Geophysical Monograph Series

Geophysical Monograph 240

Chemostratigraphy Across Major Chronological Boundaries

Alcides N. Sial
Claudio Gaucher
Muthuvairavasamy Ramkumar
Valderez Pinto Ferreira
Editors

This Work is a co-publication of the American Geophysical Union and John Wiley and Sons, Inc.

AGU
100
ADVANCING EARTH
AND SPACE SCIENCE

WILEY

Published under the aegis of the AGU Publications Committee

Brooks Hanson, Executive Vice President, Science
Lisa Tauxe, Chair, Publications Committee
For details about the American Geophysical Union visit us at www.agu.org.

Wiley Global Headquarters
111 River Street, Hoboken, NJ07030, USA

For details of our global editorial offices, customer services, and more information about Wiley products visit us at www.wiley.com.

Library of Congress Cataloging-in-Publication data is available.

ISBN: 9781119382485

Cover design: Wiley
Cover image: The Permian-Triassic boundary beds in the Ali Bashi Mountains, 9 km west of the town Julfa (East Azerbaijan province, Iran) (© Dieter Korn, February 2018)

Set in 10/12pt Times New Roman by SPi Global, Pondicherry, India

V10006167_021219

CONTENTS

Part IV: Mesozoic

Part V: Cenozoic

CONTRIBUTORS

Jhon Willy Lopes Afonso
Programa de Pós-Graduação em Geologia e
Geoquímica
Faculdade de Geologia
Instituto de Geociências
Universidade do Pará
Belém, PA, Brazil

José Antonio Barbosa
LAGESE
Department of Geology
Federal University of Pernambuco
Recife, PE, Brazil

Stig M. Bergström
School of Earth Sciences
Division of Earth History
The Ohio State University
Columbus, OH, USA

Farid Chemale Junior
Programa de Pos-graduação em Geologia
Universidade Vale do Rio dos Sinos
São Leopoldo, Rio Grande do Sul, Brazil

Jiubin Chen
State Key Laboratory of Environmental Geochemistry
Institute of Geochemistry
Chinese Academy of Sciences
Guiyang, China

Simonetta Cirilli
Department of Physics and Geology
University of Perugia
Perugia, Italy

Fábio Henrique Garcia Domingos
Programa de Pós-Graduação em Geologia e
Geoquímica
Faculdade de Geologia
Instituto de Geociências
Universidade do Pará
Belém, PA, Brazil

Karol Tatiana Dussan-Tapias
Universidad de Caldas
Manizales, Colombia

Valderez Pinto Ferreira
NEG–LABISE
Department of Geology
Federal University of Pernambuco
Recife, PE, Brazil

Robert Frei
Department of Geosciences and Natural Resource
Management
University of Copenhagen
Copenhagen, Denmark;
Nordic Center for Earth Evolution (NordCEE)
University of Southern Denmark
Odense, Denmark

Claudio Gaucher
Instituto de Ciencias Geológicas
Facultad de Ciencias
Universidad de la República
Montevideo, Uruguay

Abbas Ghaderi
Department of Geology
Faculty of Sciences
Ferdowsi University of Mashhad
Mashhad, Iran

Nova Giovanny
Corporación Geológica Ares
Bogotá, Colombia

Daniel Goldman
Department of Geology
University of Dayton
Dayton, OH, USA

Felipe Guadagnin
Universidade Federal do Pampa
Caçapava do Sul
Rio Grande do Sul, Brazil

Stephen Peter Hesselbo
Camborne School of Mines and Environment and
Sustainability Institute
University of Exeter
Cornwall, United Kingdom

Franziska Heuer
Museum für Naturkunde - Leibniz Institute for
Evolution and Biodiversity Science
Berlin, Germany

John A. Higgins
Department of Geosciences
Princeton University
Princeton, NJ, USA

Alan J. Kaufman
Department of Geology
Earth System Science Interdisciplinary Center
University of Maryland
College Park, MD, USA

Dieter Korn
Museum für Naturkunde - Leibniz Institute for
Evolution and Biodiversity Science
Berlin, Germany

Christoph Korte
Department of Geosciences and Natural Resource
Management
University of Copenhagen
Copenhagen, Denmark

Jean Michel Lafon
Programa de Pós-Graduação em Geologia e
Geoquímica
Faculdade de Geologia
Instituto de Geociências
Universidade do Pará
Belém, PA, Brazil;
Research Productivity of CNPq
Brasília, Brazil

Luis Drude Lacerda
LABOMAR, Institute of Marine Sciences
Federal University of Ceará
Fortaleza, Brazil

Ramasamy Nagarajan
Department of Applied Geology
Curtin University
Sarawak, Malaysia

Afonso César Rodrigues Nogueira
Programa de Pós-Graduação em Geologia e
Geoquímica
Faculdade de Geologia
Instituto de Geociências
Universidade do Pará
Belém, PA, Brazil;
Research Productivity of CNPq
Brasília, Brazil

József Pálfy
Department of Geology
Eötvös University
Budapest, Hungary;
Research Group for Paleontology
Hungarian Academy of Sciences-Hungarian Natural
History Museum-Eötvös University
Budapest, Hungary

Natan Silva Pereira
NEG–LABISE
Department of Geology
Federal University of Pernambuco
Recife, PE, Brazil;
Department of Biology
State University of Bahia
Paulo Afonso, Brazil

Roberto Vizeu Lima Pinheiro
Programa de Pós-Graduação em Geologia e
Geoquímica
Faculdade de Geologia
Instituto de Geociências
Universidade do Pará
Belém, PA, Brazil

Muthuvairavasamy Ramkumar
Department of Geology
Periyar University
Salem, TN, India

Danielle Santiago Ramos
Department of Geosciences
Princeton University
Princeton, NJ, USA

Guilherme Raffaeli Romero
Programa de Pós-Graduação em Geologia e
Geoquímica
Faculdade de Geologia
Instituto de Geociências
Universidade do Pará
Belém, PA, Brazil

Priyadarsi Debajyoti Roy
Instituto de Geología
Universidad Nacional Autónoma de México
Ciudad de México, México

Isaac Daniel Rudnitzki
Departamento de Geologia
Universidade Federal de Ouro Preto
Ouro Preto, MG, Brazil

Micha Ruhl
Department of Geology
Trinity College Dublin
The University of Dublin
Dublin, Ireland;
Department of Earth Sciences
University of Oxford
Oxford, United Kingdom

Evelyn Aparecida Mecenero Sanchez
Faculty of Geological Engineering
Instituto de Ciência e Tecnologia
Universidade Federal dos Vales do Jequitinhonha e
Mucuri
Diamantina, MG, Brazil

Hudson Pereira Santos
Programa de Pós-Graduação em Geologia e
Geoquímica
Faculdade de Geologia
Instituto de Geociências
Universidade do Pará
Belém, PA, Brazil

Iara Maria dos Santos
Programa de Pós-Graduação em Geologia e
Geoquímica
Faculdade de Geologia
Instituto de Geociências
Universidade do Pará
Belém, PA, Brazil

Martin Schobben
School of Earth and Environment
University of Leeds
Leeds, United Kingdom;
Museum für Naturkunde - Leibniz Institute for
Evolution and Biodiversity Science
Berlin, Germany

Alcides Nobrega Sial
NEG–LABISE
Department of Geology
Federal University of Pernambuco
Recife, PE, Brazil

José Bandeira
Programa de Pós-Graduação em Geologia e
Geoquímica
Faculdade de Geologia
Instituto de Geociências
Universidade do Pará
Belém, PA, Brazil

Joelson Lima Soares
Programa de Pós-Graduação em Geologia e
Geoquímica
Faculdade de Geologia
Instituto de Geociências
Universidade do Pará
Belém, PA, Brazil

Juan Carlos Silva-Tamayo
Antonio Nariño University
Bogotá, Colombia;
Tetslab Geoambiental
Medellin, Colombia

Vinod Chandra Tewari
Department of Geology
Sikkim University
Gangtok, SK, India

Melanie Tietje
Museum für Naturkunde - Leibniz Institute for
Evolution and Biodiversity Science
Berlin, Germany

Clemens Vinzenz Ullmann
Camborne School of Mines and Environment and
Sustainability Institute
University of Exeter
Cornwall, United Kingdom

Helmut Weissert
Department of Earth Sciences
ETH Zurich
Zurich, Switzerland

Paul B. Wignall
School of Earth and Environment
University of Leeds
Leeds, United Kingdom

PREFACE

Multiple global changes marked the major (chrono) stratigraphic boundaries in the geological history of Earth. Accordingly, these changes are documented through geochemical and stable isotopic proxies/chemostratigraphic events across the Neoproterozoic-Cambrian, Permian-Triassic, Cretaceous-Paleogene, and many other boundaries from different continents. Study of these past geological-chronological boundary records holds the key for understanding the multiple proxies and diverse consequences of these changes. This book focuses on global studies from Archean-Paleoproterozoic, Proterozoic-Paleozoic, Paleozoic-Mesozoic, and Mesozoic-Cenozoic transitions using major, trace, and platinum-group elements (PGE), REE, and elemental and stable and radiogenic isotope variations. The aim of these studies is a better understanding of causes and effects of the changes that mark these important boundaries, within the lithosphere, atmosphere, biosphere, and hydrosphere. In addition, the knowledge of past positions of continents, global sea-level changes, volcanism, and mass extinction events across these boundaries are essential clues to unravel the history of our planet.

Recent studies have demonstrated that geochemical and stable isotope changes at the end-Permian mass extinction are due to abrupt climate change induced by CO_2 emission. Catastrophic end-Permian and end-Cretaceous volcanism may have released large amounts of CO_2 and other toxic gases into the atmosphere contributing to the mass extinction at these two major boundaries. Therefore, oceanic and terrestrial records of elemental and isotope chemostratigraphy are valuable tools in establishing major tectonic and climatic changes. A global paleogeographic and paleoclimatic picture of the Earth will emerge from exploring this theme.

Chemostratigraphy, an interdisciplinary discipline, has made rapid strides and promises to provide solutions to some intriguing problems of Earth processes on microscales and global scales. This book focuses on the application of chemostratigraphy to the study of major chronostratigraphical boundaries and on how it can contribute to broaden the knowledge on these boundaries. It comprises thirteen chapters, which deal with different geological units around the world. It aims at providing a concise and updated view of major chronostratigraphical boundaries from the chemostratigraphical viewpoint, highlighting (i) chemostratigraphy as an important stratigraphical tool of wide interest, as attested by growing popularity and expanding application to many

geological problems, despite the absence of textbooks on this field; (ii) it supplements other lines of evidence for analyzing and documenting geological phenomena; (iii) it is important in unraveling the intriguing nature of chronostratigraphical boundaries; (iv) it helps to make a more accurate determination of boundaries and more robust correlations; and (v) high-resolution chemostratigraphy along available biostratigraphy of these boundaries helps in determining the cause of extreme biotic turnover.

With this book, our intention is to provide students and researchers a comprehensive review of major turnovers and global changes at chronostratigraphical boundaries from the chemostratigraphic viewpoint. Thirteen chapters in this volume embody relevant issues and conclusions on nature and possible causes of the major chronostratigraphic boundaries and are grouped into five sections: In Part I, Alcides Sial and others propose that chemostratigraphy should be a formal stratigraphic method, and Mu Ramkumar and others present a glossary of chemostratigraphy, including key phrases and the terminology used in this field. Part II encompasses five chapters on the major Precambrian boundaries, while two chapters on Paleozoic chronostratigraphical boundaries are found in Part III. Four chapters cover the major Mesozoic boundaries in Part IV, and a summary on the chemostratigraphy of the most recent era of Earth's history, the Cenozoic, is found in Part V.

Claudio Gaucher and Robert Frei focus on the Archean-Proterozoic boundary (2500 Ma) and the Great Oxygenation Event, the most dramatic change on Earth's history. They discuss three different proposals for the placement of this boundary and a corresponding Global Boundary Stratotype Section and Point: (i) to keep it at 2500 Ma, aided by prominent BIF units, Mo abundance, and Mo isotopes; (ii) to place it at the base of the second Huronian glaciation (ca. 2.35–2.40 Ga), thought to represent a "snowball" event; and (iii) to use the termination of the mass-independent fractionation of sulfur and the increase in the $\delta^{34}S$ amplitude of sulfides as the main criteria.

Farid Chemale Jr. and Felipe Guadagnin review the chronochemostratigraphy of some platform sequences across the Paleoproterozoic-Mesoproterozoic boundary. The Paleoproterozoic era is known to be an interval of major changes in the Earth's atmosphere, biosphere, oceans, and lithosphere. In contrast, the Mesoproterozoic era is considered for some as a "boring interval" due to the paucity of changes, especially in life forms. Carbonate platforms in basins of this interval exhibit essentially flat carbon isotope signature (around a mean of −0.6‰, with extreme $\delta^{13}C$ values seldom lying further than 1‰ from the mean) suggesting a stable paleoclimate, implying that the global ocean reached a state of equilibrium in the mid-Paleoproterozoic and remained stable for much of the following billion years.

Juan Carlos Silva Tamayo and others have used geochronological and C and Sr chemostratigraphic data from late Neoproterozoic to early Mesoproterozoic marine carbonate successions to propose reference $\delta^{13}C$ and $^{87}Sr/^{86}Sr$ chemostratigraphic pathways for the Mesoproterozoic-Neoproterozoic transition. While late Mesoproterozoic marine carbonates display $\delta^{13}C$ decrease from 4‰ to −2‰, carbonates across the Mesoproterozoic-Neoproterozoic transition exhibit a positive $\delta^{13}C$ shift, from −2‰ to +2‰, followed by subsequent decrease to values around −1‰. This decrease of $\delta^{13}C$ values is followed by a new increase to predominantly positive ones in the early Neoproterozoic. The reference chemostratigraphic pathways obtained also suggest that late Mesoproterozoic carbonate successions display predominantly higher $^{87}Sr/^{86}Sr$ values than early Neoproterozoic carbonates.

Afonso C. R. Nogueira and others review the status of knowledge of the Cryogenian-Ediacaran transition and focused on the southern margin of the Amazon Craton, an important area for studying evidence of Neoproterozoic glaciations. They examine four outcrops of cap carbonate that overlie Marinoan diamictites and perhaps record the best preserved boundary between Cryogenian (850–635 Ma) and Ediacaran (635–541 Ma) in South America. The new data discussed and the review of previous geological, geochemical, and isotopic information provide a robust stratigraphic framework that confirms unequivocally the record of Cryogenian-Ediacaran boundary in the Southern Amazon Craton.

Alan J. Kaufman assesses the state of knowledge of the Precambrian-Phanerozoic boundary, discussing in detail the progress in resolving several major issues of this transitional period. He also provides a review of profound changes in the carbon and sulfur cycles across this critical transition in order to better understand climatic and biological events and further proposes a novel resource-based hypothesis for the rise and fall of the Ediacaran biota.

Stig Bergström and Daniel Goldman focus on the Ordovician-Silurian interval making a comprehensive summary from C isotope chemostratigraphy and conclude that this boundary cannot be defined in terms of $\delta^{13}C$ chemostratigraphy. A comparison between biostratigraphy and chemostratigraphy indicates that the graptolite-defined base of the Silurian is located at a stratigraphic level only a little higher than the end of the Hirnantian carbon isotopic excursion (HICE).

Martin Schobben and others discuss the effect of sampling strategies on stratigraphic carbonate-carbon isotope trends using chemostratigraphy across the Permian-Triassic boundary as an example. They assess how much bed-internal carbon isotope variation of rock sequences can bias carbon isotope frameworks, as well as how much anomalous signals can be introduced to carbon isotope records by polymorph assemblages and/or microbially mediated precipitates. They propose that bulk-rock sampling strategies can improve the reliability of recording primary chemical signals.

Christoph Korte and others review the Triassic-Jurassic transition, marked by one of the biggest mass extinctions in Earth's history, coeval with early stages of the Central Atlantic magmatic province (CAMP) volcanism, showing strong perturbation of the global carbon and major fluctuations in carbon isotope ratios. Changes in magnitude and rate of change in $\delta^{13}C$, coincident with the end-Triassic mass extinction interval, differ between substrates (organics vs. calcite) and depositional environments. Thus, fluxes of carbon release at this time and links to the emplacement of CAMP are poorly understood.

Helmut Weissert reports on the Jurassic-Cretaceous carbon isotope geochemistry as a proxy for paleoceanography and tool for stratigraphy. He concludes that oceanography explains why C isotope stratigraphy may not be very useful as a tool when defining GSSP of the Jurassic-Cretaceous boundary. Alcides Sial and others made an extensive review on the Cretaceous-Paleogene boundary focusing on elemental and isotope chemostratigraphy from apparently continuous sections and testing the impact versus volcanism hypotheses using Hg chemostratigraphy and Hg isotopes.

Priyadarsi Roy and others review the geological, climatic, and paleobiotic events of the Cenozoic era using chemostratigraphic markers to identify gaps in our understanding. They suggest further subdivisions of the Cenozoic, namely, the early and late Paleocene; the early, middle, and late Eocene; the early and late Oligocene; the early, middle, and late Miocene; the Pliocene; the Pleistocene; and the Holocene. Through the review, these authors found the chemostratigraphic trends of the Cenozoic to be essentially of a continuum of Mesozoic trends.

Alcides N. Sial
Claudio Gaucher
Muthuvairavasamy Ramkumar
Valderez Pinto Ferreira

ACKNOWLEDGMENTS

This book project started with the idea of organizing a session on "Elemental and Isotopic Chemostratigraphy Across Major Chronostratigraphical Boundaries" within the framework of the 35th International Geological Congress at Cape Town, South Africa (2016). Rituparna Bose acting as books editor of the American Geophysical Union (AGU) inquired us as leaders of that session on our willingness of contributing/compiling a special volume in the form of a major reference work on a related research topic. Upon our positive reply, she formally extended us an invitation to prepare this research work to the prestigious American Geophysical Union Book Series. Technical assistance at the AGU/Wiley, especially by Kathryn Corcoran, and the editorial team at AGU are gratefully acknowledged for their professional, yet timely handling of many requests/tasks since inception of this work.

All the contributions presented in this book were reviewed by internationally renowned experts. The nature of this book required some manuscripts to be long and full of detailed information which made their reviewing a time-consuming effort. We are especially grateful to a number of active researchers and experts who shared their expertise and time and made important contributions to the success of this book by providing critical, constructive, and, in some cases, thought-provoking reviews. They are listed below in alphabetical order:

Thierry Adatte (Institute of Earth Sciences, University of Lausanne, Switzerland)

José Antônio Barbosa (Federal University of Pernambuco, Recife, Brazil)

Michael Bau (Jacobs Universität, Bremen, Germany)

Paulo Cesar Boggiani (University of São Paulo, São Paulo, Brazil)

Ana-Voica Bojar (Universitat Salzburg, Salzburg, Austria)

Zhong-Qiang Chen (China University of Geosciences, Wuhan, China)

Sean Crowe (University of British Columbia, Vancouver, Canada)

Milene Figueiredo (Center of Research and Development Leopoldo Américo Miguez de Mello, Petróleo Brasileiro S.A., Brazil)

Karl B. Föllmi (Institute of Earth Sciences, University of Lausanne, Switzerland)

Robert Frei (University of Copenhagen, Copenhagen, Denmark)

Reinhardt Adolf Fuck (University of Brasília, Brasília, Brazil)

Francesca Furlanetto (Simon Fraser University, Vancouver, British Columbia, Canada)

Leo Afraneo Hartmann (Federal University of Rio Grande do Sul, Porto Alegre, Brazil)

Jens Herrle (Goethe-Universitaet, Frankfurt am Main, Germany)

Wolfram M. Kürschner (Department of Geosciences, University of Oslo, Norway)

Aroldo Misi (Federal University of Bahia, Salvador, Bahia, Brazil)

Ramasamy Nagarajan (Curtin University, Sarawak, Malaysia)

Manoj Kumar Pandit (Department of Geology, University of Rajasthan, Jaipur, India)

Gustavo Paula-Santos (Institute of Geosciences, State University of Campinas, São Paulo, Brazil)

Gregory Price (School of Geography, Earth and Environmental Sciences, University of Plymouth, United Kingdom)

Claudio Riccomini (University of São Paulo, São Paulo, Brazil)

Priyadarsi D. Roy (National Autonomous University of Mexico, Mexico)

Isaac Rudnitzki (Federal University of Ouro Preto, Minas Gerais, Brazil)

Finn Surlyk (University of Copenhagen, Copenhagen, Denmark)

Vinod Chandra Tewari (Geology Department, Sikkim University, Sikkim, India)

Manish Tiwari (National Centre for Antarctic and Ocean Research, Vasco da Gama, Goa, India)

Paul B. Wignall (School of Earth and Environment Sciences, University of Leeds, United Kingdom)

ANS and VPF acknowledge the continuous financial support from Brazilian agencies (CNPq, FINEP, CAPES, VITAE, FACEPE) through funds to defray costs with traveling, visiting scientists, scholarships, field trips, maintenance of laboratories, and chemical and isotope analyses.

CG gratefully acknowledges continued support from the Sistema Nacional de Investigadores (SNI) (Uruguay). Involvement of MR on this subject and scientific collaboration with national and international academic and research institutions have been supported by research grants from various organizations, namely, the Alexander von Humboldt Foundation and the German Research Foundation (Germany), the University Grants Commission, the Council of Scientific and Industrial Research, the Department of Science and Technology, the Oil Industry Development Board, and the Oil and Natural Gas Corporation Limited (India), for which MR is thankful.

Part I
Introduction

1

Chemostratigraphy as a Formal Stratigraphic Method

Alcides Nobrega Sial[1], Claudio Gaucher[2], Muthuvairavasamy Ramkumar[3], and Valderez Pinto Ferreira[1]

ABSTRACT

Elemental and isotope chemostratigraphies are used as tracers for glacial events, buildup of volcanic gases during glaciations (e.g., CO_2), role of volcanism in mass extinction, salinity variation, redox state of the ocean and atmosphere, and provenance, among other applications. The use of isotope systems (C, O, S, N, Sr, Nd, Os), nontraditional stable isotope systems (e.g., Ca, Mg, B, Mo, Fe, Cr, Li), and elemental composition or elemental ratio (e.g., V, Ir, Mo, P, Ni, Cu, Hg, Rb/K, V/Cr, Zr/Ti, Li/Ca, B/Ca, Mg/Ca, I/Ca, Sr/Ca, Mn/Sr, Mo/Al, U/Mo, Th/U) in chemostratigraphy, especially across major chronological boundaries, are reviewed in this chapter. Furthermore, it is discussed what validates chemostratigraphy as a formal stratigraphic method.

1.1. INTRODUCTION

The use of elemental and isotope chemostratigraphy in interpretation and correlation of global events was established with the pioneer work of *Emiliani* [1955] on oxygen isotope composition of foraminifers from deep-sea cores. *Shackleton and Opdyke* [1973] established the first 22 oxygen isotope stages, which was effectively the first formal application of chemostratigraphy. *Williams et al.* [1988] extended the oxygen isotope stage zonation to the rest of the Quaternary and *Lisiecki and Raymo* [2005] to the whole Pliocene. The success of oxygen isotope chemostratigraphy encouraged researchers to use stable isotope stratigraphy in ancient sedimentary successions.

Precambrian chemostratigraphy followed the pioneer research by William T. Holser on ancient ocean water chemistry [*Kaufman et al.*, 2007a]. Long-term fluctuations

in the chemistry of the seawater have been examined from the C isotope record across thick successions [e.g., *Veizer et al.*, 1980; *Magaritz et al.*, 1986], and, in spite of potential effects of late diagenesis on isotope record, important isotope events were demonstrated on a global scale [e.g., *Knoll et al.*, 1986; *Magaritz*, 1989; *Holser*, 1997]. Since then, it became evident that contemporaneous, geographically widely separated marine strata registered similar isotopic compositions. Thereafter, chemostratigraphy became an important technique/tool of intrabasinal and interbasinal stratigraphic correlation to help assemble Precambrian stratigraphic record from fragments preserved in different successions [*Kaufman et al.*, 2007b; *Karhu et al.*, 2010; *Sial et al.*, 2010a], compensating for poor biostratigraphic resolution of Precambrian fossils [*Veizer et al.*, 1980; *Knoll et al.*, 1986; *Magaritz et al.*, 1986; *Knoll and Walter*, 1992; *Kaufman et al.*, 1997; *Corsetti and Kaufman*, 2003; *Halverson et al.*, 2005]. Correlations established through chemostratigraphy can be used to comment on biogeochemical and climate changes through time although the paucity of radiometric constraints on the absolute age of few of the extreme isotope excursions have led to debates on their temporal equivalence [e.g., *Kaufman et al.*, 1997; *Kennedy et al.*, 1998; *Calver et al.*, 2004; *Allen and Etienne*, 2008].

[1] NEG–LABISE, Department of Geology, Federal University of Pernambuco, Recife, PE, Brazil

[2] Instituto de Ciencias Geológicas, Facultad de Ciencias, Universidad de la República, Montevideo, Uruguay

[3] Deparment of Geology, Periyar University, Salem, TN, India

Chemostratigraphy Across Major Chronological Boundaries, Geophysical Monograph 240, First Edition.
Edited by Alcides N. Sial, Claudio Gaucher, Muthuvairavasamy Ramkumar, and Valderez Pinto Ferreira.

1.2. BASIS AND DEVELOPMENT OF CHEMOSTRATIGRAPHY

High-resolution chemostratigraphy provides records that are multidimensional and that may yield climatic, stratigraphic, biologic, environmental, oceanographic, and, last but not least, tectonic information. Hence, the number of studies relying on isotope stratigraphy has grown substantially. In the case of C isotope stratigraphy, it can be even applied to sedimentary rocks diagenetically altered or that have undergone up to amphibolite facies metamorphism but that may have retained the original isotope signal [*Melezhik et al.*, 2005; *Nascimento et al.*, 2007; *Kaufman et al.*, 2007b; *Chiglino et al.*, 2010].

There are a myriad of isotope systems that have been successfully used in chemostratigraphy: carbon, oxygen, sulfur, nitrogen, calcium, boron, chromium, molybdenum, lithium, strontium, neodymium, osmium, iron, and zinc. In order to apply the isotope record of any of these systems for chemostratigraphy of sedimentary sequences, it is essential to have good knowledge of the secular and other variations of marine isotope ratios. As carbon isotopes have higher resilience against postdepositional alteration, they are measured in carbonates and organic matter that led to the establishment of a larger database than other isotope systems. Therefore, $\delta^{13}C$ on carbonates are more widely used in chemostratigraphy, except in carbonate-poor successions characterized by black shales [e.g., *Johnston et al.*, 2010] in which one can measure organic carbon isotopes or carbonate carbon isotopes on fossils (bivalves, ammonites, belemnites, ostracods, etc.). An attempt to compile carbon isotope data to determine a secular variation curve of $\delta^{13}C$ has revealed remarkable $\delta^{13}C$ anomalies in the Proterozoic and Phanerozoic [e.g., *Veizer et al.,* 1980, 1999; *Karhu and Holland,* 1996; *Hoffman et al.,* 1998b; *Kah et al.,* 1999; *Melezhik et al.,* 1999, 2007; *Zachos et al.,* 2001; *Lindsay and Brasier,* 2002; *Halverson et al.,* 2005, 2010a, 2010b; *Saltzman,* 2005; *Bekker et al.,* 2006; *Saltzman and Thomas,* 2012], and it became apparent that $\delta^{13}C$ minima, perhaps, follow main extinction events [e.g., *Magaritz,* 1989]. The Hirnantian and Frasnian-Famennian episodes, however, are characterized by a positive excursion, and negative excursions are known where extinction was only minor (e.g., early Aptian). A compilation of global secular variation curves of $\delta^{13}C$, $\delta^{18}O$, $\delta^{34}S$, and $^{87}Sr/^{86}Sr$, together with major anoxic events, glaciations, and sea-level variation, can be found in *Sial et al.* [2015a].

The use of chemostratigraphy as a stratigraphic tool requires a careful examination of the diagenetic history of rocks. Petrographic, elemental (e.g., Mn/Sr, Sr, and Rb/Sr vs. $\delta^{13}C$), and isotopic ($\delta^{18}O$ vs. $\delta^{13}C$) proxies are fundamental for the assessment of the nature of C isotope signals [e.g., *Marshall,* 1992; *Jacobsen and Kaufman,* 1999; *Melezhik et al.,* 2001]. In doing so, dolostones and limestones have to be dealt with separately due to their different capacity to retain primary isotopic compositions [e.g., *Kah et al.,* 1999; *Gaucher et al.,* 2007].

Two special issues focusing Precambrian chemostratigraphy were published in *Chemical Geology* [*Kaufman et al.,* 2007a] and *Precambrian Research* [*Karhu et al.,* 2010]. In these special issues, results of some cutting-edge research on traditional (C, Sr, S) isotope chemostratigraphy, few nontraditional isotope systems (Ca), and Hg chemostratigraphy have been reported. These publications encompass studies that highlighted chemical events from the Paleoproterozoic (Africa, South America, Europe, and India), Mesoproterozoic (South America), and Cryogenian-Ediacaran (North America, South America, and India) and a special focus to the atmospheric, climatic, and biogeochemical changes in both ends of the Proterozoic eon. In addition, a comprehensive synthesis on the basis and use of chemostratigraphy is presented in the book by *Ramkumar* [2015].

1.2.1. Hydrogen Isotopes

Hydrogen isotopes are relatively little used in chemostratigraphy except in studies of ice and snow stratigraphy, but deuterium has proved to be important isotope in defining the Holocene Global Stratotype Section and Point (GSSP) [*Walker et al.,* 2009]. Quaternary scientists have always sought a boundary stratotype for the Holocene in terrestrial sedimentary records, but it was within the NorthGRIP (NGRIP) ice core, Greenland, that the Holocene GSSP at 1492.45 m depth has been ratified by the International Union of Geological Sciences (IUGS). Physical and chemical parameters within the ice

enable the base of the Holocene, marked by the first signs of climatic warming at the end of the Younger Dryas/Greenland Stadial 1 cold phase, located with a high degree of precision [*Walker et al.,* 2009]. This climatic event is reflected in an abrupt shift in deuterium excess values, accompanied by more gradual changes in $\delta^{18}O$, dust concentration, a range of chemical species, and annual layer thickness.

1.2.2. Carbon Isotopes

Carbon isotope investigation on Paleoproterozoic carbonate rocks of the Lomagundi province in Africa revealed much larger $\delta^{13}C$ variation [*Schidlowski et al.,* 1983] than previously known from the Phanerozoic carbonate successions [*Veizer et al.,* 1980]. This observation led to the assumption that $\delta^{13}C$ stratigraphic variation could be a tool in stratigraphic correlation. The pioneer work of *Scholle and Arthur* [1980] is one of the first to use carbon isotopes as stratigraphic tool, and *Berger and Vincent* [1981] recognized chemostratigraphy as a valid stratigraphic method. The potential use of $\delta^{13}C$ trends and excursions of marine carbonates to date and correlate rocks relies on the fact that their $^{13}C/^{12}C$ ratios varied over time as the result of partitioning of carbon between C_{org} and C_{carb} reservoirs in the lithosphere [e.g., *Shackleton and Hall,* 1984; *Berner,* 1990; *Kump and Arthur,* 1999; *Falkowski,* 2003; *Sundquist and Visser,* 2004; *Saltzman and Thomas,* 2012]. The knowledge of the C isotope record is very important not only in stratigraphic correlation but also because of its potential to help understand the development of Earth's climate, evolution of its biota, and CO_2 levels in the atmosphere.

The compilations of the secular $\delta^{13}C_{carb}$ variation for the entire Phanerozoic [*Veizer et al.,* 1999] and the Cenozoic [*Zachos et al.,* 2001] were important steps to enable carbon isotope chemostratigraphy to be routinely used as a stratigraphic tool. Currently, the most complete available curve on the $\delta^{13}C_{carb}$ fluctuations through geologic time has been compiled from multiple literature sources by *Saltzman and Thomas* [2012]. Difficulties faced in constructing such a curve reside on the fact that materials analyzed for curve construction, available in the literature, differ between authors and geological time periods, as cautioned by *Saltzman and Thomas* [2012]. In an attempt to use these compiled curves, one should carefully consider whether skeletal carbonate secreted by specific organisms or bulk carbonate has been used in evaluating or comparing C isotope stratigraphic records. Apparently, the most accepted carbonate $\delta^{13}C_{carb}$ record spanning the Neoproterozoic era is found in *Halverson et al.* [2010a, 2010b].

Covariation between $\delta^{13}C_{carb}$ and $\delta^{13}C_{org}$ helps find out whether variations in the $\delta^{13}C_{carb}$ record reflect changes in the isotopic composition of the ancient dissolved inorganic carbon (DIC) pool [e.g., *Oehlert and Swart*, 2014]. Covariant $\delta^{13}C_{carb}$ and $\delta^{13}C_{org}$ records attest that both carbonate and organic matter were originally produced in the ocean surface waters and have retained their original $\delta^{13}C$ composition [e.g., *Korte and Kozur*, 2010; *Meyer et al.*, 2013] as no secondary process is able to shift $\delta^{13}C_{carb}$ and $\delta^{13}C_{org}$ in the same direction at the same rate [*Knoll et al.*, 1986]. Conversely, the decoupled $\delta^{13}C_{carb}$ and $\delta^{13}C_{org}$ records point to diagenetic alteration [e.g., *Grotzinger et al.*, 2011; *Meyer et al.*, 2013] or denounce that noise in the $\delta^{13}C_{org}$ record resulted from local syn-sedimentary processes [*Maloof et al.*, 2010]. One should remember, however, that the organic carbon isotope record is very much dependent on the source of the organic matter (terrestrial vs. marine) and terrestrial records may retain the secular variations known from the marine records.

Carbon isotopes can also be used as a pCO_2 proxy. Stratigraphic variation in the offset between the $\delta^{13}C_{carb}$ and $\delta^{13}C_{org}$ expressed by $\Delta^{13}C$ offers a potential tool for tracing paleo-pCO_2 change [*Kump and Arthur*, 1999; *Jarvis et al.*, 2011]. Increased burial of organic carbon leads to a fall in atmospheric pCO_2 and a positive excursion in both inorganic and organic carbon. The peak in $\delta^{13}C_{org}$ may postdate that of inorganic carbon and may be larger in magnitude, because $\Delta^{13}C$ decreases as atmospheric pCO_2 falls. This difference in response is tied to a drawdown in atmospheric pCO_2 [*Kump and Arthur*, 1999]. The "robust voice" of carbon isotopes has the potential to tell us about Earth's history [*Knauth and Kennedy*, 2009], but some postdepositional alteration of carbonate rocks may alter the story [*Bristow and Kennedy*, 2008]. However, indiscriminate use of C isotope stratigraphy to correlate Neoproterozoic carbonates ("blind dating") has been cautioned by *Frimmel* [2008, 2009, 2010] from his studies on REE + Y distribution in Neoproterozoic carbonates from different settings in Africa. These studies have raised some doubt on the usefulness of cap carbonates for stratigraphic correlation of Neoproterozoic sediment successions based on carbon isotopes. They deserve further investigation, although one can argue that rare earth elements (REEs) and DIC behave differently in seawater and are affected by diagenesis in a complete different way.

The application of carbon isotope chemostratigraphy to the study of oceanic anoxic events (OAEs) which record profound global climatic and paleoceanographic changes and disturbance of the carbon cycle, is one of the best examples of use of chemostratigraphy as a stratigraphic tool. The OAEs resulted from abrupt global warming induced by rapid influx of CO_2 into the atmosphere from volcanogenic or methanogenic sources and were accompanied by accelerated hydrological cycle, increased weathering, nutrient discharge to oceans, intensified upwelling, and increase in organic productivity [*Jenkyns*,

2010]. Nine major OAEs are known, the oldest in the Jurassic (Toarcian, called T-OAE, around 183 Ma), seven in the Cretaceous, and the youngest one in the Cenozoic (corresponding to the Paleocene-Eocene Thermal Maximum (PETM), around 55.8 Ma).

An OAE event implies very high burial rates of marine organic carbon (^{12}C), resulting in an increase in $\delta^{13}C$ values of marine and atmospheric carbon, as observed in the pronounced regionally developed positive carbon isotope excursion in $\delta^{13}C_{carb}$ across the Cenomanian-Turonian boundary [*Scholle and Arthur*, 1980]. However, the carbon isotope signatures of the early Toarcian, early Albian, and early Aptian OAEs are more complicated as signals from $\delta^{13}C_{carb}$, $\delta^{13}C_{org}$, and specific biomarkers exhibit both positive and pronounced negative excursions [*Jenkyns and Clayton*, 1986; *Herrle et al.*, 2003; *Jenkyns*, 2003, 2010]. This observation suggests that besides carbon burial driving to global $\delta^{13}C$ heavier values, input of light carbon implies movement in the opposite direction.

The selection of a section at El Kef, Tunisia, to be the GSSP for the Cretaceous-Paleogene boundary (K-Pg; 66.02; *Molina et al.*, 2006, 2009), and of one at Dababiya, Egypt, to be the one for the Paleocene-Eocene boundary (PETM; 58.8 ± 0.2 Ma; *Aubry et al.*, 2007), is the best example of use of carbon isotope chemostratigraphy in boundary definition. A $\delta^{13}C$ negative shift in the section at El Kerf was one of the five marker criteria to define the K/Pg boundary, while the Paleocene-Eocene boundary was defined based on global $\delta^{13}C_{org}$ and $\delta^{13}C_{carb}$ isotope excursions (CIE).

1.2.3. Nitrogen Isotopes

The use of $\delta^{15}N$ variations in organic matter (kerogen, $\delta^{15}N_{org}$) has proved to be a valuable tool in the investigation of the evolution of the ocean chemistry, bioproductivity, and chemostratigraphic correlation, especially where biostratigraphy is of limited usefulness [*Beaumont and Robert*, 1999; *Papineau et al.*, 2005; *Algeo et al.*, 2008; *Cremonese et al.*, 2009]. Nitrogen isotope values for bulk samples ($\delta^{15}N_{bulk}$) from sections across the Ediacaran-Cambrian boundary in South China display positive values in the uppermost Ediacaran strata and strong negative shift in the Cambrian strata, especially in black shales, testifying to the changes in the biogeochemical cycle of the ancient ocean [*Cremonese et al.*, 2009, 2013, 2014]. Nitrate and nitrite are reduced to nitrogen gas by denitrification, as part of the global nitrogen cycle in modern oceans [*Algeo et al.*, 2008].

The hypothesis that transition from anoxic to oxygenated deep ocean took place at the end of the Neoproterozoic era (Neoproterozoic Oxygenation Event) is relatively well accepted [e.g., *Canfield et al.*, 2008; *Och and Shields-Zhou*, 2012]. Some of the available geochemical data for the age interval of this transition, however, allow the interpretation of possibly full oxygenation in the early Ediacaran and preservation of deep ocean anoxia up to as late as the Early Cambrian [*Ader et al.,* 2014].

Changes in marine redox structure are related to changes in the nitrogen nutrient cycling in the global ocean, implying that $\delta^{15}N_{sed}$ probably reflects deep ocean redox transition [*Ader et al.,* 2014]. Nitrogen isotope data from Canada, Svalbard, Amazonia, and China, spanning the 750–580 Ma interval, together with other available $\delta^{15}N_{sed}$ data, show no apparent change between the Cryogenian and Ediacaran, revealing a $\delta^{15}N_{sed}$ distribution that closely resembles modern marine sediments, ranging from −4 to +11, with a $\delta^{15}N$ mode close to +4 [*Ader et al.,* 2014]. $\delta^{15}N$ data from the earlier Proterozoic show distribution relatively similar to this, but shifted slightly toward more negative $\delta^{15}N$ values and with a wider range. A possible explanation for similarity of this $\delta^{15}N$ distributions is that as in the modern ocean, nitrate (and hence O_2) was stable in most of the middle to late Neoproterozoic ocean and possibly much of the Proterozoic eon [*Ader et al.,* 2014].

Global climate over Quaternary glacial-interglacial time scales may have affected fluctuations of denitrification intensity whose rates varied over time, especially during OAEs (e.g., T-OAE; *Jenkyns et al.,* 2001). Some Upper Carboniferous black shales display C_{org}/N ratios and nitrogen isotope data that attest to fluctuations in the intensity of denitrification associated with glacially driven sea-level changes [*Algeo et al.,* 2008]. Sedimentary $\delta^{15}N$ increases during rapid sea-level rise in each cycle, with intensified denitrification, returning to background levels as sea level stabilized during the interglacial phase.

Bulk $^{15}N_{tot}$ data from early Toarcian black carbon-rich shales from British Isles and northern Italy (T-OAE; *Jenkyns et al.,* 2001, 2010) and from the Toarcian-Turonian OAE [*Jenkyns et al.,* 2007] have revealed a pronounced positive $\delta^{15}N_{tot}$ excursion that broadly correlates with a relative maximum in weight percent TOC and, in some sections, with a negative $\delta^{13}C_{org}$ excursion. Perhaps, the upwelling of a partially denitrified, oxygenated water mass is the explanation for the relative enrichment of $\delta^{15}N_{tot}$, and the development of early Toarcian suboxic water masses and partial denitrification is attributed to increases in organic productivity [*Jenkyns et al.,* 2001]. A negative $\delta^{15}N_{org}$ peak to near 0‰ air/N_2 occurs at the Permian-Triassic (P-T) boundary parallel to a negative $\delta^{13}C$ excursion. It has been interpreted as the result of a diminished biomass of eukaryotic algae due to mass extinction, which were replaced by microbial N_2 fixers such as cyanobacteria [*Fio et al.,* 2010]. An analogous negative $\delta^{15}N_{org}$ and $\delta^{15}N_{bulk}$ excursion has been reported from the Ordovician-Silurian boundary [*Luo et al.,* 2016] and from the Ediacaran-Cambrian boundary [*Kikumoto*

et al., 2014]. Thus, nitrogen isotopes are valuable for the definition of major chronostratigraphic boundaries.

1.2.4. Oxygen Isotopes

Oxygen isotope chemostratigraphy has become an important tool for Mesozoic and Cenozoic stratigraphic correlation of marine sediments [e.g., *Friedrich et al.,* 2012]. For such studies, $\delta^{18}O$ is usually measured on benthic foraminifera to avoid isotopic gradient effects [e.g., *Emiliani,* 1955; *Shackleton and Opdyke,* 1973; *Lisiecki and Raymo,* 2005]. The demonstration of primary nature of $\delta^{18}O$ values in older successions, however, is often difficult, although oxygen isotopes have been successfully used in carbonates from belemnites and brachiopods and phosphates from shark teeth and conodonts [e.g., *Vennemann and Hegner,* 1998; *Joachimski and Buggisch,* 2002; *Puceat et al.,* 2003; *Price and Mutterlose,* 2004; *Bodin et al.,* 2009; *Dera et al.,* 2009; *Van de Schootbrugge et al.,* 2013].

Oxygen isotope ratios in foraminifera from deep-sea cores have shown a consistent pattern representing changes in the ocean-atmosphere system through time. *Emiliani* [1955], based on the major swings in his data, has recognized the "marine isotope stages" (MIS). *Shackleton* [1969] has subdivided Emiliani's stage 5 into lettered substages, and since then, Quaternary time is divided into marine isotope stages and substages. The MIS scheme was the first attempt to use oxygen isotope chemostratigraphy in the Quaternary. *Railsback et al.* [2015] have proposed the scheme of marine isotope substages currently in use.

A general increase from −8 to 0‰ VPDB in the Phanerozoic, punctuated by positive excursions coincident with cold intervals, has been recognized by *Veizer et al.* [1999] who have suggested that $\delta^{18}O$ analyses of carefully screened, well-preserved brachiopods and mollusks can still retain a primary signal even in Paleozoic samples. Nevertheless, similar consideration is not possible for the Precambrian due to the absence of calcified metazoans, except for the Ediacaran. $\delta^{18}O$ analyses of whole rock samples of Precambrian successions usually reflect diagenetic conditions, although primary trends have been reported in rare/limited occasions [*Tahata et al.,* 2012].

According to *Bao et al.* [2008, 2009], triple oxygen isotope evidence proved to be an important tool in the discrimination of early-Cryogenian from end-Cryogenian cap carbonates. Sulfate from ancient evaporites and barite shows variable negative ^{17}O isotope anomalies over the past 750 million years. An important difference in ^{17}O isotope anomalies of barite at top of the dolostones from the Marinoan cap carbonates (negative spike ∼ −0.70‰) suggests that by the time this mineral was precipitated, P_{CO2} was highest for the past 750 million years (CO_2 levels reached 0.01–0.08 bar during and just after ∼635 Ma glacial event; *Bao et al.,* 2008, 2009].

Oxygen isotopes of dissolved inorganic phosphate (δ^{18}Op) are a powerful stable isotope tracer for biogeochemical research, offering insights into the relative importance of different sources of phosphorus within natural ecosystems [*Davies et al., 2014*]. Besides, the isotope fractionations alongside the metabolism of phosphorus allow δ^{18}Op to be used to better understand intracellular/extracellular reaction mechanisms that control phosphorus cycling.

An organic paleothermometer based upon the membrane lipids of mesophilic marine Thaumarchaeota, the tetraether index of lipids, with 86 carbon atoms (TEX86) has been used for over a decade when attempting to reconstruct sea surface temperatures (SSTs). This thermometer is particularly useful when other SST proxies are diagenetically altered (e.g., planktic foraminifera; *Pearson et al., 2007*) or absent (e.g., alkenones; *Bijl et al., 2009*).

The oldest TEX$_{86}$ record is from the Middle Jurassic (~160Ma) and indicates relatively warm SST [*Jenkyns et al., 2012*]. It has been also used to reconstruct SST throughout the Cenozoic era (66–0Ma) [e.g., *Sluijs et al., 2009; Zachos et al., 2006*] and particularly to reconstruct the Eocene (55.8–34Ma) SST. During the early Eocene, TEX$_{86}$ values indicate warm high southern hemisphere latitude SSTs (20–25 °C) in agreement with other independently derived proxies (e.g., alkenones, Mg/Ca). During the middle and late Eocene, high southern latitude sites cooled, while the tropics remained stable and warm.

The field of clumped isotopes is concerned with how the various isotopes of carbon and oxygen are distributed in the lattice of the carbonate crystal, allowing distinction of the "isotopologues," that is, molecules of similar chemical composition but different isotopic composition [*Eiler, 2007*] This field is concerned with measuring an isotopologue of CO_2 gas with a mass of 47, that is, where the two "heavy" rare isotopes (^{13}C and ^{18}O) are substituted in the CO_2 molecule. This is representative of the amount of "clumping" of the heavy isotopes in the crystal lattice of the carbonate. As $\Delta 47$ is measured, the amount of clumping at a known temperature can be determined [e.g., *Ghosh et al., 2006*]. *Guo et al.* (2009b) provided a theoretical Δ_{47} calibration for a number of different mineralogies, making clumped isotopes to be one of the most promising paleothermometer for paleoclimate and diagenesis [e.g., *Eagle et al., 2010; Tripati et al., 2010; Petrizzo et al., 2014*]. The great advantage is that it is unnecessary to know the oxygen isotope composition of the water with which carbonates have isotopically equilibrated. The growing interest on use of this technique is reflected in a rapid increase in the number of laboratories equipped to perform routine analyses of clumped isotope and by the organization of a series of international workshops focusing on its development and general applications.

1.2.5. Sulfur Isotopes

A secular δ^{34}S variation curve for evaporites (1.0Ga to present) was reported by *Claypool et al.* [1980], and since then sulfur isotope chemostratigraphy has been largely used for marine evaporite sulfate, in terrains ranging from 1.0Ga to recent. Extensive critical review on sedimentary sulfur through time and on potential use of sulfur isotopes in the investigation of time boundaries is found in *Strauss* [1997], while detailed discussion on the use of sulfur isotopes on Neoproterozoic chemostratigraphy can be found in *Halverson et al.* [2010a]. *Halverson et al.* [2010b] have subdivided Neoproterozoic sulfur isotope data into two kinds: one recording seawater sulfate (δ^{34}S$_{sulph}$) and the other recording epigenic or authigenic pyrite (δ^{34}S$_{pyr}$). The former is recovered from evaporites, barites, phosphorites, and carbonates (as carbonate-associated sulfate (CAS)). Fractionation that occurs during bacterial sulfate reduction (BSR) plus additional fractionation effects of reactions during oxidative recycling of sulfides is recorded by the pyrite data [*Canfield and Teske*, 1996], while the sulfur isotope data from barite, phosphorite, and CAS depict seawater sulfate (δ^{34}S$_{sulph}$). Due to BSR, δ^{34}S$_{pyr}$ is usually lower (lighter) than δ^{34}S$_{sulph}$. Two important exceptions to this rule have been reported [*Ries et al., 2009*]: (i) Archean successions usually yield similar values for pyrite and CAS, because the ocean was anoxic, and therefore BSR was negligible. (ii) Superheavy pyrites, that is, with δ^{34}S values exceeding that of coeval sulfides, occur in late Neoproterozoic successions and were interpreted as the result of very low sulfate concentrations and ferruginous conditions in the ocean and intense aerobic reoxidation of pyrite [*Ries et al., 2009*].

Mass-independent fractionation (MIF) is observed in O, S, and Hg, linked to photochemical reactions in the atmosphere, and in the case of sulfur, it can be observed in ancient sediments [*Farquhar et al., 2000; Guo et al., 2009b*] where it preserves a signal of the prevailing environmental conditions which makes sulfur isotopes as a tracer of early atmospheric oxygenation up to the formation of the ozone shield. The method implies measurements of multiple sulfur isotopes (δ^{33}S, δ^{34}S, and δ^{36}S) on CAS and sulfides. The creation and transfer of the mass-independent (MI) signature into minerals would be unlikely in an atmosphere containing abundant oxygen, constraining the Great Oxygenation Event (GOE) and the establishment of an ozone shield to sometime after 2.45 Ga ago. Prior to this time, the MI sulfur record implies that sulfate-reducing bacteria did not play a significant role in the global sulfur cycle and that the MI sulfur signal is due primarily to changes in volcanic activity [*Halevy et al., 2010*]. After 2.3 Ga, the MIF signal disappears, attesting to the continued existence of an ozone layer since the Paleoproterozoic [*Guo et al., 2009a*].

Therefore, sulfur isotopes are important in the study of the Archean-Paleoproterozoic boundary and the fundamental biotic and environmental changes that took place during the GOE.

Biological and abiotic reactions in the sulfur biogeochemical cycle show distinctive stable isotopic fractionation and are important in regulating the Earth's surface redox state [*Pasquier et al.,* 2017]. The $\delta^{34}S$ composition of sedimentary sulfate-bearing phases reflects temporal changes in the global sulfur cycle and can be used to infer major changes in the Earth's surface environment, including rise of atmospheric oxygen.

Sulfur isotope pyrite-based records have been less explored. *Pasquier et al.* [2017] have studied Mediterranean sediments deposited over 500,000 y which exhibit stratigraphic variations >76‰ in the $\delta^{34}S_{pyr}$ data. These authors have demonstrated the relationship between the stratigraphic isotopic variation and phases of glacial-interglacial sedimentation rates. Their results suggest that the control of the sulfur isotope record can be associated with strong sea-level variations. Besides, they provided an important perspective on the origin of variability in such records and suggested that meaningful paleoenvironmental information can be derived from pyrite $\delta^{34}S$ records.

1.2.6. Calcium, Magnesium, and Boron Isotopes

Technological advances in analytical procedures and sophistication of equipment (e.g., micro-SIMS, nano-SIMS, MC-ICPMS) for few nontraditional stable isotopes, mainly Li, B, Mg, Cl, Ca, Cr, Fe, Ni, Cu, Zn, Ge, Se, Mo, Os, Hg, and Th [*Johnson et al.,* 2004; *Baskaran,* 2012; *Teng et al.,* 2017], have opened new avenues, some still to be explored in terms of isotope chemostratigraphy. In particular, Ca, Mo, and Fe have received more attention in Precambrian isotope chemostratigraphy [*Kasemann et al.,* 2005; *Arnold et al.,* 2004; *Siebert et al.,* 2003; *Johnson and Beard,* 2006; *Staubwasser et al.,* 2006, among others], and Cr isotopes have proven to be an important tool in this regard [*Frei et al.,* 2009, 2011, among others].

It is not known exactly how Ca isotopes work in modern carbonate rocks or the extension on how diagenesis affects them. A fairly updated review on the global calcium cycle is found in *Fantle and Tipper* [2014] and *Gussone et al.* [2016].

The global Ca isotope signal from end-Cryogenian carbonate successions suggests that Ca isotope chemostratigraphy can be an additional tool for the correlation of postglacial Neoproterozoic carbonate successions [*Higgins and Schrag,* 2010; *Kasemann et al.,* 2005; *Silva Tamayo et al.,* 2007, 2010a, 2010b]. These authors have claimed that the Neoproterozoic Ca isotopic record is, perhaps, an archive of changes in the oceanic Ca isotopic composition.

Rapid glacier melting and significant increase in the Ca input to the ocean immediately after deglaciation, followed by progressive increase in carbonate precipitation and burial compensating for the large initial Ca input, have been depicted from Ca isotope behavior. Post-Sturtian and post-Marinoan global $\delta^{44/40}Ca$ patterns seem to differ from each other, probably because of the difference in Ca mass balance evolution among these two deglaciation events as a consequence of contrasting glacier melting regimes [*Silva Tamayo et al.,* 2010a, 2010b]. This divergent behavior of the Ca isotopic evolution makes Ca isotope stratigraphy a promise, perhaps, to discriminate and correlate Neoproterozoic postglacial carbonate successions. Possibly, there is a close connection between Ca isotopic cycling in the Phanerozoic, seawater chemistry, carbonate sedimentation, and evolutionary trends [*Blättler et al.,* 2012]. MI isotope fractionation effects as observed in O, S, and Hg isotopes were not so far observed in Ca isotopes [*Gussone et al.,* 2016].

Use of magnesium isotope to understand geological phenomenon/processes has been on the rise during recent times [e.g., *Tipper et al.,* 2006a, 2006b, 2006c; *Higgins and Schrag,* 2010; *Wombacher et al.,* 2011; *Azmy et al.,* 2013; *Geske et al.,* 2015]. *Chang et al.* [2003], *Tipper et al.* [2008], and *Wombacher et al.* (2009) presented the systematics and analytical protocols in Mg isotope analyses, and accuracy of Mg isotope determination in MC-ICPMS was discussed by *Tipper et al.* [2008]. *Brenot et al.* (2008) examined the Mg isotope variability within a lithologically diverse river basin. The relationships between continental weathering, riverine influx of Mg into the oceans, and global Mg isotope budgets of modern oceans were examined by *Tipper et al.* [2006a, 2006b, 2006c]. *Higgins and Schrag* [2010] demonstrated the utility of constraining Mg cycle in marine sediments through the use of Mg isotope. As magnesium is part of the C cycle and dolomite is a major sink for Mg and a main control for $\delta^{26}Mg_{seawater}$, *Geske et al.* [2015] studied Mg isotope and suggested its use as a vital proxy. *Azmy et al.* [2013] are also of the similar opinion. Nevertheless, use of Mg isotopes in truly stratigraphic context has been scarce, for example, *Strandmann et al.* [2014] and *Pokrovsky et al.* [2011], to name a few. Despite this scarcity, the information that the Mg isotope system follows that of Sr and Ca isotopic systems [*Fantle and Tipper,* 2014] and the fact that the Mg isotopic composition of the oceans is relatively constant ($\delta^{26}Mg_{seawater} = -0.82 \pm 0.01$‰, *Foster et al.,* 2010) and Mg has a long residence time in the ocean ($\approx$10 Myr; *Berner and Berner,* 1987; 14–16 Myr, *Lécuyer et al.,* 1990) could suggest its utility in establishing chemostratigraphic curve similar to that of Sr isotopic curve; however, the potential remains yet to be tapped and tested. It was *Galy et al.* [2002] who have reported a latitudinal gradient of Mg isotopic fractionation in calcites of speleothems. *Li et al.* [2012] precipitated calcite in a wide range of temperature (4–45°C) and reported a feeble gradient between $\delta^{26}Mg_{calcite\ in\ solution}$ and temperature (0.011±0.002‰ °C^{-1}). This finding could help establish Mg isotope as a proxy to temporal trends of paleotemperature and paleolatitudinal variations.

There is fair agreement on that the aftermath of the Cryogenian glaciations has been marked by cap dolostone deposition that have followed intense continental chemical weathering. *Huang et al.* [2016] have explored the behavior of Mg isotopes to demonstrate that this was the picture in the deposition of the terminal Cryogenian-age Nantuo Formation and the overlying cap carbonate of the basal Doushantuo Formation, South China. They observed a $\delta^{26}Mg$ positive excursion, with values ranging from +0.56 to +0.95‰, in the top of the Nantuo Formation that likely resulted from an episode of intense chemical weathering. The siliciclastic component of the overlying Doushantuo cap carbonate, on the contrary, has yielded much lower $\delta^{26}Mg$ values (<+0.40‰), suggesting low-intensity chemical weathering during the cap carbonate deposition. *Huang et al.* [2016] concluded that such a behavior of Mg isotopes confirms an intense chemical weathering at the onset of deglaciation and that it has reached its maximum before the cap carbonate deposition.

There are a growing number of publications that have applied boron isotopes as a paleo-pH proxy although boron isotope analyses are complex [e.g., *Palmer et al.*, 1998; *Sanyal et al.*, 2001; *Joachimski et al.*, 2005; *Hemming and Hönisch*, 2007; *Hönisch et al.*, 2012; *Foster and Rae*, 2016]. A secular change in the boron isotope geochemistry of seawater over the Phanerozoic is found in *Joachimski et al.* [2005], based on the boron isotope geochemistry of brachiopod calcite.

It is known that oceanic uptake of CO_2 decreases ocean pH [*Kasemann et al.*, 2005]. Calcium and boron isotopes have been used to estimate paleoenvironmental conditions in the aftermath of the two major Neoproterozoic glaciations in Namibia. *Kasemann et al.* [2005] presented a record of Cryogenian interglacial ocean pH based on boron isotopes in marine carbonates. Their B isotope data suggest a largely constant ocean pH and no critically elevated pCO_2 throughout the older postglacial and interglacial periods. Marked ocean acidification event, in contrast, marks the younger deglaciation period and is compatible with elevated postglacial pCO_2 concentration. Negative $\delta^{11}B$ excursions in postglacial carbonates have been interpreted as an indication of temporary decrease in seawater pH.

It has been proposed that during the PETM, thousands of petagrams of carbon (Pg C) were released as methane or CO_2 into the ocean-atmosphere system for about 10 kyr, concomitant to a carbon isotope excursion, widespread

dissolution of deep-sea carbonates, and global warming, leading to possible severe acidification of the ocean surface [*Penman et al.*, 2014]. Using boron-based proxies for ocean carbonate chemistry, these authors demonstrated that there is evidence for a pH drop of surface and seawater thermocline during the PETM. They have observed a decrease of 0.8‰ in $\delta^{11}B$ at the onset of the PETM event and a reduction of almost 40% in shell B/Ca, at a drill site in the North Pacific and similar trends in the South Atlantic and Equatorial Pacific, consistent with global acidification of the surface of the ocean.

1.2.7. Chromium, Iron, Molybdenum, and Thallium Isotopes

Widespread deepwater anoxia predominated in the Archean and Paleoproterozoic oceans, while the Neoproterozoic was transitional between anoxic and largely oxygenated Phanerozoic oceans. Stratified, ferruginous oceans have characterized the Archean-Paleoproterozoic and Neoproterozoic ocean chemistries, while during the Mesoproterozoic, sulfidic (euxinic) marine conditions prevailed in contrast with Phanerozoic oxygenated conditions [*Canfield et al.*, 2008]. Investigation on the Fe, Cr, and Mo isotope behavior has provided further insights into the question of surface ocean oxygenation [*Scott et al.*, 2008; *Frei et al.*, 2009].

It is well known that Cr is very sensitive to the redox state of the surface environment, oxidative weathering processes producing the oxidized hexavalent Cr. Positive isotopic fractionation of up to 5‰ accompanies the oxidation of the reduced Cr(III) on land [*Frei et al.*, 2009 and references therein]. *Lyons and Reinhard* [2009] and *Døssing et al.* [2011] have discussed in detail the isotopic systematic of the Cr cycle, including incorporation into banded iron formation (BIF). From Cr isotopes in BIFs, one can track the presence of hexavalent Cr in Precambrian oceans to understand the oxygenation history of the Earth's atmosphere-hydrosphere system [*Frei et al.*, 2009]. *Frei et al.* [2011] applied for the first time Cr isotope systematics to ancient carbonates, representing a useful tracer for climate change and for reconstructing the redox state of ancient seawater and atmosphere. Cr and C isotope curves in carbonates are virtually parallel [*Frei et al.*, 2011], and therefore, coupled $\delta^{13}C$-$\delta^{53}Cr$ chemostratigraphy of mixed BIF/carbonate/chert successions may provide more continuous curves than C isotopes alone. This method is suitable for chemical sediments (BIF, chert, and carbonates), with low amounts of terrigenous material; otherwise Cr isotopic composition of the rock will predominate [*Frei et al.*, 2013]. Cr isotopes enabled the detection of early oxygenation pulses at 2.7 Ga [*Frei et al.*, 2009] and even at 2.95 Ga [*Crowe et al.*, 2013], long before the GOE. If one compares the $\delta^{53}Cr$

values of BIF of different ages, Archean BIFs are the less fractionated and Neoproterozoic BIFs show the largest positive values, in accordance with the progressive oxygenation of surface environments [*Frei et al.*, 2009, 2013, 2017].

Iron (Fe) isotopes are a tool in the study of iron cycling due to its large isotopic fractionation attending to redox transformations in near-surface environment. Before the impossibility of applying the traditional stable or radiogenic isotope systems, the Fe isotope system has been largely applied to BIF [*Halverson et al.*, 2011]. Archean and Paleoproterozoic BIFs have revealed an extraordinary variability in Fe isotope compositions, from the stratigraphic [e.g., *Beard et al.*, 2003; *Johnson et al.*, 2008; *Heimann et al.*, 2010] to the mineral [e.g., *Johnson et al.*, 2003; *Frost et al.*, 2007] and microscale [e.g., *Steinhoefel et al.*, 2010]. These variations are usually ascribed to the large fractionation resulting from reduction/oxidation of Fe and the isotopic differences between mineral phases [e.g., *Johnson et al.*, 2008].

In Neoproterozoic iron formations (IF), Fe occurs almost predominantly as hematite [*Klein and Beukes*, 1993] in contrast to some Archean-Paleoproterozoic BIFs in which iron occurs as both Fe^{2+} and Fe^{3+} in a range of different minerals [*Klein and Beukes*, 1993]. Therefore, primary isotope signatures are easier to obtain from the Neoproterozoic BIFs which are usually associated with episodes of global glaciation (Rapitan-type BIF) as their Fe isotope composition reflects the chemistry of the glacial ocean [*Halverson et al.*, 2011].

The evolution of the redox state of the oceans can be also investigated using Mo concentrations in black shales [*Scott et al.*, 2008]. Its isotopic composition, in turn, allows differentiation between euxinic (i.e., sulfidic) and oxygenated environments [*Arnold et al.*, 2004]. Three oxygenation events at 2.65 Ga, ca. 2.5 Ga, and 550 Ma were recognized, with the late Paleoproterozoic and Mesoproterozoic (1.8–1.0 Ga) being characterized by euxinic conditions ("Canfield Ocean"; *Canfield*, 1998; *Arnold et al.*, 2004; *Scott et al.*, 2008). This is consistent with other proxies, such as MIF of sulfur and chromium isotopes.

High-precision measurements of thallium (Tl) isotope ratios were only made possible in the late 1990s, and, therefore, one has only limited knowledge of its isotopic behavior. Despite of their heavy masses of 203 and 205 a.m.u., it is known that thallium isotopes can be fractionated substantially in the marine environment [*Nielsen et al.*, 2017].

Thallium isotopes have been applied to investigate paleoceanographic processes in the Cenozoic, and a compilation of the Tl ($\varepsilon^{205}Tl_{sw}$) isotope composition of seawater over the last 75 Myrs is found in *Nielsen et al.* [2009, 2017], together with contemporaneous $\delta^{34}S_{sw}$ variation curve. These two curves show relatively similar behavior,

with the lowest values within the 55–70 Ma range, the $\delta^{34}S_{sw}$ curve displaying minimum values around 55 Ma and $\varepsilon^{205}Tl_{sw}$ around 66 Ma. Thallium isotopes may be utilized as a proxy for changes in Fe and Mn supply to the water column over million year time scales according to *Nielsen and Rehkämper* [2012] to monitor changes in marine Mn sources and/or Mn oxide precipitation rates back in time.

1.2.8. Strontium and Neodymium Isotopes

As $^{87}Sr/^{86}Sr$ ratios and $\delta^{13}C$ fluctuate independently from each other, their combined use through the application of high-resolution chemostratigraphy represents a powerful tool to resolve geological problems. The radiogenic nature of ^{87}Sr, which forms as a result of radioactive decay of ^{87}Rb, implies that the $^{87}Sr/^{86}Sr$ ratio of the mantle, crust, and surface environments rises with time [e.g., *Shields*, 2007a, 2007b]. The $^{87}Sr/^{86}Sr$ seawater variation curve is better known for the Phanerozoic [e.g., *Burke et al.*, 1982; *Veizer et al.*, 1999; *McArthur et al.*, 2001; *Leckie et al.*, 2002; *McArthur*, 2010] showing long-term variations of about 500–550 Ma from the Upper Cambrian (0.709; *Montañez et al.*, 2000) gradually decreasing with a nadir of 0.7068 at 250 Ma and rising again to values of 0.7092 in the present-day ocean [*Macdougall*, 1991; *McArthur et al.*, 2001]. This makes Sr isotopes a pretty straightforward and precise method for dating marine carbonates and calcareous fossils in the upper half of the Cenozoic, for example, because $^{87}Sr/^{86}Sr$ ratios rise continuously from 0.7077 in the Bartonian (ca. 40 Ma) to 0.7092 in the Holocene [*McArthur et al.*, 2001]. Sr isotope chemostratigraphy is equally well feasible in the other periods of the Phanerozoic, depending on the morphology of the Sr isotope record. The close correlation in time between the strontium isotope excursions and the major OAEs (Jurassic and Cretaceous) is compatible with a causal linkage [e.g., *Jones and Jenkyns*, 2001].

The main source of ^{87}Sr is the weathering of Rb-rich granitic rocks. Hydrothermal vents near mid-ocean ridges are enriched in non-radiogenic ^{86}Sr [*Shields*, 2007a], and therefore, high $^{87}Sr/^{86}Sr$ ratios are considered as an indication of periods of enhanced orogenesis, while low ratios characterize periods of continental breakup and enhanced hydrothermal activity. *Flament et al.* [2011], however, have pointed out that $^{87}Sr/^{86}Sr$ is influenced by the area of emerged land rather than by orogenic processes alone, something especially important for calculations of continental growth in the Archean, when maybe <4% of Earth's area was emerged [*Shields*, 2007b; *Flament et al.*, 2011].

Efforts to compile Sr isotope data aimed at determining the secular $^{87}Sr/^{86}Sr$ seawater curve for the Proterozoic have been made [e.g., *Jacobsen and Kaufman*, 1999;

Melezhik et al., 2001; *Halverson et al.*, 2007, 2010a, 2010b; *Kuznetsov et al.*, 2010]. The use of strontium isotopes in chemostratigraphy, however, is limited by the paucity of limestone in many successions. Another difficulty is posed by the likelihood of alteration in samples with low strontium contents through the incorporation of ^{87}Sr from the decay of ^{87}Rb in coexisting clay minerals [*Kaufman et al.*, 2009]. Therefore, it is advisable to consider only analyses of high-Sr limestones, less prone to postdepositional alteration. Geochemical screens (Rb/Sr, Mn/Sr, Sr concentration, and δ^{18}O) have been widely adopted to evaluate the degree of postdepositional alteration of strontium isotope ratios [*Veizer et al.*, 1983; *Kaufman et al.*, 1992, 1993; *Marshall*, 1992; *Jacobsen and Kaufman*, 1999; *Melezhik et al.*, 2001]. Dolostones are usually not suitable for Sr isotope studies due to the lower Sr concentrations of usually a few tens of ppm [*Kah et al.*, 1999; *Gaucher et al.*, 2007], although a few exceptions have been reported [*Sawaki et al.*, 2010].

Another problem of the method is the differing laboratory procedures, which yield different results for the same samples. The use of pre-leaching with ammonium acetate removes adsorbed Sr and yields lower ^{87}Sr/^{86}Sr ratios than a more aggressive one-step HCl leaching method [*Melezhik et al.*, 2001; *Rodler et al.*, 2017]. An intermediate approach for limestones is the use of 0.5 M acetic acid for a short time (5–10 min), which predominantly liberates calcite-associated Sr, thereby yielding lower Sr isotope ratios [e.g., *Frei et al.*, 2011].

Details on the neodymium isotope geochemistry are found in *DePaolo* [1988]. A "global average" εNd curve for the oceans since 800 Ma has been constructed [*Keto and Jacobsen*, 1988; *Macdougall*, 1991], although neodymium isotopes have been seldom used as a chemostratigraphic tool. Similar to Sr isotope secular curve, this curve shows ε_{Nd} values at the end of the Precambrian oceans not substantially different from those at present ones. There is a remarkable decrease of the average ε_{Nd} values (−5 to −15) in the time interval between 700 and 550 Ma. Despite the precision of modern instruments, the scatter in measured values is substantial, limiting the use of ε_{Nd} in chemostratigraphic studies. Even so, secular ε_{Nd} variations coupled with δ^{13}C and δ^{53}Cr have been reported from Ediacaran rocks from Uruguay [*Frei et al.*, 2011, 2013] and Brazil [*Dantas et al.*, 2009], yielding valuable information regarding the tectonic evolution of the basin.

1.2.9. Osmium and Lithium Isotopes

The temporal variations of the ^{187}Os/^{188}Os ratio are preserved in several marine depositional environments, where osmium is an ultra trace element [*Peckeur-Ehrenbrink and Ravizza*, 2000]. Several developments over the last three decades have allowed direct measuring of ^{187}Os/^{188}Os

ratio and osmium concentration in seawater, river water, and rain, improving the knowledge on the surficial cycle of osmium [*Sharma et al.*, 1997; *Peckeur-Ehrenbrink and Ravizza*, 2000]. *Ravizza and Peucker-Ehrenbrink* [2003] have observed a decline of about 25% in the marine ^{187}Os/^{188}Os record that predated the Cretaceous-Paleocene transition (K-Pg) and that coincides with a warming in the late Maastrichtian. They have interpreted this osmium isotope ratio decline as a chemostratigraphic marker of the Deccan volcanism which was responsible for a transient global warming event (3–5 °C) and likely one of the causes of the K-Pg mass extinction.

Precambrian-to-Pleistocene marine osmium isotope records, particularly the Cenozoic and Mesozoic ones, and interpretations of their temporal variations have been reviewed by *Peckeur-Ehrenbrink and Ravizza* [2012]. Although the Cenozoic seawater ^{187}Os/^{188}Os mimics the marine ^{87}Sr/^{86}Sr record and suggests that both reflect continental weathering linked to climatic or tectonic processes, these two marine isotope systems differ fundamentally from each other [*Peckeur-Ehrenbrink and Ravizza*, 2000]. The marine residence time of osmium is distinctly shorter, allowing to record short-term fluctuations (e.g., glacial-interglacial periods), something that escapes to the buffered marine strontium isotope system. This difference between these two systems allows discrimination between climatic and tectonic forcings. Besides, large-amplitude changes in the marine ^{187}Os/^{188}Os record can be useful as chemostratigraphic event markers [*Peckeur-Ehrenbrink and Ravizza*, 2012].

The decline of atmospheric CO_2 has a potential role in initiating glaciation and its increase of terminating it [*Vandenbroucke et al.*, 2010]. Both cases involve changes in silicate weathering rates [*Lenton et al.*, 2012; *Ghienne et al.*, 2014]. The change of ^{187}Os/^{188}Os ratios during glacial periods may represent a response to change in silicate weathering, but does not help in tracing the weathering rate or processes involved [*Finlay et al.*, 2010]. The behavior of Li isotopes, however, is solely controlled by silicate weathering processes and, therefore, gives a unique insight into CO_2 drawdown and climate stabilization [*Pogge von Strandmann et al.*, 2017].

Biological processes do not lead to lithium isotope fractionation [*Pogge von Strandmann et al.*, 2017], and carbonate weathering does not affect Li isotope signals [*Dellinger et al.*, 2017]. The δ^7Li of primary silicate rocks have a narrow range [*Sauzeat et al.*, 2015] if compared to the high variability of modern rivers which reflects weathering processes, particularly the extent of preferential uptake of ^{6}Li into secondary minerals [*Dellinger et al.*, 2017]. Marine carbonates have a negligible sink of Li [*Marriott et al.*, 2004; *Pogge von Strandmann et al.*, 2013].

A comprehensive review on lithium isotope geochemistry is found in *Tomascak et al.* [2016] and *Penniston-Dorland et al.* [2017] in which the possibility of use of Li isotope in chemostratigraphy has been overlooked. Lithium isotope chemostratigraphy of Late Ordovician bulk carbonate sections and brachiopods in Anticosti Island, Canada [*Achab et al.,* 2013] (Pointe Laframboise Ellis Bay West), and of an equivalent shale section at Dob's Linn, United Kingdom [*Finlay et al.,* 2010; *Melchin et al.,* 2013], was presented by *Pogge von Strandmann et al.* [2017]. In all sections in that study, the relative timings of δ^7Li and the Hirnantian carbon isotope excursion (HICE) are similar, suggesting that Li isotope excursions occur contemporaneously, consistent with the Li residence time in the ocean (1 Myr). The positive δ^7Li excursion during the Hirnantian cooling event compares well to negative δ^7Li during warming events [*Pogge von Strandmann et al.,* 2013; *Lechler et al.,* 2015].

1.2.10. Elemental Chemostratigraphy

Elemental chemostratigraphy (element and element ratios) is a supplementary, useful tool in stratigraphy, and Mo, Ir, V, Ni, Cu, P, Hg, REEs, and Fe are among the most used elements, while Mo/Al, U/Mo, Rb/K, V/Cr, Zr/Ti, I/Ca, Li/Ca, B/Ca, Sr/Ca, Mg/Ca, Mo/Th, V/Th, and Th/U ratios seem to be particularly interesting. Paleoceanographic applications of trace-metal concentration data have been reviewed by *Algeo and Rowe* [2012].

Iron speciation has been widely used to determine the redox state of ancient basins. The method involves sequential extraction procedures to extract highly reactive iron (oxide, carbonates, and sulfide) and compare their concentration to total iron (Fe_{HR}/Fe_T; *Canfield,* 1989; *Shen et al.,* 2003; *Poulton and Canfield,* 2005). Sediments deposited in an oxygenated water column yield Fe_{HR}/Fe_T lower than 0.38 [*Canfield,* 1989]. Furthermore, the sulfide-bound iron (Fe_P) can be compared to highly reactive iron (Fe_P/Fe_{HR}), with values higher than 0.8 characterizing sulfidic (euxinic) basins [*Canfield et al.,* 2008]. Iron speciation chemostratigraphy has been applied successfully to sedimentary units of different ages, from the Archean to recent [e.g., *Shen et al.,* 2003; *Poulton et al.,* 2004; *Canfield et al.,* 2008; *Lyons et al.,* 2009; *Johnston et al.,* 2010; *Scott et al.,* 2011; *Hammarlund et al.,* 2012; *Frei et al.,* 2013].

In sediments deposited immediately after major glacial events, Hg tends to concentrate as a result from leaching of volcanogenic Hg from land surface and accumulation along argillaceous sediments [*Santos et al.,* 2001]. This element is usually found in low geological background concentrations, and this makes this trace element suitable for identifying accumulation pulses in sediments that can be tentatively related to weathering processes and thus to climatic changes.

Carbon dioxide buildup in the atmosphere during the Neoproterozoic glacial events resulted from volcanism that led to enhanced greenhouse effect, ice melting, and cap carbonate deposition [e.g., *Hoffman et al.,* 1998a; *Hoffman,* 2011]. Besides, intense volcanism may have witnessed the P-T and Cretaceous-Paleogene transition (K-Pg) and was, perhaps, co-responsible for dramatic climatic changes and thus for the decrease in biodiversity and mass extinction [e.g., *Keller,* 2005; *Archibald et al.,* 2010]. *Sial et al.* [2010b] demonstrated the use of Hg chemostratigraphy of the cap carbonates to document intense volcanism and resultant CO_2 buildup in the atmosphere, following the Neoproterozoic snowball events. Moreover, Hg chemostratigraphy was applied to investigate the relationships between large igneous province (LIP) activity, abrupt environmental changes, and mass extinctions [e.g., *Nascimento-Silva et al.,* 2011, 2013; *Sanei et al.,* 2012; *Sial et al.,* 2013a, 2014, 2016, 2017, this volume; *Adatte et al.,* 2015; *Grasby et al.,* 2013, 2015, 2017; *Percival et al.,* 2015, 2017; *Font et al.,* 2016, 2018; *Thibodeau et al.,* 2016; *Charbonnier et al.,* 2017; *Jones et al.,* 2017; *Thibodeau and Bergquist,* 2017; *Keller et al.,* 2018). To assure that the measured Hg contents result from true Hg loading to the environment, it is necessary to examine Hg/TOC ratios for chemostratigraphy [e.g., *Grasby et al.,* 2015; *Percival et al.,* 2015]. Mercury enrichments in sedimentary successions that recorded the mid-Cenomanian Event and Oceanic Anoxic Event 2 (OAE2) in the Late Cretaceous have been regarded by *Scaife et al.* [2017] as a marker for submarine LIP volcanism, and Hg enrichment recorded in the PETM is assumed to be related to volcanic activity of the North Atlantic Igneous Province (NAIP) (e.g., *Keller et al.,* 2018). Hg is doubtless a good benchmark for high volcanic activity, but normalization by TOC is in some cases problematic if TOC values are <0.2, leading to exaggerated peaks.

Mo and V chemostratigraphy may be useful in the investigation of the redox state of deep ocean water. Mo is a redox-sensitive element, scavenged from seawater into sediments in the form of $MoSxO_{4-}x^{2-}$, under anoxic conditions [*Wen et al.,* 2015]. The transfer of aqueous Mo to the sediment can be increased by means of metal-oxyhydroxide particulate shuttles, but aqueous U is not affected by this process [*Tribovillard et al.,* 2012]. Therefore, an increase in U/Mo ratio may suggest oxic conditions [e.g., *Sosa-Montes et al.,* 2017]. According to *Scheffler et al.* [2003], certain elemental ratios can be useful as proxies for investigation of salinity variation (Rb/K), redox state (V/Cr), or provenance (Zr/Ti). Mo and V can be normalized with Th (Mo/Th and V/Th) and, together with other redox-sensitive trace elements such as Ni, Zn, and Pb, can be used to determine redox variations in ancient sedimentary successions [*Spangenberg et al.,* 2014].

REE has been extensively used in different types of sedimentary rocks, often in combination with yttrium (REEY). The most widely used proxies are Ce, Eu, and Pr anomalies (Ce/Ce*, Eu/Eu*, Pr/Pr*), Y/Ho, La/Yb, and ΣREE, which can be applied to shales, carbonates, BIF, cherts, phosphorites, and other fine-grained rocks [*Elderfield and Greaves*, 1982; *Liu et al.*, 1988; *Bau and Dulski*, 1996; *Kato et al.*, 2006; *Lawrence and Kamber*, 2006; see *Sial et al.*, 2015b for an overview of proxies]. REE chemostratigraphy has been applied to Archean [*Kamber et al.*, 2014], Paleoproterozoic [*Bau and Dulski*, 1996], Mesoproterozoic [*Azmy et al.*, 2009], Neoproterozoic [*Tribovillard et al.*, 2006; *Frimmel*, 2009; *Sansjofre et al.*, 2014; *Spangenberg et al.*, 2014; *Gaucher et al.*, 2015; *Sial et al.*, 2015b; *Hu et al.*, 2016; *Rodler et al.*, 2016], and Phanerozoic successions [*Schmitz et al.*, 1988; *Lécuyer et al.*, 2004; *Fio et al.*, 2010].

The redox behavior of iodine is well known [*Broecker et al.*, 1982]. Besides, it is also known that there is a linear covariation between carbonate-associated iodine (CAI) and IO_3^- during calcite precipitation, but I^- is completely excluded [*Lu et al.*, 2010]. This trait, coupled with the residence time of iodine in seawater (300 ky; *Broecker et al.*, 1982) and concentration near 450 nM in modern ocean, makes I/Ca (or I/Ca + Mg) ratios in carbonates a robust indicator of the presence of IO_3^- and hence oxygen in the water column. Therefore, surface ocean oxygenation has been investigated using I/Ca ratios as a paleoredox indicator [e.g., *Hardisty et al.*, 2014].

Li/Ca and B/Ca in carbonates are regarded as proxies for carbonate saturation state [*Hall and Chan*, 2004; *Hall et al.*, 2005; *Lear and Rosenthal*, 2006; *Yu and Elderfield*, 2007; *Foster*, 2008], and Mg/Ca ratios of foraminiferal shells have been regarded as useful paleothermometer to determine ocean temperature. The difference in the Mg/Ca ratio of the foraminiferal shell and that from a baseline value (defined by the global ocean Mg and Ca concentration) when calibrated for the vital effects of the organism is a function of temperature [e.g., *Lea et al.*, 2000; *Lear et al.*, 2000]. The baseline composition of seawater is relatively simple to infer, once both Mg and Ca have long residence times in the oceans (>1 Ma) and are major components of ocean salts.

1.3. CHEMOSTRATIGRAPHY AND CHRONOSTRATIGRAPHIC BOUNDARIES

The International Commission on Stratigraphy (ICS) recognizes the existence of one hundred fourteen chronostratigraphic boundaries. Sixty-seven sections straddling chronostratigraphic boundaries were internationally agreed upon as reference points to define the lower boundaries of stages on the geologic time scale, the Global Boundary Stratotype Section and Point (GSSP), and a golden spike is placed precisely at the boundary defined. Accessibility and degree of representativity of the same boundary on sections worldwide are among the most

important criteria in the GSSP selection. Since GSSPs require well-preserved sections of rock without interruptions in sedimentation, and since most are defined by different biozones, defining them becomes more difficult as one goes further back in time in the Precambrian.

So far, chemostratigraphy has been overlooked as a formal criterion on GSSP selection. Carbon isotope excursions (CIE) have been reported only from seven of the established GSSPs [*Cooper et al.*, 2001; *Dupuis et al.*, 2003; *Peng et al.*, 2004; *Knoll et al.*, 2006; *Xu et al.*, 2006; *Aubry et al.*, 2007; *Goldman et al.*, 2007; *Schmitz et al.*, 2011; *Keller et al.*, 2018], probably due to the absence of carbonate rocks in several chronostratigraphic boundary sections. Only in the selection of the Cretaceous-Paleogene (K-Pg; *Molina et al.*, 2006, 2009) and the Paleocene-Eocene (PETM; *Aubry et al.*, 2007) GSSPs was carbon isotope chemostratigraphy one of the criteria, and hydrogen isotopes in the Pleistocene–Holocene GSSP [*Walker et al.*, 2009]. In addition, oxygen isotopes have been reported from two other GSSPs [*Steininger et al.*, 1997; *Hilgen et al.*, 2009]. Heavy element (e.g., Ir, Os, Hg) enrichments at the Cretaceous-Tertiary boundary are well known since the seminal paper of *Alvarez et al.* [1980] and later studies [e.g., *Schmitz et al.*, 1988; *Frei and Frei*, 2002; *Sial et al.*, 2016; *Keller et al.*, 2018, and references therein].

1.4. CHEMOSTRATIGRAPHY AS FORMAL STRATIGRAPHIC METHOD

The stratigraphic record shows changes of the concentration of certain elements with time [*Morante et al.*, 1994], as a function of geological conditions including, but not limited to, tectonic, climatic, redox, oceanographic, biotic, and other processes. Chemostratigraphy enables not only apparently uniform thick successions to be subdivided and correlated with coeval strata located elsewhere [*Ramkumar*, 1999] but also thinner and more heterogeneous sedimentary records. Initially, chemostratigraphy was applied to recognize unique geochemical compositions for characterizing depositional units and correlating them with coeval strata elsewhere and found its use in the stratigraphic location of boundaries and later expanded to examination of specific causes to the stratigraphic variations of geochemical compositions [*Ramkumar et al.*, 2010, 2011]. The utility of chemostratigraphy for age determination was demonstrated through documentation of stratigraphic variations of isotopic trends, beginning with oxygen isotopes. Linear, secular, cyclic, and perturbed trends have been recognized, which are utilized for stratigraphic classification and spatial correlation [e.g., *Zachos et al.*, 2001; *Ramkumar*, 2014, 2015]. In addition, the chemozones, calibrated with absolute time, are in use as chemochrons. Although chemostratigraphy is firmly recognized as a valid stratigraphic method since more than thirty years [e.g., *Berger and Vincent*, 1981], given to its sensitivity and wide applications, time has come to recognize this technique as an individual method of stratigraphy. The International Stratigraphic Commission (ISC) defines stratigraphy as "the description of all rock bodies forming the Earth's crust and their organization into distinctive, useful, mappable units based on their inherent properties or attributes… in order to establish their distribution and relationship in space and their succession in time, and to interpret geologic history" [*Salvador*, 1994]. Chemostratigraphy "recognizes and organizes" rock bodies into useful units based on their inherent chemical properties. Most importantly, it helps in establishing the distribution and relationships of these units in space and time and, especially, interpreting geological history. In many cases, the resolution of chemostratigraphy proves to be better than other conventional methods of stratigraphy and can be refined by improving sampling resolution, although cyclostratigraphy can be even better, providing a resolution on Milankovitch time scales, reaching back far into the Cenozoic. In this regard, it can be stated that chemostratigraphy as an independent stratigraphic method can serve well even where other methods fail or have limitations. Chemostratigraphy is complementary to other types of stratigraphic units, such as lithostratigraphy, biozones, and magnetostratigraphy. With these attributes in mind, we are of the opinion that chemostratigraphy can be recognized as an independent standard method of stratigraphic classification.

ACKNOWLEDGMENTS

We express our thankfulness to the American Geophysical Union (AGU)/Wiley for the invitation to compile a special publication on "chemostratigraphy across major chronological eras." Three anonymous reviewers are thanked for thorough, critical analysis of the original manuscript. This is the contribution n. 287 from the Nucleus for Geochemical Studies–Stable Isotope Laboratory (NEG-LABISE), Department of Geology, Federal University of Pernambuco, Brazil.

REFERENCES

Achab, A., Asselin, E., Desrochers, A., Riva, J.F., 2013. The end-Ordovician chitinozoan zones of Anticosti Island, Quebec: Definition and stratigraphic position. Review of Palaeobotany and Palynology *198*, 92–109.

Adatte, T., Keller, G., Schoene, B., Samperton, K.M., Font, E., Sial, A.N., Lacerda, L.D., Punekar, J., Fantasia, A., Khadri, S., 2015. Paleoenvironmental influence of Deccan volcanism relative to the KT extinction. Geological Society of America Abstracts with Programs *47*(7), 210, Baltimore.

Ader, A., Sansjofre, P., Halverson, G.P., Busigny, V., Trindade, R.I.F., Kunzmann, M., Nogueira, A.C.R., 2014. Ocean redox structure across the Late Neoproterozoic Oxygenation Event: A nitrogen isotope perspective. Earth and Planetary Science Letters *396*, 1–13.

Algeo, T.J., Rowe, H., 2012. Paleoceanographic applications of trace-metal concentration data. Chemical Geology *324–325*, 6–18.

Algeo, T., Rowe, H., Hower, J.C., Schwark, L., Herrmann, A., Heckel, P., 2008. Changes in ocean denitrification during late carboniferous glacial interglacial cycles. Nature Geoscience *1*, 709–714.

Allen, P.A., Etienne, J.L., 2008. Sedimentary challenge to snowball Earth. Nature Geoscience *1*, 817–825.

Alvarez, L.W., Alvarez, W., Asaro, F., Michel, H.V., 1980. Extraterrestrial cause for the Cretaceous–Tertiary extinction. Science *208*, 1095–1108.

Archibald, J.D., Clemens, W.A., Padian, K., Rowe, T., Macleod, N., Barrett, P.M., Gale, A., Holroyd, P., Sues, H.D., Arens, N.C., Horner, J.R., Wilson, G.P., Goodwin, M.B., Brochu, C.A., Lofgren, D.L., Hurlbert, S.H., Hartman, J.H., Eberth, D.A., Wignall, P.B., Currie, P.J., Weil, A., Prasad, G.V., Dingus, L., Courtillot, V., Milner, A., Milner, A., Bajpai, S., Ward, D.J., Sahni, A., 2010. Cretaceous extinctions: Multiple causes. Science *328*, 973–976.

Arnold, G.L., Anbar, A.D., Barling, J., Lyons, T.W., 2004. Molybdenum isotope evidence for widespread anoxia in mid-Proterozoic oceans. Science *304*, 87–90.

Aubry, M.-P., Ouda, K., Dupuis, C., Berggren, W.A., Van Couvering, J.A., Working Group on the Paleocene/Eocene Boundary, 2007. The Global Standard Stratotype-section and Point (GSSP) for the base of the Eocene Series in the Dababiya section (Egypt). Episodes *30*, 271–286.

Azmy, K., Sylvester, P., de Oliveira, T.F., 2009. Oceanic redox conditions in the Late Mesoproterozoic recorded in the upper Vazante Group carbonates of São Francisco Basin, Brazil: Evidence from stable isotopes and REEs. Precambrian Research *168*(3), 259–270.

Azmy, K., Lavoie, D., Wang, Z., Brand, U., Al-Aasm, I., Jackson, S., Girard, I., 2013. Magnesium isotope and REE composition of Lower Ordovician carbonates from eastern Laurentia: Implications for the origin of dolomites and limestones. Chemical Geology *356*, 64–75.

Bao, H., Lyons, J.R., Zhou, C., 2008. Triple oxygen isotope evidence for elevated CO_2 levels after a Neoproterozoic glaciations. Nature *453*, 504–506.

Bao, H., Fairchild, I.J., Wynn, P.M., Spötl, C., 2009. Stretching the envelope of past surface environments: Neoproterozoic glacial lakes from Svalbard. Science *323*, 119–122.

Baskaran, M., 2012. Handbook of Environmental Isotope Geochemistry. Advances in Isotope Geochemistry. Springer-Verlag, Berlin, 951 pages.

Bau, M., Dulski, P., 1996. Distribution of yttrium and rare-earth elements in the Penge and Kuruman iron-formations, Transvaal Supergroup, South Africa. Precambrian Research *79*(1–2), 37–55.

Beard, B.L., Johnson, C.M., Skulan, J.L., Nealson, K.H., Cox, L., Sun, H., 2003. Application of Fe isotopes to tracing the geochemical and biological cycling of Fe. Chemical Geology *195*, 87–117.

Beaumont, V., Robert, F., 1999. Nitrogen isotope ratios of kerogens in Precambrian cherts: A record of the evolution of atmosphere chemistry? Precambrian Research *96*, 63–82.

Bekker, A., Karhu, J.A., Kaufman, A.J., 2006. Carbon isotope record for the onset of the Lomagundi carbon isotope excursion in the Great Lakes area, North America. Precambrian Research *148*, 145–180.

Berner, R.A., 1990. Atmospheric carbon dioxide levels over Phanerozoic time. Science *249*, 1382–1386.

Berger, W.H., Vincent, E., 1981. Chemostratigraphy and biostratigraphic correlation: Exercises in systemic stratigraphy. Oceanologica Acta *1981*(SP), 115–127.

Berner, E.K., Berner, R.A., 1987. The Global Water Cycle: Geochemistry and Environment. Prentice Hall, New York, 397 p.

Bijl, P.K., Schouten, S., Sluijs, A., Reichart, G.-J., Zachos, J.C., Brinkhuis, H., 2009. Early Palaeogene temperature evolution of the southwest Pacific Ocean. Nature *461*, 776–779.

Blättler, C.L., Henderson, G.M., Jenkys, H.C., 2012. Explaining the Phanerozoic Ca isotope history of seawater. Geology *40*, 843–846.

Bodin, S., Fiet, N, Godet, A., Matera, V., Westermann, S., Clement, A., Janssen, N.M.M., Stille, P, Foellmi, K.B., 2009. Early Cretaceous (late Berriasian to early Aptian) palaeoceanographic change along the northwestern Tethyan margin (Vocontian Trough, southeastern France): $\delta^{13}C$, $\delta^{18}O$ and Sr-isotope and whole-rock records. Cretaceous Research *30*, 1247–1262.

Brenot, A., Cloquet, C., Vigier, N., Carignan, J., France-Lanord, C., 2008. Magnesium isotope systematics of the lithologically varied Moselle river basin, France. Geochimica et Cosmochim Acta *72*, 5070–5089.

Bristow, T.F., Kennedy, M.J., 2008. Carbon isotope excursions and the oxidant budget of the Ediacaran atmosphere and ocean. Geology *36*, 863–866.

Broecker, W.S., Peng, T.H., Beng, Z., 1982. Tracers in the Sea. Lamont-Doherty Geological Observatory, Columbia University, New York, 690 p.

Burke, W.H., Denison, R.E., Hetherington, E.A., Koepnick, R.B.,Nelson, H.F., Otto, J.B., 1982. Variation of seawater $^{87}Sr/^{86}Sr$ throughout Phanerozoic time. Geology *10*, 516–519.

Calver, C.R., Black, L.P., Everard, J.L., Seymour, D.B., 2004. U-Pb zircon age constraints on late Neoproterozoic glaciation in Tasmania. Geology *32*, 892–896.

Canfield, D.E., 1989. Reactive iron in marine sediments. Geochimica et Cosmochimica Acta *53*, 619–632.

Canfield, D.E., 1998. A new model for Proterozoic ocean chemistry. Nature *396*, 450–453

Canfield, D.E., Teske, A., 1996. Late Proterozoic rise in atmospheric oxygen concentration inferred from phylogenetic and sulphur-isotope studies. Nature *381*, 127–132.

Canfield, D.E., Poulton, S.P., Knoll, H., Narbonne, G.M., Ross, G., Goldberg, T., Strauss, H., 2008. Ferruginous conditions dominated later Neoproterozoic deep-water chemistry. Science *321*, 949–952.

Chang, V.T.-C., Masishima, A., Belshaw, N.S., Onions, R.K., 2003. Purification of Mg from low Mg biogenic carbonates for isotope ratio determination using multi-collector ICP-MS. Journal of Analytical Atomic Spectrometry *18*, 296–301.

Charbonnier, G., Morales, C., Duchamp-Alphonse, S., Westermann, S., Adatte, T., Föllmi, K.B., 2017. Mercury enrichment indicates volcanic triggering of Valanginian environmental change. Scientific Reports 7, 40808, January 2017. doi:10.1038/srep40808.

Chiglino, L., Gaucher, C., Sial, A.N., Bossi, J., Ferreira, V.P., Pimentel, M.M., 2010. Chemostratigraphy of Mesoproterozoic and Neoproterozoic carbonates of the Nico Pérez Terrane, Río de la Plata Craton, Uruguay. Precambrian Research *182*, 313–336.

Claypool, G.E., Holser, W.T., Kaplan, I.R., Sakai, H., Zak, I., 1980. The ages, curves of sulfur and oxygen isotopes in marine sulfate and their mutual interpretation. Chemical Geology *28*, 199–260.

Cooper, R.A., Nowlan, G.S., Williams, S.H., 2001. Global Stratotype Section and Point for base of the Ordovician System. Episodes *24*, 19–28.

Corsetti, F.A., Kaufman, A.J., 2003. Stratigraphic investigations of carbon isotope anomalies and Neoproterozoic ice ages in Death Valley, California. Geological Society of America Bulletin *115*, 916–932.

Cremonese, L., Struck, U., Shields-Zhou, G., Ling, H., Och, L., 2009. $\delta^{15}N$ chemostratigraphy of Ediacaran–Cambrian sections of South China. Supplement Geochimica et Cosmochimica Acta *73*(13S), A-251.

Cremonese, L., Shields-Zhou, G., Struck, U., Ling, H.F., Och, L., Chen, X., Li, D., 2013. Marine biogeochemical cycling during the early Cambrian constrained by a nitrogen and organic carbon isotope study of the Xiaotan section, South China. Precambrian Research *225*, 148–165.

Cremonese, L., Shields-Zhou, G., Struck, U., Ling, H.F., Och, L., 2014. Nitrogen and organic carbon isotope stratigraphy of the Yangtze Platform during the Ediacaran–Cambrian transition in South China. Palaeogeography, Palaeoclimatology, Palaeoecology *398*, 165–186.

Crowe, S.A., Døssing, L.N., Beukes, N.J., Bau, M., Kruger, S.J., Frei, R., Canfield, D.E., 2013. Atmospheric oxygenation three billion years ago. Nature *501*, 535–538.

Dantas, E.L., Alvarenga, C.J.S., Santos, R.V., Pimentel, M.M., 2009. Using Nd isotopes to understand the provenance of sedimentary rocks from a continental margin to a foreland basin in the Neoproterozoic Paraguay Belt, Central Brazil. Precambrian Research *170*, 1–12.

Davies, C.L., Surridge, B.W.J., Gooddy, D.C., 2014. Phosphate oxygen isotopes within aquatic ecosystems: Global data synthesis and future research priorities. Science of the Total Environment *496*, 563–575.

Dellinger, M., Bouchez, J., Gaillardet, J., Faure, L., Moureau, J., 2017. Tracing weathering regimes using the lithium isotope composition of detrital sediments. Geology *45*, 411–414.

DePaolo, D.J., 1988. Neodymium Isotope Geochemistry: An Introduction. Springer-Verlag, Berlin, New York, 187 p.

Dera, G., Puceat, E., Pellenard, P., Neige, P., Delsate, D., Joachimski, M.M., Reisberg, L., Martinez, M., 2009. Water mass exchange and variations in seawater temperature in the NW Tethys during the Early Jurassic; evidence from neodymium and oxygen isotopes of fish teeth and belemnites. Earth and Planetary Science Letters *286*, 198–207.

Døssing, L.N., Dideriksen, K., Stipp, S.L.S., Frei, R., 2011. Reduction of hexavalent chromium by ferrous iron: A process of chromium isotope fractionation and its relevance to natural environments. Chemical Geology *285*, 157–166.

Dupuis, C., Aubry, M.-P., Steurbaut, E., Berggren, W.A., Ouda, K., Magioncalda, R., Cramer, B.S., Kent, D.V., Speijer, R.P., Heilmann-Clausen, C., 2003. The Dababiya Quarry section: Lithostratigraphy, clay mineralogy, geochemistry and paleontology. Micropaleontology 49, 41–59.

Eagle, R.A., Schauble, E.A., Tripati, A.K., Tutken, T., Hulbert, R.C., Eiler, J.M., 2010. Body temperatures of modern and extinct vertebrates from ^{13}C–^{18}O bond abundances in bioapetite. Proceedings of the National Academy of Sciences 107, 10377–10382.

Eiler, J.M., 2007. "Clumped-isotope" geochemistry: The study of naturally-occurring, multiply-substituted isotopologues. Earth and Planetary Science Letters 262, 309–327. doi:10.1016/j.epsl.2007.08.020.

Elderfield, H., Greaves, M.J., 1982. The rare earth elements in seawater. Nature 296, 214–219.

Emiliani, C. 1955. Pleistocene temperatures. The Journal of Geology 63(6), 538–578.

Falkowski, P., 2003. Biogeochemistry of primary production in the sea. Treatise on Geochemistry 8, 185–213.

Fantle, M.S., Tipper, E.T., 2014. Calcium isotopes in the global biogeochemical Ca cycle: Implications for development of a Ca isotope proxy. Earth-Science Review 129, 148–177.

Farquhar, J., Bao, H., Thiemens, M., 2000. Atmospheric influence of Earth's earliest sulfur cycle. Science 289, 756–758.

Finlay, A.J., Selby, D., Grocke, D.R., 2010. Tracking the Hirnantian glaciation using Os isotopes. Earth and Planetary Science Letters 293, 339–348.

Fio, K., Spangenberg, J.E., Vlahović, I., Sremac, J., Velić, I., Mrinjek, E., 2010. Stable isotope and trace element stratigraphy across the Permian–Triassic transition: A redefinition of the boundary in the Velebit Mountain, Croatia. Chemical Geology 278(1), 38–57.

Flament, N., Coltice, N., Rey, P.F., 2011. The evolution of $^{87}Sr/^{86}Sr$ of marine carbonates does not constrain continental growth. Precambrian Research 229, 177–188. doi:10.1016/j.precamres.2011.10.009.

Font, F., Adatte, T., Sial, A.N., Lacerda, L.D., Keller, G., Punekar, J., 2016. Mercury anomaly, Deccan Volcanism and the end-Cretaceous Mass Extinction. Geology 44, 171–174. doi:10.1130/G37451.1.

Font, E., Adatte, T., Andrade, M., Keller, G., Mbabi Bitchong, A., Carvallo, C., Ferreira, J., Diogo, Z., Mirão, J., 2018. Deccan volcanism induced high-stress environment during the Cretaceous–Paleogene transition at Zumaia, Spain: Evidence from magnetic, mineralogical and biostratigraphic records. Earth and Planetary Science Letters 484, 53–66.

Foster, G.L., 2008. Seawater pH, pCO_2 and $[CO_3 2-]$ variations in the Caribbean Sea over the last 130 kyr: A boron isotope and B/Ca study of planktic foraminifera. Earth Planetary Science Letters 271, 254–66.

Foster, G.L., Rae, J.W.B., 2016. Reconstructing ocean pH with boron isotopes in foraminifera. Annual Reviews Earth Planetary Science 44, 207–37.

Foster, G.L., Pogge von Strandmann, P.A.E., Rae, J.W.B., 2010 Boron and magnesium isotopic composition of seawater. Geochemistry, Geophysics, Geosystems 11, 1–10.

Frei, R., Frei, K.M., 2002. A multi-isotopic and trace element investigation of the Cretaceous–Tertiary boundary layer at Stevns Klint, Denmark–inferences for the origin and nature of siderophile and lithophile element geochemical anomalies. Earth and Planetary Science Letters 203(2), 691–708.

Frei, R., Gaucher, C., Poulton, S.W., Canfield, D.E., 2009. Fluctuations in Precambrian atmospheric oxygenation recorded by chromium isotopes. Nature 46, 250–254.

Frei, R., Gaucher, C., Døssing, L.N., Sial, A.N., 2011. Chromium isotopes in carbonates: A tracer for climate change and for reconstructing the redox state of ancient seawater. Earth Planetary Sciences Letters 312, 114–125.

Frei, R., Gaucher, C., Stolper, D., Canfield, D.E., 2013. Fluctuations in late Neoproterozoic atmospheric oxidation: Cr isotope chemostratigraphy and iron speciation of the late Ediacaran lower Arroyo del Soldado Group (Uruguay). Gondwana Research 23, 797–811.

Frei, R., Døssing, L.N., Gaucher, C., Boggiani, P.C., Frei, K.M., Bech Árting, T., Crowe, S.A., Freitas, B.T., 2017. Extensive oxidative weathering in the aftermath of a late Neoproterozoic glaciation: Evidence from trace element and chromium isotope records in the Urucum district (Jacadigo Group) and Puga iron formations (Mato Grosso do Sul, Brazil). Gondwana Research 49, 1–20.

Friedrich, O., Norris, R.D., Erbacher, J., 2012. Evolution of middle to Late Cretaceous oceans: A 55 m.y. record of Earth's temperature and carbon cycle. Geology 40, 107–110.

Frimmel, H.E., 2008. REE geochemistry of Neoproterozoic carbonates: Deviations from normal marine signatures. Abstract, 33 International Geological Congress, Oslo, Norway.

Frimmel, H.E., 2009. Trace element distribution in Neoproterozoic carbonates as palaeoenvironmental indicator. Chemical Geology 258, 338–353.

Frimmel, H.E., 2010. On the reliability of stable carbon isotopes for Neoproterozoic chemostratigraphic correlation. Precambrian Research 182, 239–252.

Frost, C.D., von Blanckenburg, F., Schoenberg, R., Frost, B.R., Swapp, S.M., 2007. Preservation of Fe isotope heterogeneities during diagenesis and metamorphism of banded iron formation. Contributions Mineralogy Petrology 153, 211–235.

Galy, A., Bar-Matthews, M., Halicz, L., O'Nions, R.K., 2002. Mg isotopic composition of carbonate: Insight from speleothem formation. Earth Planetary Science Letters 201, 105–115.

Gaucher, C., Sial, A.N., Ferreira, V.P., Pimentel, M.M., Chiglino, L., Sprechmann, P., 2007. Chemostratigraphy of the Cerro Victoria Formation (Lower Cambrian, Uruguay): Evidence for progressive climate stabilization across the Precambrian–Cambrian boundary. Chemical Geology 237, 28–46.

Gaucher, C., Sial, A.N., Frei, R., 2015. Chemostratigraphy of Neoproterozoic banded iron formation (BIF): Types, age and origin. In: Ramkumar, M. (Ed.), Chemostratigraphy: Concepts Techniques and Applications. Elsevier, Amsterdam, pp. 433–449.

Geske, A., Goldstein, R.H., Mavromatis, V., Richter, D.K., Buhl, D., Kluge, T., John, C.M., Immenhauser, A., 2015. The magnesium isotope ($\delta^{26}Mg$) signature of dolomites. Geochimica et Cosmochimica Acta 149, 131–151.

Ghienne, J.-F., Desrochers, A., Vandenbroucke, T.R.A., Achab, A., Asselin, E., Dabard, M.-P., Farley, C., Loi, A., Paris, F., Wickson, S., Veizer, J., 2014. A Cenozoic-style scenario for the end-Ordovician glaciation. Nature Communications 5, 4485. doi:10.1038/ncomms5485.

Ghosh, P., Adkins, J., Affek, H., Balta, B., Guo, W., Schauble, E.A., Schrag, D., Eiler, J.M., 2006. ^{13}C-^{18}O bonds in carbonate minerals: A new kind of paleothermometer. Geochimica et Cosmochimica Acta 70, 1439–1456.

Goldman, D., Leslie, S.A., Nõlvak, J., Young, S., Bergström, S.M., Huff, W.D., 2007. The Global Stratotype Section and Point (GSSP) for the base of the Katian Stage of the Upper Ordovician Series at Black Knob Ridge, Southeastern Oklahoma, USA. Episodes 30, 258–270.

Grasby, S.E., Sanei, H., Beauchamp, B., Chen, Z., 2013. Mercury deposition through the Permo–Triassic Biotic Crisis. Chemical Geology 351, 209–16.

Grasby, S.E., Beauchamp, B., Bond, D.P.G., Wignall, P.B., Sanei, H., 2015. Mercury anomalies associated with three extinction events (Capitanian Crisis, Latest Permian Extinction and the Smithian/Spathian Extinction) in NW Pangea. Geological Magazine 153, 285–297.

Grasby, S.E., Shen, W., Yin, R., Gleason, J.D., Blum, J.D., Lepak, R.F., Hurley, J.P., Beauchamp, B., 2017. Isotopic signatures of mercury contamination in latest Permian oceans. Geology 45, 55–58.

Grotzinger, J.P., Fike, D.A., Fischer, W.W., 2011. Enigmatic origin of the largest-known carbon isotope excursion in Earth's history. Nature Geoscience 4, 285–291.

Guo, Q., Strauss, H., Kaufman, A.J., Schröder, S., Gutzmer, J., Wing, B., Baker, M.A., Bekker, A., Jin, Q., Kim, S.-T., Farquhar, J., 2009a. Reconstructing Earth's surface oxidation across the Archean–Proterozoic transition. Geology 3, 399–402.

Guo, W., Mosenfelder, J.L., Goddard, W.A., III, Eiler, J.M., 2009b. Isotopic fractionations associated with phosphoric acid digestion of carbonate minerals: Insights from first-principles theoretical modeling and clumped isotope measurements. Geochimica et Cosmochimica Acta 73, 7203–7225.

Gussone, N., Schmitt, A.-D., Heuser, A., Wmbacher, F., Dietzel, M., Tipper, E., Schiller, M., 2016. Calcium Isotope Geochemistry. Advances in Isotope Geochemistry. Springer, Berlin, Heidelberg, 260 p. doi:10.1007/978-3-540-68953-9.

Halevy, I., Johnston, D., Schrag, D., 2010. Explaining the structure of the Archean mass-independent sulfur isotope record. Science 329, 204–207

Hall, J.M., Chan, L.H., 2004. Li/Ca in multiple species of benthic and planktonic foraminifera: Thermocline, latitudinal, and glacial–interglacial variation. Geochimica Cosmochimica Acta 68, 529–545.

Hall, J.M., Chanb, T.L.-H., McDonough, W.F., Turekian, K.K., 2005. Determination of the lithium isotopic composition of planktic foraminifera and its application as a paleo-seawater proxy. Marine Geology 217, 255–265.

Halverson, G.P., Hoffman, P.F., Schrag, D.P., Maloof, A.C., Hugh, A., Rice, N., 2005. Towards a Neoproterozoic composite carbon-isotope record. Geological Society of America Bulletin 117, 1181–1207.

Halverson, G.P., Dudás, F.Ö., Maloof, A., Bowring, S.A., 2007. Evolution of the ^{87}Sr/^{86}Sr composition of Neoproterozoic seawater. Palaeogeography, Palaeoclimatology, Palaeoecology 256, 103–129.

Halverson, G.P., Hurtgen, M.T., Porter, S.M., Collins, A.S., 2010a. Neoproterozoic–Cambrian biogeochemical evolution. In: Gaucher, C., Sial, A.N., Halverson, G.P., Frimmel, H. (Eds.), Neoproterozoic-Cambrian Tectonics, Global Change and Evolution: A Focus on South Western Gondwana. Developments in Precambrian Geology 16. Elsevier, Amsterdam, Boston, pp. 351–365.

Halverson, G.P., Wade, B.P., Hurtgen, M.T., Barovich, K., 2010b. Neoproterozoic chemostratigraphy. In: Karhu, J., Sial, A.N., Ferreira, V.P. (Eds.), Precambrian Isotope Stratigraphy, special issue. Precambrian Research 182. Elsevier, Amsterdam, pp. 337–350.

Halverson, G.P., Poitrasson, F., Hoffman, P.F., Nedelec, A., Montel, J.M., Kirby, J., 2011. Fe isotope and trace element geochemistry of the Neoproterozoic syn-glacial Rapitan iron formation. Earth Planetary Science Letters 309, 100–112.

Hammarlund, E.U., Dahl, T.W., Harper, D.A., Bond, D.P., Nielsen, A.T., Bjerrum, C.J., Schovsbo, N.H., Schönlaub, H.P., Zalasiewicz, J.A., Canfield, D.E., 2012. A sulfidic driver for the end-Ordovician mass extinction. Earth and Planetary Science Letters 331, 128–139.

Hardisty, D.S., Lu, Z., Planavsky, N.J., Bekker, A., Philippot, P., Zhou, Z., Lyons, T.W., 2014. An iodine record of Paleoproterozoic surface ocean oxygenation. Geology 42, 619–622.

Heimann, A., Johnson, C.M., Beard, B.L., Valley, J.W., Roden, E.E., Spicuzza, M.J., Beukes, N.J., 2010. Fe, C, and O isotope compositions of banded iron formation carbonates demonstrate a major role for dissimilatory iron reduction in 2.5 Ga marine environments. Earth Planetary Science Letters 294, 8–18.

Hemming, N.G., Hönisch, B., 2007. Boron isotopes in marine carbonate sediments and the pH of the Ocean. Developments in Marine Geology 1, 717–734.

Herrle, J.O., Pross, J., Friedrich, O., Koßler, P., Hemleben, C., 2003. Forcing mechanisms for mid-Cretaceous black shale formation: Evidence from the Upper Aptian and Lower Albian of the Vocontian Basin (SE France). Palaeogeography, Palaeoclimatology, Palaeoecology 190, 399–426.

Higgins, J.A., Schrag, D.P., 2010. Constraining magnesium cycling in marine sediments using magnesium isotopes. Geochimica Cosmochimica Acta 74, 5039–5053.

Hilgen, F.J., Abels, H.A., Iaccarino, S., Krijgsman, H., Raffi, I., Sprovieri, R., Turco, E., Zachariasse, E.W., 2009. The Global Stratotype Section and Point (GSSP) of the Serravallian Stage (Middle Miocene). Episodes 32, 152–166.

Hoffman, P.F., 2011. Strange bedfellows: Glacial diamictite and cap carbonate from the Marinoan (635 Ma) glaciation in Namibia. Sedimentology 58, 57–119.

Hoffman, P.F., Kaufman, A.J., Halverson, G.P., Schrag, D.P., 1998a. A Neoproterozoic snowball Earth. Science 281, 1342–1346.

Hoffman, P.F., Kaufman, A.J., Halverson, G.P., 1998b. Comings and goings of global glaciations on a Neoproterozoic tropical platform in Namibia. GSA Today 8, 1–9.

Holser, W.T., 1997. Geochemical events documented in inorganic carbon isotopes. Palaeogeography, Palaeoclimatology, Palaeoecology *132*, 173–182.

Hönisch, B., Ridgwell, A., Schmidt, D.N., Thomas, E., Gibbs, S.J., Sluijs, A., Zeebe, R., Kump, L., Martindale, R.C., Greene, S.E., Kiessling, W., Ries, J., Zachos, J.C., Royer, D.L., Barker, S., Marchitto, T.M., Jr., Moyer, R., Pelejero, C., Ziveri, P., Foster, G.L., Williams, B., 2012. The geological record of ocean acidification. Science *335*, 1058. doi:10.1126/science.1208277.

Hu, R., Wang, W., Li, S.-K.,Yang, Y.-Z., Chen, F., 2016. Sedimentary environment of Ediacaran sequences of South China: Trace element and Sr-Nd isotope constraints. The Journal of Geology *124*, 769–789.

Huang, K.-J., Teng, F.-Z., Shen, B., Xiao, S., Lang, X., Ma, H.-R., Fu, Y., Peng, Y., 2016. Episode of intense chemical weathering during the termination of the 635 Ma Marinoan glaciation. Proceedings of the National Academy of Science *113*, 14904–14909.

Jacobsen, S.B., Kaufman, A.J., 1999. The Sr, C and O isotopic evolution of Neoproterozoic seawater. Chemical Geology *161*, 37–57.

Jarvis, I., Lignum, J.S., Gröcke, D.R., Jenkyns, H.C., Pearce, M.A., 2011. Black shale deposition, atmospheric CO_2 drawdown, and cooling during the Cenomanian-Turonian Oceanic Anoxic Event. Paleoceanography *26*, PA3201. doi:10.1029/2010PA002081.

Jenkyns, H.C., 2003. Evidence for rapid climate change in the Mesozoic–Palaeogene greenhouse world. Philosophical Transactions of the Royal Society of London *361*, 1885–1916.

Jenkyns, H.C., 2010. Geochemistry of oceanic anoxic events. Geochemistry, Geophysics, Geosystems *11*(3). doi:10.1029/2009GC002788.

Jenkyns, H.C., Clayton, C.J., 1986, Black shales and carbon isotopes from the Tethyan Lower Jurassic. Sedimentology *33*, 87–106.

Jenkyns, H.C., Gröcke, D.R., Hesselbo, S.P., 2001. Nitrogen isotope evidence for water mass denitrification during the early Toarcian (Jurassic) oceanic anoxic event. Paleoceanography *16*(6), 593–603. doi:10.1029/2000PA000558.

Jenkyns, H.C., Matthews, A., Tsikos, H., Erel, Y., 2007. Nitrate reduction, sulfate reduction, and sedimentary iron isotope evolution during the Cenomanian–Turonian oceanic anoxic event. Paleoceanography *22*, PA3208. doi:10.1029/2006PA001355.

Jenkyns, H., Schouten-Huibers, L., Schouten S., Sinninghe-Damste, J.S., 2012. Warm Middle Jurassic-early Cretaceous high-latitude sea surface temperature from the Southern Ocean. Climate of the Past *8*, 215–226.

Joachimski, M.M., Buggisch, W. 2002. Conodont apatite $\delta^{18}O$ signatures indicate climatic cooling as a trigger of the Late Devonian mass extinction. Geology *30*, 711–714.

Joachimski, M.M., Simon, L., Van Geldern, R., Lécuyer, C., 2005. Boron isotope geochemistry of Paleozoic brachiopod calcite: Implications for a secular change in the boron isotope geochemistry of seawater over the Phanerozoic. Geochimica et Cosmochimica Acta *69*, 4035–4044.

Johnson, C., Beard, L., 2006. Fe isotopes: An emerging technique for understanding modern and ancient biogeochemical cycles. GSA Today *16*, 4–10.

Johnson, C.M., Beard, B.L., Beukes, N.J., Klein, C., O'Leary, J.M., 2003. Ancient geochemical cycling in the Earth as inferred from Fe isotope studies of banded iron formations from the Transvaal Craton. Contributions to Mineralogy and Petrology *114*, 523–547.

Johnson, C.M., Beard, B.L., Albarede, F., 2004. Geochemistry of non-traditional stable isotopes. Reviews in Mineralogy and Geochemistry *55*, 454 pp.

Johnson, C.M., Beard, B.L., Roden, E.E., 2008. The iron isotope fingerprints of redox and biogeochemical cycling in modern and ancient oceans. Annual Reviews Earth Planetary Science *36*, 457–493.

Johnston, D.T., Poulton, S.W., Dehler, C., Porter, S., Husson, J., Canfield, D.E., Knoll, A.H., 2010. An emerging picture of Neoproterozoic ocean chemistry: Insights from the Chuar Group, Grand Canyon, USA. Earth and Planetary Science Letters *290*, 64–73.

Jones, C.E., Jenkyns, H.C., 2001. Seawater strontium isotopes, oceanic anoxic events, and seafloor hydrothermal activity in the Jurassic and Cretaceous. American Journal of Science *301*, 112–149.

Jones, D.S., Martini, A.M., Fike, A., Kaiho, K., 2017. A volcanic trigger for the Late Ordovician mass extinction? Mercury data from south China and Laurentia. Geology *45*(7), 631–634. doi:10.1130/G38940.1.

Kah, L.C., Sherman, A.G., Narbonne, G.M., Knoll, A.H., Kaufman, A.J., 1999. $\delta^{13}C$ stratigraphy of the Proterozoic by lot supergroup, Baffin Islands, Canada: Implications for regional lithostratigraphy correlations. Canadian Journal of Earth Sciences *36*, 313–332.

Kamber, B.S., Webb, G.E., Gallagher, M., 2014. The rare earth element signal in Archaean microbial carbonate: Information on ocean redox and biogenicity. Journal of the Geological Society *171*(6), 745–763.

Karhu, J.A., Holland, H.D., 1996. Carbon isotopes and the rise of atmospheric oxygen. Geology *24*, 867–870.

Karhu, J., Sial, A.N., Ferreira, V.P., 2010. Insights from precambrian isotope stratigraphy. Precambrian Research *182*, 239–412.

Kasemann, S.A., Hawkesworth, C.J., Prave, A.R., Fallick, A.E., Pearson, P.N., 2005. Boron and calcium isotope composition in Neoproterozoic carbonate rocks from Namibia: Evidence for extreme environmental change. Earth Planetary Sciences Letters *23*, 73–86.

Kato, Y., Yamaguchi, K.E., Ohmoto, H., 2006. Rare earth elements in Precambrian banded iron formations: Secular changes of Ce and Eu anomalies and evolution of atmospheric oxygen. Geological Society of America Memoirs *198*, 269–289.

Kaufman, A.J., Knoll, A.H., Awramik, S.M., 1992. Biostratigraphic and chemostratigraphic correlation of Neoproterozoic sedimentary successions: Upper Tindir Group, northwestern Canada, as a test case. Geology *20*, 181–185.

Kaufman, A.J., Jacobsen, S.B., Knoll, A.H., 1993. The Vendian record of C- and Sr-isotopic variations: Implications for tectonics and paleoclimate. Earth and Planetary Science Letters *120*, 409–430.

Kaufman, A.J., Knoll, A.H., Narbonne, G.M., 1997. Isotopes, ice ages, and terminal Proterozoic Earth history. Proceedings of the National Academy Science *94*, 6600–6605.

Kaufman, A.J., Sial, A.N., Ferreira, V.P., (guest editors), 2007a.Precambrian isotope chemostratigraphy. Chemical Geology 237(1/2), special Issue, 232 p.

Kaufman, A.J., Sial, A.N., Ferreira, V.P., 2007b. Preface to special issue of chemical geology on precambrian chemostratigraphy in honor of the late William T. Holser. Chemical Geology 237, 1–4.

Kaufman, A.J., Sial, A.N., Frimmel, H.E., Misi, A., 2009. Neoproterozoic to Cambrian Palaeoclimatic events in Southwestern Gondwana. In: Gaucher, C., Sial, A.N., Halverson, G.P., Frimmel, H. (Eds.), Neoproterozoic–Cambrian Tectonics, Global Change and Evolution: A Focus on Southwestern Gondwana. Developments in Precambrian Geology 16. Elsevier, Amsterdam, pp. 369–388.

Keller, G., 2005. Impacts, volcanism and mass extinction: Random coincidence or cause and effect? Australian Journal of Earth Science 52, 725–757.

Keller, G., Mateo, P., Punekar, J., Khozyem, H., Gertsch, B., Spangenberg, J., Bitchong, A.M., Adatte, A., 2018. Environmental changes during the Cretaceous–Paleogene mass extinction and Paleocene–Eocene Thermal Maximum: Implications for the Anthropocene. Gondwana Research 56, 69–89.

Kennedy, M.J., Runnegar, B., Prave, A.R., Hoffmann, K.H., Arthur, M.A., 1998. Two or four Neoproterozoic glaciations? Geology 26, 1059–1063.

Keto, L.S., Jacobsen, S.B., 1988. Nd isotopic variations of early Paleozoic oceans. Earth Planetary Science Letters 84, 27–41.

Kikumoto, R., Tahata, M., Nishizawa, M., Sawaki, Y., Maruyama, S., Shu, D., Han, J., Komiya, T., Takai, K., Ueno, Y., 2014. Nitrogen isotope chemostratigraphy of the Ediacaran and Early Cambrian platform sequence at Three Gorges, South China. Gondwana Research 25(3), 1057–1069.

Klein, C., Beukes, N.J., 1993. Sedimentology and geochemistry of the glaciogenic Late Proterozoic Rapitan Iron Formation in Canada. Economic Geology 88, 542–565.

Knauth, L.P., Kennedy, M.J., 2009. The late Precambrian greening of the Earth. Nature 460, 728–732.

Knoll, A.H., Walter, M.R., 1992. Latest Proterozoic stratigraphy and Earth history. Nature 356, 673–678.

Knoll, A.H., Hayes, J.M., Kaufman, A.J., Swett, K., Lambert, I.B., 1986. Secular variation in carbon isotope ratios from Upper Proterozoic successions of Svalbard and East Greenland. Nature 321, 832–838.

Knoll, A., Walter, M., Narbonne, G., Christie-Blick, N., 2006. The Ediacaran Period: A new addition to the geologic time scale. Lethaia 39, 13–30.

Korte, C., Kozur, H.W, 2010. Carbon-isotope stratigraphy across the Permian–Triassic boundary: A review. Journal of Asian Earth Sciences 39, 215–235.

Kump, L.R., Arthur, M.A., 1999. Interpreting carbon-isotope excursions: Carbonates and organic matter. Chemical Geology 161, 181–198.

Kuznetsov, A., Melezhik, V., Gorokhov, I., Melnikov, N., Konstantinova, G., Kutyavin, E., Turchenko, T. 2010. Sr isotopic composition of Paleoproterozoic ^{13}C-rich carbonate rocks: The Tulomozero Formation, SE Fennoscandian Shield. Precambrian Research 182, 300–312.

Lawrence, M.G., Kamber, B.S., 2006. The behaviour of the rare earth elements during estuarine mixing: Revisited. Marine Chemistry 100(1), 147–161.

Lea, D.W., Pak, D.K., Spero, H.J., 2000. Climate impact of late Quaternary equatorial Pacific sea surface temperature variations. Science 289, 1719–1724.

Lear, C.H., Y. Rosenthal, 2006. Benthic foraminiferal Li/Ca: Insights into Cenozoic seawater carbonate saturation state, Geology 34, 985–988. doi:10.1130/G22792A.1.

Lear, C.H., Elderfield, H., Wilson, P.A., 2000. Cenozoic deep-sea temperatures and global ice volumes from Mg/Ca in benthic foraminiferal calcite. Science 287, 269–272.

Lechler, M., Pogge von Strandmann, P.A.E., Jenkyns, H.C., Prosser, G., Parente, M., 2015. Lithium-isotope evidence for enhanced silicate weathering during OAE 1a (Early Aptian Selli event). Earth and Planetary Science Letters 432, 210–222.

Leckie, R.M., Bralower, T.J., Cashman, R., 2002, Oceanic anoxic events and plankton evolution: Biotic response to tectonic forcing during the mid-Cretaceous. Paleoceanography 17, 1041. doi:10.1029/2001PA000623.

Lécuyer, C., Brouxel, M., Albarède, F., 1990 Elemental fluxes during hydrothermal alteration of the Trinity ophiolite (California, USA) by seawater. Chemical Geology 89, 87–115.

Lécuyer, C., Reynard, B., Grandjean, P., 2004. Rare earth element evolution of Phanerozoic seawater recorded in biogenic apatites. Chemical Geology 204(1), 63–102.

Lenton, T.M., Crouch, M., Johson, M., Pires, N., Dolan, L., 2012. First plants cooled the Ordovician. Nature Geoscience 5, 86–89.

Li, W., Chakraborty, S., Beard, B.L., Romanek, C.S., Johnson, C.M., 2012. Magnesium isotope fractionation during precipitation of inorganic calcite under laboratory conditions. Earth Planetary Science Letters 333–334, 304–316.

Lindsay, J.F., Brasier, M.D., 2002. Did global tectonics drive early biosphere evolution? Carbon isotope record from 2.6 to1.9 Ga carbonates of Western Australian basins. Precambrian Research 114, 1–34.

Lisiecki, L.E., Raymo, M.E., 2005. A Pliocene-Pleistocene stack of 57 globally distributed benthic δ18O records. Paleoceanography 20(1), PA1003.

Liu, Y.G., Miah, M.R.U., Schmitt, R.A., 1988. Cerium: A chemical tracer for paleo-oceanic redox conditions. Geochimica et Cosmochimica Acta 52(6), 1361–1371.

Lu, Z., Jenkyns, H.C., Rickaby, R.E.M., 2010. Iodine to calcium ratios in marine carbonates as a paleo-redox proxy during oceanic anoxic events. Geology 38, 1107–1110.

Luo, G., Algeo, T.J., Zhan, R., Yan, D., Huang, J., Liu, J., Xie, S., 2016. Perturbation of the marine nitrogen cycle during the Late Ordovician glaciation and mass extinction. Palaeogeography, Palaeoclimatology, Palaeoecology 448, 339–348.

Lyons, T.W., Reinhard, C.T., 2009. Oxygen for heavy-metal fans. Nature 461,179–181.

Lyons, T.W., Anbar, A.D., Severmann, S., Scott, C., Gill, B.C., 2009. Tracking euxinia in the ancient ocean: A multiproxy perspective and Proterozoic case study. Annual Review of Earth and Planetary Sciences 37, 507–534.

Macdougall, J.D., 1991. Radiogenic isotopes in seawater and sedimentary systems In: Heaman, L., Ludden, J.N. (Eds.), Applications of Radiogenic Isotope Systems to Problems in Geology. NMCA Short Course Handbook. Mineralogical Association of Canada, Toronto, ON, pp. 337–364, (Chapter10).

Magaritz, M., 1989. δ^{13}C minima follow extinction events: A clue to faunal radiation. Geology *17*, 337–340.

Magaritz, M., Holser, W.T., Kirschvink, J.L., 1986. Carbon-isotope events across the Precambrian/Cambrian boundary on the Siberian Platform. Nature *320*, 258–259.

Maloof, A.C., Porter, S.M., Moore, J.H., Dudás, F.O., Bowring, S.A., Higgins, J.A., Fike, D.A., Eddy, M.P., 2010. The earliest Cambrian record of animals and ocean geochemical change. Geological Society of America Bulletin *122*, 1731–1774.

Marriott, C.S., Henderson, G.M., Crompton, R., Staubwasser, M., Shaw, S., 2004. Effect of mineralogy, salinity, and temperature on Li/Ca and Li isotope composition of calcium carbonate. Chemical Geology *212*, 5–15.

Marshall, J.D., 1992. Climatic and oceanographic isotopic signals from the carbonate rock record and their preservation. Geological Magazine *129*, 143–160.

McArthur, J.M., 2010. Strontium isotope stratigraphy. In: Ratcliffe K.T., Zaitlin B.A. (Eds.), Application of Modern Stratigraphic Techniques: Theory and Case Histories. SEPM Special Publication *94*. SEPM (Society for Sedimentary Geology), Tulsa, OK, pp. 129–142.

McArthur, J.M., Howarth, R.J., Bailey, T.R., 2001. Strontium isotope stratigraphy: Lowess version 3: Best fit to the marine Sr-isotope curve for 0–509 Ma and accompanying look-up table for deriving numerical age. Journal of Geology *109*, 155–170.

Melchin, M.J., Mitchell, C.E., Holmden, C., Storch, P., 2013. Environmental changes in the Late Ordovician-early Silurian: Review and new insights from black shales and nitrogen isotopes. Geological Society of America Bulletin *125*, 1635–1670.

Melezhik, V.A., Fallick, A.E., Medvedev, P.V., Marakarikhin, V.V., 1999. Extreme ^{13}Ccarb enrichment in ca. 2.0Ga magnesite–stromatolite–dolomite-'redbeds' association in a global context: A case for the world-wide signal enhanced by a local environment. Earth Science Review *48*, 71–120.

Melezhik, V.A., Gorokov, I.M., Kuznetsov, A.B., Fallick, A.E., 2001. Chemostratigraphy of Neoproterozoic carbonates: Implications for "blind dating". Terra Nova *13*, 1–11.

Melezhik, V.A., Roberts, D., Fallick, A.E., Gorokhov, I.M., Kusnetzov, A.B., 2005. Geochemical preservation potential of high-grade calcite marble versus dolomite marble: Implication for isotope chemostratigraphy. Chemical Geology *216*, 203–224.

Melezhik, V.A., Huhma, H., Fallick, A.E., Whitehouse, M.J., 2007.Temporal constraints on the Palaeoproterozoic Lomagundi-Jatuli carbon isotope event. Geology *35*, 655–658.

Meyer, K.M., Yu, M., Lehrmann, D., van de Schootbrugge, B., Payne, J.L., 2013. Constraints on early Triassic carbon cycle dynamics from paired organic and inorganic carbon isotope records. Earth Planetary Science Letters *361*, 429–435.

Molina, E., Alegret, L., Arenillas, I., Arz, J.A., Gallala, N., Hardenbol, J., Von Salis, K., Steurbaut, E., Vandenberghe, N., Zaghbib-Turki, D., 2006. The global boundary Section and Point for the base of the Danian Stage (Paleocene, Paleogene, "Tertiary", Cenozoic) at El Kef, Tunisia: Original definition and revision. Episodes *29*, 263–273.

Molina, E., Alegret, L., Arenillas, E., Arz, J.A., Gallala, N., Grajales-Nishimura, J.M., Murillo-Muñetón, G., Zaghbib-Turki, D., 2009. The Global Boundary Stratotype Section and Point for the base of the Danian Stage (Paleocene, Paleogene, "Tertiary", Cenozoic): Auxiliary sections and correlation. Episodes *32*, 84–95.

Montañez, I.P., Osleger, D.A., Banner, J., Mack, L.E., Musgrove, M., 2000. Evolution of the Sr and C isotope composition of Cambrian Oceans. GSA Today *10*, 1–7.

Morante, R., Veevers, J.J., Andrew, A.S., Hamilton, P.J., 1994. Determination of the Permian–Triassic boundary in Australia. APEA Journal *34*, 330–336.

Nascimento, R.S.C., Sial, A.N., Pimentel, M.M., 2007. C-and Sr-isotope systematics applied to Neoproterozoic marbles of the Seridó Belt, northeastern Brazil. Chemical Geology *237*, 209–228.

Nascimento-Silva, V.M., Sial, A.N., Ferreira, V.P., Neumann, V.H., Barbosa, J.A., Pimentel, M.M., Lacerda, L.D., 2011. Cretaceous–Paleogene transition at the Paraíba Basin, Northeastern, Brazil: Carbon-isotope and mercury subsurface stratigraphies. Journal of South American Earth Sciences *32*, 379–392.

Nascimento-Silva, M.V., Sial, A.N., Ferreira, V.P., Barbosa, J.A., Neumann, V.H., Pimentel, M.M., Lacerda, L.D., 2013. Carbon Isotopes, rare-earth elements and mercury behavior of Maastrichtian–Danian carbonate succession of the Paraíba Basin, Northeastern Brazil. In: Bojar, A.V., Melinte-Dobrinescu, M.C., Smit, J. (Eds.), Isotopic Studies in Cretaceous Research. Geological Society, London, Special Publications *382*. Geological Society of London, London, pp. 85–104.

Nielsen, S.G., Rehkämper, M., 2012. Thallium isotopes and their application to problems in Earth and environmental science. In: Baskaran, M. (Ed.), Handbook of Environmental Isotope Geochemistry. Advances in Isotope Geochemistry. Springer-Verlag, Berlin, pp. 247–269.

Nielsen, S.G., Mar-Gerrison, S., Gannoun, A., LaRowe, D.E., Klemm, V., Halliday, A., Burton, K.W., Hein, J.R., 2009. Thallium isotope evidence for increased marine organic carbon export in the early Eocene. Earth Planetary Science Letters *278*, 297–307.

Nielsen, S.G., Rehkämper, M., Prytulak, J., 2017. Investigation and application of thallium isotope fractionation. Reviews in Mineralogy and Geochemistry *82*, 759–798.

Och, L.M., Shields-Zhou, G.A., 2012. The Neoproterozoic oxygenation event: Environmental perturbations and biogeochemical cycling. Earth-Science Reviews *110*, 26–57.

Oehlert, A.M., Swart, P.K., 2014. Interpreting carbonate and organic carbon isotope covariance in the sedimentary record. Nature Communications *5*, 4672. doi:10.1038/ncomms5672.

Palmer, M.R., Pearson, P.N., Cobb, S.J., 1998. Reconstructing past ocean pH-depth profiles. Science *282*(5393), 1468–1471. doi:10.1126/science.282.5393.1468.

Papineau, D., Mojzsis, S.J., Karhu, J.A., Marty, B., 2005. Nitrogen isotopic composition of ammoniated phyllosilicates: Case studies from Precambrian metamorphosed sedimentary rocks. Chemical Geology *216*, 37–58.

Pasquier, V., Sansjofre, P., Rabineau, M., Revillon, S., Houghton, J., Fike, D.A., 2017. Pyrite sulfur isotopes reveal glacial–interglacial environmental changes. Proceedings of the National Academy of Science *114*, 5941–5945. doi:10.1073/pnas.1618245114.

Pearson, P.N., van Dongen, B.E., Nicholas, C.J., Pancost, R.D., Schouten, S., Singano, J.M., Wade, B.S., 2007. Stable warm tropical climate through the Eocene Epoch. Geology *35*, 211–214.

Peckeur-Ehrenbrink, B., Ravizza, G., 2000. The marine osmium isotope record. Terra Nova *12*, 205–219.

Peckeur-Ehrenbrink, B., Ravizza, G., 2012. Osmium isotope stratigraphy. In: Gradstein, F.M., Ogg, J.G., Schmitz, M.D., Ogg, G.M. (Eds.), The Geologic Time Scale 2012. Elsevier, Amsterdam, Boston, vol. 1, chapter 8, pp. 145–166.

Peng, S., Babcock, L., Robison, R., Lin, H., Rees, M., Saltzman, M., 2004. Global Standard Stratotype-section and Point (GSSP) of the Furongian Series and Paibian Stage (Cambrian). Lethaia *37*, 365–379.

Penman, D.E., Hönisch, B., Zeebe, R.E., Thomas, E., Zachos, J.C., 2014. Rapid and sustained surface ocean acidification during the Paleocene–Eocene Thermal Maximum. Paleoceanography *29*, 357–369. doi:10.1002/2014PA002621.

Penniston-Dorland, S., Liu, X.M., Rudnick, R.L., 2017. Lithium isotope geochemistry. Reviews in Mineralogy and Geochemistry *82*, 165–217.

Percival, L.M.E., Witt, M.L.I., Mather, T.A., Hermoso, M., Jenkyns, H.C., Hesselbo, S.P., Al-Suwaidi, A.H., Storm, M.S., Xu, W., Ruhl, M., 2015. Globally enhanced mercury deposition during the end-Pliensbachian extinction and Toarcian OAE: A link to the Karoo–Ferrar Large Igneous Province. Earth and Planetary Science Letters *428*, 267–280.

Percival, L.M.E., Ruhl, M., Hesselbo, S.P., Jenkyns, H.C., Mather, T.A., Whiteside, J.H., 2017. Mercury evidence for pulsed volcanism during the end-Triassic mass extinction. Proceedings of the National Academy of Sciences of the United States of America *114*(30), 7929–7934. doi:10.1073/pnas.1705378114.

Petrizzo, D.A., Young, E.D., Runnegar, B.N., 2014. Implications of high-precision measurements of $^{13}C–^{18}O$ bond ordering in CO_2 for thermometry in modern bivalved mollusc shells. Geochimica et Cosmochimica Acta *142*, 400–410.

Pogge von Strandmann, P.A.E., Jenkyns, H.C., Woodfine, R.G., 2013. Lithium isotope evidence for enhanced weathering during Oceanic Anoxic Event 2. Nature Geoscience *6*, 668–672.

Pogge von Strandmann, P.A.E., Desrochers, A., Murphy, M.J., Finlay, A.J., Selby, D., Lenton, T.M., 2017. Global climate stabilisation by chemical weathering during the Hirnantian glaciation. Geochemical Perspectives Letters *3*(2), 230–237.

Pokrovsky, B.G., Mavromatis, V., Pokrovsky, O.S., 2011. Covariation of Mg and C isotopes in late Precambrian carbonates of the Siberian Platform: A new tool for tracing the change in weathering regime? Chemical Geology *290*, 67–74.

Poulton, S.W., Canfield, D.E., 2005. Development of a sequential extraction procedure for iron: Implications for iron partitioning in continentally derived particulates. Chemical Geology *214*, 209–221.

Poulton, S.W., Fralick, P.W., Canfield, D.E., 2004. The transition to a sulphidic ocean ~1.84 billion years ago. Nature *431*, 173–77.

Price, G.D., Mutterlose, J., 2004. Isotopic signals from late Jurassic–early Cretaceous (Volgian–Valanginian)sub-Arctic belemnites, Yatria River, Western Siberia. Journal of the Geological Society *161*, 959–968.

Puceat, E., Lecuyer, C., Sheppard, S.M.F., Dromart, G., Reboulet, S., Grandjean, P., 2003. Thermal evolution of Cretaceous Tethyan marine waters inferred from oxygen isotope composition of fish tooth enamels. Paleoceanography *18*(2), 1029. doi:10.1029/2002PA000823.

Railsback, L.B., Gibbard, P.L., Head, M.J., Voarintsoa, N.R.G., Toucanne, S., 2015. An optimized scheme of lettered marine isotope substages for the last 1.0 million years, and the climatostratigraphic nature of isotope stages and substages. Quaternary Science Reviews *111*, 94–106.

Ramkumar, M., 1999. Role of chemostratigraphic technique in reservoir characterization and global stratigraphic correlation. Indian Journal of Geochemistry *14*, 33–45.

Ramkumar, M., 2014. Characterization of depositional units for stratigraphic correlation, petroleum exploration and reservoir characterization. In: Sinha, S. (Ed.), Advances in Petroleum Engineering. Studium Press L.L.C, Houston, pp. 1–13.

Ramkumar, M., 2015. Toward standardization of terminologies and recognition of chemostratigraphy as a formal stratigraphic method. In: Ramkumar, M. (Ed.), Chemostratigraphy: Concepts, Techniques and Applications. Elsevier, Amsterdam, pp. 1–21. doi:10.1016/ B978-0-12-419968-2.00001-7.

Ramkumar, M., Stüben, D., Berner, Z., 2010. Hierarchical delineation and multivariate statistical discrimination of chemozones of the Cauvery Basin, South India: Implications on Spatio-temporal scales of stratigraphic correlation. Petroleum Science *7*, 435–447.

Ramkumar, M., Stüben, D., Berner, Z., 2011. Barremian–Danian chemostratigraphic sequences of the Cauvery Basin, South India: Implications on scales of stratigraphic correlation. Gondwana Research *19*, 291–309.

Ravizza, G., Peucker-Ehrenbrink, B., 2003. Chemostratigraphic evidence of Deccan Volcanism from the marine osmium isotope record. Science *302*, 1392–1395.

Ries, J.B., Fike, D.A., Pratt, L.M., Lyons, T.W., Grotzinger, J.P., 2009. Superheavy pyrite (δ34Spyr > δ34SCAS) in the terminal Proterozoic Nama Group, southern Namibia: A consequence of low seawater sulfate at the dawn of animal life. Geology *37*, 743–746.

Rodler, A.S., Frei, R., Gaucher, C., Germs, G.J.B., 2016. Chromium isotope, REE and redox-sensitive trace element chemostratigraphy across the late Neoproterozoic Ghaub glaciation, Otavi Group, Namibia. Precambrian Research *286*, 234–249.

Rodler, A., Frei, R., Gaucher, C., Korte, C., Rosing, S.A., Germs, G.J.B. 2017. Multiproxy isotope constraints on ocean compositional changes across the late Neoproterozoic Ghaub glaciation, Otavi Group, Namibia. Precambrian Research *298*, 306–324.

Saltzman, M.R., 2005. Phosphorus, nitrogen, and the redox evolution of the Paleozoic oceans. Geology *33*, 573–576.

Saltzman, M.R., Thomas, E., 2012. Carbon isotope stratigraphy. In: Gradstein, F.M., Ogg, J.G., Schmitz, M., Ogg, G.

(Eds.), The Geologic Time Scale. Elsevier, Amsterdam, Heidelberg. doi:10.1016/B978-0-444-59425-9.00011-1.

Salvador, A., 1994. International Stratigraphic Guide: A Guide to Stratigraphic Classification, Terminology and Procedure, 2nd Edition. IUGS-GSA, Boulder, pp. 1–214.

Sanei, H., Grasby, S.E., Beauchamp, B., 2012. Latest Permian mercury anomalies. Geology 40, 63–66.

Sansjofre, P., Trindade R.I.F., Ader M., Soares, J.L., Nogueira A.C.R., Tribovillard, N., 2014. Paleoenvironment reconstruction of the Ediacaran Araras platform (Western Brazil) from the sedimentary and trace metals records. Precambrian Research 241, 185–202.

Santos, G.M., Cordeiro, R.C., Silva Filho, E.V., Turcq, B., Lacerda, L.D., Fifield, L.K., Gomes, P.R.S., Hauscaden, P.A., Sifeddine, A., Albuquerque, A.L.S., 2001. Chronology of the atmospheric mercury in Lagoa da Pata Basin, Upper Rio Negro of Brazilian Amazon. Radiocarbon 43, 801–808.

Sanyal, A., Bijma, J., Spero, H., David, H., Lea, W., 2001. Empirical relationship between pH and the boron isotopic composition of Globigerinoides sacculifer implications for the boron isotope paleo-pH proxy. Paleoceanography 16, 515–519.

Sauzeat, L., Rudnick, R.L., Chauvel, C., Garcon, M., Tang, M., 2015. New perspectives on the Li isotopic composition of the upper continental crust and its weathering signature. Earth and Planetary Science Letters 428, 181–192.

Sawaki, Y., Ohno, T., Tahata, M., Komiya, T., Hirata, T., Maruyama, S., Windley, B.F., Han, J., Shud, D., Li, Y., 2010. The Ediacaran radiogenic Sr isotope excursion in the Doushantuo Formation in the Three Gorges area, South China. Precambrian Research 176, 46–64.

Scaife, J.D., Ruhl, M., Dickson, A.J., Mather, T.A., Jenkyns, H.C., Percival, L.M.E., Hesselbo, S.P., Cartwright, J., Eldrett, J.S., Bergman, S.C., Minisini, D., 2017. Sedimentary mercury enrichments as a marker for submarine Large Igneous Province volcanism? Evidence from the Mid-Cenomanian Event and Oceanic Anoxic Event 2 (Late Cretaceous). Geochemistry, Geophysics, Geosystems 18(12), 4253–4275. doi:10.1002/2017GC007153.

Scheffler, K., Hoernes, S., Schwark, L., 2003. Global changes during Carboniferous–Permian glaciation of Gondwana: Linking polar and equatorial climate evolution by geochemical proxies. Geology 31, 505–608.

Schidlowski, M., Hayes, J.M., Kaplan, I.R., 1983. Isotopic inferences of ancient biochemistries: Carbon, sulfur, hydrogen and nitrogen. In: Schopf, J.W. (Ed.), Earth's Earliest Biosphere: Its Origin and Evolution. Princeton University Press, Princeton, NJ, pp. 149–186.

Schmitz, B., Andersson, P., Dahl, J., 1988. Iridium, sulfur isotopes and rare earth elements in the Cretaceous-Tertiary boundary clay at Stevns Klint, Denmark. Geochimica et Cosmochimica Acta 52(1), 229–236.

Schmitz, B., Pujalte, V., Molina, E., Monechi, S., Orue-Etxebarria, X., Speijer, R.P., Alegret, L., Apellaniz, E., Arenillas, I., Aubry, M.-P., Baceta, J.-I., Berggren, W.A., Bernaola, G., Caballero, F., Clemmensen, A., Dinarès-Turell, J., Dupuis, C., Heilmann-Clausen, C., Orús, A.H., Knox, R., Martín-Rubio, M., Ortiz, S., Payros, A., Petrizzo, M.R., von Salis, K., Sprong, J., Steurbaut, E., Thomsen, E., 2011. The global stratotype sections and points for the bases of the Selandian (Middle Paleocene) and Thanetian (Upper Paleocene Paleocene) stages at Zumaia, Spain. Episodes 34, 220–243.

Scholle, P.A., Arthur, M.A. 1980. Carbon isotope fluctuations in Cretaceous pelagic limestones: Potential stratigraphic and petroleum exploration tool. American Association of Petroleum Geologists, Bulletin 64, 67–87.

Scott, C., Lyons, T.W., Bekker, A., Shen, Y., Pultron, S.W., Chu, X., Anbar, A.D., 2008. Tracing the stepwise oxygenation of the Proterozoic ocean. Nature 452, 456–459.

Scott, C.T., Bekker, A., Reinhard, C.T., Schnetger, B., Krapež, B., Rumble, D., Lyons, T.W., 2011. Late Archean euxinic conditions before the rise of atmospheric oxygen. Geology 39(2), 119–122.

Shackleton, N.J., 1969. The last interglacial in the marine and terrestrial record. Proceedings of the Royal Society of London 174, 135–154.

Shackleton, N.J., Opdyke, N.D., 1973. Oxygen isotope and palaeomagnetic stratigraphy of Equatorial Pacific core V28-238: Oxygen isotope temperatures and ice volumes on a 105 year and 106 year scale. Quaternary research 3(1), 39–55.

Shackleton, N.J., Hall, M.A., 1984. Carbon isotope data from Leg 74 sediments. Initial Reports of the Deep Sea Drilling Project 74, 613–619.

Sharma, M., Papanastassiou, D.A., Wasseburg, J., 1997. The concentration and isotopic composition of osmium in the oceans. Geochimica et Cosmochimica Acta 61, 3287–3299.

Shen, Y., Knoll, A.H., Walter, M.R., 2003. Evidence for low sulphate and anoxia in a mid-Proterozoic marine basin. Nature 423, 632.

Shields, G.A., 2007a. A normalised seawater strontium isotope curve and the Neoproterozoic-Cambrian chemical weathering event. Earth Discussions 2, 69–84.

Shields, G., 2007b. The marine carbonate and chert isotope records and their implications for tectonics, life and climate on the early Earth. In: van Kranendonk, M.J., Smithies, R.H., Bennett, V.C. (Eds.), Earth's Oldest Rocks. Developments in Precambrian Geology 15. Elsevier, Amsterdam, pp. 971–983.

Sial, A.N., Karhu, J., Ferreira, V.P., 2010a. Insights from isotope stratigraphy. Preface to the special issue on precambrian isotope stratigraphy. Precambrian Research 182(4), v–viii.

Sial, A.N., Gaucher, C., Silva Filho, M.A., Ferreira, V.P., Pimentel, M.M., Lacerda, L.D., Silva Filho, E.V., Cezario, W., 2010b. C-, Sr-isotope and Hg chemostratigraphy of Neoproterozoic cap carbonates of the Sergipano Belt, Northeastern Brazil. In: Karhu, J., Sial, A.N., Ferreira, V.P. (Eds.), Precambrian Isotope Stratigraphy. Precambrian Research 182. Elsevier, Amsterdam, pp. 351–372.

Sial, A.N., Lacerda, L.D., Ferreira, V.P., Frei, R., Marquillas, R.A., Barbosa, J.A., Gaucher, C., Windmöller, C.C., Pereira, N.S., 2013a. Mercury as a proxy for volcanic activity during extreme environmental turnover: The Cretaceous–Paleogene transition. Palaeogeography, Palaeoclimatology, Palaeoecology 387, 153–164.

Sial, A.N., Chen, J.-B., Lacerda, L.D., Peralta, S., Gaucher, C., Frei, R., Cirilli, S., Ferreira, V.P., Marquillas, R.A., Barbosa, J.A., Pereira, N.S., Belmino, I.K.C., 2014. High-resolution

Hg chemostratigraphy: A contribution to the distinction of chemical fingerprints of the Deccan volcanism and Cretaceous–Paleogene Boundary impact event. Palaeogeography, Palaeoclimatology, Palaeoecology 414, 98–115. doi:10.1016/j.palaeo.2014.08.013.

Sial, A.N., Gaucher, G., Ferreira, V.P., Pereira, N.S., Cezario, W.S., Chiglino, L., Monteiro, H., 2015a. Isotope and elemental chemostratigraphy. In: M. Ramkumar (Ed.), Chemostratigraphy, Concepts, Techniques and Applications. Elsevier, Amsterdam, pp. 23–64.

Sial, A.N., Campos, M.S., Gaucher, C., Frei, R., Ferreira, V.P., Nascimento, R.C., Pimentel, M.M., Pereira, N.S., Rodler, A., 2015b. Algoma-type Neoproterozoic BIFs and related marbles in the Seridó Belt (NE Brazil): REE, C, O, Cr and Sr isotope evidence. Journal of South American Earth Sciences 61, 33–52.

Sial, A.N., Chen, J., Lacerda, L.D., Frei, R., Tewari, V.C., Pandit, M.K., Gaucher, C., Ferreira, V.P., Cirilli, S., Peralta, S., Korte, C., Barbosa, J.A., Pereira, N.S., 2016. Mercury enrichment and mercury isotopes in Cretaceous–Paleogene boundary successions: Links to volcanism and palaeoenvironmental impacts. Cretaceous Research 66, 60–81. doi:10.1016/j.cretres.2016.05.006.

Sial, A.N., Chen, J., Lacerda, L.D., Frei, R., Tewari, V.C., Pandit, M.K., Gaucher, C., Ferreira, V.P., Cirilli, S., Peralta, S., Korte, C., Barbosa, J.A., Pereira, N.S., 2017. Reply to comments by Sanjay K. Mukhopadhyay, Sucharita Pal, J. P. Shrivastava on the paper by Sial et al. (2016) Mercury enrichments and Hg isotopes in Cretaceous–Paleogene boundary successions: Links to volcanism and palaeoenvironmental impacts. Cretaceous Research 66, 60–81. Cretaceous Research 78, 84–88.

Sial, A.N., Chen, J., Lacerda, L.D., Frei, R., Higgins, J., Tewari, V.C., Gaucher, C., Ferreira, V.P., Cirilli, S., Peralta, S., Korte, C., Barbosa, J.A., Pereira, N.S., Ramos, D.S., this volume. Chemostratigraphy across the Cretaceous–Paleogene boundary. In: Sial, A.N., Gaucher, C., Ramkumar, M., Ferreira, V.P. (guest editors), Chemostratigraphy Across Major Chronological Eras. AGU/Wiley, United States.

Siebert, C., Nägler, T.F., von Blanckenburg, F., Kramers, J.D., 2003. Molybdenum isotope records as a potential new proxy for paleoceanography. Earth Planetary Science Letters 211, 159–171.

Silva-Tamayo, J.C., Nägler, T.F., Villa, I.M., Kyser, K., Narbonne, G., James, N.P., Sial, A.N., Silva Filho, M.A., 2007. The aftermath of Snowball Earth: Ca- and Mo- isotope constraints on post-glacial ocean conditions. Geophysical Research Abstracts, 9, 01980.

Silva Tamayo, J.C., Nägler, T., Villa, I.M., Kyser, K., Vieira, L.C., Sial, A.N., Narbonne, G.M., James, N.P., 2010a. Global Ca isotope variations in Post-Sturtian carbonate successions. Terra Nova 22, 188–194.

Silva Tamayo, J.C., Nägler, T.F., Nogueira, A., Kyser, K., Villa, I., Riccomini, C., Sial, A.N., Narbonne, G.M., James, N.P., 2010b. Global perturbation of the marine Ca-isotopic composition in the aftermath of the Marinoan global glaciations. Precambrian Research 182(4), 373–381.

Sluijs, A., Schouten, S., Donders, T.H., Schoon, P.L., Rohl, U., Reichart, G.-J., Sangiorgi, F., Kim, J.-H., Sinninghe Damste, J.S., Brinkhuis, H., 2009. Warm and wet conditions in the Arctic region during Eocene Thermal Maximum 2. Nature Geosciences 2, 777–780.

Sosa-Montes, C., Rodríguez-Tovar, F.J., Martínez-Ruiz, F., Monaco, C.P., 2017. Paleoenvironmental conditions across the Cretaceous–Paleogene transition at the Apennines sections (Italy): An integrated geochemical and ichnological approach. Cretaceous Research 71, 1–13. doi:10.1016/j.cretres.2016.11.005.

Spangenberg, J.E., Bagnoud-Velásquez, M., Boggiani, P.C., Gaucher, C., 2014. Redox variations and bioproductivity in the Ediacaran: Evidence from inorganic and organic geochemistry of the Corumbá Group, Brazil. Gondwana Research 26, 1186–1207.

Staubwasser, M., Von Blackenburg, F., Shoenberg, R., 2006. Iron isotopes in the early marine diagenetic iron cycle. Geology 34, 629–632.

Steinhoefel, G., von Blanckenburg, F., Horn, I., Konhauser, K.O., Beukes, N.J., Gutzmer, J., 2010. Deciphering formation processes of banded iron formations from the Transvaal and the Hamersley successions by combined Si and Fe isotope analysis using UV femtosecond laser ablation. Geochimica et Cosmochimica Acta 74, 2677–2696.

Steininger, F.F., Aubry, M.P., Berggren, W.A., Biolzi, M., Borsetti, A.M., Cartlidge, J.E., Cati, F., Corfield, R., Gelati, R., Iaccarino, S., Napoleone, C., Ottner, F., Rögl, F., Roetzel, R., Spezzaferri, S., Tateo, F., Villa, G., Zevenboom, D., 1997. The Global Stratotype Section and Point (GSSP) for the base of the Neogene. Episodes 20, 23–28.

Strandmann, P.A.E., Forshaw, J., Schmidt, D.N., 2014. Modern and Cenozoic records of seawater magnesium from foraminiferal Mg isotopes. Biogeosciences 11, 5155–5168.

Strauss, H., 1997. The isotopic composition of sedimentary sulfur through time. Paleogeography, Palaeoclimatology, Palaeoecology 132, 97–118.

Sundquist, E.T., Visser, K., 2004. The geologic history of the carbon cycle. Treatise on Geochemistry 8, 425–472.

Tahata, M., Ueno, Y., Ishikawa, T., Sawaki, Y., Murakami, K., Han, J., Shu, D., Li, Y., Guo, J., Yoshida, N., Komiya, T., 2013. Carbon and oxygen isotope chemostratigraphies of the Yangtze platform, South China: Decoding temperature and environmental changes through the Ediacaran. Gondwana Research 23(1), 333–356. doi:10.1016/j.gr.2012.04.005.

Teng, F.-Z., Watkins, J.M., Dauphas, N., 2017. Non-traditional stable isotopes. Reviews in Mineralogy and Geochemistry 82, 885 pp.

Thibodeau, A.M., Bergquist, B.A., 2017. Do mercury isotopes record the signature of massive volcanism in marine sedimentary records? Geology 45, 95–96.

Thibodeau, A.M., Ritterbush, K., Yager, J.A., West, A.J., Ibarra, Y., Bottjer, D.J., Berelson, W.M., Bergquist, B.A., Corsetti, F.A., 2016. Mercury anomalies and the timing of biotic recovery following the end-Triassic mass extinction. Nature Communications 7, 1–8.

Tipper, E.T., Bickle, M.J., Galy, A., West, J., Pomiés, C., Chapman, H.J., 2006a. The short term sensitivity of carbonates and silicate weathering fluxes: Insight from seasonal variations in river chemistry. Geochimica et Cosmochimica Acta 70, 2737–2754.

Tipper, E.T., Galy, A., Bickle, M.J., 2006b. Riverine evidence for a fractionated reservoir of Ca and Mg on the continents: Implications for the oceanic Ca cycle. Earth Planetary Science Letters *247*, 267–279.

Tipper, E.T., Galy, A., Gaillardet, J., Bickle, M.J., Elderfield, H., Carder, E.A., 2006c. The magnesium isotope budget of the modern ocean: Constraints from riverine magnesium isotope ratios. Earth Planetary Science Letters *250*, 241–253.

Tipper, E.T., Louvat, P., Capmas, F., Galy, A., Gaillardet, J., 2008. Accuracy of stable Mg and Ca isotope data obtained by MC-ICP-MS using the standard addition method. Chemical Geology *257*, 65–75.

Tomascak, P.B., Magna, T.S., Dohmen, R., 2016. Advances in Lithium Isotope Geochemistry. Springer International Publishing, Cham, 195 pp.

Tribovillard, N., Algeo, T.J., Lyons, T., Riboulleau, A., 2006. Trace metals as paleoredox and paleoproductivity proxies: An update. Chemical Geology *232*, 12–32.

Tribovillard, N., Algeo, T.J., Baudin, F., Riboulleau, A., 2012. Analysis of marine environmental conditions based on molybdenum-uranium covariation-applications to Mesozoic paleoceanography. Chemical Geology *324/325*, 46–58.

Tripati, A., Eagle, R., Thiagarajan, N., Gagnon, A., Bauch, H., Halloran, P., Eiler, J., 2010. ^{13}C–^{18}O isotope signatures and "clumped isotope" thermometry in foraminifera and coccoliths. Geochimica et Cosmochimica Acta *74*, 5697–5717.

Vandenbroucke, T.R.A., Armstrong, H.A., Williams, M., Paris, F., Zalasiewicz, J.A., Sabbe, K., Nolvak, J., Challandsa, T.J., Verniers, J., Servais, T., 2010. Polar front shift and atmospheric CO$_2$ during the glacial maximum of the Early Paleozoic Icehouse. Proceedings of the National Academy of Sciences of the United States of America *107*, 14983–14986.

Van de Schootbrugge, B., Bachan, A., Suan, G., Richoz, S., Payne, J.L., 2013. Microbes, mud and methane: Cause and consequence of recurrent Early Jurassic anoxia following the end-Triassic mass extinction. Paleontology *56*, 1–25. doi:10.1111/pala.120341.

Veizer, J., Holser, W.T., Wilgus, C.K., 1980. Correlation of ^{13}C/^{12}C and ^{34}S/^{32}S secular variations. Geochimica Cosmochimica Acta *44*, 579–588.

Veizer, J., Compston, W., Clauer, N., Schidlowski, M., 1983. ^{87}Sr/^{86}Sr in late Proterozoic carbonates: Evidence for a "mantle" event at approximately 900 Ma ago. Geochimica et Cosmochimica Acta *47*, 295–302.

Veizer, J., Ala, D., Azmy, K., Bruckschen, P., Buhl, P., Bruhn, F., Carden, G.A.F., Diener, A., Ebneth, S., Godderis, Y., Jasper, T., Korte, C., Pawellek, F., Podlaha, O.G., Strauss, H., 1999. ^{87}Sr/^{86}Sr, δ^{13}C and δ^{18}O evolution of Phanerozoic seawater. Chemical Geology *161*, 59–88.

Vennemann, T.W., Hegner, E., 1998. Oxygen, strontium, and neodymium isotope composition of fossil shark teeth as a proxy for the palaeoceanography and palaeoclimatology of the Miocene northern Alpine Paratethys. Palaeogeography, Palaeoclimatology, Palaeoecology *142*, 107–121.

Walker, M., Johnsen, S., Rasmussen, S.O., Popp, T., Steffensen, J.-P., Gibbard, P., Hoek, W., Lowe, J., Andrews, J., Rck, S.B., Cwynar, L.C., Hughen, K., Kershaw, P., Kromer, B., Litt, T., Lowe, D.J., Nakagawa, T., Newnham, R., Schwander, J., 2009. Formal definition and dating of the GSSP (Global Stratotype Section and Point) for the base of the Holocene using the Greenland NGRIP ice core, and selected auxiliary records. Journal of Quaternary Science *24*, 3–17.

Wen, H., Fan, F., Zhang, Y., Cloquet, C., Carignan, J., 2015. Reconstruction of early Cambrian ocean chemistry from Mo isotopes. Geochimica et Cosmochimica Acta *164*, 1–16.

Williams, D.F., Thunell, R.C., Tappa, E., Rio, D., Raffi, I. 1988. Chronology of the Pleistocene oxygen isotope record: 0–1.88 my BP. Palaeogeography, Palaeoclimatology, Palaeoecology *64*(3–4), 221–240.

Wombacher, F., Eisenhauer, A., Heuser, A., Weyer, S., 2009. Separation of Mg, Ca and Fe from geological reference materials for stable isotope ratio analyses by MC-ICP-MS and double-spike TIMS. Journal Analytical Atomic Spectrometry *24*, 627–636.

Wombacher, F., Eisenhauer, A., Böhm, F., Gussone, N., Regenberg, M., Dullo, W.-C., Rüggeberg, A., 2011. Magnesium stable isotope fractionation in marine biogenic calcite and aragonite. Geochimica Cosmochica Acta *75*, 5797–5818.

Xu, C., Rong, J., Fan, J., Zhan, R., Mitchell, C.E., Harper, D.A.T., Melchin, M.J., Peng, P., Finney, S.C., Wang, X., 2006. The Global Boundary Stratotype Section and Point (GSSP) for the base of the Hirnantian Stage (the uppermost of the Ordovician System). Episodes *2*, 183–196.

Yu, J., Elderfield, H., 2007. Benthic foraminiferal B/Ca ratios reflect deep water carbonate saturation state. Earth and Planetary Science Letters *258*, 73–86.

Zachos, J., Pagani, M., Sloan, L., Thomas, E., Billups, K., 2001. Trends, rhythms, and aberrations in global climate 65 Ma to present. Science *292*, 686–693.

Zachos, J.C., Schouten, S., Bohaty, S., Quattlebaum, T., Sluijs, A., Brinkhuis, H., Gibbs, S.J., Bralower, T.J., 2006. Extreme warming of mid-latitude coastal ocean during the Paleocene-Eocene Thermal Maximum: Inferences from TEX86 and isotope data. Geology *34*, 737–740.

2

Glossary of Chemostratigraphy

Muthuvairavasamy Ramkumar[1], Alcides Nobrega Sial[2], Claudio Gaucher[3], and Valderez Pinto Ferreira[2]

ABSTRACT

Owing to the interdisciplinary nature and practitioners from multitudes of science disciplines, chemostratigraphy, one of the younger branches of geosciences, finds itself in one of the most frequently published subjects. Analysis of two major scientific literature databases suggests that the number of publications including books and journal articles dealing with chemical stratigraphy or isotopic stratigraphy grows at an exponential rate. Published geoscientific articles/books dealing with chemostratigraphy have unequivocally established that this technique can be relied upon when other conventional stratigraphic methods failed and/or have limitations. Among the different time slices of Earth's history, the Phanerozoic, especially the Cenozoic, was intensively studied chemostratigraphically, as revealed by the number of publications. The plethora of literature has also resulted in highly varied terminologies that hinder clear understanding and development of the subject. A glossary of commonly used terminologies and intended meanings is presented in this chapter. The glossary is to be considered as a platform for further discussion and improvement rather than an exhaustive list.

2.1. INTRODUCTION

Chemostratigraphy has traveled a long way from its simple beginning of comparison between geochemical and biostratigraphic subdivisions of sedimentary strata [*Scholle and Arthur*, 1980; *Berger and Vincent*, 1981], understanding oceanographic parameters through sediment geochemistry [*Renard*, 1986; *Oppo et al.*, 1990], and recognizing unique geochemical compositions of sedimentary deposits with a sole objective of comparing and contrasting the depositional environmental settings across major chronological boundary [*Romein and Smit*, 1981; *Keller*, 1988; *Ramkumar et al.*, 2004, 2005] to the interpretation of planetary habitats [*Wiens et al.*, 2015] and to its current position as a recognized independent accurate stratigraphic method [*Ramkumar*, 2015].

Notwithstanding its limitations [*Ramkumar*, 1999], spatial and temporal resolution of chemostratigraphy is dependent upon the sampling frequency, which makes this a preferred method over others and also serves as complementary to conventional methods of stratigraphic study [*Ramkumar*, 2014]. In this chapter, we present the commonly used terminologies based on a thorough review.

2.2. CHEMOSTRATIGRAPHY: TRANSCENDING BOUNDARIES AND EXPANDING POSSIBILITIES

Recently, *Ramkumar* [2015] and *Sial et al.* [2015] presented detailed reviews on the development, systematics, and applications of chemostratigraphy. The growing popularity, wider acceptance, and application of this technique can be gauged from few simple metrics as culled from the commonly used scientific data sources such as www.sciencedirect.com and www.geoscienceworld.org. For example, a search with the keyword "chemostratigraphy" in ScienceDirect website returns a total of 1453 books and journal articles,

[1] *Department of Geology, Periyar University, Salem, TN, India*
[2] *NEG-LABISE, Department of Geology, Federal University of Pernambuco, Recife, PE, Brazil*
[3] *Instituto de Ciencias Geológicas, Facultad de Ciencias, Universidad de la República, Montevideo, Uruguay*

Chemostratigraphy Across Major Chronological Boundaries, Geophysical Monograph 240, First Edition.
Edited by Alcides N. Sial, Claudio Gaucher, Muthuvairavasamy Ramkumar, and Valderez Pinto Ferreira.
© 2019 the American Geophysical Union. Published 2019 by John Wiley & Sons, Inc.

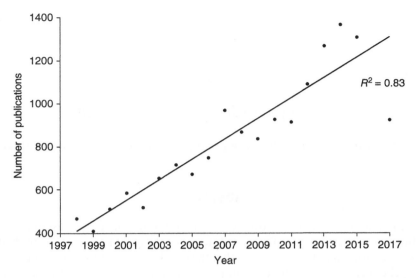

Figure 2.1 Yearly increase of number of publications using geochemistry for stratigraphy. Based on data displayed in www.sciencedirect.com, as retrieved on 18 June 2017.

whereas the keywords "chemical stratigraphy" and "isotope stratigraphy" return a total of 280 and 26,210 books and journal articles, while the GeoScienceWorld returns 715, 6298, 7157 books and articles, respectively. There were about 9059 articles and books until the year 1997. The number of publications vs. time since the year 1998 shows a linear trend with a R^2 value of 0.83 (Fig. 2.1). The statistics also show that halfway through the year 2017, there were 923 articles and books published, evidencing the rapid strides made by this field of geosciences. About 30 major geoscientific societies have so far published either articles or books or both on chemostratigraphy during these years. Among these, the Geological Society of America leads with 2029 publications, followed by Geological Society of London (1378 publications). Out of the total publications listed in the GeoScienceWorld website, the eon Phanerozoic was intensively reported/studied in 2760 publications. Among the eras, the Cenozoic era leads with 1480 publications, while the Paleozoic (607) and the Mesozoic (532) follow not so closely in terms of number of publications. Similarly, among the individual periods, the Quaternary leads with 631 publications. When individual epochs are considered, the list is led by the Paleocene with 1745 publications, while among the ages, the Danian leads with 1471 publications. The interdisciplinary nature and applicability of chemostratigraphy in a variety of fields could be gauged from the fact that as many as 46 subject fields are covered by the publications listed in the GeoScienceWorld website. These subject fields range from applied geophysics; geomorphology; igneous, sedimentary, and metamorphic petrology; engineering geology; petroleum exploration; archaeology; oceanography; limnology; and paleontology. A cursory examination of many of the articles listed in these websites suggest that though there is an agreement among the researchers on the utility of chemostratigraphy as a tool for precise characterization, fingerprinting, and correlating strata and of geogenic and anthropogenic objects of interest, there appears to be absence of synchronicity of many terminologies. Many of the terminologies are used interchangeably, and some of them are used in a contradicting, if not confusing, context, leading to anarchy. This state of affair is due to many reasons including, but not limited to, absence of standardized terms and explanations on their precise usage. In the following part, we attempt listing the terminologies as available in the chemostratigraphy publications and provide a list of most important terms. The list may be commented, modified, expanded, and added with new terminologies.

2.3. GLOSSARY

Note that these are the terms that are in use in chemostratigraphic literature. While efforts are made by the authors to synchronize their meanings in the context of standardization, as could be appreciated, this is a continuing and evolving exercise. The contextual explanations provided herein may be added, expanded, and modified. In addition, newer terms may also be introduced befitting evolutionary development of this subject and practice. Thus, our objective is to provide a platform from where discussion among the practitioners of this field may be initiated.

Anomaly	Unique/distinctly different geochemical composition from adjoining strata on a spatial and temporal scale; more frequently used in a temporal sense. Depending on increment/decrease or sign of isotopic ratios, it is termed as positive or negative anomaly. The positive "shift" is termed as "spike"; however, this term is also used less commonly to indicate negative shift also.
Background value	Same as the normal value, but more often used while analyzing the trends of geochemical curve and/or average values of specific rock types, strata, or stratigraphic section.
Chemochron	Chemozone with well-defined absolute age constraints. However, well-defined chronological boundaries in terms of conventional biostratigraphy or magnetostratigraphy can also serve to define the upper and lower limits of a chemochron.
Chemofacies	Unique rock type/strata defined exclusively by elemental composition. While chemozone has temporal meaning, chemofacies defines the strata/portion of stratigraphic record/bed itself without temporal connotation.
Chemozone	A body of rock strata or other stratified materials (i.e., ice) defined or characterized by its geochemical composition (elemental, isotopic, or both). Examples of widely used chemozones are oxygen isotope stages in the Cenozoic. The term has a marked temporal connotation. This is characterized and recognized in geochemical curve, defined by unique geochemical index, or its specific range of value or its unique pattern, often recognizable at specific spatial and/or temporal scale.
Chemostratigraphy	Documentation, characterization, analysis, and correlation of stratigraphic variation of geochemical composition (absolute values of elemental or isotopic composition, ratios and their combinations) of geogenic and anthropogenic materials.
Chemostratigraphic index/indices	Elemental composition/ratio or isotopic composition used to characterize a body of rock/strata. Only those element/isotope values or ratios that show distinct variation between rocks/strata/rock components qualify to be geochemical signature/index.
Correlation	Correspondence that could be established with the help of geochemical trends and signatures at a spatial and temporal scale between geographically separated noncontiguous strata.
Cycle, cyclic	Repetition of geochemical properties in a stratigraphic sequence often with a fixed duration/period. The cycles are hierarchical such that higher-order cycles (short term) occur within lower-order cycles (long term).
Duration	Refers to the temporal scale of cycle, anomaly, etc., measured and/or perceived from the geochemical trends. Used in a relative scale with reference to the context as "short" or "long." Also referred as "short term" or "long term."
Excursion	Change of geochemical composition (elemental or isotopic, absolute values or ratios) that depart from normal or background values. The change can be spatial or temporal or both, but this term is normally used in a temporal sense. Depending on the change toward increment or decrease from the background reference values, excursion is defined as either positive or negative. The increment is referred to as "spike" although "spike" is also used less commonly to denote negative excursion. Excursions that occur in geographically widely separated noncontiguous sedimentary successions and exhibit remarkable isotope or elemental composition are termed anomalies (see "Anomaly").
Fingerprinting	Recognition of unique geochemical trait of rock/rock component/sediment/objects under study. While the process of recognition of a certain horizon or rock body by a set of geochemical characteristics is termed fingerprinting, the "fingerprint" is represented by the geochemical signature, index, or proxy.
Fluctuation	A general description of geochemical changes on a temporal scale.
Geochemical curve	A line connecting the absolute or ratio or isotopic values arranged in stratigraphic order.
Geochemical marker	Unique geochemical character with which rock bodies are distinguished at a spatial and more often temporal scale.
Geochemical signature	Unique geochemical characteristic of body of rock under study.
Magnitude	The extent of change of compositional values or elemental or isotopic ratios from established/recognized trend or background/reference value, usually given in a relative scale (high, low, moderate, extreme) or in absolute measures (percent, ppm, per mil).
Normal value	Mean compositional (elemental or isotopic) values usually encountered in a strata/sediment/rock/rock component.
Order of cycle	Classification of cyclic geochemical trends depending on hierarchy (temporal duration) of the considered cycle.
Perturbation	Unique or distinct change of geochemical composition from background/normal value. Here, the emphasis is made on the "change" rather than the direction of change.

Proxy	Often used in paleoceanography, paleolimnology, and paleoclimatology to designate "indicators" used as an indirect measure of the processes/events studied.
Rate	Used in the context of defining or describing the rate at which the change of composition in a temporal scale, either positively or negatively or on a relative context.
Resolution	Duration of the shortest time interval resolvable by means of chemostratigraphy in a particular succession. Though relative, there are many differentiations of the scale in terms of low, high, very high, ultra high, etc.
Sampling resolution/ interval	Stratigraphic distance between samples in a chemostratigraphic study.
Secular	Monotonous and/or uniform stratigraphic variation of elemental composition or isotopic ratio values.
Shift	Used in the same context of the term excursion, but less frequently. Specifically used while describing the direction of change of geochemical composition from background value.
Spatial scale	A measure of geographically separated regions; the units of which are absolute and relative.
Stage	Normally used as "isotopic stage" to indicate distinct shift of isotopic trend as a result of climatic and/or major geological event.
Stratigraphy	A branch of geology concerned with the study of the original succession, age relations, distribution, lithologic composition, fossil content, and geophysical and geochemical properties of rock bodies, be it sedimentary, igneous, or metamorphic [*Salvador*, 1994].
Temporal scale	A measure of separation of strata in terms of time; the units of which are absolute and relative.
Term	Refer to "Duration."
Trend	The nature of the geochemical curve, constructed on a stratigraphic scale. Also used in the context of positive and negative change from background/reference values; sometimes used interchangeably with shift and excursion.
Trendline	Indicates the curve depicting geochemical values in stratigraphic scale. Also representing statistical measure of the geochemical curve, in terms of linear, polynomial, logarithmic, and exponential functions. Based on these statistical measurements, the trends are determined as secular, cyclic, and positive or negative anomaly.

2.4. SUMMARY

Chemostratigraphy is one of the young and rapidly growing fields of geosciences that has made rapid strides in multiple disciplines and has practitioners in many branches of science. Owing to these, a plethora of terminologies and their interchangeable usages thwart precise understanding and impede the development of this fascinating subject. Based on a thorough literature survey, a standard list of terminologies and explanations is presented in this chapter that may serve as a platform for further improvement and standard usage.

ACKNOWLEDGMENTS

This glossary was made possible by previous publications that are listed in this chapter and many others. Authors thank those pioneers who have contributed to the development of this subject. A few of the journal metrics presented in this chapter are collected from the scientific websites www.sciencedirect.com and www.geoscienceworld. org. The reviewers are thanked for their suggestions that have improved clarity of certain definitions.

REFERENCES

Berger, W.H. and Vincent, E., 1981. Chemostratigraphy and biostratigraphic correlation: exercises in systematic stratigraphy. Oceanologica Acta NP, *4*, 115–127.

Keller, G., 1988. Extinction, survivorship and evolution of planktonic foraminifera across the Cretaceous/Tertiary boundary at El Kef, Tunisia. Marine Micropalaeontology *13*, 239–263.

Oppo, D.W., Fairbanks, R.G. and Gordon, A.L., 1990. Late Pleistocene southern ocean $\delta^{13}C$ variability. Paleoceanography *5*, 43–54.

Ramkumar, M., 1999. Role of chemostratigraphic technique in reservoir characterisation and global stratigraphic correlation. Indian Journal of Geochemistry *14*, 33–45.

Ramkumar, M., 2014. Characterization of depositional units for stratigraphic correlation, petroleum exploration and reservoir characterization. In: Sinha, S., (Ed.), Advances in Petroleum Engineering. Studium Press L.L.C, Houston, pp. 1–13.

Ramkumar, M., 2015. Toward standardization of terminologies and recognition of chemostratigraphy as a formal stratigraphic method. In: Ramkumar, M., (Ed.). Chemostratigraphy: Concepts, Techniques and Applications. Elsevier, Amsterdam, pp. 1–21.

Ramkumar, M., Stüben, D., Berner, Z. and Schneider, J., 2004. Isotopic and geochemical anomalies preceding K/T boundary in the Cauvery Basin, south India: timing of events in the context of global scenario. Current Science *87*, 1738–1747.

Ramkumar, M., Harting, M. and Stüben, D., 2005. Barium anomaly preceding K/T boundary: possible causes and implications on end Cretaceous events of K/T sections in Cauvery Basin (India), Israel, NE-Mexico and Guatemala. International Journal of Earth Sciences *94*, 475–489.

Renard, M., 1986 Pelagic carbonate chemostratigraphy (Sr, Mg, ^{18}O, ^{13}C). Marine Micropaleontology *10*, 117–164.

Romein, A.J.T. and Smit, J., 1981. Carbon–Oxygen stable isotope stable isotope stratigraphy of the Cretaceous–Tertiary boundary interval: data from the Biarritz section (SW France). Geologie en Mijnbouw *60*, 514–544.

Salvador, A., 1994. International Stratigraphic Guide: A Guide to Stratigraphic Classification, Terminology and Procedure. 2nd Edition, IUGS-GSA, Boulder, pp. 1–214.

Scholle, P.A. and Arthur, M.A., 1980. Carbon isotope fluctuations in Cretaceous pelagic limestones: potential stratigraphic and petroleum exploration tool. American Association of Petroleum Geologists Bulletin *64*, 67–87.

Sial, A.N., Gaucher, C., Ferreira, V.P., Pereira, N.S., Cezario W.S., Chiglino, L. and Lima, H.M., 2015. Isotope and elemental chemostratigraphy. In: Ramkumar, M., (Ed.). Chemostratigraphy: Concepts, Techniques and Applications. Elsevier, Amsterdam, pp. 23–64.

Wiens, R.C., Maurice, S. and MSL Science Team, 2015. ChemCam: chemostratigraphy by the first Mars microprobe. Elements *11*, 33–38.

Part II
Precambrian

3

The Archean-Proterozoic Boundary and the Great Oxidation Event

Claudio Gaucher[1] and Robert Frei[2,3]

ABSTRACT

The Archean-Proterozoic boundary is placed at 2500 Ma and marks possibly the most dramatic change in Earth's history. The Great Oxidation Event (GOE) took place between 2.45 and 2.32 Ga as part of an Archean-Proterozoic transition rather than sharp boundary and postdates the currently established boundary. Apart from geological proxies of atmospheric oxygenation, such as banded iron formation (BIF) abundance and paleosol mineralogy, isotope chemostratigraphy provides the most powerful tool for studying the GOE and establishing the boundary. Sulfur isotopes, and especially mass-independent fractionation of sulfur, are the best studied proxy, indicating the formation of an ozone layer at ca. 2.33 Ga. Cr and Mo isotopes are more sensitive indicators of surface environment oxygenation, reflecting the redox state of surface seawater. We discuss three different proposals for the placement of the Archean-Proterozoic boundary and a corresponding Global Stratotype Section and Point: (i) to keep it at 2500 Ma, aided by prominent BIF units, Mo abundance, and Mo isotopes; (ii) to place it at the base of the second Huronian glaciation (ca. 2.35–2.40 Ga), thought to represent a "snowball" event; and (iii) to use the termination of the mass-independent fractionation of sulfur and the increase in the $\delta^{34}S$ amplitude of sulfides as the main criteria.

3.1. INTRODUCTION

The Archean-Proterozoic boundary marks what can be described as the most dramatic global change that our planet experienced since the formation of the Moon by a giant impact some 60 Myr after the formation of the solar system [*Touboul et al.*, 2007]. Early indications of a major change in seawater and atmospheric composition and redox state were provided by geological evidence. Anoxic Archean atmosphere and oceans were deduced from the occurrence of detrital pyrite [*England et al.*, 2002] and uraninite in conglomerates and the reduced nature of mineral phases in Archean paleosols and uranium enrichments in marine shales [*Holland*, 1994; Fig. 3.1]. The deposition of the largest banded iron formation (BIF), the first occurrence of red beds, and a shift to oxidized paleosols (Fig. 3.1) showed that the redox state of the atmosphere changed at or near the Archean-Proterozoic boundary, exceeding the threshold oxygen concentration of 1% present atmospheric level (PAL).

Macgregor [1927] proposed a photosynthetic origin for atmospheric oxygen, a model that was later developed by *Cloud* [1965, 1968, 1972] mainly to explain BIF deposition. The model has so far remained unchallenged, because alternative abiotic mechanisms of oxidant production, such as CO_2 photolysis to O_2 and photochemically produced H_2O_2, are inefficient and were relatively unimportant even in the high-CO_2 Archean

[1] *Instituto de Ciencias Geológicas, Facultad de Ciencias, Universidad de la República, Montevideo, Uruguay*

[2] *Department of Geosciences and Natural Resource Management, University of Copenhagen, Copenhagen, Denmark*

[3] *Nordic Center for Earth Evolution (NordCEE), University of Southern Denmark, Odense, Denmark*

Chemostratigraphy Across Major Chronological Boundaries, Geophysical Monograph 240, First Edition.
Edited by Alcides N. Sial, Claudio Gaucher, Muthuvairavasamy Ramkumar, and Valderez Pinto Ferreira.
© 2019 the American Geophysical Union. Published 2019 by John Wiley & Sons, Inc.

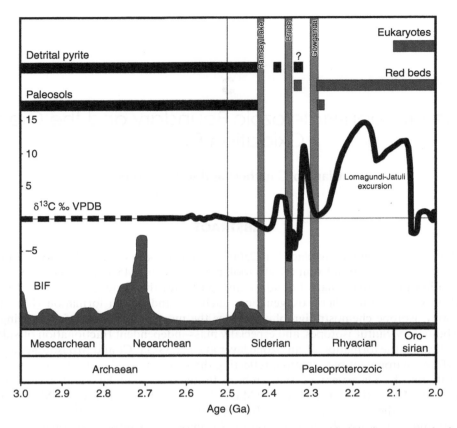

Figure 3.1 Summary of geological evidence of atmospheric oxygenation and glacial events in the later Archean and Paleoproterozoic. Glacial events named after the type Huronian glacials in Canada [*Papineau et al.*, 2007]. BIF abundance curve according to *Isley and Abbott* [1999]; carbon isotope curve compiled from *Lindsay and Brasier* [2002], *Melezhik et al.* [2007], and *Maheshwari et al.* [2010]; and the other proxies from *Holland* [1994] and *Bekker et al.* [2006].

atmosphere [*Canfield*, 2005; *Haqq-Misra et al.*, 2011]. This extraordinary event, during which organisms irreversibly changed Earth's surface environments and even lithospheric composition, has been named Great Oxidation Event (GOE) [*Holland*, 2002].

The GOE took place, even in geological terms, over a long period of time, between ca. 2.45 and 2.32 Ga [*Farquhar et al.*, 2011 and references therein] as part of an Archean-Proterozoic transition rather than sharp boundary. This actually postdates the currently established age of the Archean-Proterozoic boundary at 2.5 Ga [*Cohen et al.*, 2013] and prompted *Gradstein et al.* [2012] to propose that the boundary be established at 2.42 Ga and to include the Siderian as the last period of the Neoarchean era. In their informal proposal, *Gradstein et al.* [2012] proposed the name "Oxygenian" for the oldest Paleoproterozoic period, between 2.42 Ga and the onset of the Lomagundi carbon isotope anomaly at ca. 2.25 Ga [*Melezhik et al.*, 2007; *Maheshwari et al.*, 2010; Fig. 3.1]. However, several lines of evidence show significant "whiffs of oxygen" already at ca. 2.7 Ga [*Anbar*

et al., 2007; *Frei et al.*, 2009] and even as far back as 3.0 Ga [*Crowe et al.*, 2013], meaning that the oxygenation process and the Archean-Proterozoic transition took probably more time than currently accepted.

In this chapter, we focus on the main chemostratigraphic proxies that allow us to understand and perhaps better establish the Archean-Proterozoic boundary and the GOE. As we show below, this is possibly the first major boundary that can be defined mainly on the basis of chemostratigraphy.

3.2. SULFUR ISOTOPES

Sulfur has four naturally occurring isotopes, namely, ^{32}S (95.02%), ^{33}S (0.75%), ^{34}S (4.21%), and ^{36}S (0.02%), which can be quantified by three ratios, such as $\delta^{33}S$, $\delta^{34}S$, and $\delta^{36}S$, corresponding, respectively, to $^{33}S/^{32}S$, $^{34}S/^{32}S$, and $^{36}S/^{32}S$ isotopic ratios[1] [*Mojzsis*, 2007].

---

[1] *The delta value for sulfur (in per mil) is defined as $\delta^x S = 1000 \times (^x Rsample/^x R_{VCDT} - 1)$, where $^x R$ is the isotopic ratio $^x S/S$ with $x = 33$, 34, or 36 and VCDT is the Vienna-Canyon Diablo Troilite standard.*

Most geological and biological isotopic fractionation processes are mass dependent and thus proportional to the relative mass difference between the isotopes considered. Because of this and the fact that the three sulfur isotope ratios are relative to ^{32}S concentration, there is a simple linear relationship between the different sulfur isotope systems [*Hulston and Thode*, 1965]:

$$\delta^{33}S = 0.515 \times \delta^{34}S \qquad (3.1)$$

$$\delta^{36}S = 1.90 \times \delta^{34}S \qquad (3.2)$$

Thus, the magnitudes $\Delta^{33}S$ and $\Delta^{36}S$ were defined to measure the difference of isotopic ratios with Equations 3.1 and 3.2. They equal zero for samples that follow those relationships and can be mathematically defined as [*Farquhar et al.*, 2000]

$$\Delta^{33}S = 1000 \times \left[\left(\delta^{33}S/1000 + 1 \right) - \left(\delta^{34}S/1000 + 1 \right)^{0.518} - 1 \right]$$
$$(3.3)$$

$$\Delta^{36}S = 1000 \times \left[\left(\delta^{36}S/1000 + 1 \right) - \left(\delta^{34}S/1000 + 1 \right)^{1.91} - 1 \right]$$
$$(3.4)$$

When $\Delta^{33}S$ and $\Delta^{36}S$ are different from zero or, expressed in other terms, when isotopic ratios do not follow Equations 3.1 and 3.2, it is evidence of mass-independent fractionation (MIF). Processes that have been observed to cause MIF in sulfur are gas-phase reactions such as H_2S photolysis and SO_2 photooxidation [*Farquhar et al.*, 2000; *Mojzsis*, 2007]. These reactions involve ultraviolet (UV) radiation, which is blocked by the ozone layer in the stratosphere. Present-day geological samples, such as marine and lacustrine sulfates and sulfides, yield $\Delta^{33}S = 0$ and $\Delta^{36}S = 0$, indicating the dominance of mass-dependent fractionation.

Farquhar et al. [2000] reported for the first time MIF for Archean and earliest Proterozoic (>2450 Ma) sulfides and sulfates, with $\Delta^{33}S$ reaching 2‰ VCDT (Fig. 3.2a). Samples between 2450 and 2090 Ma yielded weaker MIF signals with $\Delta^{33}S$ between 0.02 and 0.34‰ VCDT. On the other hand, samples younger than 2090 Ma consistently yielded $\Delta^{33}S = 0$ [*Farquhar et al.*, 2000; Fig. 3.2a].

The explanation put forward for the observed pattern is related to the absence of an ozone layer prior to ca. 2.45 Ga [*Farquhar et al.*, 2000]. Without ozone, UV radiation could penetrate deeply into Earth's atmosphere, and sulfur aerosols exhibiting MIF ($\Delta^{33}S \neq 0$ and $\Delta^{36}S \neq 0$) were formed. The absence of an effective ozone layer implies atmospheric pO_2 lower than 1% PAL before 2.45 Ga and may be even lower than 10^{-5} PAL [*Pavlov and Kasting*, 2002]. After the GOE, higher oxygen levels in the atmosphere (>1% PAL) enabled oxidative weathering of sulfides to overwhelm the diminished sulfur MIF signal in surface environments, effectively driving $\Delta^{33}S$ and $\Delta^{36}S$ to zero for the last 2.1 Ga [*Farquhar et al.*, 2000].

Guo et al. [2009] proposed that the establishment of an effective ozone layer and the consequent shutdown of MIF of sulfur took place during deposition of the middle Duitschland Formation of the Transvaal Supergroup (South Africa). Whereas the lower Duitschland Formation and its equivalent Rooihoogte Formation exhibit $\Delta^{33}S$ of up to 1.5‰, the upper part of the unit is characterized by the absence of an MIF signal ($\Delta^{33}S = 0$; *Bekker et al.*, 2004; *Guo et al.*, 2009; Fig. 3.2b). The age of the upper Duitschland Formation is constrained by a Re-Os isochron on carbonaceous, pyritic shales of 2316±7 Ma for the correlative Rooihoogte Formation [*Hannah et al.*, 2004]. Thus, the buildup of an effective ozone layer and the oxygenation of the atmosphere beyond 1% PAL took place around 2.32 Ga. It is worth noting that a positive $\delta^{34}S$ excursion of up to 42‰ VCDT is recorded concomitantly to the termination of MIF [*Guo et al.*, 2009; Fig. 3.2b], which provides further evidence for a more oxygenated atmosphere and marine waters richer in sulfate. Finally, a positive $\delta^{13}C$ excursion in carbonates reaching 8‰ VPDB is consistent with the described evolution of sulfur isotopes and strongly suggests that enhanced bioproductivity and/or organic carbon burial was responsible for this event [*Guo et al.*, 2009].

Sulfur isotope composition of authigenic sulfides from the Huronian Supergroup in Canada, which hosts evidence of three glaciations, mimics that reported for the Transvaal Supergroup. Whereas samples from the older McKim and Pecors formations yielded a clear MIF signal of up to 0.88‰, sulfides of the La Española and Gordon Lake formations exhibit $\Delta^{33}S$ values near zero [*Papineau et al.*, 2007]. Concomitantly, $\delta^{34}S$ values rise from near zero to 32‰ VCDT. The transition between an anoxic and an oxygenated atmosphere took place between the middle (Bruce Formation) and the youngest (Gowganda Formation) Huronian glaciations [*Papineau et al.*, 2007]. *Hoffman* [2013] correlated the Bruce Formation with a prominent erosional unconformity in the middle Duitschland/Rooihoogte Formation of the Transvaal Supergroup, and estimated an age of ca. 2.38 Ga for the MIF termination, to take into account the necessary global nature of the GOE. This age is some 50–60 Myr older than the one estimated by *Guo et al.* [2009]. The glacial diamictite at the base of the Duitschland Formation in South Africa is correlated by *Hoffman* [2013] with the oldest Huronian glaciation, the ca. 2.4 Ga Ramsay Lake Formation.

More recently, *Luo et al.* [2016] were able to better constrain the timing of the sulfur MIF termination

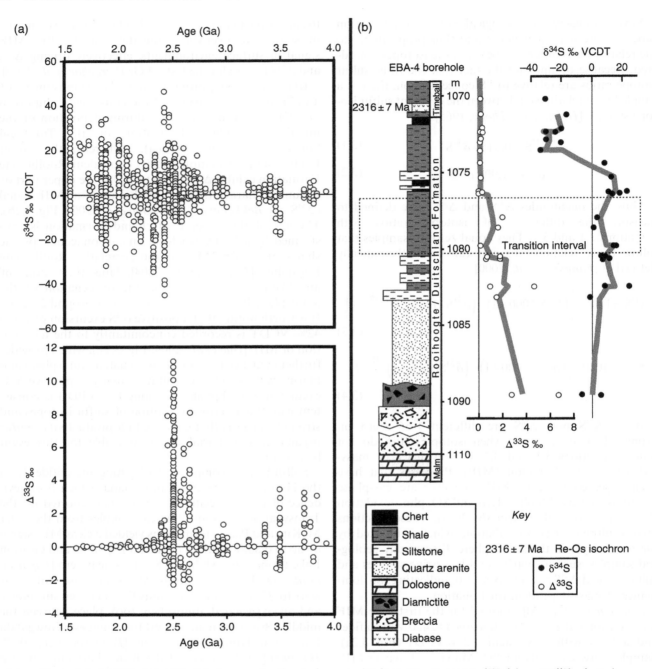

Figure 3.2 (a) $\delta^{34}S$ and $\Delta^{33}S$ global curves for the Archean and Paleoproterozoic. Modified from Williford *et al.* [2011 and references therein]. Reproduced with permission of Elsevier. (b) Termination of the MIF sulfur signal ($\Delta^{33}S$) in the Transvaal Supergroup. From *Luo et al.* [2016 and references therein]. Note that the transition, and thus the formation of a functional ozone layer, takes place in <4 m stratigraphically. Malm, Malmani Formation; Timeball, Timeball Hill Formation. Re-Os age according to *Hannah et al.* [2004].

within the Duitschland/Rooihoogte Formation with drill core material, which occurs 5 m above the middle Duitschland unconformity (Fig. 3.2b). *Luo et al.* [2016] calculated an age for the transition interval of 2.33 Ga, which is 50 Myr younger than the one proposed by

Hoffman [2013]. The transition, which involves a drop of $\Delta^{33}S$ from 8‰ down to essentially 0‰ (Fig. 3.2b), occurs in only 5 m of section and therefore represented a rather fast event (<10 Myr), equated by *Luo et al.* [2016] with the GOE.

3.3. CHROMIUM ISOTOPES

Chromium is a redox-sensitive transition metal and has four stable isotopes: ^{50}Cr (4.345%), ^{52}Cr (83.789%), ^{53}Cr (9.501%), and ^{54}Cr (2.365%). Of these, the two more abundant isotopes ^{52}Cr and ^{53}Cr are currently used in geochemistry and are quantified using the $\delta^{53}Cr$ ratio (The delta notation for Cr (in per mil) is defined as $\delta^{53}Cr = [(^{53}Cr/^{52}Cr)_{sample}/(^{53}Cr/^{52}Cr)_{SRM979}) - 1].$)., with the SRM 979 standard as a reference.

Mantle and magmatic rocks exhibit a narrow range of $\delta^{53}Cr$ values of $-0.12 \pm 0.10‰$, meaning that they range between -0.22 and $-0.02‰$ (mean: $-0.124‰$; *Schoenberg et al.*, 2008). These high-temperature materials represent the ultimate source of chromium for surface processes.

From a redox point of view, Cr has two oxidation states: Cr (III) and Cr (VI). The former is immobile and is the one encountered in high-temperature minerals. Upon oxidation to Cr(VI) by oxidative weathering, chromium becomes mobile and is transported in solution by rivers [e.g., *D'arcy et al.*, 2016]. The oxidation process is catalyzed by manganese oxides and involves a positive isotopic fractionation, with enrichment of the mobile hexavalent chromium in the heavier ^{53}Cr [*Frei et al.*, 2009]. Present-day river waters yield positive $\delta^{53}Cr$ values [*Frei et al.*, 2014; *D'arcy et al.*, 2016, and references therein], and reported seawater values range between 0.13 and 1.5‰ [*Bonnand et al.*, 2013; *Scheiderich et al.*, 2015; *Paulukat et al.*, 2016]. The lower seawater values are encountered in restricted basins, such as the Baltic Sea, which are strongly influenced by local river discharge [*Paulukat et al.*, 2016]. Positive fractionation is thus confirmed in present-day oxygenated atmosphere. Soils and altered rocks, on the other hand, typically yield negative $\delta^{53}Cr$ values [*Frei et al.*, 2014; *D'arcy et al.*, 2016], because chromium remaining in the weathered rocks is impoverished in ^{53}Cr, which is preferentially lixiviated and transported.

Frei et al. [2009] reported $\delta^{53}Cr$ values for Precambrian BIFs and cherts (Fig. 3.3). Archean samples yielded mostly unfractionated values within the range of high-temperature Cr ($-0.12 \pm 0.10‰$), with the notable

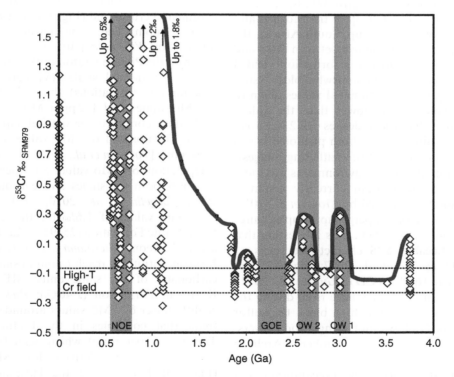

Figure 3.3 Cr isotope evolution of seawater in the Precambrian and present values. Modified from *Frei et al.* [2009]. Shaded areas represent notable oxygenation events: OW1, Mesoarchean Oxygen Whiff; OW2, Neoarchean Oxygen Whiff; GOE, Great Oxidation Event; NOE, Neoproterozoic Oxidation Event. Gray line represents the envelope of the most fractionated data and approximates the evolution of atmospheric oxygen. Present-day seawater values from *Paulukat et al.* [2016]. Other sources of data: Isua BIF: *Frei et al.* [2016], Ijzermyn iron formation (ca. 3 Ga): *Crowe et al.* [2013], Mesoproterozoic-Tonian carbonates: *Gilleaudeau et al.* [2016], Otavi Group carbonates (Cryogenian): *Rodler et al.* [2016]; Ediacaran iron formations: *Frei et al.* [2013, 2017].

exception of four BIFs deposited between 2.8 and 2.45 Ga and showing positive $\delta^{53}Cr$ values of up to 0.29‰. Samples immediately predating the GOE, with an age between 2.48 and 2.45 Ga, were found to be unfractionated [*Frei et al.,* 2009; Fig. 3.3]. Due to the scarcity of BIF in the 2.45–2.1 Ga period (see Fig. 3.1), which includes the entire GOE, there is a gap in the data of *Frei et al.* [2009]. BIF samples in the 2.1–2.0 Ga range yield weakly fractionated values around 0.1‰. A substantial positive $\delta^{53}Cr$ excursion up to 5‰ characterizes the later Neoproterozoic and marks a second oxygenation event [*Frei et al.,* 2009, 2013, 2017].

More recently, chromium isotopes were for the first time successfully measured on Neoproterozoic carbonates [*Frei et al.,* 2011], meaning that the gap in the $\delta^{53}Cr$ curve spanning the GOE can be filled using carbonate samples.

The above $\delta^{53}Cr$ data imply that transient "whiffs" of oxygen occurred in the Neoarchean between 2.7 and 2.5 Ga (Fig. 3.3), as also indicated by the redox-sensitive trace elements Mo and Re [*Anbar et al.,* 2007; see below]. Even more tantalizing are positively fractionated values recently reported for the ca. 3.8 Ga Isua BIF [*Frei et al.,* 2016; Fig. 3.3]. Evidences of oxidative weathering were also reported by *Crowe et al.* [2013] for the Mozaan and Nsuze groups (Pongola Supergroup, South Africa), the age of which is tightly constrained between 2.98 and 2.92 Ga. Whereas the Ijzermyn iron formation yielded $\delta^{53}Cr$ values as high as 0.28‰, the somewhat older Nsuze paleosol yielded negatively fractionated values down to −1‰, which are considerably lower than the values observed in the parent basaltic andesites (−0.2‰; *Crowe et al.,* 2013). As explained above, both phenomena can be explained as a result of oxidative weathering, suggesting that the oxidation of surface environments was more complicated and protracted than currently assumed. Oxygen concentrations calculated by *Crowe et al.* [2013] on the basis of the Pongola Supergroup samples range between 6×10^{-5} and 3×10^{-4} PAL. It is interesting that lower but still noticeable $\Delta^{33}S$ characterize rocks of comparable age [*Farquhar et al.,* 2007; Fig. 3.2] and $\delta^{34}S_{Pyrite}$ values reach 16.5‰ VCDT [*Guy et al.,* 2012], suggesting that oxygen levels may have risen during the late Mesoarchean, but not enough to block the sulfur MIF signal. If so, the GOE was not a single and irreversible event, but only the strongest of a number of Archean and Proterozoic oxygenation events. Oxygen levels may have not risen steadily during the Precambrian, but rather fluctuated up and down, with an overall trend of increased oxygenation.

Chromium isotopes are a still young but promising tool to trace atmospheric oxygenation at the Archean-Proterozoic boundary. Further studies of continuous, well-preserved sections encompassing the boundary are needed, especially those dominated by carbonates.

3.4. MOLYBDENUM ISOTOPES AND CONCENTRATIONS

Molybdenum exhibits a similar chemical behavior as Cr, both of them belonging to group 6 in the periodic table. It has seven stable isotopes, the isotopic ratio used in chemostratigraphy being $\delta^{98/95}Mo$, which is reported relative to the Johnson Matthey ICP standard solution (J&M standard, *Siebert et al.,* 2003) or the CPI Mo ICPMS standard [*Goldberg et al.,* 2009]. The redox behavior of Mo is similar to Cr: the oxidized Mo (VI) is mobile, and after being released by oxidative weathering, it is transported by rivers to the ocean. Relative to the J&M standard, present-day river waters exhibit average $\delta^{98/95}Mo$ values of 0.9‰, seawater has a tight average value of 2.3 ± 0.2‰, and the mantle/crust reservoir ranges from −0.2 to 0.4‰ [*Siebert et al.,* 2003; *Goldberg et al.,* 2009, and references therein].

Wille et al. [2007] studied the $\delta^{98/95}Mo$ chemostratigraphy of the 2.64–2.15 Ga Transvaal Supergroup and the older 3.23 Ga Fig Tree Group in South Africa. Shale and chert samples of the Fig Tree Group yielded, except for one sample, unfractionated $\delta^{98/95}Mo$ values within the continental crust band (Fig. 3.4) and Mo concentrations around 2 ppm. The same unfractionated results were reported by *Wille et al.* [2013] for black shales >2.8 Ga in age from the Pilbara Craton in Western Australia.

Lower Transvaal shales (Vryburg, Boomplaas, and Lokammona formations) show similar unfractionated $\delta^{98/95}Mo$ values and 1–2 ppm Mo. Carbonate data and shale data corrected for detrital input, however, show clearly fractionated $\delta^{98/95}Mo$ values of up to 1.3‰ for these units [*Voegelin et al.,* 2010; Fig. 3.4].

Upsection, $\delta^{98/95}Mo$ values are fractionated, and [Mo] increases both for shales and carbonates [*Wille et al.,* 2007; *Voegelin et al.,* 2010], with $\delta^{98/95}Mo$ reaching maximum values of 1.6‰ for the upper Nauga/lower Klein Naute Formation (2549 ± 7 Ma, U-Pb SHRIMP on ash bed zircon; *Altermann and Nelson,* 1998; Fig. 3.4). Mo concentrations reach 6 ppm in shales and 1.5 ppm in carbonates of the same units. BIF of the overlying Kuruman Formation (2.46 Ga; *Pickard,* 2003) yielded slightly lower $\delta^{98/95}Mo$ values around 0.7‰ (Fig. 3.4), a trend that continues in the Pretoria Group (upper Transvaal Supergroup), which yielded mostly unfractionated values and relatively low Mo concentrations (Fig. 3.5). For instance, the Timeball Hill Formation, whose lower part is well constrained by a Re-Os age of 2316 ± 7 Ma [*Hannah et al.,* 2004], yielded $\delta^{98/95}Mo$ values between −0.34 and 0.35‰ and [Mo] between 0.56 and 3.7 ppm [*Wille et al.,* 2007].

These results coincide in part with the Neoarchean "whiffs" of oxygen revealed by Cr isotopes, especially for the 2.8–2.48 Ga period [*Frei et al.,* 2009].

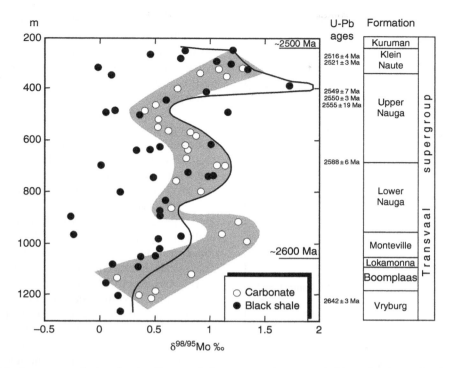

Figure 3.4 Mo isotope evolution in the Transvaal Supergroup between 2.64 and ca. 2.5 Ga. Modified after *Voegelin et al.* [2010], including shale data of *Wille et al.* [2007]. The curve in gray represents the $\delta^{98/95}$Mo evolution of marine carbonates, and the black curve the isotopic evolution of shales. U-Pb ages according to Altermann and Nelson [1998 and references therein].

Interestingly, the same pattern is revealed by Mo elemental chemostratigraphy of the Mount McRae Shale (Australia), which shows enrichments of up to 40 ppm [*Anbar et al., 2007*]. The beds where this Mo enrichment is recorded yielded a Re-Os isochron of 2501 ± 8 Ma [*Anbar et al., 2007*], correlating them with the upper Klein Naute/lower Kuruman Formation in South Africa [*Altermann and Nelson, 1998*], which also shows Mo enrichments and a fractionated $\delta^{98/95}$Mo signature [*Wille et al., 2007*]. A possible explanation of this pattern suggested by *Voegelin et al.* [2010] is that Mo isotopes may be more sensitive to low oxygen concentrations ($<10^{-5}$ PAL) than rare earth data (Ce anomaly) or sulfur MIF.

On the other hand, the low $\delta^{98/95}$Mo values reported for the upper Transvaal Supergroup are difficult to reconcile with sulfur isotope data for the same units (Fig. 3.5), which record the shutdown of the MIF signal already in the middle Duitschland/Rooihoogte Formation [*Luo et al., 2016*]. *Wille et al.* [2007] suggested that near-global glaciations [e.g., *Hoffman, 2013*] and consequent anoxic conditions in the ocean during this period may account for the unfractionated, low $\delta^{98/95}$Mo values (Fig. 3.5) during a time when the

atmosphere was already oxygenated, as indicated by the loss of the sulfur MIF signature.

3.5. DISCUSSION AND CONCLUSIONS

3.5.1. Significance of Geochemical Proxies for Atmospheric Oxygenation

It becomes clear that the different oxygenation proxies behave differently. MIF of sulfur is dependent on (i) the absence of an effective ozone layer and (ii) the absence of widespread mass-dependent fractionation processes. Both conditions are met in a reducing atmosphere with less oxygen than 10^{-5} PAL [*Pavlov and Kasting, 2002*], conditions that undoubtedly existed before 3.2 Ga. At the other end, there is consensus that none of those conditions were met after 2.33 Ga, that is, an ozone layer existed, and mass-dependent processes, mainly bacterial sulfate reduction, dominated the sulfur cycle, due to oxygen concentrations larger than 10^{-2} PAL [*Farquhar et al., 2007; Luo et al., 2016*]. As for the time in between, it is reasonable to consider them a transitional period. A strong MIF signal with $\Delta^{33}S > 3‰$ is observed for samples older than 3.2 Ga, but it is attenuated between 3.2–3.0

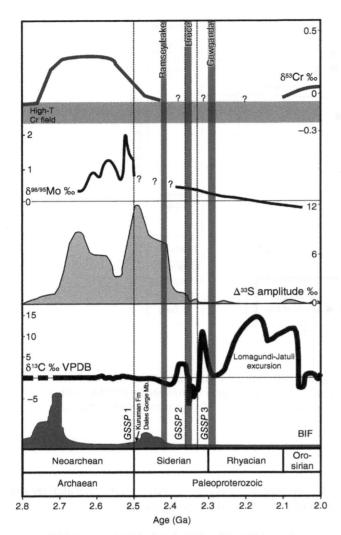

Figure 3.5 Summary of chemostratigraphic curves, BIF deposition, and glacial events (blue bars) near the Archean-Proterozoic boundary. Three different proposals of Global Stratotype Section and Point (GSSP) for the boundary are shown. GSSP 1: current boundary at 2500 Ma. GSSP 2: base of the second and possibly near-global Huronian glaciation [*Gradstein et al.*, 2012]. GSSP 3: termination of the sulfur MIF (Δ^{33}S) at ca. 2.33 Ga [*Luo et al.*, 2016]. Sources of data: δ^{53}Cr: *Frei et al.* [2009], δ$^{98/95}$Mo of shales: *Wille et al.* [2007], Δ^{33}S amplitude: *Williford et al.* [2011 and references therein], *Luo et al.* [2016]; δ^{13}C, BIF abundance and glacial events: same as for Figure 3.1. *(See insert for color representation of the figure.)*

and 2.7–2.9 Ga to less than ±1‰ [*Farquhar et al.*, 2007; *Guy et al.*, 2012; Fig. 3.2a]. *Farquhar et al.* [2007] suggested that the transparency of the atmosphere to certain UV wavelengths may have changed due to the formation of an organic haze. Alternatively, *Ono et al.* [2006] proposed that the low amplitude of the signal could have been caused by an incipient atmospheric oxidation

between 10^{-5} and 10^{-2} PAL. More recently, *Eickmann et al.* [2018] reported negative Δ^{33}S in pyrite, δ^{34}S as large as 30‰, and iron isotope evidence from the 2.97 Ga Nsuze Group, all pointing to oxygenated shallow waters (at least 5 mM sulfate) despite the existence of a reducing atmosphere.

The 2.7–2.45 period, on the other hand, is more puzzling. Large-amplitude sulfur MIF of up to 12‰ is characteristic [*Farquhar et al.*, 2007; *Guy et al.*, 2012] and coincides with larger-amplitude δ^{34}S excursions (Fig. 3.2a) and with evidence for transient oxidation events from Cr and Mo isotopes and Mo concentrations in shales [*Anbar et al.*, 2007; *Wille et al.*, 2007; *Frei et al.*, 2009; Fig. 3.5]. These observations are difficult to reconcile, but one possibility is a greater sensitivity of Cr and Mo to low oxygen concentrations than sulfur MIF [*Voegelin et al.*, 2010]. The puzzle has been solved recently by *Fakhraee et al.* [2018] by chemical modeling, showing that large-amplitude Δ^{33}S values require micromolar levels of O$_2$ in seawater in order to be preserved during early diagenesis. Present-day oxygen concentration in surface seawater at 20 °C is 225 μM, increasing with depth. Therefore, seawater oxygen concentrations modeled for the 2.7–2.45 period are at least 1% of present-day concentrations, but does not mean that the atmosphere had reached 1% PAL yet [*Fakhraee et al.*, 2018].

Summing up, while sulfur MIF records the redox state of the atmosphere, Mo concentrations, δ$^{98/95}$Mo, and possibly δ^{53}Cr trace the redox state of surface seawater (Fig. 3.5). Levels of up to 10 μM O$_2$ in surface seawater can coexist with an atmosphere devoid of oxygen, due to the relatively small diffusion coefficient of O$_2$ that allows significant disequilibrium [*Olson et al.*, 2013].

3.5.2. Where to Place the Archean-Proterozoic Boundary?

The Archean-Proterozoic boundary is currently placed at 2500 Ma, but only on the basis of round-number chronometric divisions and not necessarily following important events in Earth's history [*Gradstein et al.*, 2012]. Indeed, there are few geological events that occurred at 2500 Ma (Fig. 3.5). One of them is the base of the massive iron deposits (up to 750 m thick) of the Kuruman and Penge formations in the Transvaal Basin (South Africa), calculated at 2500 Ma by *Altermann and Nelson* [1998] on the basis of U-Pb SHRIMP ages of interbedded ash beds (Fig. 3.5). Another event identified at 2501 ± 8 Ma is represented by Mo and Re enrichments in the Mount McRae Shale (Australia), which have been interpreted as an ephemeral oxygenation event [*Anbar et al.*, 2007]. A possible Global Boundary Stratotype Section and Point (GSSP) could be located at the base of the Kuruman Formation, but the question arises as to whether this boundary could be easily recognized elsewhere. A possible correlate in the Hamersley Group (Australia) may be the boundary between the Mount McRae Shale (2501 ± 8 Ma Re-Os; *Anbar et al.,* 2007) and overlying BIF of the Dales Gorge Member of the Brockman Iron Formation, the base of which has been dated by U-Pb SHRIMP on tuff zircon at 2495 ± 16 Ma [*Trendall et al.,* 2004]. Chemostratigraphic criteria which may aid in establishing the boundary would be the mentioned Mo enrichment in shales and the coeval positive $\delta^{98/95}$Mo excursion, which reaches its maximum just beneath the Kuruman Formation [*Wille et al.,* 2007; Fig. 3.5]. Establishing the GSSP for the Archean-Proterozoic boundary in this way would maintain the currently used boundary at 2.5 Ga, regardless of the GOE, which is clearly younger.

A different boundary has been suggested by *Gradstein et al.* [2012] that emphasizes the importance of Siderian glacial events, which they include in the Archean. Their proposal for a new GSSP is at the base of glacial deposits in Australia, which are assigned to the Meteorite Bore Member of the Kungarra Formation, Turee Creek Group [*Williford et al.,* 2011; Fig. 3.5]. These glacial diamictites are stratigraphically ca. 2 km on top of an ash bed dated U-Pb SHRIMP at 2445 ± 5 Ma and are intruded by a sill dated at 2208 ± 15 Ma [*Trendall et al.,* 2004; *Müller et al.,* 2005]. The Meteorite Bore glacials are correlated with the middle Duitschland Formation in the Transvaal Supergroup, on the basis of the termination of the Δ^{33}S signal [*Williford et al.,* 2011]. Using the same criteria, the Meteorite Bore Formation would be correlative to the Bruce Formation of the Huronian Supergroup [*Hoffman,* 2013]. Therefore, rather than the first appearance of Siderian glacial deposits as stated by *Gradstein et al.* [2012], this GSSP would be at the base of the second Siderian glaciation (Fig. 3.5), which seems to be the only one with a cap carbonate and possible near-global extent [*Hoffman,* 2013 and references therein].

In any case, if this approach is preferred, the GSSP is better located in the Huronian Supergroup in Canada, which records the three glacial events and the termination of the sulfur MIF [*Papineau et al.,* 2007]. The most important chemostratigraphic aides for the placement of the boundary in this proposal are the disappearance of Δ^{33}S and the expansion of δ^{34}S values, although both occur above the base of the second glacial event [*Papineau et al.,* 2007; Figs. 3.2 and 3.5]. One disadvantage of a boundary defined on the basis of the second Huronian glaciation is the occurrence of a prominent erosional unconformity at the same level in the Transvaal Supergroup [*Guo et al.,* 2009; *Hoffman,* 2013; *Luo et al.,* 2016], which represents a hiatus of unknown duration. It is likely that the same situation will be encountered in other coeval successions elsewhere.

Another approach would be to establish the GSSP mainly on the basis of chemostratigraphy. Atmospheric oxygenation beyond a threshold of ca. 1% PAL (the GOE) is a very significant event in Earth's history, and was of global nature, at around 2.33 Ga (Fig. 3.5). It is recorded in several geological units in different continents, such as the Transvaal Supergroup in South Africa [*Luo et al.,* 2016], the Huronian Supergroup in North America [*Papineau et al.,* 2007], and the Turee Creek Group in Australia [*Williford et al.,* 2011]. The transition takes place in 5 m of section in the Transvaal Supergroup [*Luo et al.,* 2016; Fig. 3.2b] and in >2 m in the Turee Creek Group, which highlights the feasibility of placing a GSSP by means of high-resolution chemostratigraphy of the transitional interval. The boundary layer could be identified in continuous sections worldwide, including those dominated by shales and carbonates.

The three possible boundaries have their own different advantages. The conservative proposal to keep the boundary at 2.5 Ga would not require any revisions to existing geological maps and stratigraphic tables, something that the other two proposals do. Placing the boundary at the base of the second Huronian glaciation (ca. 2.4 Ga) would be analogous to the definition of the Ediacaran-Cryogenian boundary, which has proven successful since its establishment [*Knoll et al.,* 2006]. Finally, choosing the termination of sulfur MIF as the Archean-Proterozoic boundary would make atmospheric oxygenation the real and global event that marks the boundary, thereby permitting easier correlations on a global scale.

ACKNOWLEDGMENTS

Financial support to the first author by the Polo de Desarrollo Universitario "Geología y Recursos Minerales" (CURE, Universidad de la República, Uruguay) and from the Sistema Nacional de Investigadores (ANII, Uruguay) is gratefully acknowledged. The useful comments of two anonymous reviewers helped improve an earlier version of this manuscript.

REFERENCES

Altermann, W., Nelson, D.R., 1998. Sedimentation rates, basin analysis and regional correlations of three Neoarchaean and Palaeoproterozoic sub-basins of the Kaapvaal craton as inferred from precise U–Pb zircon ages from volcaniclastic sediments. Sedimentary Geology *120,* 225–256.

Anbar, A.D., Duan, Y., Lyons, T.W., Arnold, G.L., Kendall, B., Creaser, R.A., Kaufman, A.J., Gordon, G.W., Scott, C., Garvin, J., Buick, R., 2007. A whiff of oxygen before the great oxidation event? Science *317,* 1903–1906.

Bekker, A., Holland, H.D., Wang, P.L., Rumble, D., III, Stein, H.J., Hannah, J.L., Coetzee, L.L., Beukes, N.J., 2004. Dating the rise of atmospheric oxygen. Nature 427, 117–120.

Bekker, A., Karhu, J.A., Kaufman, A.J., 2006. Carbon isotope record for the onset of the Lomagundi carbon isotope excursion in the Great Lakes area, North America. Precambrian Research 148, 145–180.

Bonnand, P., James, R.H., Parkinson, I.J., Connelly, D.P., Fairchild, I.J., 2013. The chromium isotopic composition of seawater and marine carbonates. Earth and Planetary Science Letters 382, 10–20.

Canfield, D.E., 2005. The early history of atmospheric oxygen: homage to Robert M. Garrels. Annual Review of Earth and Planetary Sciences 33, 1–36.

Cloud, P.E., 1965. Significance of the Gunflint (Precambrian) microflora. Science 148, 27–35.

Cloud, P.E., 1968. Atmospheric and hydrospheric evolution on the primitive Earth. Science 160, 729–736.

Cloud, P.E., 1972. A working model of the primitive Earth. American Journal of Science 272, 537–548.

Cohen, K.M., Finney, S.C., Gibbard, P.L., Fan, J.-X., 2013. The ICS International Chronostratigraphic Chart. Episodes 36, 199–204.

Crowe, S.A., Døssing, L.N., Beukes, N.J., Bau, M., Kruger, S.J., Frei, R., Canfield, D.E., 2013. Atmospheric oxygenation three billion years ago. Nature 501, 535–538.

D'Arcy, J., Babechuk, M.G., Dossing, L.N., Gaucher, C., Frei, R., 2016. Processes controlling the chromium isotopic composition of river water: constraints from basaltic river catchments. Geochimica et Cosmochimica Acta 186, 296–315.

Eickmann, B., Hofmann, A., Wille, M., Bui, T.H., Wing, B.A., Schoenberg, R., 2018. Isotopic evidence for oxygenated Mesoarchaean shallow oceans. Nature Geoscience 11, 133–138.

England, G.L., Rasmussen, B., Krapez, B., Groves, D.I., 2002. Palaeoenvironmental significance of rounded pyrite in siliciclastic sequences of the Late Archaean Witwatersrand Basin: oxygen-deficient atmosphere or hydrothermal alteration? Sedimentology 49, 1133–1156.

Fakhraee, M., Crowe, S.A., Katsev, S., 2018. Sedimentary sulfur isotopes and Neoarchean ocean oxygenation. Science Advances 4(1), e1701835.

Farquhar, J., Bao, H., Thiemens, M.H., 2000. Atmospheric influence of Earth's earliest sulfur cycle. Science 289, 756–758.

Farquhar, J., Peters, M., Johnston, D.T., Strauss, H., Masterson, A., Wiechert, U., Kaufman, A.J., 2007. Isotopic evidence for Mesoarchaean anoxia and changing atmospheric sulphur chemistry. Nature 449, 706.

Farquhar, J., Zerkle, A.L., Bekker, A., 2011. Geological constraints on the origin of oxygenic photosynthesis. Photosynthesis Research 107, 11–36.

Frei, R., Gaucher, C., Poulton, S.W., Canfield, D.E., 2009. Fluctuations in Precambrian atmospheric oxygenation recorded by chromium isotopes. Nature 461, 250–254.

Frei, R., Gaucher, C., Døssing, L.N., Sial, A.N., 2011. Chromium isotopes in carbonates: a tracer for climate change and for reconstructing the redox state of ancient seawater. Earth and Planetary Science Letters 312, 114–125.

Frei, R., Gaucher, C., Stolper, D., Canfield, D.E., 2013. Fluctuations in late Neoproterozoic atmospheric oxidation: Cr isotope chemostratigraphy and iron speciation of the late Ediacaran lower Arroyo del Soldado Group (Uruguay). Gondwana Research 23, 797–811.

Frei, R., Poiré, D., Frei, K.M., 2014. Weathering on land and transport of chromium to the ocean in a subtropical region (Misiones, NW Argentina): a chromium stable isotope perspective. Chemical Geology 381, 110–124.

Frei, R., Crowe, S.A., Bau, M., Polat, A., Fowle, D.A., Døssing, L.N., 2016. Oxidative elemental cycling under the low O2 Eoarchean atmosphere. Scientific reports 6, 21058.

Frei, R., Døssing, L.N., Gaucher, C., Boggiani, P.C., Frei, K.M., Bech Árting, T., Crowe, S.A., Freitas, B.T., 2017. Extensive oxidative weathering in the aftermath of a late Neoproterozoic glaciation: evidence from trace element and chromium isotope records in the Urucum district (Jacadigo Group) and Puga iron formations (Mato Grosso do Sul, Brazil). Gondwana Research 49, 1–20.

Gilleaudeau, G.J., Frei, R., Kaufman, A.J., Kah, L.C., Azmy, K., Bartley, J.K., Chernyavskiy, P., Knoll, A.H., 2016. Oxygenation of the mid-Proterozoic atmosphere: clues from chromium isotopes in carbonates. Geochemical Perspectives Letters 2, 178–187.

Goldberg, T., Archer, C., Vance, D., Poulton, S.W., 2009. Mo isotope fractionation during adsorption to Fe (oxyhydr) oxides. Geochimica et Cosmochimica Acta 73, 6502–6516.

Gradstein, F.M., Ogg, J.G., Hilgen, F.J., 2012. On the geologic time scale. Newsletters on Stratigraphy 45, 171–188.

Guo, Q., Strauss, H., Kaufman, A.J., Schröder, S., Gutzmer, J., Wing, B., Baker, M.A., Bekker, A., Jin, Q., Kim, S.-T., Farquhar, J., 2009. Reconstructing Earth's surface oxidation across the Archean-Proterozoic transition. Geology 37, 399–402.

Guy, B.M., Ono, S., Gutzmer, J., Kaufman, A.J., Lin, Y., Fogel, M.L., Beukes, N.J., 2012. A multiple sulfur and organic carbon isotope record from non-conglomeratic sedimentary rocks of the Mesoarchean Witwatersrand Supergroup, South Africa. Precambrian Research 216, 208–231.

Hannah, J.L., Bekker, A., Stein, H.J., Markey, R.J., Holland, H.D., 2004. Primitive Os and 2316 Ma age for marine shale: implications for Paleoproterozoic glacial events and the rise of atmospheric oxygen. Earth and Planetary Science Letters 225, 43–52.

Haqq-Misra, J., Kasting, J.F., Lee, S., 2011. Availability of O_2 and H_2O_2 on pre-photosynthetic Earth. Astrobiology 11, 293–302.

Hoffman, P.F., 2013. The Great Oxidation and a Siderian snowball Earth: MIF-S based correlation of Paleoproterozoic glacial epochs. Chemical Geology 362, 143–156.

Holland, H.D., 1994. Early Proterozoic atmospheric change. In: Bengtson, S. (ed.): Early Life on Earth, Nobel Symposium No. 84. Columbia University Press, New York, p. 237–244.

Holland, H.D., 2002. Volcanic gases, black smokers, and the Great Oxidation Event. Geochimica et Cosmochimica Acta 66, 3811–3826.

Hulston, J.R., Thode, H.G., 1965. Variations in the S33, S34, and S36 contents of meteorites and their relation to chemical and nuclear effects. Journal of Geophysical Research 70, 3475–3484.

Isley, A.E., Abbott, D.H., 1999. Plume-related mafic volcanism and the deposition of banded iron formation. Journal of Geophysical Research: Solid Earth *104*, 15461–15477.

Knoll, A., Walter, M., Narbonne, G., Christie-Blick, N., 2006. The Ediacaran Period: a new addition to the geologic time scale. Lethaia *39*, 13–30.

Lindsay, J.F., Brasier, M.D., 2002. Did global tectonics drive early biosphere evolution? Carbon isotope record from 2.6 to 1.9 Ga carbonates of Western Australian basins. Precambrian Research *114*, 1–34.

Luo, G., Ono, S., Beukes, N.J., Wang, D.T., Xie, S., Summons, R.E., 2016. Rapid oxygenation of Earth's atmosphere 2.33 billion years ago. Science Advances *2*, e1600134.

Macgregor, A.M., 1927. The problem of the Precambrian atmosphere. South African Journal of Science *24*, 155–172.

Maheshwari, A. Sial, A.N., Gaucher, C., Bossi, J., Bekker, A., Ferreira, V.P., Romano, A.W., 2010. Global nature of the Paleoproterozoic Lomagundi carbon isotope excursion: a review of occurrences in Brazil, India, and Uruguay. Precambrian Research *182*, 274–299.

Melezhik, V.A., Huhma, H., Condon, D.J., Fallick, A.E., Whitehouse, M.J., 2007. Temporal constraints on the Paleoproterozoic Lomagundi-Jatuli carbon isotope event. Geology *35*, 655–658.

Mojzsis, S., 2007. Sulphur on the early Earth. In: van Kranendonk, M.J., Smithies, R.H., Bennett, V.C. (eds.): *Earth's Oldest Rocks*, Developments in Precambrian Geology *15*. Elsevier, Amsterdam, p. 923–970.

Müller, S.G., Krapez, B., Barley, M.E., Fletcher, I.R., 2005. Giant iron-ore deposits of the Hamersley province related to the breakup of Paleoproterozoic Australia: new insights from in situ SHRIMP dating of baddeleyite from mafic intrusions. Geology *33*, 577–580.

Olson, S.L., Kump, L.R., Kasting, J.F. 2013. Quantifying the areal extent and dissolved oxygen concentrations of Archean oxygen oases. Chemical Geology *362*, 35–43.

Ono, S., Beukes, N.J., Rumble, D., Fogel, M.L., 2006. Early evolution of atmospheric oxygen from multiple-sulfur and carbon isotope records of the 2.9 Ga Mozaan Group of the Pongola Supergroup, Southern Africa. South African Journal of Geology *109*, 97–108.

Papineau, D., Mojzsis, S.J., Schmitt, A.K., 2007. Multiple sulfur isotopes from Paleoproterozoic Huronian interglacial sediments and the rise of atmospheric oxygen. Earth and Planetary Science Letters *255*, 188–212.

Paulukat, C., Gilleaudeau, G.J., Chernyavskiy, P., Frei, R., 2016. The Cr-isotope signature of surface seawater: a global perspective. Chemical Geology *444*, 101–109.

Pavlov, A.A., Kasting, J.F., 2002. Mass-independent fractionation of sulfur isotopes in Archean sediments: strong evidence for an anoxic Archean atmosphere. Astrobiology *2*, 27–41.

Pickard, A.L., 2003. SHRIMP U–Pb zircon ages for the Palaeoproterozoic Kuruman Iron Formation, northern Cape Province, South Africa: evidence for simultaneous BIF deposition on Kaapvaal and Pilbara cratons. Precambrian Research *125*, 275–315.

Rodler, A., Frei, R., Gaucher, C., Germs, G.J.B., 2016. Chromium isotope, REE and redox-sensitive trace element chemostratigraphy across the late Neoproterozoic Ghaub glaciation, Otavi Group, Namibia. Precambrian Research *286*, 234–249.

Scheiderich, K., Amini, M., Holmden, C., Francois, R., 2015. Global variability of chromium isotopes in seawater demonstrated by Pacific, Atlantic, and Arctic Ocean samples. Earth and Planetary Science Letters *423*, 87–97.

Schoenberg, R., Zink, S., Staubwasser, M., von Blanckenburg, F., 2008. The stable Cr isotope inventory of solid earth reservoirs determined by double spike MC-ICP-MS. Chemical Geology *249*, 294–306.

Siebert, C., Nägler, T.F., von Blanckenburg, F., Kramers, J.D., 2003. Molybdenum isotope records as a potential new proxy for paleoceanography. Earth and Planetary Science Letters *211*, 159–171.

Touboul, M., Kleine, T., Bourdon, B., Palme, H., Wieler, R., 2007. Late formation and prolonged differentiation of the Moon inferred from W isotopes in lunar metals. Nature *450*, 1206–1209.

Trendall, A.F., Compston, W., Nelson, D.R., De Laeter, J.R., Bennett, V.C., 2004. SHRIMP zircon ages constraining the depositional chronology of the Hamersley Group, Western Australia. Australian Journal of Earth Sciences *51*, 621–644.

Voegelin, A.R., Nägler, T.F., Beukes, N.J., Lacassie, J.P., 2010. Molybdenum isotopes in late Archean carbonate rocks: implications for early Earth oxygenation. Precambrian Research *182*, 70–82.

Wille, M., Kramers, J.D., Nägler, T.F., Beukes, N.J., Schröder, S., Meisel, Th., Lacassie, J.P., Voegelin, A.R., 2007. Evidence for a gradual rise of oxygen between 2.6 and 2.5 Ga from Mo isotopes and Re-PGE signatures in shales. Geochimica et Cosmochimica Acta *2007*, 2417–2435.

Wille, M., Nebel, O., Van Kranendonk, M.J., Schoenberg, R., Kleinhanns, I.C., Ellwood, M.J., 2013. Mo–Cr isotope evidence for a reducing Archean atmosphere in 3.46–2.76 Ga black shales from the Pilbara, Western Australia. Chemical Geology *340*, 68–76.

Williford, K.H., Van Kranendonk, M.J., Ushikubo, T., Kozdon, R., Valley, J.W., 2011. Constraining atmospheric oxygen and seawater sulfate concentrations during Paleoproterozoic glaciation: in situ sulfur three-isotope microanalysis of pyrite from the Turee Creek Group, Western Australia. Geochimica et Cosmochimica Acta *75*, 5686–5705.

4

Chronochemostratigraphy of Platform Sequences Across the Paleoproterozoic-Mesoproterozoic Transition

Farid Chemale Junior[1] and Felipe Guadagnin[2]

ABSTRACT

The Paleoproterozoic era was an interval of major changes in Earth, whereas the Mesoproterozoic era comprised the "boring billion" due to the paucity of changes, especially in life forms, from 1.7 to 0.75 Ga. By the end of the Paleoproterozoic era, most of the continents were adjoined to form the Columbia or Nuna supercontinent. A series of Statherian to Calymmian rift, intracratonic, and passive margin basins developed in the Columbia recording the major accretionary and extensional events with dominant older ages rather stratigraphic depositional ages. These basins are dominated by siliciclastic successions with high abundance of red bed sandstones, but subordinate carbonate sequences are also preserved. The Paleoproterozoic-Mesoproterozoic transition is recognized in carbonate successions of Australia, China, India, and Laurentia, which are characterized by negative $\delta^{13}C$ values (down to −6‰) and slightly $\delta^{34}_{Sulfate}$ positive anomalies (up to 30‰) close to the boundary (1650–1600 Ma). Between 1600 and 1400 Ma, carbonate platforms in these basins exhibit relatively uniform $^{87}Sr/^{86}Sr$ values around 0.705, low concentration of V and Mo, very regular and flat carbon isotope signature (mostly $\delta^{13}C$ equals 0 ± 2‰), and the first occurrence of complex organisms (eukaryotes), suggesting low organic productivity and stable paleoclimate in these periods.

4.1. INTRODUCTION

The Paleoproterozoic era (from 2.5 to 1.6 Ga) was a time of dramatic changes in the Earth systems [e.g., *Wicander and Monroe*, 2009; *Levin*, 2013]. Based on tectonic, sedimentary, or environmental characteristics, this era encompasses 900 Ma and is subdivided into four periods: (i) Siderian (2.5–2.3 Ga), characterized by the high abundance of banded iron formations; (ii) Rhyacian (2.3–2.05 Ga), marked by injection of layered complexes such as the Bushveld Complex; (iii) Orosirian (2.05–1.8 Ga), marked by widespread orogenies; and (iv) Statherian (1.8–1.6 Ga), with stabilization of cratons [e.g., *Ogg*, 2004]. The presence of red beds and lateritic paleosols at

2.3 Ga suggests a modification of atmospheric conditions. At this time, production of reduced gases (CH_4, H_2, SO_2, and H_2S) declined with concomitant accumulation of O_2 in the atmosphere [*Zhanle et al.*, 2006]. This period marked the Great Oxygenation Event (GOE) (2.3–2.2 Ga) when a significant increase in atmospheric O_2 level was related to global cooling by at least three main glaciations (2.45–2.22 Ga) included in the Makganyene snowball Earth [*Kopp et al.*, 2005]. As a result, the Earth's atmospheric chemistry before and after GOE is defined as anoxic and oxic, respectively. In comparison, anoxic late Archean concentrations of atmospheric O_2 were <1 ppmv, atmospheric $CO_2 \leq 4\%$, atmospheric CH_4 ca. 10^3 ppmv, and marine $SO_4 = 0.2$ mM. In comparison, the oxic Mesoproterozoic atmosphere had estimated concentrations of O_2 between 0.2 and 2 vol. %, atmospheric $CO_2 \leq 0.4\%$, atmospheric $CH_4 = 10^2$ ppmv, and marine SO_4 between 1 and 4 mM [*Catling*, 2014]. Oxic atmosphere record is well defined after the GOE because of lack of

[1] *Programa de Pos-graduação em Geologia, Universidade Vale do Rio dos Sinos, São Leopoldo, Rio Grande do Sul, Brazil*

[2] *Universidade Federal do Pampa, Caçapava do Sul, Rio Grande do Sul, Brazil*

Chemostratigraphy Across Major Chronological Boundaries, Geophysical Monograph 240, First Edition.
Edited by Alcides N. Sial, Claudio Gaucher, Muthuvairavasamy Ramkumar, and Valderez Pinto Ferreira.
© 2019 the American Geophysical Union. Published 2019 by John Wiley & Sons, Inc.

mass-independent fractionation (MIF) of S isotopes in sedimentary rocks. Also, the Lomagundi carbon isotope excursion is registered in C-rich sediments with the production of a significant amount of oxygen in the atmosphere [*Karhu and Holland*, 1996].

From Rhyacian to Statherian times, global orogenies accreted most Archean terranes (Fig. 4.1). The register of such tectonic collage is found in all continents as orogens, attesting to global-scale orogenic processes [e.g., *Rogers and Santosh*, 2002; *Zhao et al.*, 2002]. These processes led to the formation of the one of the most relevant supercontinent in the Earth's history, the Columbia (or Nuna) supercontinent [e.g., *Bradley*, 2011; *Meert*, 2012; *Cawood and Hawkesworth*, 2014; Fig. 4.2a]. Despite these profound changes in atmosphere and lithosphere, life remained unmodified in Archean being restricted to stromatolites and bacteria.

The following Mesoproterozoic era (from 1.6 to 1.0 Ga) was characterized by expansion of existing platform covers in the Calymmian (1.6–1.4 Ga) and Ectasian periods (1.4–1.2 Ga) and by formation of narrow polymetamorphic belts in the Stenian period (1.2–1.0 Ga; *Ogg*, 2004). Some authors consider the time from 1.7 to 0.75 Ga spanning all Mesoproterozoic and part of Paleoproterozoic and Neoproterozoic eras as a "boring billion" (or the "Earth's middle age"; e.g., *Brasier and Lindsay*, 1998; *Cawood and Hawkesworth*, 2014). Indeed, anomalies in climate, tectonic processes, environments, and mineral deposits are well recognized in the geological record preceding Paleoproterozoic and succeeding Neoproterozoic eras [*Cawood and Hawkesworth*, 2014]. Nonetheless, some anomalies are characteristic of this interval from Statherian (late Paleoproterozoic) to Tonian (early Neoproterozoic). The anomalies are (i) significant occurrences of abundant intraplate magmatism [*Ashwal*, 2010] and reduced continental and oceanic arc volcanism [*Cawood and Hawkesworth*, 2014]; (ii) lack of positive $\varepsilon_{Hf(t)}$ in the detrital zircon record [*Belousova et al.*, 2010]; (iii) warm and stable climate with absence of glacial and iron deposits [e.g., *Condie et al.*, 2001; *Bekker et al.*, 2010; *Bradley*, 2011]; (iv) low abundance of passive margins [e.g., *Bradley*, 2008]; (v) lack of phosphate deposits [*Papineau*, 2010] and Sr anomaly in the paleo-seawater record [*Shields*, 2007]; (vi) low atmospheric pO_2 and generally anoxic deep oceans [e.g., *Cox et al.*, 2016]; (vii) low chemical index of alteration [*Condie et al.*, 2001]; (viii) limited orogenic gold, volcanic-hosted massive sulfide, and sedimentary rock-hosted manganese deposits [*Goldfarb et al.*, 2001; *Huston et al.*, 2010; *Maynard*, 2010]; and (ix) no expressive glacial events.

Oxic atmospheric conditions in late Paleoproterozoic (Statherian period) and early Mesoproterozoic (Calymmian period) provided conditions to produce more complex organism in Earth. The appearance of the earliest complex cell morphologies (eukaryotes) occurred in the late Paleoproterozoic era [e.g., *Butterfield*, 2015; *Agić et al.*, 2017]. The organisms are described as acathomorphic acritarchs in the Deonar Formation of India [*Prasad et al.*, 2005; *Ray*, 2006] and Ruyang Group of North China Craton (NCC) [*Yin*, 1997; *Su et al.*, 2012; *Pang et al.*, 2013; *Lan et al.*, 2014]. Additionally, the prolonged stability of the carbon cycle during this time interval could have led to the evolutionary radiation occurred in the Neoproterozoic era [*Brasier and Lindsay*, 1998].

Complex evolution of the Columbia supercontinent occurred from assembly, at ca. 1.8 Ga, to breakup, at ca. 1.3 Ga [e.g., *Bradley*, 2011; *Pehrsson et al.*, 2015; *Meert and Santosh*, 2017]. Subsequent Earth's crustal evolution led to the assembly of Rodinia at ca. 1.1 Ga and breakup at 0.8 Ga [e.g., *Evans*, 2009]. A series of intracontinental rift-sag basins and epicratonic basins deposited within and along the margins of Columbia after the Orosirian period. Some sequences were later deformed, whereas others preserved inside cratonic continental masses [*Guadagnin et al.*, 2015a; *Furlanetto et al.*, 2016]. The Paleoproterozoic-Mesoproterozoic boundary is preserved within the platform covers such as Calymmian and Ectasian deposits [*Ogg*, 2004]. These sequences are widespread in the continents and can be correlated based on chronostratigraphy, lithostratigraphy, chemostratigraphy, and sequence stratigraphy. All the craton fragments preserve Paleoproterozoic to Neoproterozoic platform successions including Australia, Baltica, Congo-São Francisco, India, Laurentia, North China, Siberia, and South China. The chronostratigraphic framework of those sequences is a work in progress, whereas in the last decade vast improvement has been achieved through in situ dating of detrital zircon grains to define maximum depositional ages especially using LA-ICPMS [*Gehrels*, 2014] and SIMS [*Stern*, 1997; *Ireland and Williams*, 2003] techniques. These procedures also allow to obtain ages of source rocks used in provenance analysis, improving the knowledge of sources and sinks [e.g., *Guadagnin et al.*, 2015a, 2015b; *Khudoley et al.*, 2015; *Furlanetto et al.*, 2016]. Eventually, minimum depositional ages are obtained by dating intruded dikes and sills, younger volcanic units, or diagenetic or metamorphic minerals [e.g., *Eriksson et al.*, 2001], instead of using the fossil record. The determination of maximum and minimum depositional ages allowed to elaborate the chronostratigraphic framework of Paleoproterozoic to Mesoproterozoic cratonic basins [e.g., *Guadagnin et al.*, 2015a, 2015b; *Furlanetto et al.*, 2016]. The chronostratigraphic framework based on age range for unfossiliferous successions is heterogeneous because of a variable degree of knowledge of those sequences. In spite of this, it is now possible to use the record from Paleoproterozoic to Mesoproterozoic cratonic and epicontinental basins to unravel isotopic composition of paleo-seawater, paleo-source rocks, paleodrainage dispersion, and

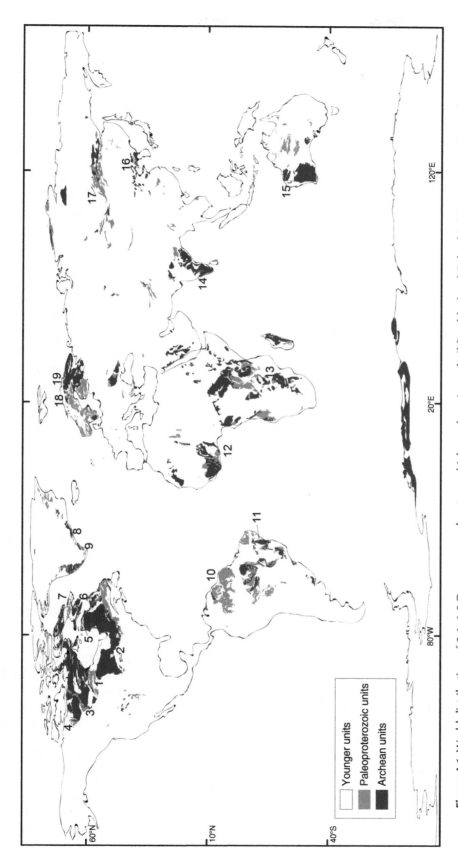

Figure 4.1 World distribution of 2.1–1.8 Ga orogens and cratons which are the primary building blocks of Columbia/Nuna supercontinent shown in Figure 4.2. 1. Trans-Hudson orogen; 2. Penokean orogen; 3. Taltson–Thelon orogen; 4. Wopmay orogen; 5. Cape Smith–New Quebec orogen; 6. Torngat orogen; 7. Foxe orogen; 8. Nagssugtoqidian orogen; 9. Makkovikian-Ketilidian orogen; 10. Transamazonian orogen; 11. Itabuna-Salvador-Curaçá orogen; 12. Eburnean orogen; 13. Limpopo belt; 14. Moyar belt; 15. Capricorn orogen; 16. Trans–North China orogen; 17. Central Aldan belt; 18. Svecofennian orogen; 19. Kola-Karelian orogen; 20. Transantarctic orogen. Paleoproterozoic to Mesoproterozoic basins occur in most cratons. Modified from *Bouysse* [2010].

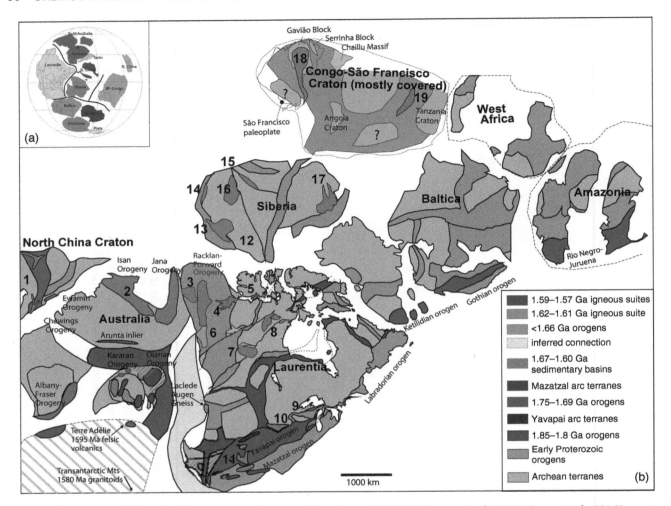

Figure 4.2 (a) Columbia/Nuna supercontinent configuration at 1.4 Ga [*Pehrsson et al.,* 2015; *Evans et al.,* 2016]. (b) Reconstruction of the Columbia supercontinent at the Paleoproterozoic-Mesoproterozoic boundary showing the location of main depocenters and orogens. 1. Changcheng-Jixian systems; 2. Mount Isa and McArthur super-basins; 3. Wernecke Supergroup; 4. Fort Simpson basin; 5. Coppermine basin; 6. Muskwa basin; 7. Athabasca Basin; 8. Thelon basin; 9. Baraboo sequence; 10. Sioux sequence; 11. Ortega sequence; 12. Kureika-Anabar basin; 13. Teya-Chapa basin; 14. Turukhansk basin; 15. Udzha and East Anabar basins; 16. Kotuy basin; 17. Uchur and Aimchan groups; 18. Espinhaço Supergroup; 19. Akanyaru and Kibara supergroups. Adapted from *Furlanetto et al.* [2016]. Reproduced with permission of Elsevier. *(See insert for color representation of the figure.)*

paleocontinent position [e.g., *Guadagnin et al.,* 2015a; *Khudoley et al.,* 2015; *Bállico et al.,* 2017].

In this chapter, we review the available chronostratigraphic and chemostratigraphic data from Paleoproterozoic to Mesoproterozoic basins distributed worldwide in the Columbia supercontinent paleogeographical framework (Figs. 4.2b and 4.3) and discuss its implications for the chronostratigraphic boundary and potential Global Boundary Stratotype Section and Point (GSSP). Evaluated chrono-correlated successions are Barren, Arid, Leichardt, Calvert, and Isan basins in the North Australian Craton [*Lindsay and Brasier,* 2000; *Spaggiari et al.,* 2015]; Onega and Vilcha basins and Vestfjordalen Supergroup in the Baltica

paleocontinent [*Köykkä,* 2011]; Kagera, Muva, and Espinhaço supergroups in the Congo-São Francisco craton [e.g., *Fernandez-Alonso et al.,* 2012; *Guadagnin et al.,* 2015b]; Cuddapah, Pranhita-Godavari, Dalma-Chandil, Kolhan, and Delhi basins in the Indian Shield [e.g., *Miall et al.,* 2015]; the Wernecke Supergroup, the Baker Lake, Wharton, Athabasca, Barrensland, and Horny Bay groups, the Fort Simpson basin, and the Muskwa assemblage in the Laurentia paleocontinent [e.g., *Furlanetto et al.,* 2016]; Changcheng and Jixian systems in the NCC [e.g., *Peng,* 2015]; the Ulkan, Uyan, Uchur, and Mukun groups in the Siberian platform [*Didenko et al.,* 2015]; and the Sin Quyen and Dongchuan groups in the South China Craton [*Wang et al.,* 2016].

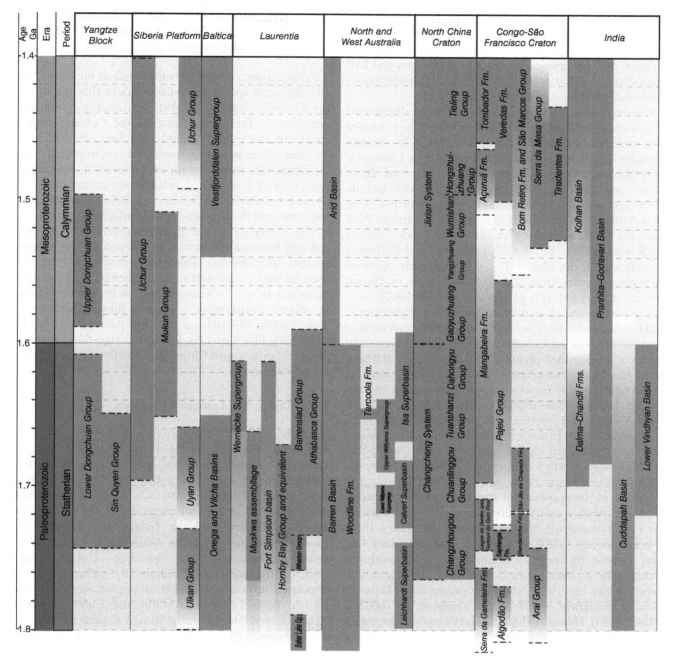

Figure 4.3 Chronostratigraphic framework of the preserved stratigraphic record of the Columbia supercontinent between the Statherian and Calymmian periods. Modified from *Lindsay and Brasier* [2000], *Davidson* [2008], *Hall et al.* [2008], *Fernandez-Alonso et al.* [2012], *Chemale et al.* [2012], *Hahn et al.* [2013], *Wang and Zhou* [2014], *Didenko et al.* [2015], *Guadagnin and Chemale* [2015], *Khudoley et al.* [2015], *Miall et al.* [2015], *Spaggiari et al.* [2015], *Zhai et al.* [2015], and *Furlanetto et al.* [2016].

4.2. COLUMBIA SUPERCONTINENT AND POTENTIAL GSSP FOR PALEOPROTEROZOIC-MESOPROTEROZOIC BOUNDARY

The Columbia supercontinent was proposed by *Rogers and Santosh* [2002] based on the supposed connection between Columbia basalts in North America and eastern India [*Meert and Santosh*, 2017]. Since the definition, the concept of the Columbia supercontinent has been debated by the use of one or a combination of the following information: (i) geological correlation of Archean terranes and Paleoproterozoic orogens [*Zhao et al.*, 2002], (ii) tectonic and sedimentary record within Paleoproterozoic to Mesoproterozoic sedimentary basins [e.g., *Furlanetto et al.*, 2016], (iii) paleomagnetism [e.g., *Pisarevsky et al.*, 2014; *Meert and Santosh*, 2017], (iv) correlation of age and strike of large-scale radiating dike swarms and location of large igneous province (LIP) centers and LIP barcode [e.g., *Ernst et al.*, 2016], and (v) correlation of mineral deposits [e.g., *Pehrsson et al.*, 2015].

Paleomagnetic data are inhomogeneously distributed in space and time, and current data support the existence of a large landmass of ca. 1.5–1.4 Ga (Fig. 4.2a), despite probable long-lived connection between Siberia, Laurentia, and Baltica at the core of Columbia [e.g., *Ernst et al.*, 2016; *Meert and Santosh*, 2017]. Remaining blocks include Amazonia, Congo-São Francisco, India, North China, West Africa, East Antarctica, Gawler, and North and West Australia paleocontinents and the smaller blocks such as the Rio de la Plata, South China (Yangtze and Cathaysia blocks), and Tarim blocks (Fig. 4.2). Assembly occurred through orogenies from 2.2 to 1.8 Ga with maximum packing at ca. 1.5–1.4 Ga [*Meert and Santosh*, 2017].

The occurrence of significant long-lived intracratonic and passive margin basins within some of the paleocontinents involved in the Columbia supercontinent is an indication of an extended period of stability (ca. 1.6–0.9 Ga). Most basins are intracratonic, whereas some occur at paleoplate margins (epicontinental and passive margin basins). LIP or plume-related events at 1.8–1.7, ~1.5, ca. 1.4–1.3, ~1.2, and ~0.9 Ga actively control basins [e.g., *Ernst et al.*, 2016]. In some segments of Columbia, paleocontinents stabilized earlier, such as in the Congo-São Francisco and North China cratons, where intracratonic and rift-related basins occur back to 1.8–1.7 Ga.

In spite of the exposures of chrono-correlated intracratonic and passive margin successions deposited through the Paleoproterozoic-Mesoproterozoic boundary, there is not a formal GSSP at 1600 Ma. The boundary was first defined by the Subcommission on Precambrian Stratigraphy as Proterozoic I-Proterozoic II [*Plumb and James*, 1986], and later, the Paleoproterozoic-Mesoproterozoic boundary was formally approved at 1600 Ma [*Plumb*, 1991]. However, this boundary was not adequately defined, but just selected, because it records approximate ending of complex events [*Plumb and James*, 1986], nowadays characterized as the completion of Columbia supercontinent as described above (Fig. 4.2a). Nonetheless, Paleoproterozoic-Mesoproterozoic carbonate successions exposed in Australia, China, India, and Laurentia paleocontinents are the best preserved successions of the Paleoproterozoic-Mesoproterozoic boundary. Based on the stratigraphic and isotopic data, the successions of the Changcheng and Jixian systems (China) and of the Isa and McArthur superbasins (Australia; Figs. 4.2b and 4.3) deposited through the Paleoproterozoic-Mesoproterozoic transition would be potential sections for a GSSP, as described in the next sections.

4.3. CHRONOSTRATIGRAPHIC CORRELATIONS FROM LATE PALEOPROTEROZOIC TO EARLY MESOPROTEROZOIC ERAS

Paleoproterozoic and Mesoproterozoic basins occur in all continents within the crustal blocks that constituted the Columbia supercontinent. In fact, these sequences are evidence for the existence of the Columbia supercontinent itself [e.g., *Rogers and Santosh*, 2002]. For decades, these sequences were chrono-correlated based on similarities in maximum and minimum depositional ages, ages of intercalated volcanic sequences, and other age constraints [e.g., *Eriksson et al.*, 2001]. The paleoenvironments and the depositional, chemical, and provenance trends preserved in the Paleoproterozoic to Mesoproterozoic basins are essential to reconstruct the paleogeography of Columbia. This information is also primarily used to understand tectonic, sedimentary, and climatic processes in the Proterozoic eon. A review on chrono-correlated stratigraphic sequences formed between 1.8 and 1.4 Ga, distributed in the paleoplates of the Columbia, like Australia, Baltica, Congo-São Francisco, India, Siberia, South China, and North China cratons, is found in Figure 4.3.

4.3.1. Congo-São Francisco Craton

The preserved sedimentary record of Paleoproterozoic and Mesoproterozoic basins in the Congo-São Francisco craton is assigned to several units: Kagera and Muva supergroups in the eastern Congo craton; Espinhaço Supergroup at the São Francisco craton interior; Araí, Serra da Mesa, and Paranoá groups in the western São Francisco Craton; and the Tiradentes, Carandaí, and Prados formations and the Andrelândia Group in the south-southwestern São Francisco craton margin (Fig. 4.3; *Guadagnin and Chemale*, 2015). These units were deposited in three tectonic-stratigraphic megasequences formed from 1.8 to 0.9: Statherian, Calymmian-early Ectasian, and Stenian-early Tonian sequences (Fig. 4.3; *Chemale et al.*, 2012; *Guadagnin et al.*, 2015a). Intracratonic covers are deposited between late Paleoproterozoic (Statherian) and early Neoproterozoic (Tonian; Marshak and Alkmim, 1989; *Chemale et al.*, 1993; *Uhlein et al.*, 1998) and overlay the Archean-Paleoproterozoic basement of the São Francisco Craton (*Teixeira and Figueiredo*, 1991; *Barbosa and Sabaté*, 2004; *Oliveira et al.*, 2013; *Teixeira et al.*, 2015; *Cruz et al.*, 2016).

The Statherian megasequence (1.8–1.6 Ga) comprises a sedimentary rift succession represented by alluvial-fluvial-lacustrine-deltaic deposits of the Bandeirinhas and São João da Chapada (southern Espinhaço); Serra da Gameleira, Ouricuri do Ouro, and Lagoa de Dentro (Chapada Diamantina); Algodão and Sapiranga formations and Pajeú

Group (northern Espinhaço); and Araí Group (external zone of the Brasília Belt). A widespread magmatic event occurred from 1.78 to 1.70 Ga in the Araí, Conceição do Mato Dentro, and Rios dos Remédios stratigraphic units. This magmatism occurred as lava flows and sills associated with the sedimentary strata and as basic and acid bodies intruding the São Francisco Craton basement adjacent to the Proterozoic cover. It is represented by the 1.73–1.68 Ga intraplate continental basic dikes and sills and acidic peralkaline granites, such as the Borrachudos and Lagoa Real suites [*Turpin et al.*, 1988; *Silva et al.*, 1995; *Chemale et al.*, 1998; *Cederberg et al.*, 2016].

The Calymmian-early Ectasian megasequence (ca. 1.6–1.38 Ga) corresponds to a typical intracratonic rift-sag basin represented by alluvial, fluvial, deltaic, and shallow marine siliciclastic strata represented by the Mangabeira, Açuará, and Tombador formations (Chapada Diamantina), Bom Retiro Formation and São Marcos Group (northern Espinhaço), Serra da Mesa Group (external zone of the Brasília Belt), and Tiradentes Formation (southern Brasília Belt; *Guadagnin and Chemale*, 2015 and references therein). Acidic tuffs and lavas are recognized in the lower and intermediate sections and were dated at 1582 ± 8 and 1569 ± 14 Ma (northern Espinhaço; *Danderfer et al.*, 2009) and 1601 ± 22 Ma (Chapada Diamantina; *M. Bállico, personal communication*]. Intraplate tholeiitic mafic dikes intruded the Mangabeira Formation at 1514 ± 22 Ma [*Babinski et al.*, 1999] and 1501 ± 9.1 Ma [*Silveira et al.*, 2013], establishing the minimum age of deposition for the intermediate section of the Calymmian-Ectasian megasequence. Furthermore, crystal-rich volcaniclastic rocks dated at 1436 ± 26 Ma occur in the upper section of the Calymmian-early Ectasian megasequence [*Guadagnin et al.*, 2015b].

4.3.2. Siberian Platform

The Siberian platform consists of Archean-Paleoproterozoic basement and Paleoproterozoic-Phanerozoic cover units [e.g., *Didenko et al.*, 2015; *Khudoley et al.*, 2015; *Priyatkina et al.*, 2016]. The basement of the Siberian platform was formed through the Paleoproterozoic collision of Archean granulite-gneiss and granite-greenstone associations [e.g., *Gladkochub et al.*, 2010]. Two main cratons compose the basement of the Siberian platform, the Aldan and Stanovoy provinces which collided with the Anabar, Tunguska, and Olenek provinces [*Didenko et al.*, 2015]. These terranes are sutured by the Paleoproterozoic Akitkan and Bilyakchan-Ulkan belts [*Didenko et al.*, 2015]. Basement units are mostly covered, whereas they are exposed in the Aldan-Stanovoy and Anabar shields and the Olenek (north), Kan, Biryusa, Sharyzhalgai, and Baikal (southwest) uplifts [*Gladkochub et al.*, 2010].

In the Siberian platform, the sedimentary cover was deposited in shallow marine to continental environments [e.g., *Priyatkina et al.*, 2016]. Late Paleoproterozoic and Mesoproterozoic sequences are classified as Riphean in Russian stratigraphy, whereas Neoproterozoic sequences are Vendian [e.g., *Khudoley et al.*, 2015]. Riphean cover occurs in several depocenters in the Siberian platform, some exposed at the margins and others in its interior. Main depocenters (Fig. 4.2b) are Turukhansk, Teya-Chapa, Kotuy, Udzha, Kureika-Anabar, and Mukun basins, preserved and partially exposed in the Anabar, Tunguska, and Olenek provinces [*Petrov*, 2014; *Priyatkina et al.*, 2016], and the Uchur and Aimchan groups, in the Aldan-Stanovoy province (Fig. 4.3; *Khudoley et al.*, 2007, 2015).

In the SE Siberian platform (Bilyakchan-Ulkan belt), the Paleoproterozoic and Mesoproterozoic sequences comprised the Ulkan Group, deposited in the Statherian period, and the Uyan and Uchur groups, deposited in the Calymmian period (Fig. 4.3; *Didenko et al.*, 2015). The Ulkan Group is mainly composed of volcanic rocks with rare intercalation of volcaniclastic sequences. The volcanic unit was dated at ca. 1.73 Ga (U-Pb zircon age; *Didenko et al.*, 2013). The Uyan Group has volcanic units succeeded by siliciclastic and carbonatic sequences, with age of 1670 ± 40 Ma (Rb-Sr isochron; *Guryanov*, 2007 in *Didenko et al.*, 2015). The Uchur Group is mostly composed of siliciclastic and carbonatic sequences with age of 1485 and 1360 Ma [*Didenko et al.*, 2015].

In the eastern Siberian platform, Paleoproterozoic and Mesoproterozoic sequences include Uchur, Aimchan, Kerpyl, Lakhanda, and Uy groups, which consist of siliciclastic and carbonatic sequences [*Khudoley et al.*, 2007, 2015]. In the Anabar shield, the Mukun basin is the early Mesoproterozoic sequence from the Siberian platform [*Petrov*, 2014]. This basin is composed of the Mukun and Billyakh groups (Fig. 4.3) which have a maximum depositional age of 1690 ± 9 Ma [*Khudoley et al.*, 2007; *Petrov*, 2014]. Deposition of these units may have occurred from 1.58 to 1.5 Ga [*Petrov*, 2014]. The Mukun Group (Fig. 4.3) was intruded by sills dated at 1493 ± 34 Ma (U-Pb baddeleyite age; *Khudoley et al.*, 2009). Two main paleocurrent vectors are characteristic in the Mukun basin, northwestern fluvial and southwestern eolian currents, in present-day coordinates [*Petrov*, 2014].

Paleoproterozoic and Mesoproterozoic sequences were intruded by mafic dike swarms at 1513 ± 51 Ma (Fomich sills), 1473 ± 24 Ma (sills of Olenek uplift; *Veselovskiy et al.*, 2006), and 1384 ± 2 Ma (Chieress dike; *Ernst et al.*, 2000).

4.3.3. Baltica Paleoplate

The East European Craton is the result of the assembly of the Fennoscandia, Sarmatia, and Volgo-Uralia cratons during the Paleoproterozoic, to form the Baltica

Paleoplate (Fig. 4.2b; e.g., *Bogdanova et al.*, 2008; *Shumlyanskyy et al.*, 2017). The three cratons consist of Archean terranes partially reworked in the Paleoproterozoic era. The Volgo-Uralia Craton is extensively covered by sedimentary deposits [*Bogdanova et al.*, 2008]. Paleoproterozoic and Mesoproterozoic basins are preserved over the East European Craton, including Rjukan, Onega, and Vilcha basins (Fig. 4.3). The Rjukan and Onega (Vepsian Formation) basins were deposited over the Archean basement of the Fennoscandia Craton [*Köykkä*, 2011; *Kulikov et al.*, 2017; *Lubnina et al.*, 2017], whereas the Vilcha basin is preserved in the Ukrainian shield of Sarmatia [*Shumlyanskyy et al.*, 2017]. Age constraints for these sequences indicate deposition from 1.8 to 1.65 Ga in the Onega and Vilcha basins and 1.52 to 1.34 Ga in the Rjukan basin [*Köykkä*, 2011; *Kulikov et al.*, 2017; *Lubnina et al.*, 2017; *Shumlyanskyy et al.*, 2017].

4.3.4. Indian Shield

The Indian Shield comprises four primary Archean nuclei, the Dharwar, Bastar, Singhbhum, and Aravalli-Bundelkhand (Fig. 4.4; e.g., *Mazumder and Saha*, 2012), which are composed of tonalite-trondhjemite-gneiss complexes, greenstone belts, and late granites assembled

between 3.6 and 2.5 Ga [*Turner et al.*, 2014]. Cratonization of this region occurred at 2.6–2.5 Ga [*Naqvi*, 2005].

Similar to the Congo-São Francisco and Baltica cratons, Proterozoic volcano-sedimentary basins are deposited on the stabilized Indian Shield. In each of these nuclei, Paleoproterozoic-Mesoproterozoic sequences are preserved (Fig. 4.4), including Lower Vindhyan sequence of Vindhyan Basin (ca. 1.7–1.45 Ga) in the Aravalli-Bundelkhand Craton, Dalma-Chandil (ca. 1.7 to <1.6 Ga) and Kolhan basins (ca. 1.6–0.9 Ga) in the Singhbhum Craton, Cuddapah Basin (ca. >1.9 to 1.6 Ga) in the Dharwar Craton, and Pranhita-Godavari basin (>1.68 to ≲1.18 Ga) at the margins of Dharwar and Bastar cratons [*Ray*, 2006; *Turner et al.*, 2014; *Miall et al.*, 2015].

The Vindhyan Basin (Fig. 4.4) has a thick Precambrian sedimentary succession of unmetamorphosed siliciclastic and carbonate rocks with few volcaniclastic beds. The stratigraphic unit of this basin is the Vindhyan Supergroup. It is divided into Lower and Upper Vindhyan, separated by a regional angular unconformity [*Ray et al.*, 2003]. Lower Vindhyan, represented by the Semri Group, deposited between 1.72 and 1.60 Ga (Fig. 4.3). The Upper Vindhyan Supergroup, represented by Kaimur, Rewa, and Bhander groups, deposited between 1.1 and 0.65 Ga [*Ray*, 2006]. The Lower Vindhyan Group, the focus of

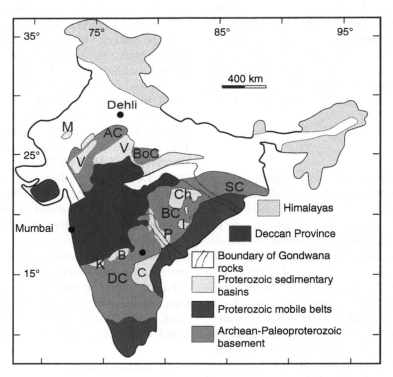

Figure 4.4 Simplified geological map of the Peninsular India showing the four main cratonic domains (Dharwar, Bastar, Aravalli-Bundelkhand, and Singhbhum) and Proterozoic basins. MB, Marwar basin; VB, Vindhyan Basin; ChB, Chhattisgarh basin; CuB, Cuddapah Basin; IB, Indravati basin; PG, Pranhita-Godavari basin; KBB, Kaladgi-Bhima basin. Modified from *Meert and Pandit* [2016] and *Saha* [2016].

this work, deposited directly on the Archean basement of the Aravalli-Bundelkhand Craton dated at 2492 ± 10 Ma [*Mondal et al.,* 2002]. From base to top, it contains the Kajrahat Limestone, Deonar Porcellanite, Sakhan Dolostone, Chorhat Sandstone, Rampur Shale, and Rohtasgarh Limestone. Pb-Pb carbonate isochrons define the minimum and maximum depositional ages for the sequence between 1721 ± 90 Ma for Kajrahat Limestone [*Sarangi et al.,* 2004] and 1601 ± 130 Ma for Rohtasgarh Limestone [*Ray et al.,* 2003]. Precise, accurate ages of Deonar Porcellanite are 1628 ± 8 (SHRIMP) and 1631 ± 0.4 Ma (ID-TIMS), and Rampur Shale at 1599 ± 8 Ma and 1602 ± 10 Ma (SHRIMP; *Rasmussen et al.,* 2002; *Ray et al.,* 2002).

Cuddapah Basin (Figs. 4.3 and 4.4) is as Paleoproterozoic intracratonic basin (<1.9 Ga; *Mazumder and Saha,* 2012) deposited on the Archean basement of the Dharwar Craton (Peninsular Gneiss). It is divided into Papaghani and Chitravati groups, which are separated by an unconformity. The Papaghani Group consists of basal siliciclastic succession (Gulcheru Formation), covered by mixed siliciclastic-carbonate package (Vempalle Formation). Mafic flows, sills, and dikes occur in the upper section of the Vempalle Formation. The Chitravati Group also comprises basal siliciclastic and upper mixed siliciclastic-carbonate sequences [*Mazumder and Saha,* 2012].

4.3.5. North China Craton

The NCC is composed of three main Precambrian associations: (i) older Neoarchean basement with crustal growth peak between 2.8 and 2.5 Ga [*Zhai and Santosh,* 2013], (ii) a lithotectonic association of three rifts that evolved into ocean basins from 2.25 to 2.0 Ga, and (iii) subducted ocean crust and final continental collisional magmatism that occurred at 1.97–1.82 Ga. The Paleoproterozoic orogeny is defined by three major collisional sutures: Inner Mongolia Suture, Central Orogenic Belt, and Jiao-Liao-Ji Belt. At 1.82 Ga, the NCC was amalgamated into two major blocks, the Western and Eastern NCC [*Zhai and Santosh,* 2013; *Peng,* 2015], coeval with the amalgamation of the Columbia supercontinent [*Zhao et al.,* 2005, 2012; *Zhai and Santosh,* 2013]. These two blocks are the basement of the late Paleoproterozoic-Neoproterozoic covers that marked a time span of stable (cratonic) conditions in the NCC [*Zhai and Santosh,* 2011, 2013; *Peng,* 2015; *Zhai et al.,* 2015].

Main dike swarms and rift associated with the intracratonic covers [*Peng,* 2015] are as follows:

1. The late Paleoproterozoic (Statherian, 1.8–1.6 Ga) rift that contains the 1.8–1.78 Ga Xiong'er igneous province and the 1.72–1.62 Ga anorogenic magmatic association [*Peng,* 2015]. This Paleoproterozoic extension started the breakup of NCC from Columbia at 1.8 Ga [*Zhai and Santosh,* 2011].

2. Mesoproterozoic rifting (Calymmian-Ectasian, 1.6–1.2 Ma). Magmatic activity was recorded in minor tuff beds formed at 1559 ± 12 Ma [*Li et al.,* 2010]; bentonite layers dated at 1379 ± 12, 1380 ± 36, and 1368 ± 12 Ma [*Su et al.,* 2008]; 1366 ± 9 and 1370 ± 11 Ma [*Gao et al.,* 2008a, 2008b]; and 1437 ± 21 and 1443 ± 21 [*Su et al.,* 2010]; and widespread dike and sill swarm intruded mainly in Yan-Liao rift (situated to the north of Beijing) at 1320 Ma [*Peng,* 2015].

3. Neoproterozoic rift (Tonian, 1000–800 Ma). The late Paleoproterozoic-Neoproterozoic fluvial-deltaic-marine sedimentary sequences of the NCC correspond to the filling of rift-sag basins formed in a stable platform. These basins point to stable lithosphere from 1.7 to 0.7 Ga because of lack of orogenic event in the period [*Zhai et al.,* 2015]. The sedimentary sequences are divided into four stratigraphic units: Statherian Changcheng, Calymmian Jixian, Ectasian to Stenian Await Name (includes Xiamanling Group), and Tonian Qingbaikou systems [*Zhai et al.,* 2015]. In this work, we focus on the Changcheng and Jixian systems, formed between 1.8 and 1.4 Ga (Figs. 4.2b and 4.3).

The Changcheng System overlies post-orogenic volcanic rocks of the Xiong'er igneous province and comprises a siliciclastic dominated package [*Xiao et al.,* 1997]. It comprises the Changzhougou, Chuanlinggou, Tuanshanzi, and Dahongyu groups [*Zhai et al.,* 2015] and equivalent, Gaoshanhe Formation, Ruyang Group, and Luoyu Group in the Xiong'er Rift System, deposited between 1.78 and 1.60 Ga [*Peng,* 2015]. The Chancheng System consists of dominant siliciclastic units and some dolomite and stromatolite layers in the upper section (Tuanshanzi and Dahongyu groups; *Xiao et al.,* 1997). Volcanic rocks of the Dahongyu Group were dated at 1625 ± 5 Ma [*Lu and Li,* 1991]. Coeval units of the Dahongyu Group are the Luoyu Group which preserves a tuff layer dated at 1611 Ma [*Su et al.,* 2012 in *Peng,* 2015].

The Jixian System succeeded the dominant siliciclastic section of the Changcheng System and formed between 1600 and 1400 Ma [*Peng,* 2015]. It is intruded by intraplate tholeiitic diabase sills with zircon ages of 1325 ± 5 and 1316 ± 37 Ma [*Zhai et al.,* 2015]. The section consists of carbonate platform divided into the Gaoyuzhuang, Yangzhuang, Wumishan, Hongshuizhuang, and Tieling groups. The basal Gaoyuzhuang Group contains limestone with nodules of Mg and Si. U-Pb zircon ages of 1559 ± 12 and 1560 ± 5 Ma for tuff layers link deposition of Gaoyuzhuang Group to Paleoproterozoic-Mesoproterozoic transition [*Li et al.,* 2010]. SHRIMP ages of 1439 ± 14 Ma obtained from bentonites in the shallow marine carbonate Tieling Group constrained the younger age of the Jixian System [*Li et al.,* 2014].

4.3.6. South China Craton

The Yangtze block has Archean basement aged 2.95–2.90 Ga, with tonalite-trondhjemite-granodiorite (TTG) suites cut by ~1.85 Ga felsic and mafic intrusions [*Peng et al.*, 2009]. The Paleoproterozoic continental collision of the Yangtze block occurred between 2.03 and 1.97 Ga as part of Columbia supercontinent assembly [e.g., *Zhang et al.*, 2006; *Xiong et al.*, 2009; *Zhao et al.*, 2010]. Proterozoic rocks are represented by metasedimentary rocks of the Dongchuan Group, and sedimentary and rare volcanic rocks of the Kunyang Group [*Greentree and Li*, 2008; *Zhao et al.*, 2010].

The late Paleoproterozoic-early Mesoproterozoic Dongchuan Group formed during fragmentation of Columbia [*Wang et al.*, 2014] and comprised conglomerates and fine-grained siliciclastic, carbonate, and volcanic rocks metamorphosed from greenschist to low amphibolite facies [*Greentree and Li*, 2008]. Dongchuan Group formed in an intracontinental rift in continental and shallow marine sedimentary environment. Metatuffs dated at 1740 ± 15 Ma [*Zhao et al.*, 2010], 1675 ± 8 Ma [*Greentree and Li*, 2008], and 1503 ± 17 Ma [*Sun et al.*, 2009] establish the depositional age for the Dongchuan Group between the Statherian and Calymmian.

4.3.7. Laurentia

Laurentia (ancestral North America) consists of a mosaic of Archean cratons (Superior, Hearne, Rae, Slave, Nain, Wyoming) sutured by Paleoproterozoic orogenic belts (Snowbird Tectonic Zone, Thelon-Taltson, Wopmay, Torngat, Trans-Hudson) and rimmed to the south (present coordinates) by accretionary orogens (Yavapai and Mazatzal; *Hoffman*, 1988, 1989; *Whitmeyer and Karlstrom*, 2007). Several Paleoproterozoic-Mesoproterozoic sedimentary sequences are preserved within Laurentia, deposited in intracratonic, rift, and passive margin settings. Statherian sequences (Figs. 4.2b and 4.3) include the Wernecke Supergroup, Athabasca Group, Baker Lake Group, Barrensland Group, Hornby Bay Group, Wharton Group, Aston Formation, Ellice Formation, Parry Bay Formation, Sims Formation, Muskwa assemblage, and their metamorphic correlatives Baraboo, Barron, and Sioux quartzites. Calymmian-early Ectasian sequences include the Belt-Purcell supergroup, Dismal Lakes Group, and Sibley Group. Ectasian-early Stenian sequences are the Bylot Supergroup, Coppermine River Group, Letitia Lake Group, Pinguicula Group, Aston Formation, Fury and Hecla Formation, and Thule Basin. Stenian-Tonian sequences include the Seal Lake Group and Midcontinent rift [*Wheeler et al.*, 1996; *Ross et al.*, 2001; *Rainbird et al.*, 2003, 2006; *Davidson*, 2008].

Laurentia records a protracted depositional history, from late Paleoproterozoic to Neoproterozoic. In northwestern Laurentia, sequences deposited between ~1.7 and 1.2 Ga are part of sequence A of *Young et al.* [1979]. Sequence A is further divided into subsequences A1 and A2, deposited between ca. 1.7 and 1.6 Ga; subsequence A3, deposited between 1.6 and ca. 1.3 Ga; and subsequence A4, deposited between 1.3 and 1.0 Ga [*MacLean and Cook*, 2004]. The analysis of the complete depositional record of Laurentia is beyond the scope of the present work, and reference is made to the work of *Ross et al.* [2001], *Rainbird et al.* [2003, 2006], *Davidson* [2008], and *Furlanetto et al.* [2016].

4.3.8. Australia

Paleoproterozoic-Mesoproterozoic sedimentary cover is widely distributed in Australia. In Northern Australia, shallow marine Paleoproterozoic-Mesoproterozoic sequences are distributed in different basins (Fig. 4.3). In this region, there are two major basins, the McArthur and Mount Isa basins (Fig. 4.2b), deposited between 1700 and 1575 Ma [*Brasier and Lindsay*, 1998; *Lindsay and Brasier*, 2000; *Page et al.*, 2000; *Southgate et al.*, 2000]. Both basins are carbonate-dominated successions of the McArthur and Nathan groups in the McArthur Basin and McNamara Group in the Mount Isa Basin. Some siliciclastic (siltstone, quartzite) and tuff layers are frequent in these basins [e.g., *Lindsay and Brasier*, 2000]. These carbonate-dominated basins have peculiar characteristics, for example, wide Pb-Zn-Ag mineral deposits, development of autotrophic eukaryotes [*Brasier and Lindsay*, 1998], and several volcanic layers. These layers are constrained by U-Pb zircon SHRIMP dating, ranging from 1708 ± 5 to 1575 Ma for the McArthur Basin, from 1694 ± 3 to 1595 ± 6 Ma for the Mount Isa Basin, and from 1729 ± 4 to 1595 ± 6 Ma for the Murphy Inlier of the Mount Isa Basin [*Brasier and Lindsay*, 1998; *Lindsay and Brasier*, 2000]. Volcanic layers were probably connected or produced by volcanic activity at the eastern margin of Columbia. These authors interpret the Paleoproterozoic to Mesoproterozoic Northern Australian basins as intracratonic marine sedimentary basins with some open ocean connection during their evolution [*Lindsay and Brasier*, 2000].

In Southern Australia, there are metasedimentary rocks of the Barren and Arid basins deposited between 1815 and 1600 Ma and between 1600 and 1305 Ma, respectively (Fig. 4.3). These Paleoproterozoic to Mesoproterozoic basins are distributed in the Archean Yilgarn Block and Mesoproterozoic Albany-Fraser Orogen, interpreted to be an earlier rift-sag basin (Barren basin) that evolved to a passive margin basin (Arid basin), strongly affected by Mesoproterozoic Albany-Fraser Orogen [*Spaggiari et al.*, 2015].

4.4. CHEMOSTRATIGRAPHY

Isotope chemostratigraphy is a powerful tool for global chronostratigraphic correlation, paleoclimate analysis, and analysis of the secular variation in Earth's processes. Fluctuations of $\delta^{13}C$ values have been used, for example, in the investigation of significant Precambrian glaciations that were followed by deposition of carbonate sequences (cap carbonates), as well as in the chronostratigraphic correlations of the Neoproterozoic era [Knoll et al., 1986]. The values of $\delta^{18}O$ are widely applied to the definition of glacial cycles in the Quaternary, while in Precambrian sequences they are used with lower resolution. Thus, the Sturtian, Marinoan, and Gaskier glaciations in the Neoproterozoic have distinct $\delta^{13}C$ and $\delta^{18}O$ anomalies [Shields and Veizer, 2002]. Similar to stable isotopes, variations in Sr isotopes such as $^{87}Sr/^{86}Sr$ ratios reflect changes related to Sr input into seawater derived from weathering of continental rocks or Earth's mantle [Shields, 2007]. $^{87}Sr/^{86}Sr$ variation curve reflects a secular variation of seawater composition.

However, the sedimentary record between 1.8 and 1.4 Ga includes the Paleoproterozoic-Mesoproterozoic transition and is dominated worldwide by siliciclastic sequences, mostly deposited in intracratonic basins. Some marine carbonate sequences are preserved from this period and can be used for isotope chemostratigraphy (C, O, and Sr), allowing chemostratigraphic correlations and definition of the paleoclimate conditions and paleo-seawater composition. Four carbonate sequences deposited between 1.7 and 1.5 Ga bring essential clues on the Paleoproterozoic-Mesoproterozoic transition (Fig. 4.3): (i) the most extensive carbonate succession is preserved in McArthur and Mount Isa basins in Northern Australia, deposited in an intracratonic basin with connection to open sea [Lindsay and Brasier, 2000]; (ii) carbonate platform of the Hornby Basin in Laurentia interpreted as epicontinental sea connected to shelf margin [Hahn et al., 2013]; (iii) carbonate platforms of the Vindhyan Basin in India [Ray et al., 2003]; and (iv) carbonate platform of Luanchuan, Lushan-Ruyang, and Jixian basins, NCC, as part of intracratonic basin [Zhao et al., 2010].

The McArthur and Mount Isa basins (~1.7–1.57 Ga) contain a well-developed carbonate platform with several tuff layers along the whole section (Fig. 4.5a) dated by U-Pb SHRIMP zircon method [Page and Sweet, 1998; Page et al., 2000]. The stratigraphic charts of these basins are supported by integrated studies of stratigraphy, sedimentology, geochronology, and geophysics [Southgate et al., 2000] and C-O isotopes [Brasier and Lindsay, 1998; Lindsay and Brasier, 2000]. This information documents high-resolution stratigraphy and secular variation in carbon isotopes for the period 1700–1575 Ma [Lindsay and Brasier, 2000]. Tuff layers encompass not only the most significant contribution of the zircon formed between 1.8 and 1.5 Ga (42% of analyzed zircon grains) in the McArthur and Mount Isa basins but offer a unique opportunity to investigate the boundary in the high-resolution isotope study. Southgate et al. [2000] recognize nine supersequences formed in an intracratonic tectonic setting connected to the Proterozoic Ocean. On the other hand, volcanic activity in these superbasins represents mostly far-field volcanism linked to a long-lived magmatic arc peripheral to the superbasins [Southgate et al., 2000], related to Kaharan and Olarian orogenies [Myers et al., 1996]. The $\delta^{13}C_{carb}$ data for these basins are the most complete and dated late Paleo-proterozoic to the early Mesoproterozoic section in the continents, ranging from −2 to +1‰ (Fig. 4.5a). However, dominant $\delta^{13}C$ values are slightly negative, with means around −0.6‰ for both basins [Brasier and Lindsay, 1998; Lindsay and Brasier, 2000]. $\delta^{18}O$ data have mean values of −8.7 to −5.9‰ in different carbonate sections of Northern Australia basins [Brasier and Lindsay, 1998; Lindsay and Brasier, 2000]. These well-constrained data and the flat pattern of the carbon isotope curve for the time span of 1.7–1.5 Ga, with low level of carbon balance (average −0.6‰ and ranging in the interval of −2 and +1), suggest a stable Proterozoic ocean during the Paleoproterozoic-Mesoproterozoic boundary [Lindsay and Brasier, 2000].

In Laurentia, most Statherian to Calymmian units are represented by fluvial, deltaic, and marine sediments dominated by siliciclastic sequences. Some basins of Laurentia preserve carbonate strata in between the siliciclastic deposits. In the Hornby Bay Basin, the Hornby Group is correlated to sequence A of Young et al. [1979] from which the Wernecke Supergroup units (Wernecke Mountains, Yukon) represent the distal deep marine equivalent of the shallow marine to continental units of Hornby Bay Group [Young et al., 1979]. Later, Hahn et al. [2013] established the Hornby Bay Group units as part of the subsequence A2, deposited between 1740 ± 5 Ma [Irving et al., 2004] and 1590 ± 3 Ma [Hamilton and Buchan, 2010]. The group was divided into two systems tracts, the basal transgressive (TST) and the upper high stand (HST), where the carbonate platform developed. Sm-Nd isotope studies of the Hornby Group and Wernecke Supergroup units suggest an essential contribution of sediments with the input of early Paleoproterozoic and Archean sources [Hahn et al., 2013; Furlanetto et al., 2016]. Volcanic tuff layers dated at 1663 ± 8 Ma [Bowring and Ross, 1985] have a mantle-derived signature (positive ε_{Nd} values at the time of crystallization; Hahn et al., 2013), suggesting a direct link with late Paleoproterozoic orogeny.

C-O isotopic studies were done by Hahn et al. [2013] on the East River Formation of the Hornby Bay Group.

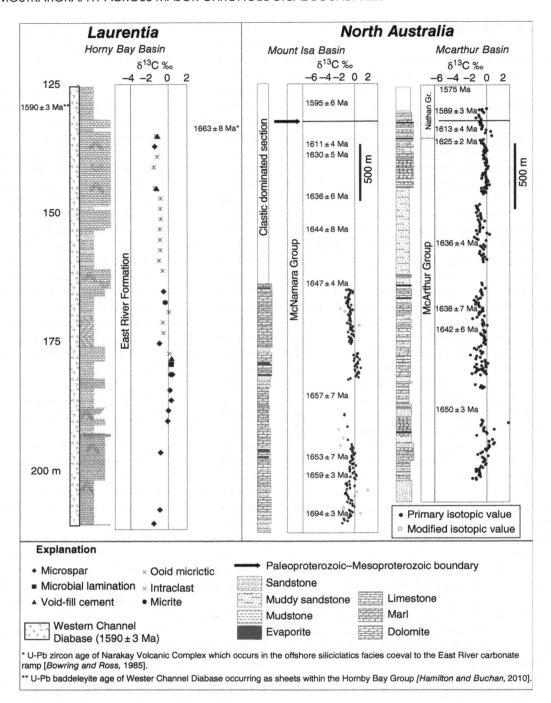

Figure 4.5 Stratigraphic sections and stable C isotope data for carbonate sequences of Laurentia (Modified after *Hahn et al.* [2013]) and North Australia [*Lindsay and Brasier*, 2000].

This section is a subtidal to peritidal carbonate platform deposited in HST, which may have a connection to the Statherian open ocean. Despite the thickness of 75 m of the carbonate sequence, $\delta^{13}C$ values vary little ranging from 1.5 to 0.5‰ and $\delta^{18}O$ from −12.55 to 6.68‰ (Fig. 4.5a). $\delta^{13}C$ values increase from δC^{13} −1.1 to −1.54‰

(VPDB) at the base to slightly positive (0.04‰) in the upper section. In the uppermost section, C isotopes decrease to −1.5‰ (Fig. 4.5a). Indeed, carbon isotope variation curves are flat and similar to those for Australia carbonate rocks, suggesting a stable environment of low nutrient availability.

The Vindhyan Basin succession overlies the basement of the Aravalli-Bundelkhand Craton (Fig. 4.4) and consists of 1.72 to ~1.6 Ga Lower Vindhyan Supergroup and 1.10–0.65 Ga Upper Supergroup [*Rasmussen et al.*, 2002; *Ray et al.*, 2003]. Lower Vindhyan units are also described as Semri Group (Fig. 4.6b) and focused in this work. They occur along the basin in the Son Valley (southern margin), Rajasthan (western margin), and Chitrakut (northern margin; *Kumar et al.*, 2002; *Ray et al.*, 2002). The Semri Group (Lower Vindhyan) is an 1170 m thick sequence formed in a shallow marine depositional environment with some deltaic-fluvial contribution [*Bose et al.*, 2001]. The group consists of siliciclastic rocks, volcanic ash (porcellanite), and three limestone layers. The limestones are (i) basal limestones, such as the Bhagwanpura Limestone (Rajasthan), Kajrahat Limestone (Son Valley), and Pellet Limestone (Chitrakut); (ii) intermediate limestone, such as Salkhan (Son Valley); and (iii) upper limestone, such as Nimbahera (Rajasthan), Rohtasgarh (Son Valley), and Tirthon (Chitrakut) [*Ray et al.*, 2003; Fig. 4.6b]. Vindhyan Basin, India, exhibits $^{87}Sr/^{86}Sr$ values of 0.70479 (Rohtasgarh profile) and 0.70460 (Kajrahat profile), based on carbonate sequences with the best preserved original $^{87}Sr/^{86}Sr$ signal [*Ray et al.*, 2003]. The obtained ratios are in agreement with a significant radiogenic contribution in the marine carbonates in the Paleoproterozoic-Mesoproterozoic boundary compared to mantle-influenced ocean water of the Archean and early Paleoproterozoic [*Shields*, 2007]. The $\delta^{13}C$ isotope values for Paleoproterozoic to Mesoproterozoic in Vindhyan Basin are also regular, varying mostly from −2 to +2‰ [*Ray et al.*, 2003]. The $\delta^{13}C$ values of basal limestones in the Rajasthan and Son Valley sections (Kajrahat and Bhagwanpura limestones) are between −1 and 0 in the basal section and −0 and −2.0‰ in the upper section (one sample has a higher $\delta^{13}C$ value of +2.4‰) (Fig. 4.6b). Upper limestone units of the Semri Group have distinct values for the Son Valley and Rajasthan sections, ranging from ca. −1‰ for the Rohtasgarh Limestone in the Son Valley to 2.1–3.1‰ in Rajasthan. $\delta^{18}O$ values for different carbonate sections in the Lower Vindhyan Supergroup range from −8.7 to −5.9‰ [*Ray et al.*, 2003].

In the NCC, the Statherian and Calymmian periods are represented by the Changcheng and Jixian systems

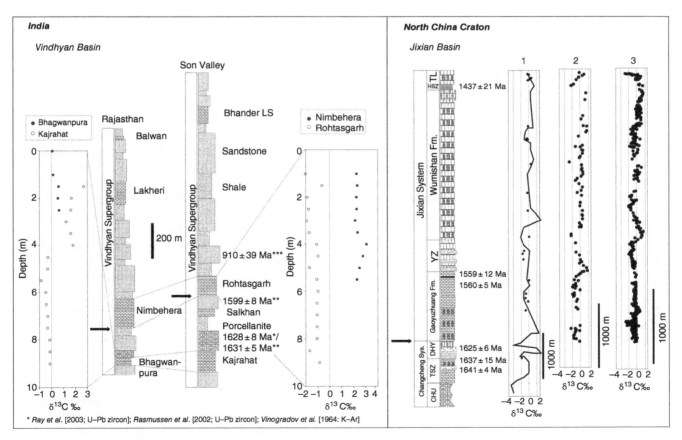

Figure 4.6 Stratigraphic sections and stable C isotope data for carbonate sequences of India [*Ray et al.*, 2003] and North China Craton. Curves after 1. *Xiao et al.* [1997] as dot; *Chu et al.* [2007] as line; 2. *Hongwei et al.* [2011]; 3. *Guo et al.* [2013].

(Figs. 4.2b and 4.3; *Meng et al.*, 2011; *Peng*, 2015; *Zhai et al.*, 2015). Basal Changcheng System overlies the 1.80–1.78 Ga Xiong'er Igneous Province and comprises a dominant siliciclastic sequence with some carbonate and tuff layers (Fig. 4.6b), deposited between 1.78 and 1.60 Ga. The Changcheng Group started with alluvial-fluvial sedimentation of late Proterozoic rifting, followed by shallow marine sedimentation [*Meng et al.*, 2011]. The Jixian System contains thick basal dolostones with intercalations of shale and mudstones (Gaoyuzhuang Group; *Zhai et al.*, 2015). The overlying units are composed of carbonate and mudstone with some tuff dated at 1559 ± 12 Ma (intermediate section; *Lu and Li*, 1991) and a bentonite layer at 1437 ± 21 Ma (top section; *Su et al.*, 2010). The Jixian System comprises coastal to shallow marine sequences [*Meng et al.*, 2011] formed between 1.4 and 1.6 Ga [*Peng*, 2015; *Zhai et al.*, 2015]. *Xiao et al.* [1997] investigated the C and O isotope composition of carbonates interlayered with siliciclastic units and few tuff and ash (bentonite) layers in both systems in three different sections in Jixian, Luonan, and Shuiyougou. The Jixian section occurs in the Jixian Rift, on northeastern margin of the NCC, and comprises the basal Changcheng System [*Xiao et al.*, 1997] deposited over the 1.78 Ga volcanic rocks of the Xiong'er Group. The group consists of a siliciclastic sequence interlayered with carbonates and tuffs [*Zhai et al.*, 2015]. The carbonate succession of Jixian System overlies these units, with some interlayered tuff and ash layers, and siliciclastic sequences and mafic dikes and sills [*Peng*, 2015] at the section top. One sample from the Chancheng System preserved original signal of $\delta^{13}C = -1.9‰$ and $\delta^{18}O = -6.7‰$ [*Xiao et al.*, 1997]. In the overlying Jixian System, 20 analyses in carbonate with the original depositional composition are used by *Xiao et al.* [1997] to construct the isotope curve for sedimentary rocks deposited between 1.6 and 1.4 Ga. The $\delta^{13}C$ values of the Jixian System range from −1.8 to 1.0‰ [*Xiao et al.*, 1997] and $\delta^{18}O$ between −14.2 and −3.3‰. The $\delta^{13}C$ isotopic values for the Jixian Basin and correlative units in NCC range from +1 to −2‰ (Fig. 4.6b). Other isotope analyses in the same section (see Section 4.1 of the Figure 4.6b for NCC) and other sections of the Jixian Basin [*Chu et al.*, 2007; *Hongwei et al.*, 2011; *Guo et al.*, 2013] confirmed the dominant $\delta^{13}C$ values between +2 and −2‰, similar to those for the studied sections of Laurentia and North Australia. The only exceptions are (i) some data of the Changcheng System in the Jixian section, Jixian Basin, which is between −4.0 and −8.0‰, and (ii) at the base of Gaoyuzhuang Formation, close to the Paleoproterozoic-Mesoproterozoic boundary, values diminishing down to −3.0‰ (Fig. 4.7).

Chu et al. [2007] and *Guo et al.* [2015] determined the composition of carbonate-associated sulfate (CAS) from

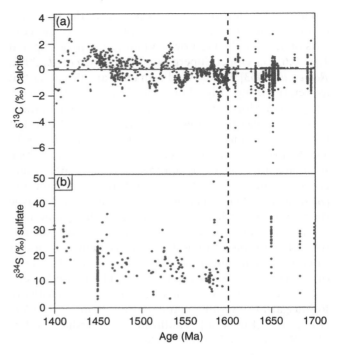

Figure 4.7 Isotope composition of $\delta^{13}C‰$ and $\delta^{34}S‰$ in carbonate sequences deposited from 1.7 and 1.4 Ga. (a) $\delta^{13}C‰$ for marine carbonates deposited from late Paleoproterozoic to early Mesoproterozoic transition in Australia, India, Laurentia, and North China. Data from *Shields and Veizer* [2002], *Ray et al.* [2003], *Xiao et al.* [1997], *Hongwei et al.* [2011], *Guo et al.* [2013], and *Hahn et al.* [2013]. (b) $\delta^{34}S‰$ data of seawater sulfate [*Guo et al.*, 2015].

the Changcheng and Jixian systems (or groups). *Chu et al.* [2007] recognized positive $\delta^{34}S$ values before and after the Paleoproterozoic-Mesoproterozoic transition in the Tuanshanzi Formation (exceeding 30‰) and basal sequences of the Gaoyuzhuang Formation (rising to +49‰), respectively. In the overlying carbonate section deposited between 1.6 and 1.4 Ga, the $\delta^{34}S$ values oscillate around +20‰ or even lower than +20 as those samples of the Yangzhuang Formation. This lower $\delta^{34}S_{sulfate}$ signature marked the Paleoproterozoic-Mesoproterozoic transition (Fig. 4.7).

Carbon isotope signature of all investigated carbonate sections in the NCC, Aravalli-Bundelkhand Craton, North Australia, and Laurentia paleoplates shows $\delta^{13}C$ values mostly between −2 and +2‰ (Fig. 4.7) with some more negative ones [*Sial et al.*, 2015 and references therein]. At the transition, there are some more negative $\delta^{13}C$ values close to −8‰ in the late Paleoproterozoic (late Statherian period) and up to −3‰ in the early Mesoproterozoic (early Calymmian period; Fig. 4.7). These flat data yield critical clues on the Paleoproterozoic-Mesoproterozoic boundary which is marked by an extensive presence of red bed sedimentary rocks, restricted development of carbonate platforms, and first appearance of complex eukaryotic characters of Ruyang biota (chrono-correlate to Changcheng System; Fig. 4.3) in the late Paleoproterozoic-early Mesoproterozoic transition [*Agić et al.*, 2017]. Deposition of large red beds during Statherian and Calymmian suggests a modification of climate conditions linked with oxidative weathering flux and increasing O_2 in the atmosphere after the GOE at 2.3 Ga. However, there were no significant changes in the Earth's climate conditions during the Statherian and Calymmian, as low diversification of biota occurred in this time interval. Significant climate and life changes occurred later in the Neoproterozoic era, linked to the presence of low-latitude glaciations in the Cryogenian period [*Kaufman et al.*, 1997; *Hoffman and Schrag*, 2002; *Halverson and Shields-Zhou*, 2011].

Variations in Sr isotopes such as the $^{87}Sr/^{86}Sr$ ratio reflect changes related to the Sr input into seawater derived from weathering of continental rocks or Earth's mantle [*Shields*, 2007]. $^{87}Sr/^{86}Sr$ curve reflects a secular variation of seawater composition. The Vindhyan Basin exhibits $^{87}Sr/^{86}Sr$ values of 0.70479 (Rohtasgarh profile) and 0.70460 (Kajrahat profile), based on carbonate sequences with the best preserved original $^{87}Sr/^{86}Sr$ signal [*Ray et al.*, 2003]. The obtained ratios are in agreement with a significant radiogenic contribution in the marine carbonates in the Paleoproterozoic-Mesoproterozoic transition compared to mantle-influenced ocean water of the Archean and Early Paleoproterozoic. Between the end of the Lomagundi-Jatuli carbon isotope excursion (including the Huronian and Makganyene glaciations), just after the GOE, and beginning of the Cryogenian isotope excursion, in the 2.1–0.72 Ga span, there are also some other data on the chemostratigraphic proxies such as S, Cr isotopes, and V and Mo. $\delta^{34}S_{sulfate}$ (sulfur isotope on sulfates) anomalies appeared at the end of the Paleoproterozoic and persisted until the present day. Close to the Paleoproterozoic-Mesoproterozoic transition, $\delta^{34}S_{sulfate}$ shows some positive excursions [see *Chu et al.*, 2007; *Guo et al.*, 2015], suggesting a change of environmental conditions (Fig. 4.7). There is no anomaly of $\delta^{53}Cr$ as well as significant amount of V and Mo in the sediments from 1.7 to 0.7 Ga, suggesting organic production [*Shields and Och*, 2011 and references therein] from the beginning of the Paleoproterozoic (~2.5 Ga) through the end of the Mesoproterozoic (~1.0 Ga). Based on the O_2 content estimation expressed as % of the present atmospheric level (PAL), a deep anoxic ocean with widespread sulfidic bottom water was suggested by Canfield [2005] and Shield and Och [2011]. From the end of the Lomagundi anomaly (~2.0–2.1 Ga) to the Paleoproterozoic-Mesoproterozoic boundary, there is a decrease in the O_2 content from 10% PAL to ~0.3%. At the boundary, there was a significant increment in the O_2 availability in the atmosphere (PAL ~10%), probably related to extensive deposition of red beds as we described above for the most paleocontinents.

4.5. DETRITAL ZIRCON RECORD

Zircon is the most common mineral used in provenance analysis due to its resistance to abrasion in the sedimentary cycle and its property of withholding age and geochemical/isotopic information of source rocks [e.g., Fedo et al., 2003; Dickinson and Gehrels, 2009; Andersen, 2014; Gehrels, 2014]. Provenance analysis using zircon data allows to establish the maximum depositional ages of Precambrian sequences as well as the stratigraphic correlations, paleogeographic reconstruction, and unraveled source areas [e.g., Gehrels, 2014]. The use of large U-Pb and Lu-Hf zircon database reproduces indistinguishable signature for a broad geologic period, as characterized by Andersen [2014] for units of Fennoscandia and Laurentia. On the other hand, zircon dataset from a short period can reproduce distinct zircon distribution patterns and is useful for paleogeographic reconstruction, source area determination, and stratigraphic correlation. This study integrates all available information of concordant U-Pb ages and Lu-Hf isotopes in detrital zircon grains within the sedimentary and low-grade metasedimentary sequences deposited between 1.8 and 1.4 Ga. We compiled information available for Baltica, Laurentia, North China, Congo, South Australia (Albany-Fraser), India, Siberia, South China (Yangtze and Cathaysia blocks), Northern Australia, and São Francisco Craton. These sequences deposited in rift, rift-sag, and epicontinental basins (e.g., São Francisco Craton, NCC, Laurentia, India, North and South Australia), formed during the period of consolidation of the Columbia supercontinent. Zircon age distribution is dominated by ages that are older than depositional or stratigraphic age (Fig. 4.8).

Two main groups of zircon age distribution in different paleoplates can be distinguished: (i) paleoplates with significant contribution of the Orosirian (2.0–1.8 Ga; peak at ca. 1.9 Ga) and early Siderian-late Neoarchean (2.4–2.7 Ga; peak at 2.56 Ga) zircon ages, which comprise NCC, Congo, Baltica, Siberia, and India, and (ii) a group marked by detrital zircon ages of 1.55–1.8 Ga, suggesting some contribution of Statherian to early Calymmian sources, represented either by the orogenic belts at the margin of the Columbia or intraplate magmatism. Paleoplates of this group are Laurentia, South China (Yangtze), South and Northern Australia, and São Francisco Craton (Fig. 4.8).

However, distribution of zircon ages for each region has some particularities. Siberia and Baltica have mainly Orosirian detrital zircon grains, whereas India has a

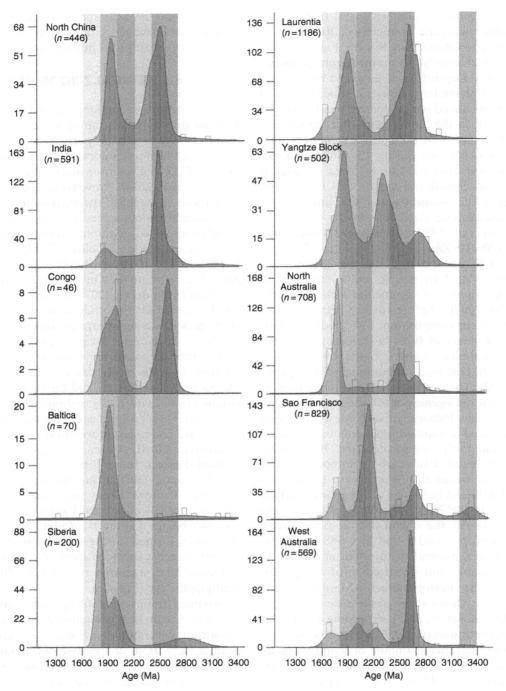

Figure 4.8 Detrital zircon age histograms for the sedimentary basins formed between 1.8 and 1.4 Ga in Siberia [*Priyatkina et al.*, 2016], Baltica [*Bogdanova et al.*, 2015], Congo [*Fernadez-Alonso et al.*, 2012], India [*Collins et al.*, 2015; *Lancaster et al.*, 2015; *Sahoo et al.*, 2018], North China [*Ying et al.*, 2011; *Liu et al.*, 2014; *Wang et al.*, 2016], Laurentia [*Rainbird and Davis*, 2007; *Furlanetto et al.*, 2013, 2016], Yangtze [*Greentree and Li*, 2008; *Zhao et al.*, 2010; *Wang et al.*, 2012; *Chen et al.*, 2013], North Australia [*Page et al.*, 2000], West Australia [*Martin et al.*, 2008; *Spaggiari et al.*, 2015], and São Francisco Craton [*Danderfer et al.*, 2009, 2015; *Marques*, 2009; *Chemale et al.*, 2012; *Santos et al.*, 2013; *Franz et al.*, 2014; *Guadagnin et al.*, 2015a, 2015b]. See text for explanation.

central peak in late Siderian to Neoarchean and low contribution of Orosirian. North China and Congo rift-sag basins have both peaks with a similar contribution. The second group encompasses different zircon age distributions controlled by major orogenic and intraplate events. Laurentia is similar to North China and Congo Craton, but there is a significant amount of zircon grains formed in the Statherian, between 1562 ± 42 and 1793 ± 63 Ma (12% of zircon population). North Australia presents a large population of zircon formed between 1497 ± 47 and 1800 ± 14 Ma (42% of the population), which were fed by orogenic belts at the margin of Eastern Australia paleoplate. These sources include ash and tuff layers deposited in intracratonic or passive margin settings connected to far-field magmatism as in Phanerozoic belts. This is similar description made by *Schindlbeck et al.* [2015] for tephra layers from Miocene plinian eruptions, transported up to 600 km from the Galápagos hotspot, or by *Rocha-Campos et al.* [2011] and *Cagliari et al.* [2014] who described Permian ash layers deposited in intracratonic setting of the Paraná Basin (Brazil) with the source situated more than 1000 km away and the Choiyoi magmatism [*Kay et al.*, 1989] associated with Neopaleozoic Gondwanide orogeny [Keidel, 1921; Veevers; 2003; Milani and DeWitt, 2008; Ramos et al., 2014]. On the other hand, the São Francisco Craton has Statherian to Calymmian zircon grains with ages ranging from 1530 ± 23 to 1800 ± 11 Ma (11.5% of zircon population) associated with post-orogenic and intraplate magmatism [*Guadagnin and Chemale*, 2015; *Guadagnin et al.*, 2015a]. We recognized two different Rhyacian and Paleoarchean populations in São Francisco Craton, which are not found as significant contributors to other Paleoproterozoic-Mesoproterozoic sequences. South Australian paleoplate has a dominant contribution of early Neoarchean zircon grains and subordinate Siderian, Rhyacian, Orosirian, and Statherian. Yangtze block contains 12% of Statherian zircon grains (63 out of 508 analyzed zircons) as well as a primary contribution of Orosirian, Siderian, and early Neoarchean to late Mesoarchean zircon grains. A significant contribution of Siderian zircon grains is uncommon in all other paleocontinents and also has yet been recognized in the Yangtze block [*Greentree et al.*, 2006].

Hf isotopes of detrital zircon grains from Paleoproterozoic-Mesoproterozoic sequences are available for the Indian Shield, NCC, São Francisco Craton, South Australia, and Yangtze block (Fig. 4.9). India is quite distinct because zircon grains have a dominant Neoarchean source (juvenile zircon with positive ε_{Hf} when zircon is crystallized). This population also occurs in the NCC together mixed with zircon grains derived from diverse sources. On another side, NCC, South China, and South Australia (southern block of the West Australia in Fig. 4.8)

present mostly Neoarchean and Mesoarchean zircon grains with slightly negative and slight to strongly positive ε_{Hf}. Some Neoarchean and Mesoarchean zircon grains from South Australia have strong negative ε_{Hf}, representing a Paleoarchean crustal source. The São Francisco Craton preserves zircon grains with long-lived crustal residence time, having dominant negative ε_{Hf} values. Detrital zircon grains of Statherian to Calymmian sedimentary rocks in the São Francisco Craton show grains with Paleoarchean and Eoarchean Hf model ages, which is a distinct signature for this craton compared to other paleoplates (Fig. 4.9).

The 1.6–1.8 Ga detrital zircon contribution is most connected to orogens at S-SE Columbia margin, as proposed by *Furlanetto et al.* [2016]. Thus, paleogeography of intracratonic and pericratonic basins allows the interpretation of an orogenic margin in S-SE Columbia and a flexural-passive margin at N-NW at current Laurentia coordinates (Fig. 4.2).

4.6. CONCLUSIONS

The Paleoproterozoic-Mesoproterozoic transition is marked by the widespread presence of red beds with moderate levels of O_2 established in the Proterozoic, probably between 0.2 and 2% O_2 by volume and around 10 times lesser of atmospheric CO_2 and CH_4 [*Catling*, 2014] compared to the anoxic Archean and early Paleoproterozoic atmosphere. Correlation is suggested among sedimentary units deposited in the paleocontinents such as Laurentia, Baltica, Siberia, Congo-São Francisco, South and North China, and North and South Australia. In all cases, basins formed mostly in intracratonic setting or epicontinental sea with a connection to the open ocean. Indeed, the dominant sedimentary record is siliciclastic with some carbonate sequences (e.g., NCC, Laurentia, India, and Western Australia) and small syn-depositional volcanogenic contributions. Very few volcanic lavas and tuff are recognized in the sedimentary successions formed in the Statherian and Calymmian (1800–1600 Ma), hampering determination of depositional age, except the North Australian superbasins. The presence of mafic and acidic dikes in the sedimentary successions is related to continental extension (e.g., LIPs and rifting processes) and defines the minimum depositional age.

The chemostratigraphic data of carbonate platforms show mostly minor variation between −2 and 2‰ in the $\delta^{13}C$ values, positive $\delta^{34}S_{sulfate}$ with some decrease in the Mesoproterozoic, more radiogenic Sr, no $\delta^{53}Cr$ anomaly recorded, and low content of V and Mo. The earliest direct evidence of eukaryotes is observed in the late Paleoproterozoic in the Ruyang Group (Chancheng System, NCC) without considerable diversification of biota. Also, low-latitude glaciation remains unregistered

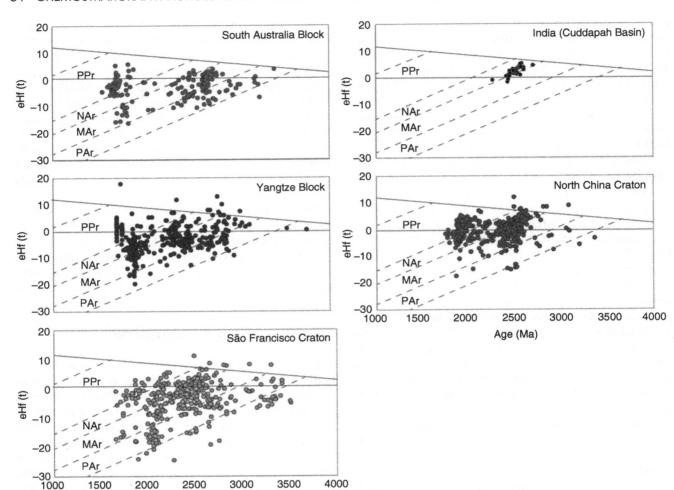

Figure 4.9 Plot of $\varepsilon_{Hf(t)}$ versus age of concordant detrital zircons from Statherian to Calymmian sequences of South Australia, India, Yangtze, North China, and Sao Francisco Craton. Depleted mantle (DM) curve after *Bodet and Scharer* [2000]. Average PPr (Paleoproterozoic), NAr (Neoarchean), MAr (Mesoarchean), and PAr (Paleoarchean) growth line for $^{176}Lu/^{177}Hf = 0.013$. Data from *Spaggiari et al.* [2015] for Australia; *Collins et al.* [2015] for India; *Zhao et al.* [2010], *Wang et al.* [2012], and *Chen et al.* [2013] for Yangtze block; *Ying et al.* [2011] and *Liu et al.* [2014] for North China Craton; and *Guadagnin et al.* [2015a, 2015b] and *Matteini et al.* [2012] for the Sao Francisco Craton.

from the Statherian to Calymmian periods. At the transition, there was a significant increment in the O_2 availability in the atmosphere (PAL ~10%), probably related to extensive deposition of red beds as described for the most paleocontinents. This establishes that in the transition between the Paleoproterozoic and Mesoproterozoic eras, little significant climatic change occurred on Earth, being quite constant, with a temperature ranging between 10 and 20 °C.

Detrital zircon age patterns of all investigated basins reflect internal processes on Columbia supercontinent, in which most the analyzed basins show dominant older zircon grains from the intracratonic or epicontinental sea. The final accretion process of the Columbia

supercontinent is recorded in the Laurentia, NCC, Baltica, Congo, Siberia, and the Yangtze blocks, where the Statherian to Calymmian sedimentary cover contains a significant contribution of 1.9–1.8 Ga zircon grains. In the basins of North Australia (McArthur and Mount Isa basins), Laurentia (Horny Bay Basin), São Francisco Craton (Espinhaço Supergroup), and Yangtze block, we also recognized the significant input of Statherian zircon grains originated from 1.8 to 1.6 Ga orogenic belts at the margin of the Columbia supercontinent (Fig. 4.2b). Ages and Lu-Hf isotopes in detrital zircon grains show two main groups of sedimentary cover with small variations. Juvenile zircon grains dominate in the North China and India cratons, whereas the São Francisco

Craton, Yangtze block, and South Australia contain dominant crustal zircon grains.

ACKNOWLEDGMENTS

We would like to thank A. N. Sial and V. P. Ferreira for the invitation to contribute this manuscript and for editorial handling. The authors thank Reinhardt Fuck and Francesca Furlanetto for their careful reviews that substantially improved the manuscript. Leo Afraneo Hartmann is thanked for English editing.

REFERENCES

Agić, H., Moczydlowska, M., Yin, L., 2017. Diversity of organic-walled microfossils from the early Mesoproterozoic Ruyang Group, North China Craton: A window into the early eukaryote evolution. Precambrian Research *297*, 101–130.

Andersen, T., 2014. The detrital zircon record: Supercontinents, parallel evolution: Or coincidence? Precambrian Research *244*, 279–287.

Ashwal, L.D., 2010. The temporality of anorthosites. Canadian Mineralogist *48*, 711–728.

Babinski, M., Pedreira, A.J., Brito Neves, B.B., Van Schmus, W.R., 1999. Contribuição à geocronologia da Chapada Diamantina. 7° Simpósio Nacional de Estudos Tectônicos. Brazilian Geological Society, Lençóis, pp. 118–120.

Bállico, M.B., Scherer, C.M.S., Mountney, N.P., Souza, E.G., Chemale, F., Jr., Pisarevsky, S.A., Reis, A.D., 2017. Wind-pattern circulation as a palaeogeographic indicator: Case study of the 1.5–1.6 Ga Mangabeira Formation, São Francisco Craton, Northeast Brazil. Precambrian Research *298*, 1–15.

Barbosa, J.S.F., Sabaté, P., 2004. Archean and Paleoproterozoic crust of the São Francisco Craton, Bahia, Brazil: Geodynamic features. Precambrian Research *133*, 1–27.

Bekker, A., Slack, J.F., Planavsky, N., Krapež, B., Hofmann, A., Konhauser, K.O., Rouxel, O.J., 2010. Iron formation: The sedimentary product of a complex interplay among mantle, tectonic, oceanic, and biospheric process. Economic Geology and the Bulletin of the Society of Economic Geologists *105*, 467–508.

Belousova, E.A., Kostitsyn, Y.A., Griffin, W.L., Begg, G.C., O'Reilly, S.Y., Pearson, N.J., 2010. The growth of the continental crust: Constraints from zircon Hf-isotope data. Lithos *119*, 457–466.

Bodet, F., Schärer, U., 2000. Evolution of the SE-Asian continent from U-Pb and Hf isotopes in single grains of zircon and baddeleyite from large rivers. Geochimica et Cosmochimica Acta *64*(12), 2067–2091.

Bogdanova, S.V., Bingen, B., Gorbatschev, R., Kheraskova, T.N., Koslov, V.I., Puchkov, V.N., Volozh, Yu.A., 2008. The East European Craton (Baltica) before and during the assembly of Rodinia. Precambrian Research *160*, 23–45.

Bogdanova, S., Gorbatschev, R., Skridlaite, G., Soesoo, A., Taran, L., Kurlovich, D., 2015. Trans-Baltic Palaeoproterozoic correlations towards the reconstruction of supercontinent Columbia/Nuna. Precambrian Research *259*, 5–33.

Bose, P.K., Sarkar, S., Chakrabarty, S., Banerjee, S., 2001. Overview of the Meso to Neoproterozoic evolution of the Vindhyan basin, central India. Sedimentary Geology *141*, 395–419.

Bouysse, P., 2010. GIS Database of the Geological Map of the World at 1/35,000,000 scale. Commission for the Geological Map of the World, 3rd revised edition.

Bowring, S.A., Ross, G.M., 1985. Geochronology of the Narakay volcanic complex: Implications for the age of the Coppermine homocline and Mackenzie igneous events. Canadian Journal of Earth Sciences = Revue Canadienne des Sciences de la Terre *22*, 774–781.

Bradley, D.C., 2008. Passive margins through Earth history. Earth-Science Reviews *91*, 1–26.

Bradley, D.C., 2011. Secular trends in the geologic record and the supercontinent cycle. Earth-Science Reviews *108*, 16–33.

Brasier, M.D., Lindsay, J.F., 1998. A billion years of environmental stability and the emergence of eukaryotes: New data from northern Australia. Geology *26*, 555–558.

Butterfield, N.J. 2015. Early evolution of the Eukaryota. Paleontology *58*(1), 5–17.

Cagliari, J., Lavina, E.L.C., Philipp, R.P., Tognoli, F.M.W., Basei, M.A.S., Faccini, U.F., 2014. New Sakmarian ages for the Rio Bonito formation (Parana Basin, southern Brazil) based on LA-ICP-MS U-Pb radiometric dating of zircons crystals. Journal of South American Earth Sciences *56*, 265–277.

Canfield, D.E., 2005. The early history of atmospheric oxygen: Homage to Robert M. Garrels. Annual Review of Earth and Planetary Sciences *33*(1), 1–36.

Catling, D.C., 2014. The great oxidation event transition. In: Turekian, K., Holland, H. (Eds.), Treatise on Geochemistry, 2nd Edition. Elsevier, Amsterdam, pp. 177–195. doi:10.1016/B978-0-08-095975-7.01307-3.

Cawood, P.A., Hawkesworth, C.J., 2014. Earth's middle age. Geology *42*, 503–506.

Cederberg, J., Söderlund, U., Oliveira, E.P., Ernst, R.E., Pisarevsky, S.A., 2016. U–Pb Baddeleyite Dating of the Proterozoic Pará de Minas Dike Swarm in the São Francisco Craton (Brazil): Implications for Tectonic Correlation with the Siberian, Congo and the North China Cratons. GFF *138*, 219–240. doi:10.1080/11035897.2015.1093543.

Chemale, F., Jr., Alkmim, F.F., Endo, I., 1993. Late Proterozoic tectonism in the interior of the São Francisco Craton. In: Findlay, R.H., Unrug, R., Banks, M.R., Veevers, J.J. (Eds.), Gondwana Eight: Assembly, Evolution and Dispersal. Balkema, Rotterdam, pp. 29–41.

Chemale, F., Jr., Quade, H., Van Schmus, W.R., 1998. Petrography, geochemistry and geochronology of the Borrachudo and Santa Bárbara metagranites, Quadrilatero Ferrífero, Brazil. Zentralblatt fur Geologie und Palaontologie Teil I *3–6*, 739–750.

Chemale, F., Jr., Dussin, I.A., Alkmim, F.F., Martins, M.S., Queiroga, G., Armstrong, R., Santos, M.N., 2012. Unravelling a Proterozoic basin history through detrital zircon geochronology: The case of the Espinhaço Supergroup, Minas Gerais, Brazil. Gondwana Research *22*, 200–206.

Chen, W.T., Zhou, M.-F., Zhao, X.-F., 2013. Late Paleoproterozoic sedimentary and mafic rocks in the Hekou area, SW China: Implication for the reconstruction of the Yangtze Block in Columbia. Precambrian Research *213*, 61–77.

Chu, X., Zhang, T., Zhang, Q., Lyons, T.W., 2007. Sulfur and carbon isotope records from 1700 to 800 Ma carbonates of the Jixian section, northern China: Implications for secular isotope variations in Proterozoic seawater and relationships to global supercontinental events. Geochimica et Cosmochimica Acta *71*, 4668–4692.

Collins, A.S., Patranabis-Deb, S., Alexander, E., Bertram, C.N., Falster, G.M., Gore, R.J., Mackintosh, J., Dhang, P.C., Payne, J.L., Jourdan, F., Backé, G., Halverson, G.P., Wade, B., 2015. Detrital mineral age, radiogenic isotopic stratigraphy and tectonic significance of the Cuddapah Basin, India. Gondwana Research *28*(4), 1294–1309.

Condie, K.C., Des Marais, D.J., Abbott, D., 2001. Precambrian superplumes and supercontinents: A record in black shales, carbon isotopes, and paleoclimates? Precambrian Research *106*, 239–260.

Cox, G.M., Jarret, A., Edwards, D., Crockford, P.W., Halverson, G.P., Collins, A.S., Poirier, A., Li, Z., 2016. Basin redox and primary productivity within the Mesoproterozoic Roper Seaway. Chemical Geology *440*, 101–114.

Cruz, S.C.P., Barbosa, J.S.F., Pinto, M.S., Peucat, J.-J., Paquette, J.L., Souza, J.S., Marins, V.S., Chemale, F., Jr., Carneiro, M.A., 2016. The Siderian-Orosirian magmatism in the Archean Gavião Paleoplate, Brazil: U-Pb geochronology, geochemistry and tectonic implications. Journal of South American Earth Sciences *69*, 43–79.

Danderfer, A., De Waele, B., Pedreira, A.J., Nalini, H.A., 2009. New geochronological constraints on the geological evolution of Espinhaço basin within the Sao Francisco Craton-Brazil. Precambrian Research *170*, 116–128.

Danderfer, A., Lana, C.C., Nalini Junior, H.A., Costa, A.F.O., 2015. Constraints on the Statherian evolution of the intraplate rifting in a Paleo-Mesoproterozoic paleocontinent: New stratigraphic and geochronology record from the eastern Sao Francisco Craton. Gondwana Research *28*(2): 668–688. doi:10.1016/j.gr.2014.06.012.

Davidson, A., 2008. Late Paleoproterozoic to mid-Neoproterozoic history of northern Laurentia: An overview of central Rodinia. Precambrian Research *160*, 5–22.

Dickinson, W.R., Gehrels, G.E., 2009. Use of U-Pb ages of detrital zircons to infer maximum depositional ages of strata: A test against a Colorado Plateau Mesozoic database. Earth and Planetary Science Letters *288*(1–2), 115–125. doi:10.1016/j.epsl.2009.09.013.

Didenko, A.N., Peskov, A.Y., Guryanov, V.A., Perestoronin, A.N., Kosynkin, A.V., 2013. Paleomagnetism of the Ulkan through (southeastern Siberian craton). Russian Journal of Pacific Geology *7*, 26–45.

Didenko, A.N., Vodovozov, V.Y., Peskov, A.Y., Guryanov, V.A., Kosynkin, A.V., 2015. Paleomagnetism of the Ulkan massif (SE Siberian platform) and the apparent polar wander path for Siberia in late Paleoproterozoic–early Mesoproterozoic times. Precambrian Research *259*, 58–77.

Eriksson, P.G., Martins-Neto, M.A., Nelson, D.R., Aspler, L.B., Chiarenzelli, J.R., Catuneanu, O., Sarkar, S., Altermann, W.,

Rautenbach, C.J.W., 2001. An introduction to Precambrian basins: Their characteristics and genesis. Sedimentary Geology *141–142*, 1–35.

Ernst, R.E., Buchan, K.L., Hamilton, M.A., Okrugin, A.V., Tomshin, M.D., 2000. Integrated paleomagnetism and U–Pb geochronology of mafic dikes of the eastern Anabar shield region, Siberia: Implications for Mesoproterozoic paleolatitude of Siberia and comparison with Laurentia. The Journal of Geology *108*, 381–401.

Ernst, R.E., Hamilton, M.A., Söderlund, U., Hanes, J.A., Gladkochub, A.V., Okrugin, A.V., Kolotilina, T., Mekhonoshin, A.S., Bleeker, W., LeCheminant, A.N., Buchan, K.L., Chamberlain, K.R., Didenko, A.N., 2016. Long-lived connection between southern Siberia and northern Laurentia in the Proterozoic. Nature Geoscience *9*(6): 464–469. doi:10.1038/NGEO2700.

Evans, D.A.D., 2009. The palaeomagnetically viable, long-lived and all-inclusive Rodinia supercontinent reconstruction. Geological Society of London, Special Publication *327*, 371–404.

Evans, D.A.D., Li, Z., Murphy, J.B., 2016. Four-dimensional context of Earth's supercontinents. In: Li, Z.X., Evans, D.A.D., Murphy, J.B. (Eds.), Supercontinent Cycles Through Earth History. Special Publications *424*. Geological Society, London.

Fedo, C.M., Sircombe, K.N., Rainbird, R.H., 2003. Detrital zircon analysis of the sedimentary record. In: Hanchar, J.M., Hoskin, P.W.O. (Eds.), Zircon. Reviews in Mineralogy and Geochemistry *53*, 277–303. Mineralogical Society of America, Chantilly, VA.

Fernandez-Alonso, M., Cutten, H., De Waele, B., Tack, L., Tahon, A., Baudet, D., Barrit, S.D., 2012. The Mesoproterozoic Karagwe-Ankole Belt (formerly the NE Kibara Belt): The result of prolonged extensional intracratonic basin development punctuated by two short-lived far-field compressional events. Precambrian Research *216–219*, 63–86.

Franz, G., Morteani, G., Gerdes, A., Rhede, D., 2014. Ages of protolith and Neoproterozoic metamorphism of Al-P-bearing quartzites of the Veredas Formation (Northern Espinhaço, Brazil): LA-ICPMS age determinations on relict and recrystallized zircon and geodynamic consequences. Precambrian Research *250*, 6–26.

Furlanetto, F., Thorkelson, D.J., Gibson, H.D., Marshall, D.D., Rainbird, R.H., Davis, W.J., Crowley, J.L., Vervoort, J.D., 2013. Late Paleoproterozoic terrane accretion in northwestern Canada and the case for circum-Columbian orogenesis. Precambrian Research *224*, 512–528.

Furlanetto, F., Thorkelson, D.J., Rainbird, R.H., Davis, W.J., Gibson, H.D., Marshall, D.D., 2016. The Paleoproterozoic Wernecke Supergroup of Yukon, Canada: Relationships to orogeny in northwestern Laurentia and basins in North America, East Australia, and China. Gondwana Research *39*, 14–40.

Gao L.Z., Zhang C.H., Yin C.Y, 2008a. SHRIMP zircon ages: Basis for refining the chronostratigraphic classification of the Meso- and Neoproterozoic strata in North China Old Land. Acta Geoscience Sinica *29*, 366–376.

Gao L.Z., Zhang C.H., Shi X.Y., 2008b. Mesoproterozoic age for Xiamaling Formation in North China Plate indicated by zircon SHRIMP dating. Chinese Science Bulletin 53, 2665–2671.

Gehrels, G., 2014. Detrital zircon U–Pb geochronology applied to tectonics. Annual Review of Earth and Planetary Sciences 42, 127–149.

Gladkochub, D.P., Pisarevsky, S.A., Donskaya, T.V., Ernst, R.E., Wingate, M.T.D., Söderlund, U., Mazukabzov, A.M., Sklyarov, E.V., Hamilton, M.A., Hanes, J.A., 2010. Proterozoic mafic magmatism in Siberian craton: An overview and implications for paleocontinental reconstruction. Precambrian Research 183, 660–668.

Goldfarb, R.J., Groves, D.I., Gardoll, S., 2001. Orogenic gold and geologic time: A global synthesis. Ore Geology Reviews 18, 1–75.

Greentree, M.R., Li, Z.X., 2008. The oldest known rocks in south–western China: SHRIMP U–Pb magmatic crystallization age and detrital provenance analysis of the Paleoproterozoic Dahongshan Group. Journal of Asian Earth Sciences 33, 289–302.

Greentree, M.R., Li, Z.X., Li, X.H., Wu, H., 2006. Late Mesoproterozoic to earliest Neoproterozoic basin record of the Sibao orogenesis in western South China and relationship to the assembly of Rodinia. Precambrian Research 151, 79–100.

Guadagnin, F., Chemale, F., Jr., 2015. Detrital zircon record of the Paleoproterozoic to Mesoproterozoic cratonic basins in the São Francisco Craton. Journal of South American Earth Sciences 60, 104–116.

Guadagnin, F., Chemale, F., Jr., Magalhães, A.J.C., Santana, A., Dussin, I., Takehara, L., 2015a. Age constraints on crystal-tuff from the Espinhaço Supergroup: Insight into the Paleoproterozoic to Mesoproterozoic intracratonic basin cycles of the Congo–São Francisco Craton. Gondwana Research 27, 363–376.

Guadagnin, F., Chemale, F., Jr., Magalhães, A.J.C., Alessandretti, L., Bállico, M.B., Jelinek, A.R., 2015b. Sedimentary petrology and detrital zircon U–Pb and Lu–Hf constraints of Mesoproterozoic intracratonic sequences in the Espinhaço Supergroup: Implications for the Archean and Proterozoic evolution of the São Francisco Craton. Precambrian Research 266, 227–245.

Guo, H., Du, Y., Kah, L.C., Huang, J., Hu, C, Huanga, H., Yu, W., 2013. Isotopic composition of organic and inorganic carbon from the Mesoproterozoic Jixian Group, North China: Implications for biological and oceanic evolution. Precambrian Research 224,169–183

Guo, H., Du, Y., Kah, L.C., Hu, C., Huang, J., Huang, H., Yu, W., Song, H., 2015. Sulfur isotope composition of carbonate-associated sulfate from the Mesoproterozoic Jixian Group, North China: Implications for the marine sulfur cycle. Precambrian Research 266, 319–336.

Hahn, K., Rainbird, R., Cousens, B., 2013. Sequence stratigraphy, provenance, C and O isotopic composition, and correlation of the late Paleoproterozoic-early Mesoproterozoic upper Hornby Bay and lower Dismal Lakes groups, NWT and Nunavut. Precambrian Research 232, 209–225.

Hall, C.E., Jones, S.A., Bodorkos, S., 2008. Sedimentology, structure and SHRIMP zircon provenance of the Woodline Formation, Western Australia: Implications for the tectonic setting of the West Australian Craton during the Paleoproterozoic. Precambrian Research 162, 577–598.

Halverson, G.P., Shields-Zhou, G., 2011. Chemostratigraphy and the Neoproterozoic glaciations. Geological Society of London, Memory 36, 51–66. doi:10.1144/M36.4.

Hamilton, M.A., Buchan, K.L., 2010. U/Pb geochronology of the Western Channel diabase, northwestern Laurentia: Implications for a large 1.59 Ga magmatic province, Laurentia's APWP and paleocontinental reconstructions of Laurentia, Baltica and Gawler craton of southern Australia. Precambrian Research 183–3, 463–473. doi:10.1016/j.precamres.2010.06.009.

Hoffman, P.F., 1988. United plates of America, the birth of a craton: Early Proterozoic assembly and growth of Laurentia. Annual Review of Earth and Planetary Sciences 16, 543–603.

Hoffman, P.F., 1989. Speculations on Laurentia's first gigayear (2.0 to 1.0 Ga). Geology 17(2), 135–138. doi:10.1130/0091-7613(1989)017<0135:SOLSFG>2.3.CO;2.

Hoffman, P.F., Schrag, D.P., 2002. The snowball Earth hypothesis: Testing the limits of global change. Terra Nova 14, 129–155.

Hongwei, K., Yongqing, L., Jiahua, L., Nan, P., Shunshe, L., Chao, C., 2011. Carbon and oxygen isotopic stratigraphy of mesoproterozoic carbonate sequences (1.6–1.4 Ga) from Yanshan in North China. International Journal of Oceanography 2011, 1–11. doi:10.1155/2011/410621.

Huston, D.L., Pehrsson, S., Eglington, B.M., Zaw, K., 2010. The geology and metallogeny of volcanic-hosted massive sulfide deposits: Variations through geologic time and with tectonic setting. Economic Geology and the Bulletin of the Society of Economic Geologists 105, 571–591.

Ireland, T.R., Williams, S., 2003. Considerations in zircon geochronology by SIMS. Reviews in Mineralogy and Geochemistry 53(1), 215–241. doi:10.2113/0530215.

Irving, E., Baker, J., Hamilton, M.A., Wynne, P.J., 2004. Early Proterozoic geomagnetic field in western Laurentia: Implications for paleolatitudes, local rotations and stratigraphy. Precambrian Research 129, 251–270.

Karhu, J.A., Holland, H.D., 1996. Carbon isotopes and the rise of atmospheric oxygen. Geology 24, 867–870.

Kaufman, A.J., Knoll, A.H., Narbonne, G.M., 1997. Isotopes, ice ages, and terminal Proterozoic earth history. Proceedings of the National Academy of Science (USA) 94, 6600–6605.

Kay, S.M., Ramos, V.A., Mpodozis C., Sruoga, P., 1989. Late Paleozoic to Jurassic silicic magmatism at the Gondwana margin: Analogy to middle Proterozoic in North America. Geology 17, 324–328, Boulder.

Keidel, J., 1921. Sobre la distribución de los depósitos glaciares del Pérmico conocidos em la Argentina y su significación para la estratigrafía de la serie del Gondwana y la paleogeografía del Hemisferio Austral. Academia Nacional de Ciencias, Boletín 25, 239–368.

Khudoley, A.K., Kropachev, A.P., Tkachenko, V.I., 2007. Mesoproterozoic to Neoproterozoic evolution of the Siberian craton and adjacent microcontinents: An overview with constraints for a Laurentian connection. In: Link, P.K., Lewis, R.S. (Eds.), Proterozoic Geology of Western North America and Siberia. SEPM Special Publication 86. SEPM, Tulsa, OK, pp. 209–226.

Khudoley, A.K., Chamberlain, K.R., Schmitt, A.K., Harrison, T.M., Prokopiev, A.V.,Sears, J.W., Veselovskiy, R.V., Proskurnin, V.F., 2009. New U–Pb baddeleyite ages from mafic intrusions from Taimyr, northern and southeastern Siberia: implications for tectonic and stratigraphy. Isotopic Systems and Time in Geological Processes, Transactions of the IV Russian Conference on the Isotopic Geology, vol. 2, St. Petersburg, pp. 243–245.

Khudoley, A., Chamberlain, K., Ershova, V., Sears, J., Prokopiev, A., MacLean, J., Kazakova, G., Malyshev, S., Molchanov, A., Kullerud, K., Toro, J., Miller, E., Veselovskiy, R., Li, A., Chipley, D., 2015. Proterozoic supercontinental restorations: Constraints from provenance studies of Mesoproterozoic to Cambrian clastic rocks, eastern Siberian Craton. Precambrian Research 259, 78–94.

Knoll, A.H., Hayes, J.M., Kaufman, A.J., Swett, K., Lambert, I.B., 1986. Secular variation in carbon isotope ratios from Upper Proterozoic successions of Svalbard and East Greenland. Nature 321, 831–838.

Kopp, R.E., Kirschvink, J.L., Hilburn, I.A., Nash, C.Z., 2005. The Paleoproterozoic snowball Earth: A climate disaster triggered by the evolution of oxygenic photosynthesis. PNAS, 102(32), 11131–11136.

Köykkä, J., 2011. Precambrian alluvial fan and braidplain sedimentation patterns: Example from the Mesoproterozoic Rjukan Rift Basin, southern Norway. Sedimentary Geology 234, 89–108.

Kulikov, V.S., Svetov, S.A., Slabunov, A.I., Kulikova, V.V., Polin, A.K., Golubev, A.I., Gorkovets, V.Y., Ivashchenko, V.I., Gogolev, M.A., 2017. Geological map of Southeastern Fennoscandia (scale 1:750 000): A new approach to map compilation. Transactions of Karelian Research Centre of Russian Academy of Science 2, Precambrian Geology Series 3–41.

Kumar, B., Das Sharma, S., Sreenivas, B., Dayal, A. M., Rao, M. N., Dubey, N., Chawla, B. R., 2002. Carbon oxygen and strontium isotope geochemistry of Proterozoic carbonate rocks of the Vindhyan Basin, central India. Precambrian Research 113, 43–63.

Lan, Z., Li, X., Chen, Z.-Q., Li, Q., Hofmann, A., Zhang, Y., Zhong, Y., Liu, Y., Tang, G., Ling, X., Li, J., 2014. Diagenetic xenotime age constraints on the Sanjiaotang Formation, Luoyu Group, southern margin of the North China Craton: Implications for regional stratigraphic correlation and early evolution of eukaryotes. Precambrian Research 251, 21–32.

Lancaster, P.J., Dey, S., Storey, C.D., Mitra, A., Bhunia, R.K., 2015. Contrasting crustal evolution processes in the Dharwar craton: Insights from detrital zircon U–Pb and Hf isotopes. Gondwana Research 28, 1361–1372.

Levin, H., 2013. The Earth Through Time, 10th Edition. Wiley, Hoboken, NJ.

Li, H.K., Zhu, S.X., Xiang, Z.Q., Su, W.B., Lu, S.N., Zhou, H.Y., Geng, J.Z., Li, S., Yang, F.J., 2010. Zircon U–Pb dating on tuff bed from Gaoyuzhuang Formation in Yanqing, Beijing: Further constraints on the new subdivision of the Mesoproterozoic stratigraphy in the northern North China craton. Acta Petrologica Sinica 26, 2131–2140.

Li, H.K., Su, W.B., Zhou, H.Y., Xiang, Z.Q., Tian, H., Yang, L.G., Huff, W.D., Ettensohn, F.R., 2014.The First Precise Age Constraints on the Jixian System of the Meso- and Neoproterozoic Standard Section of China: SHRIMP Zircon U-Pb dating of bentonites from the Wumishan and Teiling Formations in the Jixian Section, North China Craton. Acta Petrologica Sinica 30, 2999–3012.

Lindsay, J.F., Brasier, M.D., 2000. A carbon isotope reference curve for ca. 1700–1575 Ma, McArthur and Mount Isa Basins, Northern Australia. Precambrian Research 99, 271–308.

Liu, C., Zhao, G., Liu, F., 2014. Detrital zircon U–Pb, Hf isotopes, detrital rutile and whole-rock geochemistry of the Huade Group on the northern margin of the North China Craton: Implications on the breakup of the Columbia Supercontinent. Precambrian Research 254 (2014) 290–305

Lu, S.N., Li, H.M., 1991. A precise U–Pb single zircon age determination for the volcanics of the Dahongyu Formation, Changcheng System in Jinxian (in Chinese). Bulletin of the Chinese Academy Geological Sciences, 22, 137–145.

Lubnina, N.V., Pisarevsky, S.A., Stepanova, A.V., Bogdanova, S.V., Sokolov, S.J., 2017. Fennoscandia before Nuna/Columbia: Paleomagnetism of 1.98–1.96 Ga mafic rocks of the Karelian craton and paleogeographic implications. Precambrian Research 292, 1–12.

MacLean, B.C., Cook, D.G., 2004. Revisions to the Paleoproterozoic Sequence A, based on reflection seismic data across the western plains of the Northwest Territories, Canada. Precambrian Research 129, 271–289.

Marques, G.C., 2009. Geologia dos grupos Araí e Serra da Mesa e seu Embasamento no sul do Tocantins (Unpublished MSc. thesis). Universidade de Brasília, p. 116.

Marshak, S., Alkmim, F.F., 1989. Proterozoic contraction/extension tectonics of the southern São Francisco region, Minas Gerais, Brazil. Tectonics 8, 555–571.

Martin, D.M., Sircombeb, K.N., Thorne, A.M., Cawood, P.A., Nemchind, A.A., 2008. Provenance history of the Bangemall Supergroup and implications for the Mesoproterozoic paleogeography of the West Australian Craton. Precambrian Research 166, 93–110.

Matteini, M., Dantas, E.L., Pimentel, M.M., Alvarenga, C.J.S., Dardenne, M.A., 2012. U-Pb and Hf isotope study on detrital zircons from the Paranoa Group, Brasília Belt Brazil: Constraints on depositional age at Mesoproterozoic-Neoproterozoic transition and tectono-magmatic events in the Sao Francisco Craton. Precambrian Research 206–207, 168–181.

Maynard, J.B., 2010. The chemistry of manganese ores through time: A signal of increasing diversity of Earth-surface environments. Economic Geology and the Bulletin of the Society of Economic Geologists 105, 535–552.

Mazumder, R., Saha, D., 2012. Palaeoproterozoic of India. Geological Society of London, Special Publications, 365, 161–184. doi:10.1144/SP365.9.

Meert, J.G., 2012. What's in a name? The Columbia (Paleopangaea/Nuna) supercontinent. Gondwana Research 21, 987–993.

Meert, J.G., Pandit, M.K., 2015.The Archaean and Proterozoic history of Peninsular India: Tectonic framework for Precambrian sedimentary basins in India. In: Mazumder, R., Eriksson, P.G. (Eds.), Precambrian Basins of India: Stratigraphic and Tectonic Context. Geological Society, London, Memoirs 43. The Geological Society, London, pp. 29–54. doi:10.1144/M43.3.

Meert, J.G., Santosh, M., 2017. The Columbia Supercontinent revisited. Gondwana Research 50, 67–83.

Meng, Q.R., Wei, H.H., Qu, Y.Q., Ma, S.X., 2011. Stratigraphic and sedimentary records of the rift to drift evolution of the northern North China craton at the Paleo- to Mesoproterozoic transition. Gondwana Research 20, 205–218.

Miall, A.D., Catuneanu, O., Eriksson, P.G., Mazumder, R., 2015. A brief synthesis of Indian Precambrian basins: Classification and genesis of basin-fills. In: Mazumder, R., Eriksson, P.G. (Eds.), Precambrian Basins of India: Stratigraphic and Tectonic Context. Geological Society, London, Memoirs 43. The Geological Society, London, pp. 339–347. doi:10.1144/M43.3.

Milani, E.J., DeWitt, M.J., 2008. Correlations between the classic Paraná and Cape- Karoo sequences of South America and southern Africa and their basin infills flanking the Gondwanides: du Toit revisited. In: Pankhurst, R.J., Trouw, R.A.J., Brito Neves, B.B., De Wit, M.J. (Eds.), West Gondwana: Pre-Cenozoic Correlations Across the South Atlantic Region. Geological Society, Special Publications 294. The Geological Society, London, pp. 319–342.

Mondal, M.E.A., Goswami, J.N., Deomurari, M.P., Sharma, K.K., 2002. Ion microprobe 207Pb/206Pb ages of zircons from the Bundelkhand massif, northern India: Implications for the crustal evolution of the Bundelkhand–Aravalli protocontinent. Precambrian Research 117, 85–100.

Myers, J.S., Russel, D.S., Tyler, I.M., 1996. Tectonic evolution of Proterozoic Australia. Tectonics 15(6), 1431–1446.

Naqvi, S.M., 2005. Geology and Evolution of the Indian Plate. Capital Publishing, New Delhi, 450 pp.

Ogg, J.G., 2004. Status of divisions of the International Geologic Time Scale. Lethaia 37, 183–199.

Oliveira, E.P., Silveira, E.M., Söderlund, U., Ernst, R.E., 2013. U–Pb ages and geochemistry of mafic dike swarms from the Uauá Block, São Francisco Craton, Brazil: LIPs remnants relevant for Late Archean break-up of a supercraton. Lithos 174, 308–322.

Page, R.W., Sweet, I.O., 1998 Geochronology of basin phases in the western Mt Isa Inlier, and correlation with the McArthur Basin. Australian Journal of Earth Sciences 45, 219–232.

Page, R.W., Jackson, M.J., Krassay, A.A., 2000. Constraining sequence stratigraphy in north Australian basins: SHRIMP U–Pb zircon geochronology between Mt Isa and McArthur River. Australia Journal of Earth Sciences 47, 431–459.

Pang, K., Tang, Q., Schiffbauer, J.D., Yao, J., Yuan, X., Wan, B., Chen, L., Ou, Z., Xiao, S., 2013.The nature and origin of nucleus-like intracellular inclusions in Paleoproterozoic eukaryote microfossils. Geobiology 11, 499–510.

Papineau, D., 2010. Global biogeochemical changes at both ends of the Proterozoic: Insights from phosphorites. Astrobiology 10, 165–181.

Pehrsson, S.J., Eglington, B.M., Evans, D.A.D., Huston, D., Reddy, S.M., 2015. Metallogeny and its link to orogenic style during the Nuna supercontinent cycle. In: Li, Z.X., Evans, D.A.D., Murphy, J.B. (Eds.), Supercontinent Cycles Through Earth History. Special Publications 424. Geological Society, London.

Peng, P., 2015. Precambrian mafic dike swarms in the North China Craton and their geological implications. Science China Earth Sciences 57, 1–27. doi:10.1007/s11430-014-5026-x.

Peng, M., Wu, Y.B., Wang, J., Jiao, W.F., Liu, X.C., Yang, S.H., 2009. Paleoproterozoic mafic dike from Kongling terrane in the Yangtze Craton and its implication. Chinese Sciences Bulletin 54(6), 1098–1104.

Petrov, P.Y., 2014. The Mukun Basin: Settings, paleoenvironmental parameters, and factors controlling the early Mesoproterozoic terrestrial sedimentation (Lower Riphean section of the Anabar Uplift, Siberia). Lithology and Mineral Resources 49, 55–80.

Pisarevsky, S.A., Elming, S., Pesonen, L.J., Li, Z., 2014. Mesoproterozoic paleogeography: Supercontinent and beyond. Precambrian Research 244, 207–225.

Plumb, K.A., 1991. New Precambrian time scale. Episodes 14(2), 139–140.

Plumb, K.A., James, H.J., 1986. Subdivision of Precambrian time: Recommendations and suggestions by the Subcommission on Precambrian Stratigraphy. Precambrian Research 32, 65–92.

Prasad, B., Uniyal, S.N., Asher, R., 2005. Organic-walled microfossils from the Proterozoic Vindhyan Supergroup of Son Valley, Madhya Pradesh, India. Palaeobotanist 54, 13–60.

Priyatkina, N., Khudoley, A.K., Collins, W.J., Kuznetsov, N.B., Huang, H., 2016. Detrital zircon record of Meso- and Neoproterozoic sedimentary basins in northern part of the Siberian Craton: Characterizing buried crust of the basement. Precambrian Research 285, 21–38.

Rainbird, R.H., Davis, W.J. 2007. U-Pb detrital zircon geochronology and provenance of the late Paleoproterozoic Dubawnt Supergroup: Linking sedimentation with tectonic reworking of the western Churchill Province, Canada. Geological Society of America Bulletin 119(3/4), 314–328.

Rainbird, R.H., Hadlari, T., Aspler, L.B., Donaldson, J.A., LeCheminant, A.N., Peterson, T.D., 2003. Sequence stratigraphy and evolution of the Paleoproterozoic intracontinental Baker Lake and Thelon basins, western Churchill Province, Nunavut, Canada. Precambrian Research 125, 21–53.

Rainbird, R.H., Davis, W.J., Stern, R.A., Peterson, T.D., Smith, S.R., Parrish, R.R., Hadlari, T., 2006. Ar-Ar and U–Pb geochronology of a late Paleoproterozoic rift basin: Support for a genetic link with Hudsonian orogenesis, western Churchill Province. Journal of Geology 114, 1–17.

Ramos, V.A., Chemale, F., Jr., Naipauer, M., Pazos, P.J., 2014.A provenance study of the Paleozoic Ventania System (Argentina): Transient complex sources from Western and Eastern Gondwana. Gondwana Research 26, 719–740.

Rasmussen, B., Bose, P.K., Sarkar, S., Banerjee, S., Fletcher, I.R., McNaughton, N.J., 2002. 1.6Ga U–Pb zircon age for the Chorhat Sandstone, Lower Vindhyan, India: Possible implications for early evolution of animals; Geology 30, 103–106. doi:10.1130/0091-7613(2002)030<0103: GUPZAF>2.0.CO;2.

Ray, J., 2006. Age of the Vindhyan Supergroup: A review of recent findings. Journal of Earth System Science 115(1), 149–160.

Ray, J.S., Martin, M.W., Veizer, J., Bowring, S.A., 2002. U–Pb zircon dating and Sr isotope systematics of the Vindhyan

Supergroup, India. Geology *30*, 131–134. doi:10.1130/0091-7613(2002)030<0131:UPZDAS>2.0.CO;2.

Ray, J.S, Veizer, J., Davis, W.J., 2003. C, O, Sr and Pb isotope systematics of carbonate sequences of the Vindhyan Supergroup, India: Age, diagenesis, correlations and implications for global events. Precambrian Research *121*, 103–140.

Rocha-Campos, A.C., Basei, M.A., Nutman, A.P., Kleiman, L.E., Varela, R., Llambias, E., Canile, F.M., Rosa, O. de C.R. da, 2011. 30 million years of Permian volcanism recorded in the Choiyoi igneous province (W Argentina) and their source for younger ash fall deposits in the Parana Basin: SHRIMP U-Pb zircon geochronology evidence. Gondwana Research *19*, 509–523.

Rogers, J.J.W., Santosh, M., 2002. Configuration of Columbia, a Mesoproterozoic Supercontinent. Gondwana Research *5*, 5–22.

Ross, G.M., Villeneuve, M.E., Theriault, R.J., 2001. Isotopic provenance of the lower Muskwa assemblage (Mesoproterozoic, Rocky Mountains British Columbia): New clues to correlation and source areas. Precambrian Research *111*, 57–77.

Saha, S., Das, K., Hidaka, H., Kimura, K., Chakraborty, P.P., Hayasaka, Y., 2016. Detrital zircon geochronology (U–Pb SHRIMP and LA-ICPMS) from the Ampani Basin, Central India: Implication for provenance and Mesoproterozoic tectonics at East Indian cratonic margin. Precambrian Research *281*, 363–383.

Sahoo, D., Pruseth, K.L., Upadhyay, D., Ranjan, S., Pal, D.C., Banerjee, R., Gupta, S., 2018. New constraints from zircon, monazite and uraninite dating on the commencement of sedimentation in the Cuddapah basin, India. Geological Magazine *155*, 1230–1246.

Santos, M.N., Chemale, F., Jr., Dussin, I.A., Martins, M., Assis, T.A.R., Jelinek, A.R., Guadagnin, F., Armstrong, R., 2013. Sedimentological and paleoenvironmental constraints of the Statherian and Stenian Espinhaço rift system, Brazil. Sedimentary Geology *290*, 47–59.

Sarangi, S., Gopalan, K., Kumar, S., 2004. Pb–Pb age of earliest megascopic eukaryotic algae bearing Rohtas Formation, Vindhyan Supergroup, India: Implications for Precambrian oxygen evolution. Precambrian Research *132*, 107–121.

Schindlbeck, J.C., Kutterolf, S., Freundt, A. Straub, S.M., Wang, K.-L. Jegen, M., Hemming, S.R., Baxter, A.T., Sandoval, M.I., 2015. The Miocene Galápagos ash layer record of Integrated Ocean Drilling Program Legs 334 and 344: Ocean-island explosive volcanism during plume-ridge interaction. Geology, *43*(7), 599–602. doi:10.1130/G36645.1.

Shields, G.A., 2007. A normalized seawater strontium isotope curve: Possible implications for Neoproterozoic–Cambrian weathering rates and the further oxygenation of the Earth. eEarth *2*, 35–42.

Shields, G.A., Veizer, J., 2002. Precambrian marine carbonate isotope database: Version 1.1. Geochemistry, Geophysics, Geosystems *6*, 1–12. doi:10.1029/2001GC000266.

Shields, G.A., Och, L., 2011. The case for a Neoproterozoic oxygenation event: Geochemical evidence and biological consequences. GSA Today *21*(3), 1–11. doi:10.1130/GSATG102A.1.

Shumlyanskyy, L., Hawkesworth, C., Billström, K., Bogdanova, S., Mytrokhyn, O., Romer, R., Dhuime, B., Claesson, S., Ernst, R., Whitehouse, M., Bilan, O., 2017. The origin of the Palaeoproterozoic AMCG complexes in the Ukrainian Shield: New U-Pb ages and Hf isotopes in zircon. Precambrian Research *292*, 216–239.

Sial, A.N., Gaucher, C., Ferreira, V.P., Pereira, N.S., Cezario, W.S., Chiglino, L., Lima, H.M. 2015. Isotope and elemental chemostratigraphy. In: Ramkumar, M. (Ed.), Concepts, Techniques, and Applications. Elsevier, Netherlands, pp. 23–64.

Silva, A.M., Chemale, F., Kuyumjian, R.M., Heaman, L., 1995. Mafic dike swarms of Quadrilátero Ferrífero and Southern Espinhaço, Minas Gerais, Brazil. Brazilian Journal of Geology *25*, 124–137.

Silveira, E.M., Söderlund, U., Oliveira, E.P., Ernst, R.E., Menezes Leal, A.B., 2013. First precise U–Pb baddeleyite ages of 1500 Ma mafic dikes from the São Francisco Craton, Brazil, and tectonic implications. Lithos *174*, 144–156.

Southgate, P.N., Bradshaw, B.E., Domagala, J., Jackson, M.J., Idnurm, M., Krassay, A.A., Page, R.W., Sami, T., Scott, D., Lindsay, J.F., Scott, D., McConachie, B.A., Tarlowski, C., 2000. Chronostratigraphic basin framework for Palaeoproterozoic rocks (1730–1575 Ma) in northern Australia and implications for base-metal mineralization. Australia Journal of Earth Sciences *47*, 461–483.

Spaggiari, C.V., Kirkland, C.L., Smithies, R.H., Wingate, M.T.D., Belousova, E.A., 2015. Transformation of an Archean craton margin during Proterozoic basin formation and magmatism: The Albany–Fraser Orogen, Western Australia. Precambrian Research *266*, 440–466.

Stern, R.A., 1997. The GSC Sensitive High-Resolution Ion Microprobe (SHRIMP): Analytical techniques of zircon U-Th-Pb age determinations and performance evaluation. Current Research 1997-F: Radiogenic Age and Isotopic Studies: Report 10. (0704-2884, 0704-2884). Geological Survey of Canada, Ottawa, pp. 1–31.

Su, W.B., Zhang, S.H., Huff, W.D., Li, H.-K., Ettensohn, F.R., Chen, X.-Y., Yang, H.-M., Han, Y.-G., Song, B., Santosh, M., 2008. SHRIMP U-Pb ages of K–bentonite beds in the Xiamaling Formation: Implications for revised subdivision of the Meso- to Neoproterozoic history of the North China Craton. Gondwana Research *14*, 543–553.

Su, W.B., Li, H.K., Huff, W.D., Ettensohn, F.R., Zhang, S.H., Zhou, H.Y., Wan, Y.S., 2010. Zircon SHRIMP U-Pb ages of tuff in the Tieling Formation and their geological significance. Chinese Science Bulletin *55*, 3312–3323.

Su, W.B., Li, H.K., Xu, L., Jia, S.H., Geng, J.Z., Zhou, H.Y., Wang, Z.H., Pu, H.Y., 2012. Luoyu and Ruyang Groups at the south margin of the North China Craton (NCC) should belong in the Mesoproterozoic Changchengian System: Direct constraints from the LA-MC-ICPMS U–Pb age of the tuffite in the Luoyukou Formation, Ruzhou, Henan, China. Geological Survey Research *35*, 96–108.

Sun, Z.M., Yin, F.G., Guan, J.L., Liu, J.H., Li, J.M., Geng, Y.R., Wang, L.Q., 2009. SHRIMP U–Pb dating and its stratigraphic significance of tuff zircons from Heishan Formation of Kunyang Group, Dongchuan area, Yunnan Province, China. Geological Bulletin of China *28*(7), 896–900.

Teixeira, W., Figueiredo, M.C.H., 1991. An outline of Early Proterozoic crustal evolution in the São Francisco craton, Brazil: A review. Precambrian Research *53*, 1–22.

Teixeira, W., Ávila, C.A., Dussin, I.A., Corrêa Neto, A.V., Bongiolo, E.M., Santos, J.O.S., Barbosa, N., 2015. Zircon U-Pb-Hf, Nd-Sr constraints and geochemistry of the Resende Costa Orthogneiss and coeval rocks: New clues for a juvenile accretion episode (2.36-2.33 Ga) in the Mineiro belt and its role to the long-lived Minas accretionary orogeny. Precambrian Research *256*, 148–169.

Turner, C.C., Meert, J.G., Pandit, M.K., Kamenov, G.D., 2014. A detrital zircon U–Pb and Hf isotopic transect across the Son Valley sector of the Vindhyan Basin, India: Implications for basin evolution and paleogeography. Gondwana Research *26*, 348–364. doi:10.1016/j.gr.2013.07.009.

Turpin, L., Maruejol, P., Cuney, M., 1988. U–Pb, Rb–Sr and Sm-Nd chronology of granitic basement hydrothermal albitites and uranium mineralization (Lagoa Real, South Bahia, Brazil). Contributions to Mineralogy and Petrology *98*, 139–147.

Uhlein, A., Trompette, R.R., Egydio-Silva, M., 1998. Proterozoic rifting and closure, SE border of the São Francisco Craton, Brazil. Journal of South American Earth Sciences *11*, 191–203.

Veevers, J.J., 2003. Pan-African is Pan-Gondwanaland: Oblique convergence drives rotation during 650–500 Ma assembly. Geology *31*, 501–504. doi:10.1130/0091-7613(2003)031<0501:PIPOCD>2.0.CO;2.

Veselovskiy, R.V., Petrov, P.Y., Karpenko, S.F., Kostitsyn, Y.A., Pavlov, V.E., 2006. New paleomagnetic and isotopic data on the Mesoproterozoic igneous complex on the northern slope of the Anabar Uplift. Doklady Earth Sciences *411*, 1190–1194.

Vinogradov, A.P., Tugarinov, A.I., Zhykov, C., Stapnikova, N., Bibikova, E., Khorre, K., 1964. Geochronology of Indian Pre-Cambrian: 22nd International Geological Congress, Delhi (India), *10*, 553–567.

Wang, W., Zhou, M., 2014. Provenance and tectonic setting of the Paleo- to Mesoproterozoic Dongchuan Group in the southwestern Yangtze Block, South China: Implication for the breakup of the supercontinent Columbia. Tectonophysics *610*, 110–127.

Wang, L.J., Yu, J.H., Griffin, W.L., O'Reilly, S.Y., 2012. Early crustal evolution in the western Yangtze Block: Evidence from U-Pb and Lu-Hf isotopes on detrital zircons from sedimentary rocks. Precambrian Research *222–223*, 368–385.

Wang, W., Zhou, M., Zhao, X., Chen, W., Yan, D., 2014. Late Paleoproterozoic to Mesoproterozoic rift successions in SW China: Implication for the Yangtze Block-North Australia-Northwest Laurentia connection in the Columbia supercontinent. Sedimentary Geology *309*, 33–47.

Wang, W., Cawood, P.A., Zhou, M., Zhao, J., 2016. Paleoproterozoic magmatic and metamorphic events link Yangtze to northwest Laurentia in the Nuna supercontinent. Earth and Planetary Science Letters *433*, 269–279.

Wheeler, J.O., Hoffman, P.F., Card, K.D., Davidson, A., Sanford, B.V., Okulitch, A.V., Roest, W.R., 1996. Geological Map of Canada. Geological Survey of Canada, Map 1860A, scale 1:5 000 000.

Whitmeyer, S.J., Karlstrom, K.E., 2007. Tectonic model for the Proterozoic growth of North America. Geosphere *3*(4), 220–259. doi:10.1130/GES00055.1.

Wicander, R., Monroe, J.S., 2009. Historical Geology, 6th Edition. Cengage Learning, Australia.

Xiao, S., Knoll, A.H., Kaufmann, A.J., Yin, L., Zhang, Y., 1997. Neoproterozoic fossils in Mesoproterozoic rocks? Chemostratigraphic resolution of a biostratigraphic conundrum from North China Platform. Precambrian Research *84*, 197–220.

Xiong, Q., Zheng, J.P., Yu, C.M., Su, Y.P., Tang, H.Y., Zhang, Z.H., 2009. Zircon U–Pb age and Hf isotope of Quanyishang A-type granite in Yichang: Signification for the Yangtze continental cratonization in Paleoproterozoic. Chinese Science Bulletin *54*(3), 436–446.

Yin, L.M., 1997. Acanthomorphic acritarchs from Meso-Neoproterozoic shales of the Ruyang Group, Shanxi, China. Review of Palaeobotany and Palynology *98*, 15–25.

Ying, J.-F., Zhou, H.-H., Su, B.-X., Tang, Y.-J., 2011. Continental growth and secular evolution: Constraints from U-Pb ages and Hf isotope of detrital zircons in Proterozoic Jixian sedimentary section (1.8–0.8 Ga), North China Craton. Precambrian Research *189*, 229–238.

Young, G., Jefferson, C.W., Long, D.G.F., Delaney, G.D., Yeo, G.M., 1979. Middle and Late Proterozoic evolution of the northern Canadian Cordillera and Shield. Geology *7*, 125–128.

Zhai, M.G., Santosh, M., 2011. The early Precambrian odyssey of the North China Craton: A synoptic overview. Gondwana Research *20*, 6–25

Zhai, M.G., Santosh, M., 2013. Metallogeny of the North China Craton: Link with secular changes in the evolving Earth. Gondwana Research *24*, 275–297.

Zhai, M., Zhao, T., Peng, P., Meng, Q., 2015. Late Paleoproterozoic–Neoproterozoic multi-rifting events in the North China Craton and their geological significance: A study advance and review. Tectonophysics *662*, 153–166. doi:10.1016/j.tecto.2015.01.019.

Zhang, S.-B., Zheng, Y.-F., Wu, Y.-B., Zhao, Z.-F., Gao, S., Wu, F.-Y., 2006. Zircon U–Pb age and Hf–O isotope evidence for Paleoproterozoic metamorphic event in South China. Precambrian Research *151*(3–4), 265–288.

Zhanle, K., Claire, M., Catling, D., 2006. The loss of mass-independent fractionation in sulfur due to a Paleoproterozoic collapse of atmospheric methane. Geobiology *4*, 271–283.

Zhao, G., Cawood, P.A., Wilde, S.A., Sun, M., 2002. Review of global 2.1–1.8 Ga orogens: Implications for a pre-Rodinia supercontinent. Earth-Science Reviews *59*, 125–162.

Zhao, G.C., Sun, M., Wilde, S.A., Sanzhong, L., 2005. Late Archean to Paleoproterozoic evolution of the North China Craton: Key issues revisited. Precambrian Res *136*, 177–202.

Zhao, X.-F., Zhou, M.-F., Li, J.-W., Sun, M., Gao, J.-F., Sun, W.-H., Yang, J.-H., 2010. Late Paleoproterozoic to early Mesoproterozoic Dongchuan Group in Yunnan, SW China: Implications for tectonic evolution of the Yangtze Block. Precambrian Research *182*, 57–69.

Zhao, G.C., Cawood, P.A., Li, S.Z., Wilde, S.A., Sun, M., Zhang, J., He, Y.; Yin, C., 2012. Amalgamation of the North China Craton: Key issues and discussion. Precambrian Res *222–223*, 55–76.

5

Chemostratigraphy of the Mesoproterozoic-Neoproterozoic Transition

Juan Carlos Silva-Tamayo[1,2], Nova Giovanny[3], and Karol Tatiana Dussan-Tapias[4]

ABSTRACT

Geochronologic and C and Sr chemostratigraphic information of late Neoproterozoic-early Mesoproterozoic marine carbonate successions are used to propose a reference C and Sr chemostratigraphic framework for the Mesoproterozoic-Neoproterozoic transition. While late Mesoproterozoic marine carbonates display decreasing $\delta^{13}C$ values, from ~4 to ~−2‰, carbonates from the Mesoproterozoic-Neoproterozoic transition display a positive C isotope anomaly, from $\delta^{13}C$ values ~−2‰ to values ~+2‰, followed by a subsequent decrease to $\delta^{13}C$ values ~−1‰. This decrease in $\delta^{13}C$ values is followed by a new increase to predominantly positive $\delta^{13}C$ values in the early Neoproterozoic. The reference chemostratigraphic framework also suggests that late Mesoproterozoic carbonate successions display predominantly lower $^{87}Sr/^{86}Sr$ values than early Neoproterozoic carbonates.

The late Mesoproterozoic oceans registered an increase in the marine multicellular/sexual eukaryotic photosynthetic life, which parallels an increase in $\delta^{53}Cr$ and $\delta^{34/32}S$ values of several late Mesoproterozoic carbonates. Although these data suggest an increase in the atmospheric oxygen levels and intensified oxidative weathering, the occurrence of unfractionated $\delta^{98/95}Mo$ values found in a single carbonate succession (Vazante Group) suggests in contrast an oxygen-depleted ocean. This incongruence in the isotopic records of redox-sensitive elements may be related to a nonsteady-state late Mesoproterozoic Mo isotope cycle.

5.1. INTRODUCTION

The Mesoproterozoic-Neoproterozoic transition (1.3–0.8 Ga) was a period of major revolutions in the global biogeochemical cycles [*Kah et al.*, 2004; *Gellatly and Lyons*, 2005; *Kah and Bartley*, 2011; *Roberts*, 2013; *Sheldon*, 2013]. This time interval is characterized by a major change in the marine carbon cycle, evidenced by the onset of Phanerozoic like marine carbon isotope fluctuations, after a period of ca. 700 My of apparent quiescence in the carbon cycle dynamics [*Kah et al.*, 1999, 2001, 2012; *Frank et al.*, 2003; *Bartley and Kah*, 2004; *Bartley et al.*, 2015]. This adjustment of the marine

carbon cycle, which would have resulted from the decoupling of the marine inorganic and organic carbon pools, coincides with an increase in marine and atmospheric oxygen levels, as evidenced by the increased isotope fractionations of redox-sensitive elements in the global oceans [*Kah et al.*, 2004; *Gellatly and Lyons*, 2005; *Kah and Bartley*, 2011]. This increase, fueled by the oxygenation of the upper surface water column, paralleled the advent of marine multicellular/sexual eukaryotic photosynthetic life [*Gibson et al.*, 2018]. This important interval in the history of the Earth system also witnessed a significant decrease in the atmospheric pCO_2 concentrations, which likely resulted from an enhanced atmosphere-continent interaction, during a period of major global plate tectonic reconfiguration, that is, the agglutination of supercontinent Rodinia [*Frank et al.*, 2003].

Assessing the significance and temporality of the geologic and geochemical evidences of the abovementioned

---

[1] *Antonio Nariño University, Bogota, Colombia*

[2] *Tetslab Geoambiental, Medellin, Colombia*

[3] *Corporación Geológica Ares, Bogotá, Colombia*

[4] *Universidad de Caldas, Manizales, Colombia*

Chemostratigraphy Across Major Chronological Boundaries, Geophysical Monograph 240, First Edition.
Edited by Alcides N. Sial, Claudio Gaucher, Muthuvairavasamy Ramkumar, and Valderez Pinto Ferreira.

adjustments of the global marine biogeochemical cycles has remained challenging due to (i) the scarcity of well-dated marine sedimentary records and (ii) the limited availability of paleontologic and geochemical data. In this contribution, we summarize the available geochronologic (U-Pb, Rb-Sr, and Re-Os ages) and C and Sr chemostratigraphic information from several Mesoproterozoic-Neoproterozoic marine sedimentary records from which we propose refined C and Sr chemostratigraphic frameworks for the Mesoproterozoic-Neoproterozoic transition. We use this refined framework to discuss the available multi-proxy geochemical evidences of changes in ocean oxygen levels during the Mesoproterozoic-Neoproterozoic transition and to assess how those changes are related to the advent of sexually reproduced photosynthetic eukaryotic life. We finally report Mo isotope compositions from a well-dated marine Mesoproterozoic-Neoproterozoic succession in central Brazil to contribute to the quantification of the marine oxygen levels during the Mesoproterozoic-Neoproterozoic transition and ultimately to relate changes in ocean oxygenation with the initiation of photosynthetic eukaryotic life.

5.2. GEOCHRONOLOGIC AND CHEMOSTRATIGRAPHIC INFORMATION

Mesoproterozoic-Neoproterozoic marine sedimentary records are scarce and geographically limited to some remote cratonic areas of South America, West Africa,

Siberia, Svalbard, South China, and Australia (Fig. 5.1). These sedimentary records usually display poorly constrained age-diagnostic fossils. Thus, determinations of their depositional ages have commonly relied on the use of field relationships between the marine sedimentary records and other well-dated lithologies, that is, crosscutting intrusives, and C and Sr isotope chemostratigraphic correlations [*Bartley et al.*, 2001; *Azmy et al.*, 2008; *Rooney et al.*, 2010; *Kah et al.*, 2012; *Geboy et al.*, 2013; *Bertoni et al.*, 2014]. These age determinations and correlations have, however, remained to some extent contentious due to the scarcity of absolute (illite/glauconite Rb-Sr and organic-rich black shale Re-Os geochronology) and provenance-directed (detrital zircon U-Pb geochronology) geochronologic data.

Figure 5.1 shows the paleogeographic location of some of the Mesoproterozoic-Neoproterozoic marine sedimentary records for which direct and indirect age constraints are available. The marine sedimentary records from the Uchur-Maya region and the Turukhansk Uplift, southeastern Siberia, are probably some of the most studied Mesoproterozoic-Neoproterozoic sedimentary records to date (Figs. 5.1 and 5.2). *Bartley et al.* [2001] used field relationships, lithostratigraphy, and C and Sr isotope chemostratigraphy to date and correlate those marine sedimentary rocks and to construct the first reference C and Sr isotope chemostratigraphic framework for the Mesoproterozoic-Neoproterozoic transition (Fig. 5.2). According to these authors, the upper Mesoproterozoic

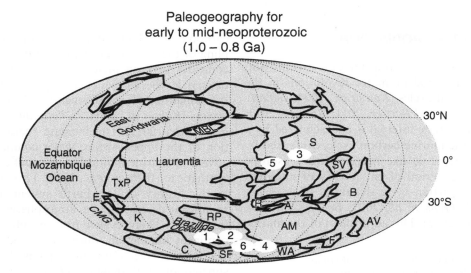

Figure 5.1 Paleogeographic map of Rodinia at 1.0 Ga and paleogeographic location of Arequipa Massif (A), Amazonian Craton (AM), Avalonia (AV), Baltica (Russian craton) (B), Congo Craton (C), Coats Land-Maudheim-Grunehogna province of East Antarctica (CMG), Ellsworth-Whitmore Mountains block (in Pangea position) (E), Florida (in pre–Pangea position within Gondwana and including the Carolina terrane) (F), Falkland-Malvinas Plateau (F/MP), Kalahari Craton (K), Marie Byrd Land (MBL), New Guinea (NG), Rockall Plateau with adjacent northwest Scotland and northwest Ireland (R), Rio de la Plata Craton (RP), Siberia (Angara Craton) (S), São Francisco Craton (SF), Svalbard block (Barentia) (SV), West African Craton (WA), and hypothetical Texas plateau (TxP). Numbers correspond to carbonate successions: (1) Vazante Group, (2) Paranoá Group, (3) Turukhansk Group, (4) Atar Group, (5) Bylot Supergroup, and (6) São Caetano Complex.

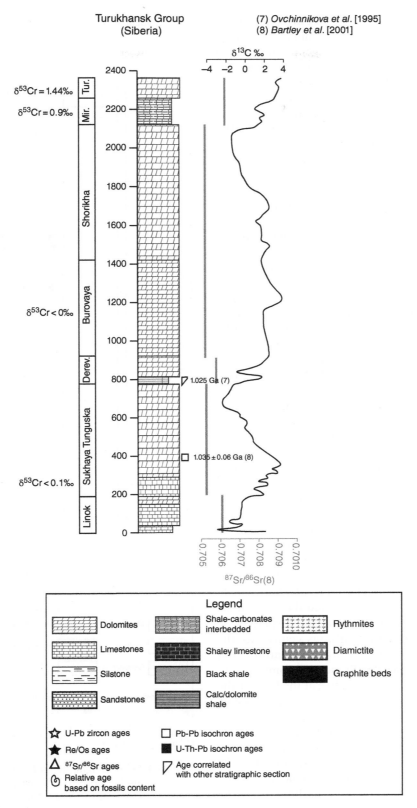

Figure 5.2 Lithostratigraphy and C isotope chemostratigraphy of the Turukhansk Group, Turukhansk Uplift, Siberia (After *Bartley et al.* [2001]). ^{87}Sr/^{86}Sr inferred ages (After *Bartley et al.* [2001]). Pb-Pb ages (After *Ovchinnikova et al.* [1995]). δ^{53}Cr data (After *Gilleaudeau et al.* [2016]).

marine carbonates (Sukhaya Tunguska Formation) from the Turukhansk Group display predominantly positive $\delta^{13}C$ values (up to +4‰). The least diagenetically altered $^{87}Sr/^{86}Sr$ isotope values displayed by carbonates from this formation average 0.70532. These values are slightly lower than those of the underlaying Linok Formation (0.70601). The positive $\delta^{13}C$ values are interrupted by a shift toward negative values (as low as −3.5‰) in the uppermost Mesoproterozoic carbonates (uppermost Sukhaya Tunguska Formation). These negative $\delta^{13}C$ values are followed by a short positive shift to $\delta^{13}C$ values as high as +2‰ in the Neoproterozoic Derevuya Formation, which display $^{87}Sr/^{86}Sr$ values averaging 0.70584. *Bartley et al.* [2001] placed the Mesoproterozoic-Neoproterozoic transition in the nadir of the $\delta^{13}C$ values at the base of the Derevuya Formation and suggested that the carbonate records from the Buroyava, Shorikha, Miroyedikha, and Turukhansk formations are early Neoproterozoic in age. These carbonates display predominantly positive $\delta^{13}C$ values (up to +4‰) which are only interrupted by a negative excursion to $\delta^{13}C$ ~−2.0‰ at the top of the Shorikha Formation. These formations display increasing $^{87}Sr/^{86}Sr$ values from 0.70523 to 0.70622 (Fig. 5.2).

Another well-studied and well-dated Mesoproterozoic-Neoproterozoic marine sedimentary record is the Vazante Group, São Francisco Basin, Brasilia Deform Belt, Central Brazil [*Azmy et al.*, 2008; *Misi et al.*, 2011, 2014; *Geboy et al.*, 2013] (Fig. 5.3). The upper Mesoproterozoic carbonates (lower half of the Serra do Poço Verde Formation) display $\delta^{13}C$ values between +1 and +4‰. These $\delta^{13}C$ values become negative (between 0 and −2.0‰) toward the top of the Serra do Poço Verde Formation and evolve to positive $\delta^{13}C$ values (~+3.5‰) in the overlaying Morro do Calcário Formation [*Azmy et al.*, 2008; *Geboy et al.*, 2013]. The Morro do Pinheiro and Pamplona formations, which are correlatives to the Serra do Poço Verde Formation, display $^{87}Sr/^{86}Sr$ values between 0.70614 and 0.714138 [*Azmy et al.*, 2001, 2008]. These authors, however, only considered the least radiogenic $^{87}Sr/^{86}Sr$ values (0.70614–0.70734) displayed by microbial aggregates and from fibrous cement as the least diagenetically altered Sr isotope signatures and thus as reflecting the seawater Sr isotope composition. The $\delta^{13}C$ values become again negative at the base of the Lapa Formation (~−8‰), right above the glaciogenic deposits that separate the Morro do Calcário and Lapa formations. The $\delta^{13}C$ values shift back to positive values (~+1.5‰) high in the stratigraphy of the Lapa Formation, before a new negative excursion in the $\delta^{13}C$ values (to ~−4.0‰) occurs. *Azmy et al.* [2008] reported $^{87}Sr/^{86}Sr$ values averaging 0.70684 for carbonates of the Lapa Formation.

The $\delta^{13}C$ pathways displayed by carbonates from the Serra do Poço Verde and Morro do Calcário formations

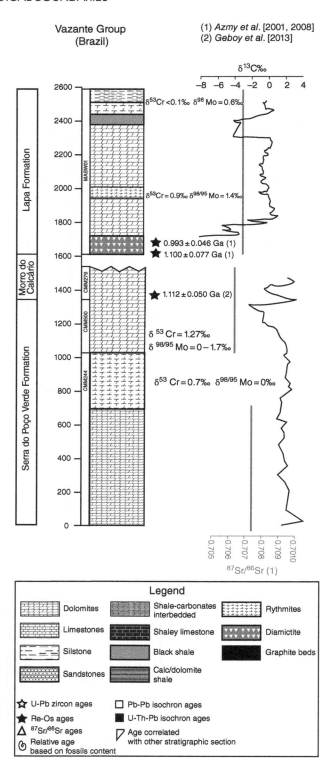

Figure 5.3 Lithostratigraphy and C isotope chemostratigraphy of the Vazante Group, São Francisco Basin, Brasilia Deformed Belt, Brazil (After *Azmy et al.* [2001, 2008]). Re-Os ages (After *Geboy et al.* [2013]). $\delta^{53}Cr$ data (After *Gilleaudeau et al.* [2016]). $\delta^{98/95}Mo$ data (this study).

are similar to those displayed by carbonates from the upper-most Mesoproterozoic Sukhaya Tunguska Formation and lower Neoproterozoic Derevuya Formation in southeastern Siberia, respectively [*Bartley et al.*, 2001]. This similarity suggests that these two records are correlative, that the Mesoproterozoic-Neoproterozoic boundary is within the Morro do Calcário Formation, and that the glaciogenic deposits that separate the Morro do Calcário and Lapa formations were deposited in the early Neoproterozoic. Although a 1.1 ± 0.05 Ga Re-Os age has been obtained for the glaciogenic deposits that overlay the Morro do Calcário Formation [*Azmy et al.*, 2008; *Geboy et al.*, 2013], the uncertainties on the Re-Os ages reported by these authors warrant caution when using this age for correlations.

To date, only one additional early Neoproterozoic glaciogenic sedimentary record has been reported. This record belongs to the Paranoá Group which discordantly overlies the basement of the São Francisco Craton (Fig. 5.4). These glaciogenic deposits are overlain by a series of dolomitic stromatolite-rich carbonates with interbedded sandstones and siltstones. U-Pb and Lu-Hf ages of detrital zircon from some sandstones of those interbedded sandstones suggest maximum depositional ages between 1.04 and 1.5 Ga for the Paranoá Group [*Alvarenga et al.*, 2014]. These maximum depositional ages along with similarities in the C isotope chemostratigraphic pathways displayed by the Lapa Formation and the Paranoá Group (Fig. 5.3) allow to propose that these units are synchronic and deposited during the early Neoproterozoic after an interval of local glaciogenic deposition along the southwestern part of the São Francisco Craton. The dolomitic stromatolitic carbonates from the Paranoá Group display $^{87}Sr/^{86}Sr$ values ranging between 0.70562 and 0.70677 [*Alvarenga et al.*, 2014]. These values are similar to those displayed by the Lapa Formation and slightly lower than those displayed by carbonates from the overlaying Bambuí Group which range between 0.707451 and 0.70839 [*Alvarenga et al.*, 2014].

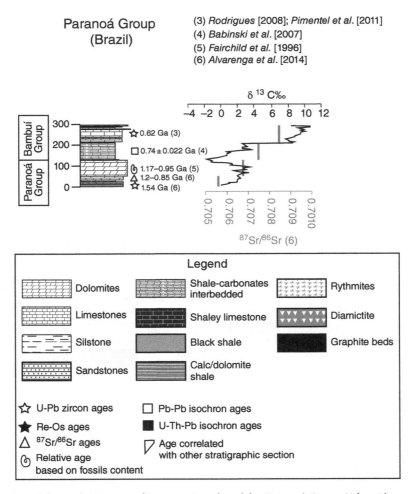

Figure 5.4 Lithostratigraphy and C isotope chemostratigraphy of the Paranoá Group (After *Alvarenga et al.* [2014]). Detrital zircon U-Pb ages (After *Rodrigues* [2008] and *Pimentel et al.* [2011]). Pb-Pb ages (After *Babinski et al.* [2007]). $^{87}Sr/^{86}Sr$ inferred ages (After *Alvarenga et al.* [2014]). Biostratigraphic ages (After *Fairchild et al.* [1996]).

Mesoproterozoic-Neoproterozoic sedimentary successions have been also reported in the Taoudeni Basin in West Africa (Atar/El Mreiti groups). The restricted marine carbonate-evaporite successions from the Atar Group are underlain by a fluvial-coastal (eolian) siliciclastic succession from the Char Group (Fig. 5.5) [*Kah et al.,* 2012]. These Atar Group carbonate-evaporite successions are overlain by shallow to deep marine siliciclastics pertaining to the Assabet el Hassiane Group. Regionally, the Atar Group is correlated with the El Mreiti Group, which mainly consists of marine carbonates and shales [*Kah et al.,* 2012; *Gilleaudeau and Kah,* 2013]. Recently, a Mesoproterozoic (1.1 Ga) depositional age has been proposed for the Atar and El Mreiti groups based on Re-Os geochronology [*Rooney et al.,* 2010]. This depositional age suggests that the Atar/El Mreiti groups are correlatives of the Serra do Poço Verde/Morro do Calcário formations (Vazante Group) and the Sukhaya Tunguska Formation (Turukhansk Group). This correlation is further supported by the similar C isotope chemostratigraphic

pathways displayed by these marine sedimentary successions (Fig. 5.6) [*Azmy et al.,* 2008; *Kah et al.,* 2012].

These late Mesoproterozoic-early Neoproterozoic C isotope chemostratigraphic pathways are also displayed by carbonates from the Bylot Supergroup in Arctic Canada (Fig. 5.7) [*Kah et al.,* 1999, 2001]. Recent Re-Os geochronologic constraints suggest that the Victor Bay Formation of the Uluksan Group, Bylot Supergroup, deposited after 1.046 ± 0.016 Ga [*Gibson et al.,* 2018]. We use this age constraint and the δ¹³C values displayed by the Bylot Supergroup to suggest that the Mesoproterozoic-Neoproterozoic transition occurs within the lowermost carbonates of the Victor Bay Formation (Figs. 5.5 and 5.6). This implies that the Society Cliffs Formation was deposited during the latest Mesoproterozoic. This depositional age is supported by the ⁸⁷Sr/⁸⁶Sr values ranging between 0.70540 and 0.70600 [*Kah et al.,* 2001]. These values fall within the range of values displayed by marine carbonates deposited within the Mesoproterozoic-Neoproterozoic transition.

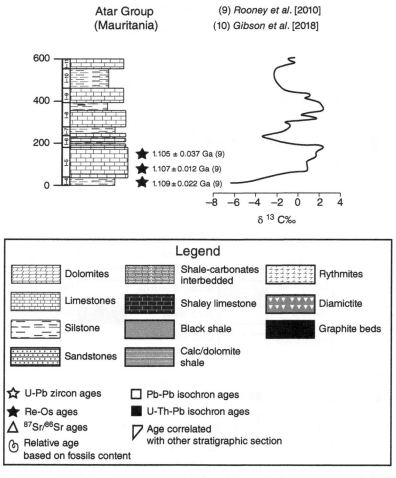

Figure 5.5 Lithostratigraphy and C isotope chemostratigraphy of the Atar Group (After *Kah et al.* [2012]). Re-Os age (After *Rooney et al.* [2010]).

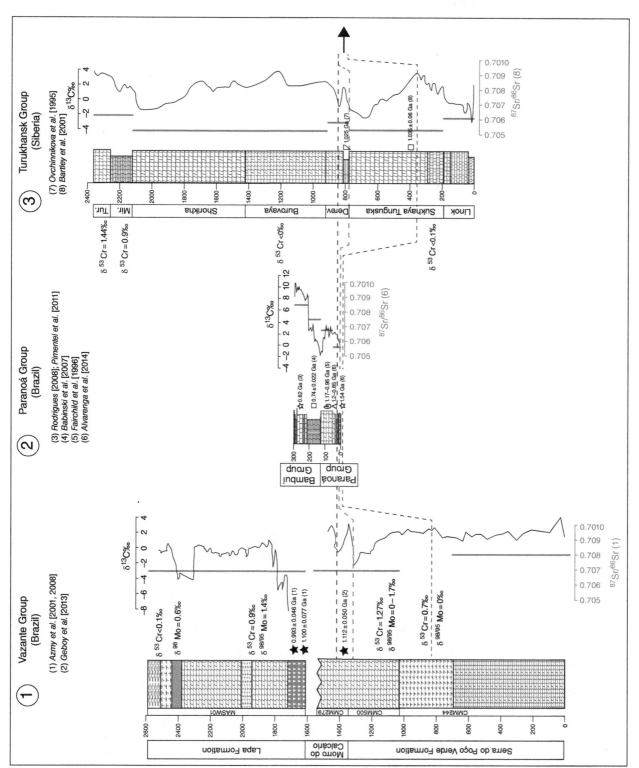

Figure 5.6 Lithostratigraphic and C isotope chemostratigraphic correlation of several Mesoproterozoic-Neoproterozoic carbonate successions. Note how the δ^{13}C values decrease from ~4‰ at 1.1 Ga to ~2‰ during the latest Mesoproterozoic. These values increase again to ~4‰ just before the Mesoproterozoic-Neoproterozoic boundary when they decrease to values below 0‰. The δ^{13}C values increase again in the Neoproterozoic to values as high as ~4‰ and remain positive. Note the occurrence of an early Neoproterozoic discrete glacial event. This event was restricted to the São Francisco Craton as evidenced by the presence of glaciogenic deposits in the Vazante and Paranoá groups.

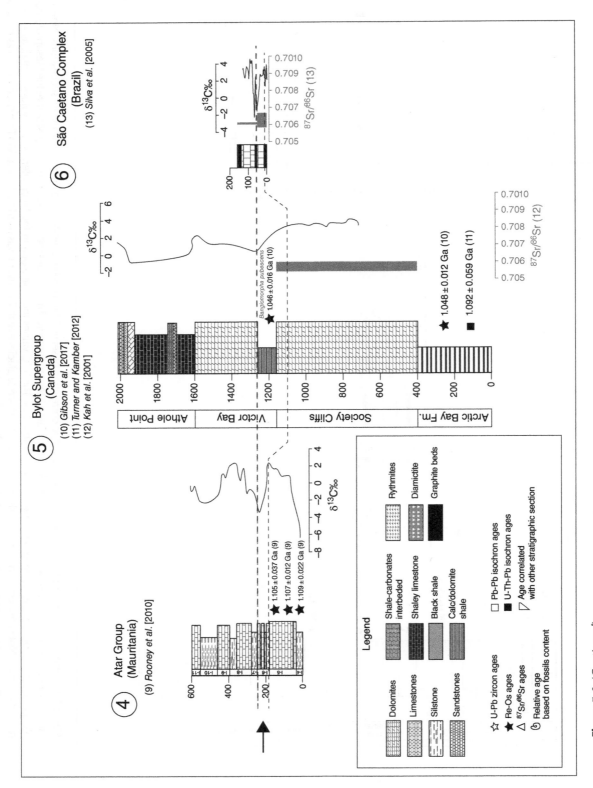

Figure 5.6 (*Continued*)

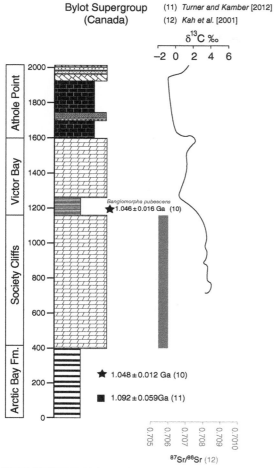

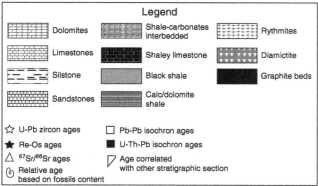

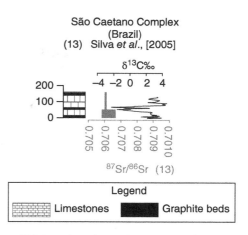

Figure 5.8 Lithostratigraphy and C isotope chemostratigraphy of the São Caetano Group, Transversal Zone, Borborema Province. After *Silva-Tamayo et al.* [2004].

Group from southeastern Siberia (Figs. 5.5 and 5.7). The Mesoproterozoic marbles from the São Caetano Complex display increasing $^{87}Sr/^{86}Sr$ values between 0.705801 and 0.706732, while the Neoproterozoic ones display $^{87}Sr/^{86}Sr$ values averaging 0.706054 (Fig. 5.8). These values further support a Mesoproterozoic-Neoproterozoic depositional age for the sedimentary protoliths of the São Caetano Complex marbles.

5.3. AVAILABLE GEOCHEMICAL EVIDENCE OF MARINE OXYGEN LEVELS

The C isotope compositions of Mesoproterozoic carbonate successions suggest two contrasting behaviors of the marine C cycle during this time interval. Before 1.3 Ga, the occurrence of carbonate $\delta^{13}C$ values near to 0‰ and the absence of isotopic fluctuations (Fig. 5.9) points toward a paucity of the marine C cycle, which was mainly favored by the occurrence of a high dissolved inorganic carbon (DIC) marine pool and thus a high carbonate supersaturation in the global ocean, both resulting from elevated pCO_2 and high alkalinity delivery from continent to oceans [*Bartley and Kah*, 2004]. After about 1.3 Ga, the average carbonate $\delta^{13}C$ value increases to a background around 4‰. This increase has been related to the decrease in the DIC pool in the ocean, the decoupling of the inorganic and organic carbon pools, and the reduction in atmospheric pCO_2 [*Bartley and Kah*, 2004].

The decrease in DIC coincides with an increase in oceanic oxygen levels [*Kah et al.,* 2004]. The increase in biospheric oxygenation has been suggested based on the rise in the $\delta^{34}S$ values of evaporites and carbonate-associated sulfates displayed by late Mesoproterozoic marine carbonate successions (Society Cliffs Formation, *Kah et al.,* 2004). The rise in $\delta^{34}S$ values implies an increase in the marine sulfate concentrations and thus in

Figure 5.7 Lithostratigraphy and C isotope chemostratigraphy of the Bylot Supergroup (After *Kah et al.* [1999, 2001]). Re-Os ages (After *Gibson et al.* [2018]). U-Pb ages (After *Turner and Kamber* [2012]). $^{87}Sr/^{86}Sr$ inferred ages (After *Kah et al.* [2001]).

Finally, the Mesoproterozoic-Neoproterozoic transition has been also reported in marble successions (São Caetano Complex) cropping out along the Transversal Zone of the Borborema Province, north of the São Francisco Craton, Brazil [*Silva-Tamayo et al.,* 2004]. These amphibolite facies marble successions display $\delta^{13}C$ values similar to those reported by *Bartley et al.* [2001] for the Turukhansk

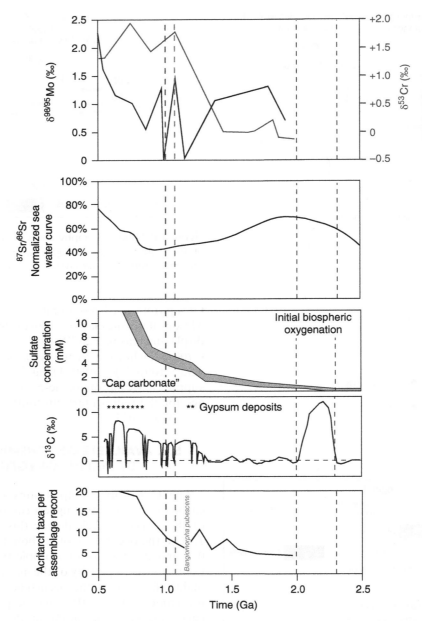

Figure 5.9 Evolution of the C, Sr, Cr, and Mo isotope composition of the seawater during the 2.5–0.5 Ga interval. Note how the increase in the variability of the seawater $\delta^{13}C$ values in the late Mesoproterozoic coincides with an increase in the seawater $\delta^{98/95}Mo$ and $\delta^{53}Cr$ values [*Gilleaudeau et al.*, 2016, this work]. This increase continues toward the Neoproterozoic, suggesting a continued increase in the ocean oxygen levels. The increase in the ocean oxygen levels implies an increase in oxidative weathering and thus an increase in atmospheric pO_2. The increase in atmospheric and oceanic oxygen levels is supported by the increase in the sulfate concentrations in the global oceans [*Kah et al.*, 2004]. This increase in biospheric oxygenation coincides with the increase in eukaryotic life as suggested by the increase in acritarch diversity [*Knoll et al.*, 2006]. Note how the first appearance of *Bangiomorpha pubescens* at 1.045 Ga [*Gibson et al.*, 2018] coincides with the highest $\delta^{53}Cr$ values in the whole Mesoproterozoic. This suggests that the appearance of marine multicellular/sexual photosynthetic eukaryotic life also coincides with an increase in marine oxygenation. Normalized $^{87}Sr/^{86}Sr$ curve after *Shields* [2007].

atmospheric oxygen concentrations [*Kah et al.*, 2004]. The late Mesoproterozoic increase in oceanic and atmospheric oxygenation is also supported by the increasing and positive $\delta^{53}Cr$ values displayed by carbonates from the Morro do Calcário Formation of the Vazante Group (Brazil) and the correlative Turukhansk (Siberia) and El Mreiti (West Africa) groups (Fig. 5.6; *Guilleaudeau et al.*, 2016).

Iron speciation data from the Victor Bay Formation of the Bylot Supergroup suggest the occurrence of predominantly anoxic bottom water redox conditions [*Scott*, 2009]. The iron speciation data from black shales interbedded with the Mesoproterozoic Atar Group (units I-4 and I-5) and the correlative El Mreiti (Tourist and En Neosoar formations) suggest, instead, the occurrence of predominantly euxinic conditions [*Gilleaudeau and Kah*, 2013]. Iron speciation data of the latest Proterozoic Morro do Calcário Formation of the Vazante Group suggest, in turn, fluctuating bottom water redox conditions, from oxic to sulfidic [*Geboy et al.*, 2013]. The differences in iron speciation data suggest that heterogeneous shallow water bottom water redox conditions occurred during the 1.2–1.04 Ga interval. These conditions contrast with the predominantly anoxic-ferruginous conditions occurring during the latest Mesoproterozoic-early Neoproterozoic (1.04–0.81 Ga) interval [*Guilbaud et al.*, 2015]. The contrasting bottom water redox conditions inferred from the available iron speciation data suggest that the global oceans were chemically stratified during 1.2–1.04 Ga interval and slightly more homogeneous and oxygenated during the 1.04–0.84 Ga interval [*Guilbaud et al.*, 2015].

5.4. MO ISOTOPE EVIDENCE OF OCEAN OXYGENATION

Although most of the available geochemical data suggest an increase in ocean oxygen levels during the Mesoproterozoic-Neoproterozoic transition, quantification of the extent of marine oxygenation during this interval of eukaryotic life revolution is lacking. Molybdenum isotope geochemistry has been extensively used to reconstruct changes in ocean Mo isotope compositions and to investigate the history of marine and atmospheric oxygen levels [*Kendall et al.*, 2017 and references therein]. The use of Mo isotopes as paleoredox proxies has generally relied on the analysis of euxinic (sulfidic) marine shales [*Kendall et al.*, 2017 and references therein]. However, Mo isotope compositions have also been successfully applied to inorganic marine carbonates to investigate changes in ocean Mo isotope compositions [*Voegelin et al.*, 2009]. Thus, these compositions can be also used, similar to euxinic black shales, to investigate the evolution of ocean oxygenation.

Voegelin et al. [2009] suggested that the $\delta^{98/95}$Mo values of inorganic carbonates can be affected if the authigenic [Mo] of the inorganic carbonate fraction is substantially lower than that of its siliciclastic fraction. In this contribution, we report detrital corrected Mo isotope compositions of carbonates from the Vazante Group and used them to further contribute to the investigation of the oceanic oxygen levels during the Mesoproterozoic-Neoproterozoic transition (see notes for details in laboratory and geochemical

methods; refer to *Voegelin et al.* [2009] for the detrital Mo isotope correction). Samples were collected in stratigraphic well MAF 4288 from Votorantim Group [*Geboy et al.*, 2013].

Inorganic marine cements from carbonates from the Serra do Poço Verde and Morro do Calcário formations display $\delta^{98/95}$Mo composition between 0 and 1.7‰ (Table 5.1). When corrected for the detrital Mo contribution, no substantial change in the $\delta^{98/95}$Mo values can be inferred. Carbonates from the base of the overlying Neoproterozoic Lapa Formation display $\delta^{98/95}$Mo increasing from 0.5 to 1.4‰ (Fig. 5.3 and Table 5.1). These values subsequently decrease to ~0.3‰. The detrital corrected $\delta^{98/95}$Mo values range from 0.5 to 1.6‰ in the lower part of the Lapa Formation and average 0.6‰ in its upper part (Fig. 5.3 and Table 5.1). The Serra do Poço Verde and Morro do Calcário formations display [Mo] below 0.6 ppm, while those of the Lapa Formation display [Mo] averaging 0.16 ppm (Table 5.1). These concentrations are higher than those of Phanerozoic and modern inorganic marine oolites [*Voegelin et al.*, 2009].

The lack of detrital Mo isotope contribution to the Mo isotope compositions of the studied carbonates suggests that their $\delta^{98/95}$Mo values reflect those from the seawater from which they precipitated. Most of the $\delta^{98/95}$Mo values of the late Mesoproterozoic Serra do Poço Verde and Morro do Calcário formations average the range of values displayed by the main unfractionated Mo isotope input to the ocean (continental crust, $\delta^{98/95}$Mo = 0). These unfractionated Mo isotope values for the inferred ancient seawater at the time of carbonate deposition can be interpreted in different ways. One possibility is that they were the result from a reduced incorporation Mo isotope into Fe-Mn oxides (oxic Mo sink) in the global oceans. This possibility would imply that the 1.2–1.04 Ga oceans were extremely depleted in dissolved oxygen. In this context, the occurrence of punctuated $\delta^{98/95}$Mo values ~1.7‰ displayed by the Morro do Calcário Formation would imply the potential occurrence of a rapid but temporary increase in ocean oxygen level during the latest Mesoproterozoic. The Morro do Calcário Formation is correlative to the Victor Bay Formation from the Bylot Supergroup in the Canadian Arctic. The latter registered the first appearance of marine multicellular/sexual photosynthetic eukaryotic life *Bangiomorpha pubescens* in the latest Mesoproterozoic [*Gibson et al.*, 2018]. The increase in photosynthetic eukaryotic life is at odds with an extremely oxygen-deficient ocean.

An alternative scenario to explain the occurrence of unfractionated marine $\delta^{98/95}$Mo values during late Mesoproterozoic is the occurrence of a low marine Mo reservoir. Under a slightly oxygenated atmosphere, the input of continental Mo into the ocean would have been extremely low and would display unfractionated $\delta^{98/95}$Mo

Table 5.1 [Mo], [Al], and $\delta^{98/95}$ Mo of Carbonates from the Vazante Group.

Stratigraphic unit	Sample	$\delta^{98/95}$ Mo	[Mo] ppm	[Al] ppm	Mo authigenic	$\delta^{98/95}$ Mo*
Lapa	MAF 4288 638.55	0.20	0.25	156	0.25	0.20
	MAF 4288 643.75	0.31	0.10	147	0.10	0.31
	MAF 4288 648.15	0.41	0.10	123	0.10	0.41
	MAF 4288 650.90	0.34	0.10	154	0.10	0.34
	MAF 4288 658.65	0.60	0.10	235	0.10	0.61
	MAF 4288 659.65	0.45	0.10	234	0.10	0.45
	MAF 4288 663.10	0.50	0.10	532	0.09	0.51
	MAF 4288 667.25	0.40	0.10	635	0.09	0.40
	MAF 4288 671.10	0.50	0.20	855	0.19	0.51
	MAF 4288 672.25	0.90	0.20	136	0.20	0.90
	MAF 4288 677.70	1.20	1.50	561	1.49	1.20
	MAF 4288 678.80	1.40	2.00	119	2.00	1.40
	MAF 4288 679.70	1.10	1.35	905	1.34	1.11
	MAF 4288 681.60	1.00	2.10	414	2.09	1.00
	MAF 4288 682.60	0.50	0.70	822	0.69	0.50
Morro do Calcário	MAF 4288 750.20	0.00	0.80	633	0.79	0.00
	MAF 4288 755.60	0.10	0.01	134	0.01	0.04
	MAF 4288 765.75	0.00	0.13	252	0.13	−0.01
	MAF 4288 801	0.00	0.62	340	0.62	0.00
	MAF 4288 806.30	0.00	0.34	289	0.34	0.00
	MAF 4288 810.86	0.80	0.49	308	0.48	0.80
	MAF 4288 814.55	1.70	0.30	194	0.29	1.71
	MAF 4288 817.30	1.50	0.45	244	0.45	1.51
Serra do Poço Verde	MAF 4288 908.05	0.00	0.09	258	0.09	−0.01
	MAF 4288 909.75	0.00	0.08	257	0.08	−0.02

Refer to *Voegelin et al.* [2009] for details on detrital correction. Reproduced with permission of Elsevier.
Asterisk denotes the detrital corrected $\delta^{98/95}$ Mo values of carbonates.

values when compared to the average continental crust (main Mo input to the ocean). The assemblage of supercontinent Rodinia resulted in the occurrence of large epicontinental seas [*Guilleaudeau and Kah*, 2013]. Iron speciation data suggest that the shallow epicontinental seas were predominantly anoxic/euxinic [*Guilleaudeau and Kah*, 2013]. The pervasive anoxic-ferruginous conditions that predominated along the late Mesoproterozoic epeiric seas resulted in a rapid and quantitative Mo removal from the seawater and thus in a global ocean characterized by $\delta^{98/95}$Mo = 0‰ and low [Mo]. This interpretation is in line not only with the $\delta^{98/95}$Mo ~0‰ values of the Serra do Poço Verde and Morro do Calcário formations but also with the low [Mo] in Mesoproterozoic euxinic/anoxic black shales from the El Mreiti Group, which has been interpreted as resulting from an enhanced marine Mo output by shallow marine anoxic shales along epeiric seas [*Guilleaudeau and Kah*, 2013].

Although the above scenario implies extremely oxygen-deficient epeiric seas, it does not necessarily imply a widespread (global) anoxic ocean. The Morro do Calcário Formation displays some punctuated $\delta^{98/95}$Mo values ~1.7‰. Under the depleted Mo ocean scenario, any rise in the input of continental Mo isotope to the

global oceans would have increased the oceanic Mo availability and favored the Mo incorporation into marine oxic sediments; this latter process potentially imparted the occurrence of fractionated seawater Mo isotope compositions. Alternatively, if the Mo input of the global oceans remained constant, the punctuated fractionated seawater $\delta^{98/95}$Mo values inferred from the Morro do Calcário Formation would imply the nonquantitative removal of oceanic Mo along the global epeiric seas. This latter possibility is most plausible and would imply punctuated increases in marine oxygen levels. The potential temporal/transient increases in marine oxygen levels would have paralleled the advent of marine multicellular/sexual photosynthetic eukaryotic life.

It is important to highlight that the low late Mesoproterozoic inferred seawater $\delta^{98/95}$Mo values (unfractionated with respect to continental crust) contrast with the positively fractionated seawater δ^{53}Cr values (Fig. 5.9; *Gilleaudeau et al.*, 2016). This difference may be explained by a nonquantitative removal of Cr along the global anoxic epeiric seas [*Frei et al.*, 2009]. This nonquantitative removal results in preferential incorporation of ^{52}Cr into anoxic sediments and a residual seawater enriched in ^{53}Cr. Although a nonquantitative removal of oceanic Mo would also result in positively fractionated seawater $\delta^{98/95}$Mo values, our data suggest that the oceanic Mo was preferentially and quantitatively removed from the oceans. More data are needed to further understand the contradictory Mo and Cr isotope data.

Finally, the $\delta^{98/95}$Mo values of the Neoproterozoic Lapa Formation are systematically positively fractionated relative to that of the continental Mo reservoir. Two potential scenarios can explain those positively fractionated $\delta^{98/95}$Mo values. A first scenario is the increase in the oxic Mo sequestration in the global ocean. This scenario implies that the main Mo source to the global ocean was similar to that of the Mesoproterozoic oceans [*Kendall et al.*, 2017]. An alternative possibility is that the Mo isotope composition of the main input to the global ocean was positively fractionated with respect to the continental crust. This scenario would imply a substantial increase in oxidative chemical weathering of the continental crust and thus an increase in atmospheric oxygen levels. The resulting associated increase in marine oxygen levels would have further favored the occurrence of positively fractionated seawater $\delta^{98/95}$Mo and δ^{53}Cr values (Fig. 5.9).

5.5. SUMMARY

Available geochronologic and C and Sr chemostratigraphic data of late Neoproterozoic-early Mesoproterozoic marine sedimentary successions allow constructing a reference C and Sr chemostratigraphic framework for the Mesoproterozoic-Neoproterozoic transition. The new chemostratigraphic framework suggests that late Mesoproterozoic marine carbonates display decreasing δ^{13}C values, from ~4 to ~−2‰. Carbonates from the Mesoproterozoic-Neoproterozoic transition display a positive C isotope anomaly, from δ^{13}C values ~−2 to values ~+2‰, followed by a subsequent decrease to δ^{13}C values ~−1‰. A new increase to predominantly positive δ^{13}C values characterizes the early Neoproterozoic carbonate successions worldwide. This increase in δ^{13}C values occurs after a period of icehouse conditions, as suggested by the occurrence of discrete glaciogenic deposits underlying the Vazante and Paranoá groups. The occurrence of these glacial deposits is restricted to the São Francisco Craton, which was paleogeographically located at highest (southern) latitudes compared to other studied Mesoproterozoic-Neoproterozoic sedimentary records. The reference Mesoproterozoic-Neoproterozoic chemostratigraphic framework also suggests that the late Mesoproterozoic carbonate successions display predominantly lower ^{87}Sr/^{86}Sr values than those characteristic of early Neoproterozoic carbonates.

The late Mesoproterozoic oceans registered an increase in the marine multicellular/sexual eukaryotic photosynthetic life. The advent of marine multicellular/sexual eukaryotic photosynthetic life parallels a rise in δ^{53}Cr and $\delta^{34/32}$S values of carbonates, which in turn suggest an intensification in oxidative fractionation of these redox-sensitive elements in the global oceans, as well as an increase in the oxidative continental weathering regime, implying an increase in the atmospheric oxygen levels. The occurrence of fractionated δ^{53}Cr and $\delta^{34/32}$S values in the late Mesoproterozoic carbonates contrasts with the unfractionated $\delta^{98/95}$Mo values. This difference may be related to the occurrence of an ocean reservoir with low [Mo] and to a nonsteady-state Mo isotope cycle.

5.6. METHODS

Samples were analyzed for their Mo isotope composition using a double spike, following the methods described by *Siebert et al.* [2001] using a Nu Instruments MC-ICPMS at the University of Bern, Switzerland. The Mo isotope compositions are reported in the delta per mil notation ($\delta^{98/95}$Mo‰) against the Johnson Matthey (JMC2 LOT # 12-12783H) standard. The external reproducibility of analyzed samples and standards was better than 0.1‰. Major and minor elements were determined using a triple quadrupole ICPMS.

ACKNOWLEDGMENTS

This work was supported by the Swiss National Science Foundation (SNSF grants 200021-107465 and 200020-115911). Juan Carlos Silva-Tamayo (JCST) is thankful to the Isotope Geochemistry Laboratory of the University of Bern for granting access to the analytical facilities. Giovanni Nova and JCST are also thankful to Colciencias for its support received through the grant 727771451027. JCST is also thankful to the University of Houston for its support through his faculty start-up grant. Authors are thankful to Votorantim Group for allowing access to core samples from the Vazante Group.

REFERENCES

Alvarenga, C.J., Santos, R.V., Vieira, L.C., Lima, B.A. and Mancini, L.H., 2014. Meso-Neoproterozoic isotope stratigraphy on carbonates platforms in the Brasilia Belt of Brazil. Precambrian Research, *251*, 164–80.

Azmy, K., Veizer, J., Misi, A., de Oliveira, T.F., Sanches, A.L. and Dardenne, M.A., 2001. Dolomitization and isotope stratigraphy of the Vazante formation, São Francisco Basin, Brazil. Precambrian Research, *112*(3–4), 303–329.

Azmy, K., Kendall, B., Creaser, R.A., Heaman, L. and de Oliveira, T.F., 2008. Global correlation of the Vazante Group, São Francisco Basin, Brazil: Re–Os and U–Pb radiometric age constraints. Precambrian Research, *164*(3–4), 160–172. doi:10.1016/j.precamres.2008.05.001.

Babinski, M., Vieira, L.C. and Trindade, R.I., 2007. Direct dating of the Sete Lagoas cap carbonate (Bambuí Group, Brazil) and implications for the Neoproterozoic glacial events. Terra Nova, *19*(6), 401–406.

Bartley, J.K. and Kah, L.C., 2004. Marine carbon reservoir, Corg-Ccarb coupling, and the evolution of the Proterozoic carbon cycle. Geology, *32*(2), 129–132. doi:10.1130/G19939.1.

Bartley, J.K., Semikhatov, M.A., Kaufman, A.J., Knoll, A.H., Pope, M.C. and Jacobsen, S.B., 2001. Global events across the Mesoproterozoic–Neoproterozoic boundary: C and Sr isotopic evidence from Siberia. Precambrian Research, *111*(1–4), 165–202. doi:10.1016/S0301-9268(01)00160-7.

Bartley, J.K., Kah, L.C., Frank, T.D. and Lyons, T.W., 2015. Deep-water microbialites of the Mesoproterozoic Dismal Lakes Group: Microbial growth, lithification, and implications for coniform stromatolites. Geobiology, *13*(1), 15–32. doi:10.1111/gbi.12114.

Bertoni, M.E., Rooney, A.D., Selby, D., Alkmim, F.F. and Le Heron, D.P., 2014. Neoproterozoic Re–Os systematics of organic-rich rocks in the São Francisco Basin, Brazil and implications for hydrocarbon exploration. Precambrian Research, *255*, 355–366.

Fairchild, T.R., Schopf, J.W., Shen-Miller, J., Guimarães, E.M., Edwards, M.D., Lagstein, A., Li, X., Pabst, M. and de Melo-Filho, L.S., 1996. Recent discoveries of Proterozoic microfossils in south-central Brazil. Precambrian Research, *80*(1–2), 125–152.

Frank, T.D., Kah, L.C. and Lyons, T.W., 2003. Changes in organic matter production and accumulation as a mechanism for isotopic evolution in the Mesoproterozoic Ocean. Geological Magazine, *140*(4), 397–420. doi:10.1017/S0016756803007830.

Frei, R., Gaucher, C., Poulton, S.W. and Canfield, D.E., 2009. Fluctuations in Precambrian atmospheric oxygenation recorded by chromium isotopes. Nature, *461*, 250–253.

Geboy, N.J., Kaufman, A.J., Walker, R.J., Misi, A., de Oliviera, T.F., Miller, K.E., Azmy, K., Kendall, B. and Poulton, S.W., 2013. Re–Os age constraints and new observations of Proterozoic glacial deposits in the Vazante Group, Brazil. Precambrian Research, *238*, 199–213. doi:10.1016/j.precamres.2013.10.010.

Gellatly, A.M. and Lyons, T.W., 2005. Trace sulfate in mid-Proterozoic carbonates and the sulfur isotope record of biospheric evolution. Geochimica et Cosmochimica Acta, *69*(15), 3813–3829. doi:10.1016/j.gca.2005.01.019.

Gibson, T.M., Shih, P.M., Cumming, V.M., Fischer, W.W., Crockford, P.W., Hodgskiss, M.S., Wörndle, S., Creaser, R.A., Rainbird, R.H., Skulski, T.M. and Halverson, G.P., 2018. Precise age of Bangiomorpha pubescens dates the origin of eukaryotic photosynthesis. Geology, *46*(2), 135–138. doi:10.1130/G39829.1.

Gilleaudeau, G.J. and Kah, L.C., 2013. Oceanic molybdenum drawdown by epeiric sea expansion in the Mesoproterozoic. Chemical Geology, *356*, 21–37.

Gilleaudeau, G.J., Frei, R., Kaufman, A.J., Kah, L.C., Azmy, K., Bartley, J.K., Chernyavskiy, P. and Knoll, A.H., 2016. Oxygenation of the mid-Proterozoic atmosphere: Clues from chromium isotopes in carbonates. Geochemical Perspectives Letters, *2*, 178–187. doi:10.7185/geochemlet.1618.

Guilbaud, R., Poulton, S.W., Butterfield, N.J., Zhu, M. and Shields-Zhou, G.A., 2015. A global transition to ferruginous conditions in the early Neoproterozoic oceans. Nature Geoscience, *8*(6), 466. doi:10.1038/NGEO2434.

Kah, L.C. and Bartley, J.K., 2011. Protracted oxygenation of the Proterozoic biosphere. International Geology Review, *53*(11–12), 1424–42. doi:10.1080/00206814.2010.527651.

Kah, L.C., Sherman, A.G., Narbonne, G.M., Knoll, A.H., and Kaufman, A.J., 1999. δ13C stratigraphy of the Proterozoic Bylot Supergroup, Baffin Island, Canada: Implications for regional lithostratigraphic correlations. Canadian Journal of Earth Sciences, *36*(3), 313–332. doi:10.1139/e98-100.

Kah, L.C., Lyons, T.W. and Chesley, J.T., 2001. Geochemistry of a 1.2 Ga carbonate-evaporite succession, northern Baffin and Bylot Islands: Implications for Mesoproterozoic marine evolution. Precambrian Research, *111*(1–4), 203–234. doi:10.1016/S0301-9268(01)00161-9.

Kah, L.C., Lyons, T.W. and Frank, T.D., 2004. Low marine sulphate and protracted oxygenation of the Proterozoic biosphere. Nature, *431*(7010), 834. doi:10.1038/nature02974.

Kah, L.C., Bartley, J.K. and Teal, D.A., 2012. Chemostratigraphy of the late Mesoproterozoic Atar Group, Taoudeni Basin, Mauritania: Muted isotopic variability, facies correlation, and global isotopic trends. Precambrian Research, *200*, 82–103. doi:10.1016/j.precamres.2012.01.011.

Kendall, B., Dahl, T.W. and Anbar, A.D., 2017. The stable isotope geochemistry of molybdenum. Reviews in Mineralogy and Geochemistry, *82*(1), 683–732. doi:0.2138/rmg.2017.82.16.

Knoll, A.H., Javaux, E.J., Hewitt, D. and Cohen, P., 2006. Eukaryotic organisms in Proterozoic oceans. Philosophical Transactions of the Royal Society B: Biological Sciences, *361*(1470), 1023–1038.

Misi, A., Kaufman, A.J., Azmy, K., Dardenne, M.A., Sial, A.N. and de Oliveira, T.F., 2011. Neoproterozoic successions of the São Francisco craton, Brazil: The Bambuí, Una, Vazante and Vaza Barris/Miaba groups and their glaciogenic deposits. Geological Society, London, Memoirs, *36*(1), 509–522.

Misi, A., Azmy, K., Kaufman, A.J., Oliveira, T.F., Sanches, A.L. and Oliveira, G.D., 2014. Review of the geological and geochronological framework of the Vazante sequence, Minas Gerais, Brazil: Implications to metallogenic and phosphogenic models. Ore Geology Reviews, *63*(October), 76–90. doi:10.1016/j.oregeorev.2014.05.002.

Ovchinnikova, G.V., Semikhatov, M.A., Gorokhov, I.M., Belyatskii, B.V., Vasilieva, I.M. and Levskii, L.K., 1995. U–Pb systematics of Pre-Cambrian carbonates: The Riphean Sukhaya Tunguska Formation in the Turukhansk Uplift, Siberia. Lithology and Mineral Resources, *30*(5), 477–487.

Pimentel, M.M., Rodrigues, J.B., DellaGiustina, M.E.S., Junges, S., Matteini, M. and Armstrong, R., 2011. The tectonic evolution of the Neoproterozoic Brasília Belt, central Brazil, based on SHRIMP and LA-ICPMS U–Pb sedimentary provenance data: A review. Journal of South American Earth Sciences, *31*(4), 345–357.

Roberts, N.M.W., 2013. The boring billion? Lid tectonics, continental growth and environmental change associated with the Columbia supercontinent. Geoscience Frontiers, *4*(6), 681–691. doi:10.1016/j.gsf.2013.05.004.

Rodrigues, J.B., 2008. Proveniência de sedimentos dos grupos Canastra, Ibiá, Vazante e Bambuí: um estudo de zircões detríticos e idades modelo Sm-Nd. PhD Thesis.

Rooney, A.D., Selby, D., Houzay, J.P. and Renne, P.R., 2010. Re–Os geochronology of a Mesoproterozoic sedimentary succession, Taoudeni basin, Mauritania: Implications for basin-wide correlations and Re–Os organic-rich sediments systematics. Earth and Planetary Science Letters, *289*(3–4), 486–496. doi:10.1016/j.epsl.2009.11.039.

Scott, C., 2009. Biogeochemical signatures in Precambrian Black shales: Window into the co-evolution of ocean chemistry and life on Earth. Doctoral dissertation, UC Riverside.

Sheldon, N.D., 2013. Causes and consequences of low atmospheric pCO_2 in the late Mesoproterozoic. Chemical Geology, *362*, 224–231.

Shields, G.A., 2007. A normalized seawater strontium isotope curve: Possible implications for Neoproterozoic-Cambrian weathering rates and the further oxygenation of the Earth. Earth, *2*(2), 35–42.

Siebert, C., Nägler, T.F., and Kramers, J.D., 2001. Determination of molybdenum isotope fractionation by double-spike multi-collector inductively coupled plasma mass spectrometry. Geochemistry, Geophysics, Geosystems, *2*, 7.

Silva Tamayo, J., Sial, A.N., Ferreira, V.P., and Pimentel, M.M., 2004. C-and Sr-isotope stratigraphy of the Sao Caetano Complex, northeastern Brazil: A contribution to the study of the Meso-Neoproterozoic seawater geochemistry. Academia Brasileira de Ciencias, *77*(1), 137–155.

Turner, E.C. and Kamber, B.S., 2012. Arctic Bay Formation, Borden Basin, Nunavut (Canada): Basin evolution, black shale, and dissolved metal systematics in the Mesoproterozoic ocean. Precambrian Research, *208*, 1–18.

Voegelin, A.R., Nägler, T.F., Samankassou, E. and Villa, I.M., 2009. Molybdenum isotopic composition of modern and Carboniferous carbonates. Chemical Geology, *265*(3–4), 488–498.

6

The Cryogenian-Ediacaran Boundary in the Southern Amazon Craton

Afonso César Rodrigues Nogueira[1,2], Guilherme Raffaeli Romero[1], Evelyn Aparecida Mecenero Sanchez[3], Fábio Henrique Garcia Domingos[1], José Bandeira[1], Iara Maria dos Santos[1], Roberto Vizeu Lima Pinheiro[1], Joelson Lima Soares[1], Jean Michel Lafon[1,2], Jhon Willy Lopes Afonso[1], Hudson Pereira Santos[1], and Isaac Daniel Rudnitzki[4]

ABSTRACT

The southern margin of the Amazon Craton is an important area regarding evidences of Neoproterozoic snowball Earth glaciations. Four outcrops of cap carbonate overlying Marinoan diamictites (~635 Ma) record the best preserved boundary between Cryogenian (720–635 Ma) and Ediacaran (635–541 Ma) in South America. Approximately 100 m thick of coastal to marine glaciogenic sediments are overlaid by ~40 m thick cap carbonate with persistent $\delta^{13}C$ negative values and $^{87}Sr/^{86}Sr$ variations consistent with an early Ediacaran age. Cap carbonate consists of (i) a basal cap dolostone, composed of shallow to moderately deepwater pinkish peloidal dolomudstone with stromatolites, tube and giant wave ripple structures, and rare crystal fans, and (ii) a cap limestone cementstone, consisting of deepwater bituminous mudstones with abundant crystal fans, subordinate shales and acritarchs. Unlike other occurrences worldwide, the base of the cap carbonate exhibits soft-sediment deformation, indicating abrupt transition from icehouse to greenhouse conditions after isostatic rebound. The postglacial transgression over craton margins generated extended shallow platforms initially with predominant dolomitic precipitation succeeded by $CaCO_3$-supersaturated deepwater carbonates. New data and review of previous geological, geochemical, and geochronological information about the Neoproterozoic deposits provide a stratigraphic framework that confirms the presence of Cryogenian-Ediacaran boundary in the southern Amazon Craton.

6.1. INTRODUCTION

The subdivision of Cryogenian (720–635 Ma) and Ediacaran (635–541 Ma) periods based on the occurrence and study of global glaciations and the Precambrian and Cambrian fossil assemblages as stratigraphic markers represent an exceptional advance in the understanding of the main events of the late Neoproterozoic [*Harland*, 1964a, 1964b]. The subdivision of Neoproterozoic successions depends on a precise lithostratigraphic description combined with facies analysis and sequential framework assisted with detailed $\delta^{13}C$ and $^{87}Sr/^{86}Sr$ data. The fossil content, mainly acanthomorphic acritarchs and late Ediacaran shelly fossils such as *Cloudina*, helped in the identification of the Ediacaran record [*Grey*, 2005; *Xiao et al.*, 2016]. Radiochronology depends on the occurrence of suitable lithologies for dating, such volcanic ash, and unconventional methods can be also used for dating such as Pb-Pb in carbonates and Re-Os in shale, despite the large uncertainty of this methods.

[1] *Programa de Pós-Graduação em Geologia e Geoquímica, Faculdade de Geologia, Instituto de Geociências, Universidade do Pará, Belém, PA, Brazil*

[2] *Research Productivity of CNPq, Brasília, Brazil*

[3] *Faculty of Geological Engineering, Instituto de Ciência e Tecnologia, Universidade Federal dos Vales do Jequitinhonha e Mucuri, Diamantina, MG, Brazil*

[4] *Departamento de Geologia, Universidade Federal de Ouro Preto, Ouro Preto, MG, Brazil*

Chemostratigraphy Across Major Chronological Boundaries, Geophysical Monograph 240, First Edition.
Edited by Alcides N. Sial, Claudio Gaucher, Muthuvairavasamy Ramkumar, and Valderez Pinto Ferreira.

The recognition of Cryogenian-Ediacaran boundary worldwide is generally associated with the presence of well-defined cap carbonates and the establishment of the Ediacaran system [*Knoll et al.*, 2004, 2006]. The boundary at the base of the Nuccaleena Formation, a typical cap carbonate that overlies the Cryogenian diamictite of the Elatina Formation, at the Enorama Creek section in South Australia is the currently accepted Global Boundary Stratotype Section and Point (GSSP). A cap carbonate comprises a postglacial succession, generally exhibiting negative isotopic excursion of ^{13}C, overlying glaciogenic diamictites formed during low-latitude glaciations, linked to the snowball Earth hypothesis [*Hoffman and Schrag*, 2002; *Hoffman et al.*, 2017]. The succession is characterized by primary pinkish dolomites and limestones, displaying diagnostic structures such as microbial laminites, tubestone structures, calcite crystal fans (after aragonite pseudomorphs), macropeloids, and giant wave ripples [*Kennedy*, 1996; *Hoffman et al.*, 1998a, 1998b; *Myrow and Kaufman*, 1999; *Kennedy et al.*, 2001; *Hoffman and Schrag*, 2002; *Nogueira et al.*, 2003, 2007; *Allen and Hoffman*, 2004; *Halverson et al.*, 2010; *Hoffman et al.*, 2017].

The Cryogenian-Ediacaran boundary in South America is well exposed to the main occurrences of Marinoan cap carbonates bordering the southern Amazon Craton, Brazil (Fig. 6.1). The Marinoan cap carbonate represents the base of a Neoproterozoic carbonate platform succession correlated for more than 900 km along the border of the southern Amazon Craton [*Nogueira et al.*, 2001, 2003, 2007; *Nogueira and Riccomini*, 2006; *Soares and Nogueira*, 2008; *Gaia et al.*, 2017; *Afonso and Nogueira*, 2018]. This succession is exposed in quarries and outcrops at the Espigão d'Oeste, Chupinguaia, Mirassol d'Oeste, and Tangará da Serra regions (Fig. 6.1). Many occurrences of Neoproterozoic carbonate rocks described in South America have been inferred as cap carbonates based mainly on δ^{13}C negative excursions and ^{87}Sr/^{86}Sr values on successions with limited facies and stratigraphic analysis [*Misi and Veizer*, 1998; *Santos et al.*, 2000; *Boggiani et al.*, 2003, 2010; *Alvarenga et al.*, 2004; *Sial et al.*, 2016]. Outcrop-based facies and stratigraphic analysis, combined with previous data of δ^{13}C and ^{87}Sr/^{86}Sr chemostratigraphy, structural geology, and geochronological data for the Ediacaran deposits, allowed to elaborate a robust stratigraphic framework

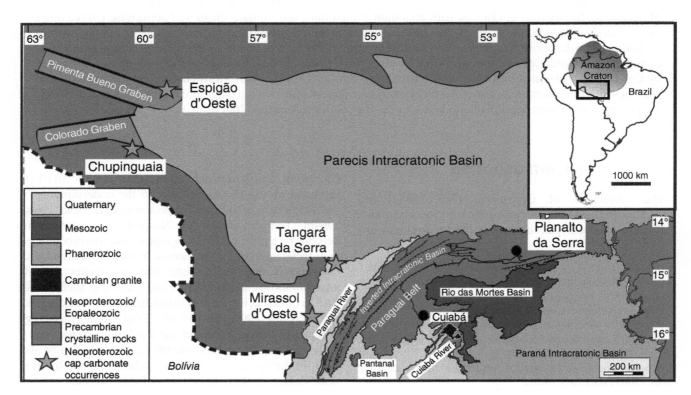

Figure 6.1 Location and simplified geologic map of southwestern Amazon Craton with Neoproterozoic cap carbonate occurrences. The Cryogenian-Ediacaran boundary is exposed in the main cap carbonate succession overlying crystalline and metasedimentary rocks of Precambrian basement in the southwest and south portion of the Amazon Craton.

that reinforces the presence of the Cryogenian-Ediacaran boundary and correlation for hundreds of kilometers along the southern Amazon Craton (Fig. 6.1). Additionally, we review the previous stratigraphic, paleomagnetic, and geochronological data, paleoenvironmental reconstructions, and global correlation of the Ediacaran deposits of the southern Amazon Craton and reassess previous geological interpretations.

6.2. GEOLOGICAL SETTING

Neoproterozoic carbonate platform deposits are discontinuously exposed overlying crystalline and metasedimentary rocks at the border of the southern Amazon Craton (Fig. 6.1). The cap carbonate outcrops are found, preferentially, at the border of the Pimenta Bueno and Colorado grabens and in the southern margin of the Amazon Craton near to the late Cryogenian-Cambrian inverted intracratonic basin (Fig. 6.1). The evolution of these grabens is associated with the implantation of the Phanerozoic intracratonic and rift-sag-type Parecis and Paraná basins, formed during an extensional event following the Rodinia breakup at ~1.0–0.75 Ga [*Siqueira and Teixeira*, 1993; *Cordani et al.*, 2000; *Pedreira and Bahia*, 2000]. The Paraguay Belt represents a folded belt belonging to the southeastern Tocantins province [*Almeida*, 1984; *Cordani et al.*, 2000], originated during a collisional event between Amazonia, São Francisco, and Paraná continental blocks [*Alkmim et al.*, 2001]. Ages obtained from metamorphic minerals (Cuiabá Group) and syntectonic plutons have suggested that the deformation took place during the Cambrian to Early Ordovician [*Basei and Brito Neves*, 1992; *Moura and Gaudette*, 1993]. This time interval contrasts with the age of the main tectonic event (~600 Ma) that formed many Brasiliano-Pan-African belts between 900 and 640 Ma [*Cordani et al.*, 2000; *Cordani et al.*, 2013]. The collisional event that originated the Paraguay Belt is recorded mainly in the metamorphosed rocks of the Cuiabá Group (Fig. 6.2). Post-collisional and non-metamorphosed intracratonic deposits are represented by Marinoan glaciogene deposits, Ediacaran carbonates, and Cambrian-Ordovician siliciclastic rocks (Fig. 6.2). These units were deformed by younger transtensional tectonics [*Santos*, 2016), marked by the emplacement of Cambrian granites with ages around 528 ± 4 to 500 ± 15 Ma [*Almeida and Mantovani*, 1975; *McGee et al.*, 2012; *Trivelli et al.*, 2017]. Other occurrences of granitoids in the southern Paraguay Belt were related to a syn-collisional continental arc magmatism [*Godoy et al.*, 2010], whereas those found in the northern Paraguay Belt are interpreted as late to post-collisional magmatism [*McGee et al.*, 2012]. These authors assume that this last magmatic event was related to the final deformational stage in the Paraguay Belt. *McGee et al.*

[2012] considered the São Vicente Granite age (518 ± 4 Ma) as constraining the minimum age of deformation and metamorphism in the Paraguay Belt and, consequently, the final assembly age of the Gondwana supercontinent in South America.

Carbonate and siliciclastic platformal settings were the main depositional sites since the early to middle Ediacaran (633–560 Ma) in the southern Amazon Craton, generally overlying metamorphic deposits of Paraguay Belt [*Boggiani et al.*, 2003; *Nogueira et al.*, 2007; *Soares and Nogueira*, 2008; *Alvarenga et al.*, 2011; *Bandeira et al.*, 2012; *Rudnitzki et al.*, 2016; *Sial et al.*, 2016; *Gaia et al.*, 2017; *Santos et al.*, 2017; *Afonso and Nogueira*, 2018]. The recurrence of epeiric sea conditions persisted until the Early Cambrian in Western Gondwana [*Torsvik and Cocks*, 2013]. The Ediacaran-Cambrian transition is confirmed by the record of the shelly fossil *Cloudina* sp. [*Warren et al.*, 2014; *Arrouy et al.*, 2016; *Becker-Kerber et al.*, 2017; *Paula-Santos et al.*, 2017] and *Skolithos* ichnofacies [*Santos et al.*, 2017]. The shelly fossil *Cloudina* sp. found in the Tamengo Formation in the southern Paraguay Belt [*Boggiani*, 1997; *Gaucher et al.*, 2003; *Becker-Kerber et al.*, 2017] was never documented in the northern Paraguay Belt.

The Cryogenian-Cambrian siliciclastic-carbonate succession in the Amazon Craton shows no evidence of metamorphism with beds dipping from 2 up to 15° to the NNW, affected by transtensional tectonics, contrasting with the transpressional dynamics in the Paraguay Belt [*Santos*, 2016]. The Ediacaran carbonate platform deposits are correlated for more than 600 km along the southern Amazon Craton (Fig. 6.1). The glaciogenic (Marinoan) deposits are represented by the Cacoal and Puga formations, comprising pelites, sandstones, and diamictites (Fig. 6.2). The platform carbonate deposits of the Araras Group, more than 700 m thick and exposed in the southern Amazon Craton, were divided into four formations, from the base to the top (Fig. 6.2): (i) the Mirassol d'Oeste Formation, a cap dolostone interpreted as shallow platform deposits; (ii) the Guia Formation, comprising limestone and shale interpreted as deep platform deposits; (iii) the Serra do Quilombo Formation, a moderately deep to shallow platform dolomitic succession; and (iv) the Nobres Formation, dolostones related to a peritidal setting. The Mirassol d'Oeste Formation is correlated with the Espigão d'Oeste Formation exposed in the southwestern Amazon Craton (Fig. 6.2). These units are occurrences of cap dolostone and together with the basal portion of the Guia Formation, a cap limestone, constitute the Marinoan cap carbonates (Fig. 6.2). The Serra Azul Formation, a putative mid-Ediacaran glacial event assigned to the Gaskiers glaciation (580 Ma), has been positioned at the base of the Alto Paraguai Group, which comprises pebbly mudstone

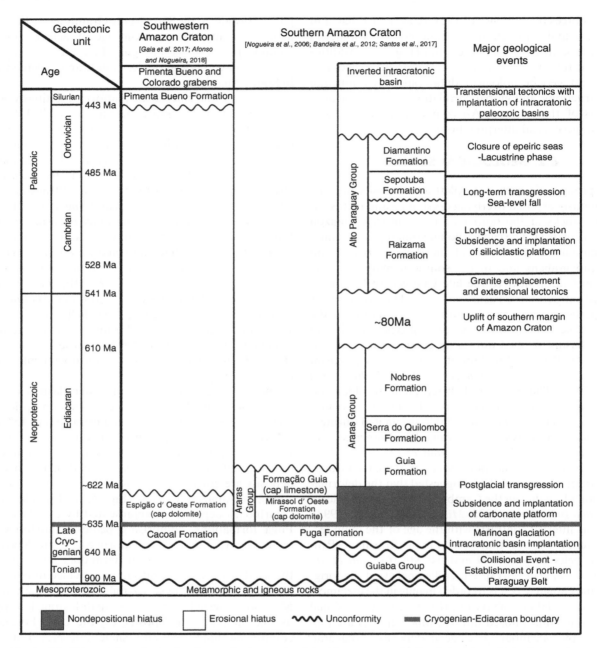

Figure 6.2 Lithostratigraphy and major geologic events of the Neoproterozoic-Cambrian deposits in the southern Amazon Craton and northern Paraguay Belt.

and diamictites with striated polymict clasts [*Alvarenga et al.*, 2007; *Figueiredo et al.*, 2008]. The contacts of the Serra Azul Formation were inferred based mainly by remote sensing and poor lithostratigraphic correlation [*Alvarenga et al.*, 2007; *Figueiredo et al.*, 2008; *McGee et al.*, 2012]. $^{87}Sr/^{86}Sr$ ratio of 0.7086 for associated carbonates and ages of 647±9 and 640±15 Ma for detrital zircon and detrital muscovite grains from the Serra Azul Formation suggest a mid-Ediacaran age [*Figueiredo and Babinski*, 2008; *Figueiredo et al.*, 2008; *McGee et al.*,

2014, 2015]. Siliciclastic deposits of the Cambrian Raizama Formation and Paleozoic deposits of intracratonic sedimentary basins unconformably overlie the Araras Group (Fig. 6.2).

The age of Neoproterozoic carbonate deposits of the Araras Group has been estimated based on chemostratigraphic and chronostratigraphic relationships and limited paleontological content [*Nogueira et al.*, 2007; *Bandeira et al.*, 2012; *Romero et al.*, 2013; *Rudnitzki et al.*, 2016]. Pb-Pb ages on carbonates of 627±32 and

622 ± 33 Ma were obtained for the underlying Mirassol d'Oeste and Guia formations, respectively [*Babinski et al.*, 2006; *Romero et al.*, 2013]. The U-Pb zircon maximum age of 615 Ma for the Sepotuba Formation [*McGee et al.*, 2015] corroborates with the record of *Skolithos* in the Raizama Formation that confers a Cambrian age for this unit [cf., *Santos et al.*, 2017]. The Cambrian age is also inferred by U-Pb detrital zircon values of 541 ± 7 and 528 ± 9 Ma and Ar-Ar detrital muscovite age of 544 Ma for the Diamantino Formation, immediately above Sepotuba deposits [*Bandeira et al.*, 2012; *McGee et al.*, 2014, 2015].

The paleomagnetic analysis of cap dolostone revealed two high-temperature components [*Trindade et al.*, 2003]. The first component comprises normal and reverse polarities attesting primary magnetization and paleolatitude of $22 \pm 6/-5°$, indicating an equatorial-tropical position for the Puga basin. The second component suggests a younger remagnetization event of ~520 Ma, correlated with remagnetized carbonates of the São Francisco Craton [*D'Agrella-Filho et al.*, 2000; *Trindade et al.*, 2004], after the tectonic deformation of the Paraguay Belt. The low latitudes measured for the Marinoan cap carbonate indicated that Amazonia was near the equator, corroborating the interpretation of a global glaciation record in South America. This interpretation agrees with sedimentologic studies that indicate a development of Ediacaran carbonate platforms with evaporitic environments in the southern Amazon Craton during the Ediacaran [*Almeida*, 1964; *Zaine and Fairchild*, 1992; *Boggiani*, 1997; *Gaucher et al.*, 2003; *Nogueira*, 2003; *Alvarenga et al.*, 2004; *Nogueira and Riccomini*, 2006; *Nogueira et al.*, 2007; *Rudnitzki et al.*, 2016].

6.3. THE MARINOAN GLACIAL DEPOSITS

Late Cryogenian glaciogenic deposits of the Puga Formation are discontinuously distributed along two thousand kilometers of the southern Amazon Craton and northern Paraguay (Figs. 6.1 and 6.2). A Marinoan age (635 Ma) for these rocks is inferred on the basis of the occurrence of the cap carbonates overlying diamictites, paleomagnetic data indicating low latitude, $\delta^{13}C$ and $^{87}Sr/^{86}Sr$ chemostratigraphy, and Pb-Pb geochronology of carbonates [*Trompette et al.*, 1998; *Nogueira et al.*, 2003, 2007; *Trindade et al.*, 2003; *Allen and Hoffman*, 2005; *Romero et al.*, 2013; *Gaia et al.*, 2017; *Afonso and Nogueira*, 2018]. The name "Puga" has its origin related to the Morro do Puga outcrop in the southern Paraguay Belt where diamictites of putatively glacial origin were described for the first time [*Maciel*, 1959]. There, the "Puga diamictites" are associated with large iron and manganese deposits, the Jacadigo Group, usually inter-

preted as glaciogenic [e.g., *Frei et al.*, 2017 and references therein], but regarded by other authors as basin margin gravitational deposits not related to glacial processes [*Freitas et al.*, 2011]. Additionally, there are only scarce evidences for the presence of cap carbonate immediately above the Puga Formation in the southern Paraguay Belt [e.g., *Boggiani et al.*, 2010]. The reddish laminated limestones of the Bocaina Formation with $\delta^{13}C$ values around −5‰ have been considered as a cap carbonate [*Boggiani et al.*, 2003, 2010]. The limestone is not micritic or dolomicritic, and except by the presence of stromatolites with tubestone, other diagnostic features of Marinoan cap carbonates were not observed (macropeloids, giant wave ripple, crystal fans, etc.). In addition, the tubestone described by these authors have been interpreted as syn-sedimentary fractures and columnar stromatolites [*Oliveira*, 2010; *Romero*, 2015; *Romero et al.*, 2016]. The intense recrystallization of these rocks may explain the pronounced negative excursion of $\delta^{13}C$ provided by diagenetic fluids, although a primary origin has been advocated by *Boggiani et al.* [2003, 2010]. Recently, U-Pb zircon values of 555 ± 0.3 Ma found in tuffs in the Bocaina Formation confirm a late Ediacaran age [*Parry et al.*, 2017].

The ^{13}C depletion pattern in the transgressive postglacial cap carbonate succession in the base of the Araras Group is compatible with other Marinoan cap carbonates worldwide. This isotopic signature has been explained as (i) the result of a fast mixture of ^{13}C-depleted deep waters with shallow waters by upwelling after glaciation [*Kaufman et al.*, 1991; *Grotzinger and Knoll*, 1995], (ii) a decrease of bioproductivity induced by ocean confinement due to the continuous ice cover as postulated by the snowball Earth hypothesis [*Hoffman et al.*, 1998a, 1998b], and/or (iii) destabilization of gas hydrates in continental permafrost succeeded by rapid warming and postglacial transgression [*Kennedy et al.*, 2001]. The temporal changes and spatial variability of $\delta^{13}C$ in a Marinoan cap dolomite in Namibia have been attributed to temperature changes, gas hydrate destabilization, and kinetic isotopic effects associated with rapid carbonate deposition [*Hoffman et al.*, 2007]. The composition of seawater could not change much on the time scale of cap carbonate deposition because the residence time of C in the postglacial ocean is relatively short [*Kump*, 1991; *Jacobsen and Kaufman*, 1999]. However, these values may vary slightly caused by local environmental conditions, such as restricted seas and water column stratification [*Ader et al.*, 2009].

The Marinoan Puga glaciogenic deposits of the northern Paraguay Belt have been included into the Cuiabá Group [*Alvarenga and Trompette*, 1992; *Alvarenga et al.*, 2008]. This interpretation is inappropriate, as Puga deposits show no evidence of metamorphism like the rocks of the Cuiabá Group. The Cuiabá Group metadiamictites are often difficult to separate from the Puga Formation matrix-supported conglomerates. Maybe this explains the arguable depositional model for the Cuiabá Group where submarine fan sediments were supplied by glaciers [*Alvarenga and Trompette*, 1992; *Alvarenga et al.*, 2008]. Diamictites are not exclusive of glacial depositional system and can also be formed by mass wasting processes, forming a gravelly, poorly sorted, matrix-rich sediment. The depositional history of the Cuiabá Group precedes the deposition of the Puga Formation, as evidenced by the regional unconformity separating these units (Figs. 6.2 and 6.3). The Puga Formation exhibits sub-horizontal dips and, locally, unconformably overlies Mesoproterozoic foliated metamorphic and crystalline rocks (Figs. 6.2, 6.3, and 6.4).

The Marinoan glacially influenced deposits represent a ~150 m thick succession recording, at least, two cycles of advance-retreat of coastal glaciers (Fig. 6.4a and b). Polymictic, matrix-supported conglomerate/diamictite exhibits rounded to subangular clasts of different compositions (gneiss, basic volcanic rock, quartzite, limestone, schist, sandstone, pelite, quartz) ranging in size from granule up to boulders, often faceted and striated (Fig. 6.4). The diamictite is greenish gray to black or red and/or purple when weathered. Gray to purple sandy-clayish matrix predominates in all outcrops of this facies, locally cemented by calcite or displaying pyrite crystals. The diamictite is generally massive, sometimes interbedded with coarse- to medium-grained sandstones with disseminated pebbles, frequently deforming the basal laminae (dropstones) (Fig. 6.4a and b). Poorly sorted fine-grained sandstone exhibits massive bedding, low-angle cross-bedding, and even parallel lamination forming up to 2 m thick layers. Rippled bedding and gradational beds occur locally. The sandstone facies occasionally occur in direct contact with dolomite. Siltstone/sandstone rhythmite overlies the sandstone forming tabular beds laterally continuous for dozens of meters. The rhythmite consists of siltstone laminae alternated with fine-grained sandstone laminae, pebbly sandstone, and locally 30–20 cm thick lenses of diamictites (dump structure). Granules and isolated, angular, faceted pebbles up to 5 cm in diameter locally deform the lamination (dropstones). The dropstones reach 20 cm in size and are mostly composed of quartzite, gneiss, and granite. Parallel lamination and undulated and convolute bedding are the most frequent sedimentary structures. These deposits are weathered with the development of intense ferruginization and reddish argillaceous soils.

The description of this glacial facies is similar to previously published works, which interpreted them as a glacio-marine platform with debris flow influence in the distal domain [*Alvarenga*, 1988; *Alvarenga and Trompette*,

1988; *Alvarenga and Trompette*, 1992; *Silva et al.*, 2015; *Gaia et al.*, 2017; *Afonso and Nogueira*, 2018]. The coastal and distal marine influence is indicated by the proximity of basement rocks, probably the margin of the basin. Likewise, the upper portion of the Puga Formation is a coastal setting due to the proximity of shallow waters and euphotic deposits represented by the cap carbonate [*Nogueira et al.*, 2003]. The diamictite clast diversity (striated and faceted clasts) suggests abrasion and reworking of basement rocks and sedimentary deposits by glaciers, which produced most of the diamictites with a sandy-pelitic matrix [*Evans*, 2006]. Liquefaction processes produced plastic adjustment between sand-silt interfaces. Subaqueous current and wave action was established after glacier retreat, when in an ice-free sea silt and fine sand were deposited by suspension or weak currents in lagoonal or marine platform environments. Ice-rafted debris from melting icebergs released granules and pebbles [*Arnaud and Etienne*, 2011]. The increase of accommodation space, during the postglacial transgression, influenced the higher degree of preservation of glacio-marine deposits.

6.4. THE CONTACT BETWEEN GLACIAL DEPOSITS AND CAP CARBONATE

In the Amazon Craton, the contact between the cap dolostone and underlying diamictites is sharp and plastically deformed [*Nogueira et al.*, 2003; *Gaia et al.*, 2017; *Afonso and Nogueira*, 2018]. The irregular dolomitic laminations are smoothly curved and pass gradually upsection to even parallel lamination (Fig. 6.5a and b). In plain view, they form smooth and disharmonic folds (limbs dipping <5°), with curved hinge lines (Fig. 6.5c). The plastic behavior of the sediments suggests a relatively fast precipitation of carbonates over partially unconsolidated glaciogenic diamictons. Water-saturated dolomitic gel generated plastic adjustments in the diamicton-dolomudstone zone, developing load-casted structures [*Nogueira et al.*, 2003]. This usually occurs in a shallow to moderately deep environment and indicates an abrupt change in the climate conditions from very cold to hot atmospheric temperatures associated with a postglacial sea-level rise. These features were formed by syn-sedimentary deformation related to postglacial isostatic rebound following the Marinoan glaciation [*Nogueira et al.*, 2003; *Soares et al.*, 2013]. The glacio-isostatic adjustment takes place after melt and retreat of a glacier, causing progressive regional uplift in continental areas [*Creveling and Mitrovica*, 2014], isostatic subsidence and relative sea-level rise in the coastal zones (forebulge), and the development of a shallow platform where the cap carbonate was precipitated.

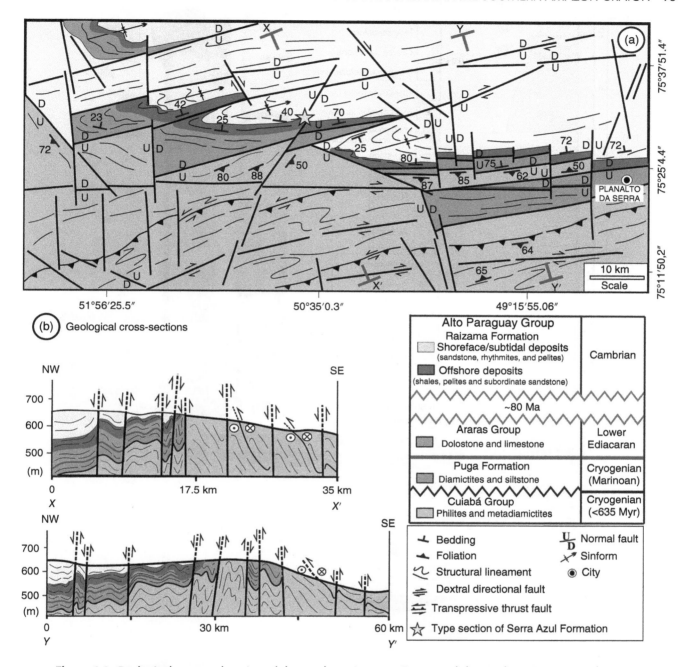

Figure 6.3 Geological-structural setting of the southern Amazon Craton and the northern Paraguay Belt, near Planalto da Serra region. Spatial and geometrical distribution of Neoproterozoic and Cambrian units in map (a) and section (b) with primary and tectonic contacts established during post-Ordovician brittle-to-brittle-ductile deformation. The type section of the Serra Azul Formation [cf., *Alvarenga et al.*, 2007] is plotted in the map for better understanding of the structural-geometric array of the region. Observed offshore deposits related to the Raizama Formation unconformably overlie the glaciogene diamictites of the Puga Formation. Reproduced with permission of Elsevier.

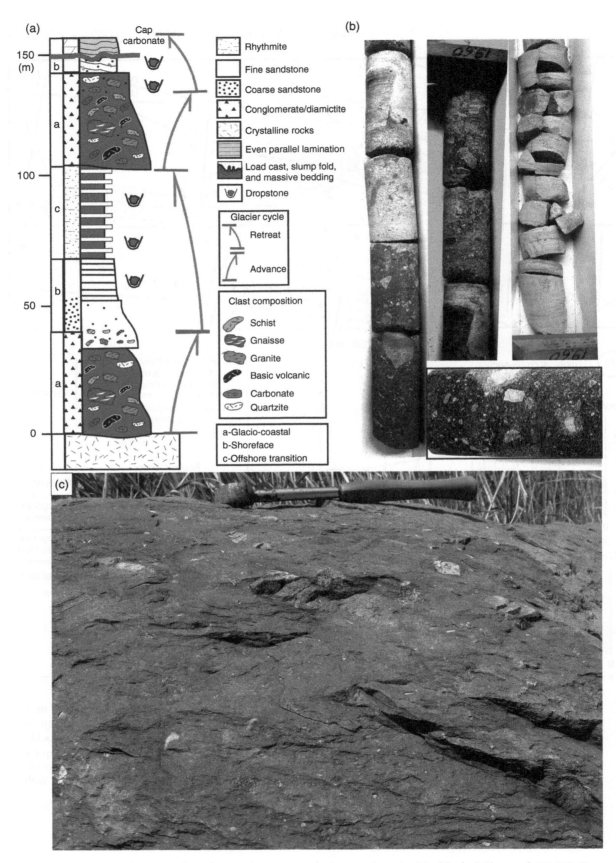

Figure 6.4 Simplified stratigraphic section of Marinoan glaciogene deposits (Modified of *Silva et al.* [2015]). Two advance-retreat cycles of coastal glaciers constitute the Puga succession (a). Core carried out by geological survey in the Rondonia Mineral Company (RMC) quarry showing the Cryogenian-Ediacaran boundary, coincident with the diamictite-dolostone contact (b). (c) Puga diamictite with disseminated clast.

Figure 6.5 The Cryogenian-Ediacaran boundary in South America. (a and b) The deformed contact between gla-ciogene diamictites and cap dolostone, respectively, in Mirassol d'Oeste and Espigão d'Oeste regions. (c) Open fold in the base of cap dolostone onlapped by subhorizontal dolostone beds, exposed in Espigão d'Oeste region. *(See insert for color representation of the figure.)*

6.5. THE MARINOAN CAP CARBONATE

The cap carbonate in the southern Amazon Craton exhibits all features related to the post-Marinoan strata worldwide [see *Plummer*, 1978; *Kaufman and Knoll*, 1995; *Kennedy*, 1996; *Hoffman et al.*, 1998a, 1998b; *Kennedy et al.*, 2001; *Hoffman and Schrag*, 2002; *Allen and Hoffman*, 2004]. The cap carbonate overlies glaciogenic diamictites (Puga and Cacoal formations) and, occasionally, basement rocks (Figs. 6.2, 6.4, and 6.6). In the Mirassol d'Oeste region, where the succession is more complete, a cap carbonate occurs at the base of the Araras Group composed of a 20 m thick cap dolostone (Mirassol d'Oeste Formation) and a 25 m thick cap limestone (lower Guia Formation) (Figs. 6.2 and 6.6). The contact between cap dolostone and cap limestone comprises a nearly horizontal surface that wraps around the megaripple morphology, with local irregularities reaching up to 20 cm depth filled by reddish siltstone. In Tangará da Serra, the siltstone is 2 m thick and overlies dolostone with syn-sedimentary deformations [*Soares et al.*, 2013]. In the Espigão d'Oeste region, the laminated

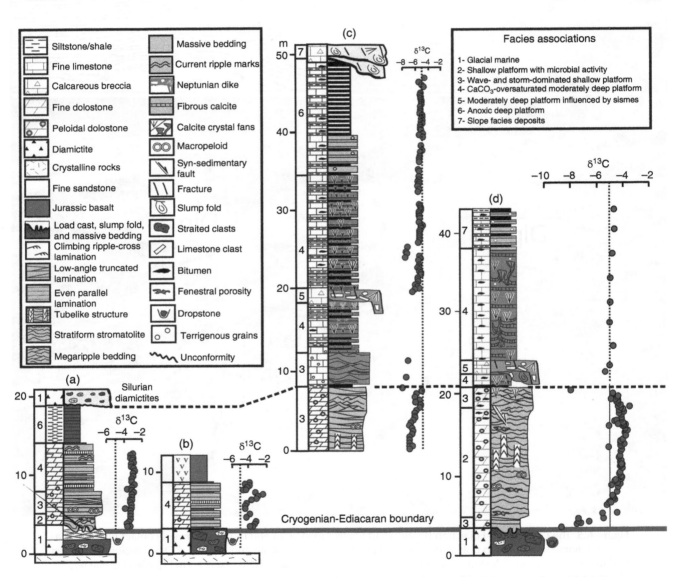

Figure 6.6 Neoproterozoic cap carbonate stratigraphic sections in the southern Amazon Craton. (a) Open pit of the RMC quarry, (b) outcrop at the Chupinguaia region, (c) open pit in the Calcario Tangará quarry, Tangará da Serra region, and (d) open pit in Terconi quarry, Mirassol d'Oeste region. The $\delta^{13}C$ curves of the carbonate platform deposits in the Amazon Craton with values in ‰ were obtained in *Nogueira et al.* [2003, 2007], *Font et al.* [2006], *Riccomini et al.* [2007], *Alvarenga et al.* [2004, 2008], *Romero et al.* [2013], and *Soares et al.* [2013].

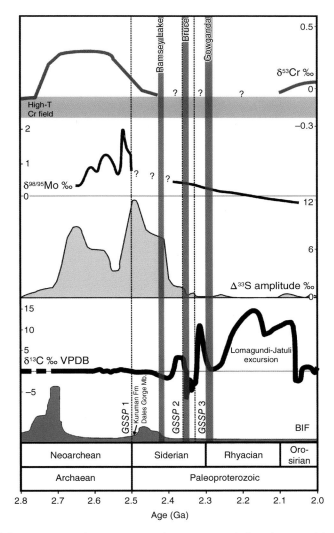

Figure 3.5 Summary of chemostratigraphic curves, BIF deposition, and glacial events (blue bars) near the Archean-Proterozoic boundary. Three different proposals of Global Stratotype Section and Point (GSSP) for the boundary are shown. GSSP 1: current boundary at 2500 Ma. GSSP 2: base of the second and possibly near-global Huronian glaciation [*Gradstein et al.,* 2012]. GSSP 3: termination of the sulfur MIF (Δ^{33}S) at ca. 2.33 Ga [*Luo et al.,* 2016]. Sources of data: δ^{53}Cr: *Frei et al.* [2009], δ$^{98/95}$Mo of shales: *Wille et al.* [2007], Δ^{33}S amplitude: *Williford et al.* [2011 and references therein], *Luo et al.* [2016]; δ^{13}C, BIF abundance and glacial events: same as for Figure 3.1.

Chemostratigraphy Across Major Chronological Boundaries, Geophysical Monograph 240, First Edition.
Edited by Alcides N. Sial, Claudio Gaucher, Muthuvairavasamy Ramkumar, and Valderez Pinto Ferreira.
© 2019 the American Geophysical Union. Published 2019 by John Wiley & Sons, Inc.

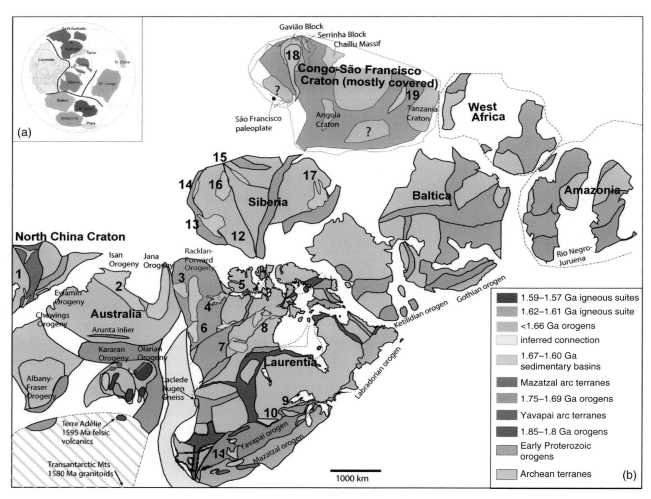

Figure 4.2 (a) Columbia/Nuna supercontinent configuration at 1.4 Ga [*Pehrsson et al., 2015; Evans et al., 2016*]. (b) Reconstruction of the Columbia supercontinent at the Paleoproterozoic-Mesoproterozoic boundary showing the location of main depocenters and orogens. 1. Changcheng-Jixian systems; 2. Mount Isa and McArthur super-basins; 3. Wernecke Supergroup; 4. Fort Simpson basin; 5. Coppermine basin; 6. Muskwa basin; 7. Athabasca Basin; 8. Thelon basin; 9. Baraboo sequence; 10. Sioux sequence; 11. Ortega sequence; 12. Kureika-Anabar basin; 13. Teya-Chapa basin; 14. Turukhansk basin; 15. Udzha and East Anabar basins; 16. Kotuy basin; 17. Uchur and Aimchan groups; 18. Espinhaço Supergroup; 19. Akanyaru and Kibara supergroups. Adapted from *Furlanetto et al.* [2016]. Reproduced with permission of Elsevier.

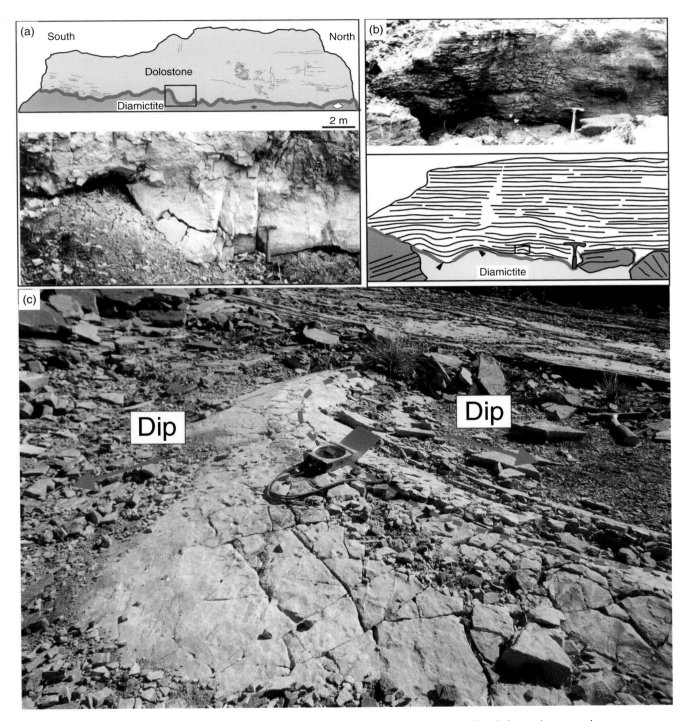

Figure 6.5 The Cryogenian-Ediacaran boundary in South America. (a and b) The deformed contact between glaciogene diamictites and cap dolostone, respectively, in Mirassol d'Oeste and Espigão d'Oeste regions. (c) Open fold in the base of cap dolostone onlapped by subhorizontal dolostone beds, exposed in Espigão d'Oeste region.

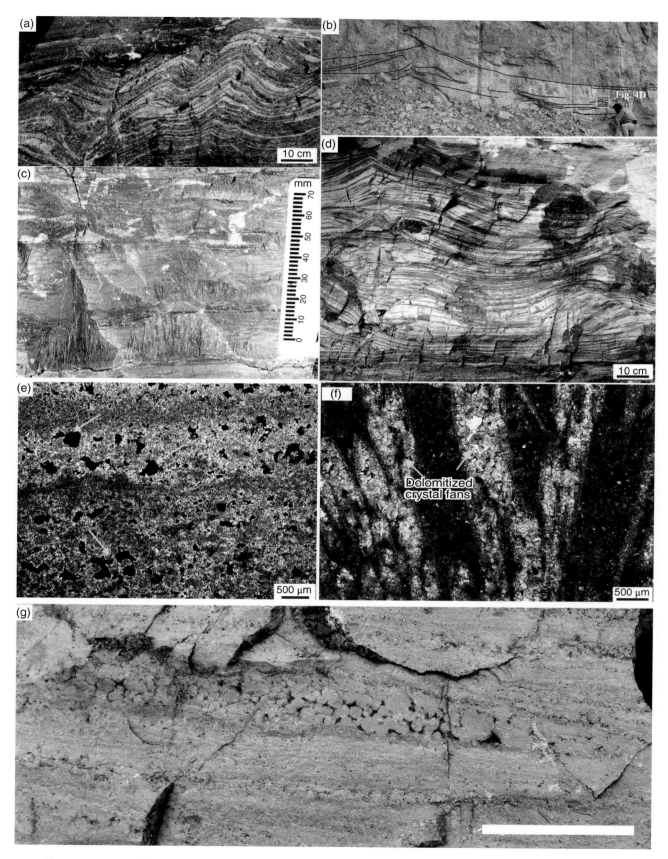

Figure 6.7 General aspects of cap carbonate in the Amazon Craton. (a) Microbialites. (b) Giant wave ripple. (c) Calcite crystal fans (pseudomorphs after aragonite) interbedded with undulated lamination. (d) Detail of (b), showing the climbing wave ripple cross-lamination and well-preserved crest of megaripple. (e) Fenestral porosity filled by bitumen. (f) Dolomitized crystal fans. (g) Macropeloids (scale bar = 2 cm). Bitumen impregnation detached all laminations of carbonates.

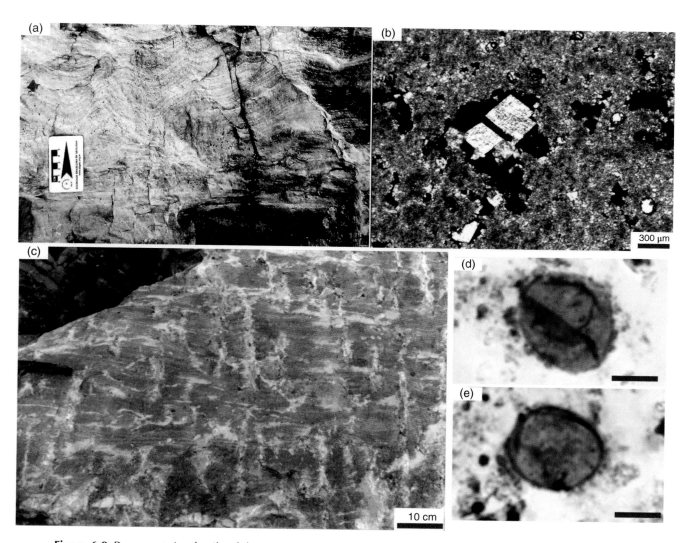

Figure 6.8 Representative fossils of the Marinoan cap carbonate in the Amazon Craton. (a) Typical microbial laminite. (b) Fenestral porosity filled by bitumen and euhedral dolomite. (c) Tubestone structures associated with microbial laminites. (d and e) *Leiosphaeridia* taxa from the intermediate part to top cap carbonate (scale bars = 20 μm).

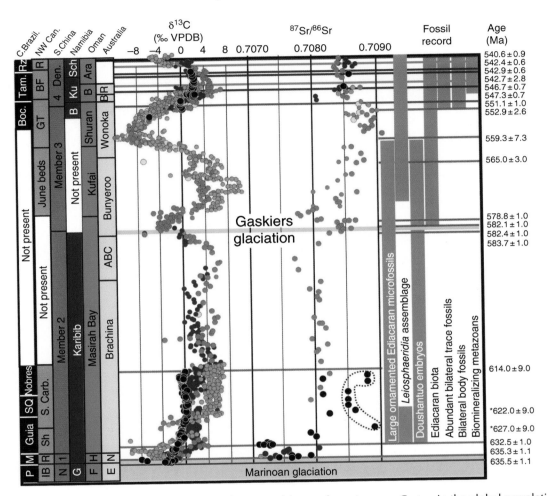

Figure 6.10 Insertion of the post-Marinoan carbonates of the southern Amazon Craton in the global correlations for the Ediacaran key locations with associated isotopic, geochronologic, and biostratigraphic data (Modified from *Macdonald et al.* [2013]). All the data are color coded for geographic location: Black, southern Amazon Craton, Central Brazil; orange, northwestern Canada; purple, South China; blue, Namibia; green, Oman; yellow, Australia; red, White Sea region of Russia; brown, Avalon Terrane of Newfoundland and the United Kingdom. The carbonates of the Araras Group and Espigão d'Oeste formations are close with lower Ediacaran age below of 614 Ma, corroborated by δ¹³C and ⁸⁷Sr/⁸⁶Sr values. Enriched positive values of ⁸⁷Sr/⁸⁶Sr in the upper portion of succession (dotted circle) represent samples probably contaminated with high content of siliciclastics. P, Puga Formation; M, Mirassol d'Oeste Formation; Guia, Guia Formation; SQ, Serra do Quilombo Formation; Nobres, Nobres Formation; Boc., Bocaina Formation; Tam., Tamengo Formation; Rz., Raizama Formation. For additional details and other abbreviation of units, see *Macdonald et al.* [2013]. Reproduced with permission of Elsevier.

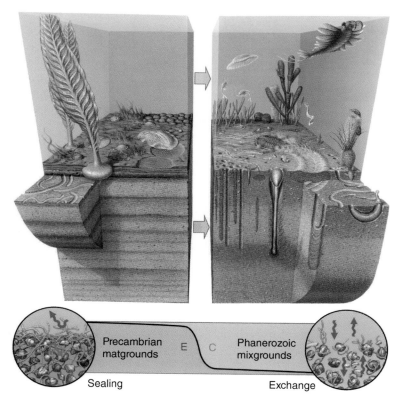

Figure 7.1 The agronomic revolution or Cambrian substrate revolution depicted in this modified illustration by Peter Trusler reveals profound changes in bioturbation across the Ediacaran (E)-Cambrian (C) transition. The enigmatic Ediacaran biota (left) is rooted in ubiquitous microbial mats that carpeted the seafloor (Precambrian matgrounds) and largely sealed the anoxic sediments beneath from the free exchange of gases. The slow diffusion of sulfate into these mats provided an oxidant for microbial sulfate reduction, leading to the buildup of toxic H_2S, a proposed critical nutrient for the Ediacaran biota. The horizontal burrows of animals within or beneath the mats and the activities of the Ediacaran organisms on its surface, which appear very late in the Ediacaran game, had little effect on sedimentary layering or the release of gases to seawater. In contrast, the deep dive of early Cambrian (right) animals into the sediments in search of food and shelter disrupted the mats and allowed the free exchange of gases across the sediment-water interface, including O_2. Ventilation and mixing of the sediments is implicated in the demise of the Ediacaran biota, if H_2S was a critical physiological resource, as well as the buildup of sulfate in the oceans. Reproduced with permission of Peter Trusler.

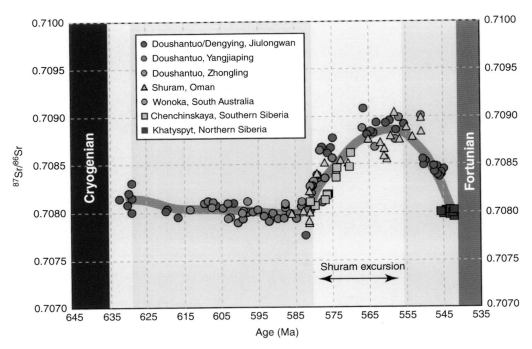

Figure 7.3 Generalized trend in the strontium isotope composition of well-preserved high-Sr marine limestones through the Ediacaran period modified from *Xiao et al.* [2016] with age constraints based on their "correlation 2" and assuming the Shuram excursion is pinned to the 551 Ma age for the Doushantuo-Dengying boundary (see discussion in text and Fig. 7.2 caption). This compilation is based on data from successions in South China [*Sawaki et al.*, 2010; *Cui et al.*, 2015], Oman [*Burns et al.*, 1994], South Australia [*Calver*, 2000], southern Siberia [*Melezhik et al.*, 2009], and northern Siberia [*Cui et al.*, 2016c]. The trend marked by the thick gray line indicates a plateau of ca. 0.7080 followed by a profound rise in $^{87}Sr/^{86}Sr$ values coincident with the Shuram excursion up to as high as 0.7090 and likely associated with intense weathering up uplifted terrains worldwide. The trend then declines back to ~0.7080 very near to the Ediacaran-Cambrian boundary.

Figure 7.4 (a) White to red colored Gaskiers diamictite overlain by a 50 cm thick white carbonate (left of the field assistant) on the shore of Conception Bay at Harbour Main, Newfoundland. Inset shows the brecciated and potentially karstified upper surface of the thin carbonate filled with green mudstone of the overlying Drook Formation. (b) Green meta-basalt of the basal Catoctin Formation with diapirs (flame structures) of Fauquier Formation carbonate injected between the chilled pillow margins on Goose Creek near Aldie, Virginia, United States. Inset illustrates a hyaloclastic texture at contact between the Catoctin meta-basalt and marble of the Fauquier cap carbonate, indicating that the sediments were water-saturated during emplacement of the volcanic rocks. Reproduced with permission of Elsevier. (c) A 20 m thick diamictite along the Tas-Yuryakh River lying unconformably above Turkut Formation dolomites in the Olenek uplift, Arctic Siberia, Russia. The freshly exposed diamictite has a green-gray sandy to clayey calcareous matrix with abundant cobble- to boulder-sized clasts in a weakly stratified pile. The randomly oriented clasts (see inset) are primarily derived from the Turkut and underlying Khatyspyt formations, but they also consist of occasional green igneous rocks and metamorphic rocks of exotic origin. Subrounded clasts plucked from the surrounding carbonate-rich matrix are notably faceted.

Figure 7.5 (a) Freshly exposed *Pteridinium* surfaces coated with yellow colored jarosite, a hydrous sulfate of potassium and iron that forms as a product of pyrite weathering, at Aarhauser on Farm Aar, near Aus, Namibia [*Hall et al.*, 2013]. Inset shows a typical iron oxidize patina on an exposed and weathered surface of *Pteridinium* at the same locality. Reproduced with permission of Elsevier. (b) "Mud chip" breccia associated with *Ernietta*-bearing sandstone from Ernietta Hill on Farm Aar. The mud chips have the iron oxide patina indicated above, and a partially exposed erniettid is exposed on the surface (see red arrow). This suggests that the chips represent the surface connection between infaunal *Ernietta* bases and their epibenthic fronds. (c) Sock-shaped sandstone concretions with flat upper surfaces likely to represent *Ernietta* specimens that have lost their tubular covering through exposure and weathering on Windy Peak, Farm Aar. These specimens are the same size and general shape to *Ernietta* preserved in situ with tubular structures (d) (yellow scale bar 1 cm).

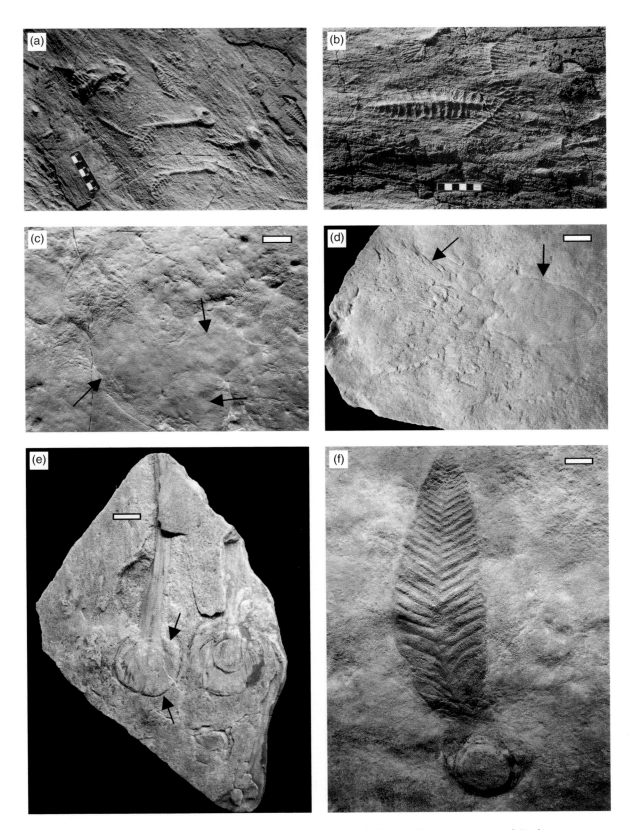

Figure 7.7 (a) Conception-style preservation [*Narbonne*, 2005] of *Charniodiscus spinosus* and *Cychrus procerus* in positive relief on a surface of the Mistaken Point Formation of Newfoundland. The fossil surface is overlain by a thin bed of volcanic ash (visible as dark layer on the upper right). (b) Spindle-shaped *Fractofusus misrai* with primary and secondary branches on the same surface in Newfoundland with visible volcanic ash. Fossils in (a) and (b) are members of the Avalon Assemblage. Images (c–f) are from the Ediacaran member in South Australia and archived at the South Australia Museum (white scale bars are 5 cm in length). These are representatives of the White Sea Assemblage. (c) *Dickinsonia costata* moving (see arrows) and resting traces. (d) *Kimberella quadrata* (right arrow) and associated trace fossil *Kimberichnus teruzzi* (left arrow). (e) *Aspidella* with holdfast and stalk, but no frond. (f) *Arborea arborea*. *Source:* Government of South Australia.

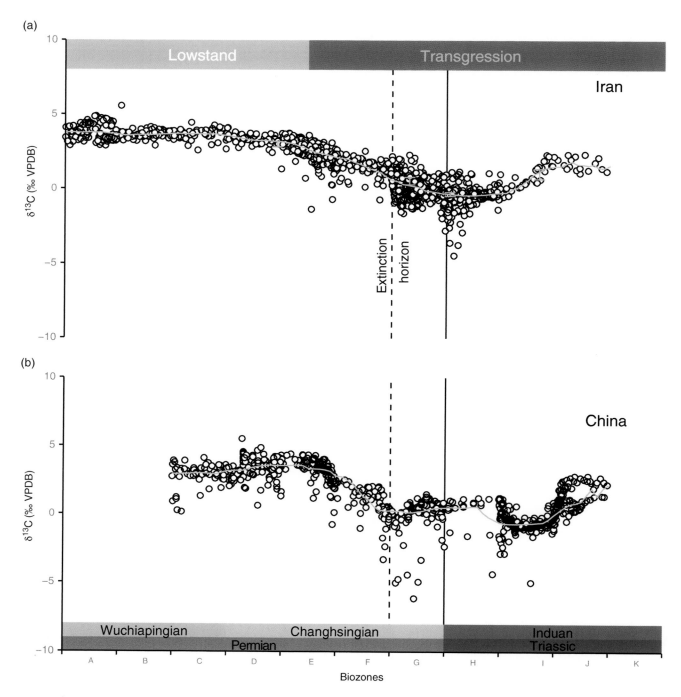

Figure 9.1 A compilation of published bulk rock δ¹³C data from multiple sites in Iran (a) and the P-Tr GSSP at Meishan in South China (b) (Modified from *Schobben et al.* [2017]. https://www.clim-past.net/13/1635/2017/. Licensed under a CC BY 3.0.). The data points are placed on a dimensionless timeline to ensure a better comparison of trends in the δ¹³C data of both regions. The dimensionless timeline is based on conodont zones. (The biostratigraphic scheme and sources regarding sea-level change and the extinction horizon are given in Supporting Information Text S1.) The trendline is based on a subsampling routine as outlined in *Schobben et al.* [2017]. The yellow error bars represent the value ranges of newly generated δ¹³C data for this study.

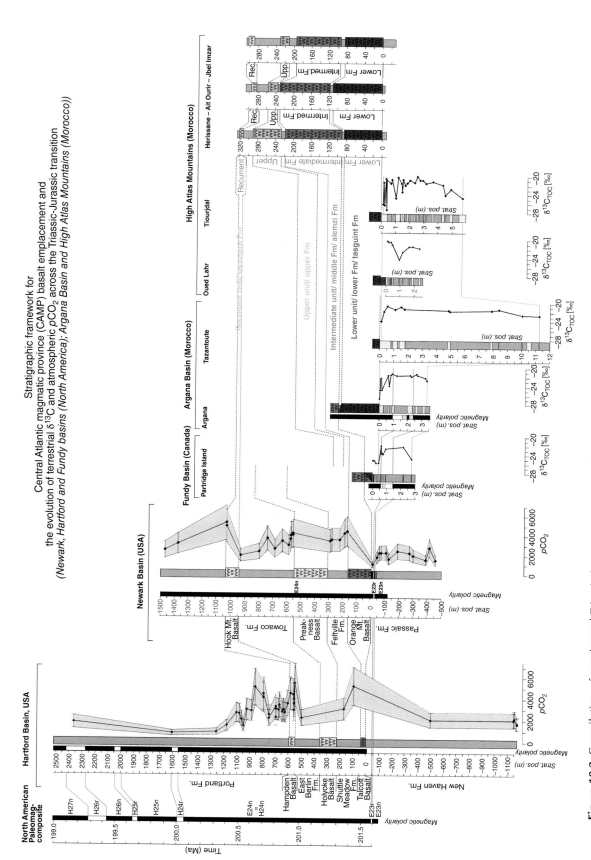

Figure 10.3 Compilation of continental Triassic-Jurassic sequences, which have been analyzed for $\delta^{13}C_{TOC}$ and $\delta^{13}C_{wood}$ and which record Central Atlantic Magmatic Province (CAMP) basalt emplacement. Data from: Hartford Basin, USA: *Schaller et al.* [2012], *Kent et al.* [2017] and references therein; Newark Basin, USA: *Schaller et al.* [2011], *Kent et al.* [2017] and references therein; Fundy Basin, Canada: *Deenen et al.* [2011]; Argana Basin, Morocco: *Deenen et al.* [2010]; High Atlas Mountains, Morocco: *Dal Corso et al.* [2014], *El Ghilani et al.* [2017].

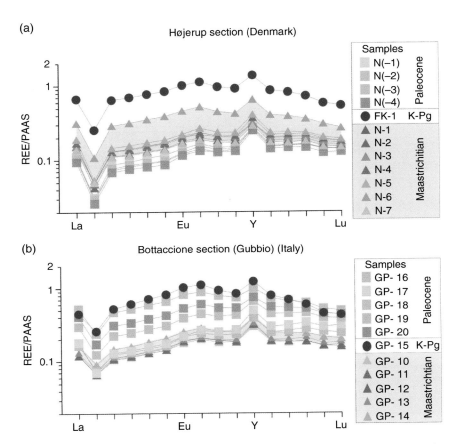

Figure 12.4 PAAS-normalized REE patterns for (a) the Højerup section and (b) Bottaccione section.

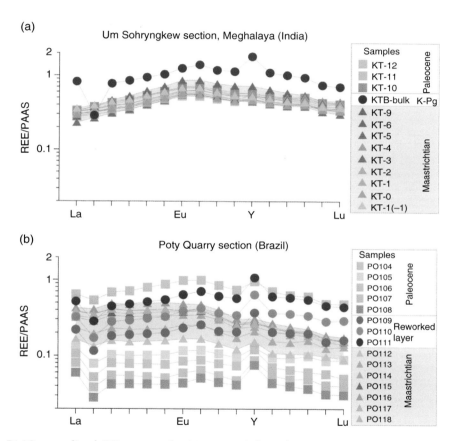

Figure 12.5 PAAS-normalized REE patterns for (a) an Um Sohryngkew River section and (b) drill core samples from a Poty drill hole.

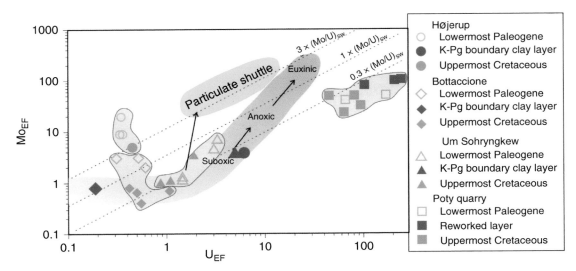

Figure 12.7 Mo$_{EF}$ versus U$_{EF}$ covariation for the Højerup, Bottaccione, Um Sohryngkew, and Poty K-Pg boundary sections (Based on *Tribovillard et al.* [2012] and *Sosa-Montes et al.* [2017]), in which Mo$_{EF}$ = [(Mo/Al)$_{sample}$/(Mo/Al)$_{PAAS}$] and U$_{EF}$ = [(U/Al)$_{sample}$/(U/Al)$_{PAAS}$]. The PAAS composition used is from *Taylor and McLennan* [1985]. The diagonal lines represent multiples of the Mo/U ratio of the present-day seawater. The orange field represents a general pattern of Mo$_{EF}$ versus U$_{EF}$ covariation in unrestricted marine trend for modern eastern tropical Pacific (From *Tribovillard et al.* [2012], modified by *Sosa-Montes et al.* [2017]), and the yellow field (From *Tribovillard et al.* [2012]) represents the particulate shuttle trend in which intense cycling of metal oxyhydroxide occurs within water column.

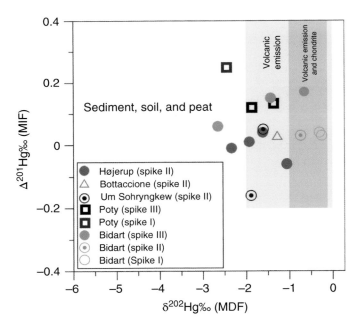

Figure 12.8 In a δ^{202}Hg (MDF)-Δ^{201}Hg (MIF) plot, modified from *Sial et al.* [2016], samples from the K-Pg layer (spike II) from the Højerup, Bottaccione, and Um Sohryngkew sections lie within the range for volcanogenic Hg. One sample from the K-Pg layer (spike II), two samples from the spike I, and one from the spike III, from the Bidart section, added for comparison, lie within the range for chondrite/volcanogenic emission. One sample from the spike III of this section lies within the volcanogenic emission field, and another within the sediment, soil, and peat range. Ranges for volcanogenic and chondritic Hg are from *Bergquist and Blum* [2009] and are shown as vertical bars.

dolostones gradually pass into dolostone/siltstone rhythmite and, subsequently, to 5 m thick laminated siltstone that are unconformably overlain by Silurian diamictites (Figs. 6.2 and 6.6). The contact corresponds to an angular unconformity where the dolostone beds dip gently (3–15°) toward SW contrasting with subhorizontal beds of overlaid Silurian deposits. In the southern border of the Colorado graben, at the Chupinguaia region, the cap dolostone is overlaid by Jurassic basalts of the Anari Formation (Fig. 6.6).

6.5.1. Cap Dolostone

Pinkish cap dolostone is composed of three peloidal microfacies (M) (Fig. 6.3): (M1) 3–5 m thick dolopackstone with low-angle to even parallel lamination, massive bedding, macropeloid lenses (Fig. 6.7g), and subordinate wavy cross-lamination; (M2) 10 m thick dolomudstone/dolopackstone, fenestral microbial laminites with tubestone structures; and (M3) 5 m thick dolopackstone showing giant wave ripple structures (megaripple bedding) with sparse calcite crystal fans (aragonite pseudomorphs) (Fig. 6.7c). The synoptic profile of megaripple bedding revealed a swing of the symmetric crest migration (Fig. 6.7b–d). At Mirassol d'Oeste, the peloidal packstone and the microbial laminites (Fig. 6.7a) are occasionally impregnated by bitumen, near the contact with bituminous cap limestone, usually associated with secondary calcite and sparry dolomite filling the porosity (Fig. 6.7e). Dolostone/siltstone rhythmite overlays the dolomite facies in the Espigão d'Oeste region. The dolomite facies generally exhibit neomorphic fabric composed also of sparry dolomite (microsparite and pseudosparite with <40 μm crystals) with some micrite (peloidal) texture preserved. Terrigenous grains (silt) occur disseminated or forming laminae and secondary calcite-filled fractures. Irregular vug and fenestral porosities (40 μm to 1.5 mm) make up to 2% of the rock. Stylolite and dissolution seams develop fitted fabric. Peloids comprise ~10–15% of the rock. Equant and subordinate bladed dolomite filling interpeloidal porosity (3–5% of the rock) also occurs. Euhedral dolomite crystals (20–50 μm) occur locally at the pore vug edges (Fig. 6.7f). In Espigão d'Oeste, dolostone rhythmically interbedded with shale overlies low-angle laminated dolomite and are underlain by 5 m thick reddish laminated siltstone. In this region, the cap carbonate is overlain unconformably by early Paleozoic diamictites and Mesozoic rocks (Fig. 6.6).

The first meters of the cap dolostone were precipitated in moderate to high-energy shallow platform environment, reworked by combined (storm) and oscillatory flows evidenced by the presence of low-angle truncations and even parallel lamination [De Raaf et al., 1977; Arnott, 1993; Dumas and Arnott, 2006]. Oscillatory flow is also indicated by megaripple-bedded dolomite whose synoptic profile revealed a swing of the symmetric crest migration. Massive bedding at the base of the cap carbonate associated with the deformed contact zone indicates plastic adjustment probably by liquefaction of dolomicritic gel. The primary (i.e., penecontemporaneous to syndepositional) to early diagenetic origin of dolomite, indicated by preservation of laminations, by fenestral (primary) porosity, and mainly by peloidal and micritic textures, is interpreted as resulting from biological activity [Vasconcelos et al., 1995; Riding, 2000; James et al., 2001; Nogueira et al., 2003; Font et al., 2005]. Macropeloids were generated by the aggregation of micropeloids on a low-energy environment substrate, indicated by exceptional preservation of these particles. The fast cementation also contributed to the preservation and aggregation of peloidal particles preserved by early cementation of the dolomitic gel [Riding, 2000; James et al., 2001; Halverson et al., 2004].

The fining upward tendency of the cap dolomite, where the chemical precipitation is successively replaced by suspension processes, indicates water depth increase. The dolomicrite precipitation occurred in conditions of intense alkalinity induced by sulfate-reducing bacteria [Nogueira et al., 2003; Font et al., 2006; Elie et al., 2007; Sansjofre et al., 2011]. Fibrous aragonite was precipitated from highly $CaCO_3$-saturated waters. Allen and Hoffman [2005] have suggested the action of extreme winds inducing the migration of giant wave ripple recorded in the top of cap dolostones. Conversely, Lamb et al. [2012] suggest that these bedforms were formed by purely oscillatory flow without extreme wind action and the high syndepositional cementation rates that allowed aggradation and preservation of these features, a mode of precipitation of carbonate without a modern analog. In fact, megaripple bedding is a complex bedform constructed by climbing wave ripple laminations above the storm wave base in a moderately deep platform.

The first meters of cap dolostone with low-angle stratification have been interpreted as storm beds [Gaia et al., 2017]. The platform carbonate succession recorded in the portion inland of the craton exhibits dolomicrite alternating with shale laminae, compatible with distal turbidites [Gaia et al., 2017]. The progressive terrigenous inflow upsection reflects the increased water depth and consequently a lower state of carbonate saturation like other cap carbonates worldwide [Hoffman and Schrag, 2002].

Tubestone structures, a sedimentary structure restricted to Marinoan cap dolostones, were also recognized at Terconi quarry in Mirassol d'Oeste. The genesis of tube structures is unknown, although fluid scape among microbial laminites has been pointed as the process responsible for these structures [Corsetti and Grotzinger,

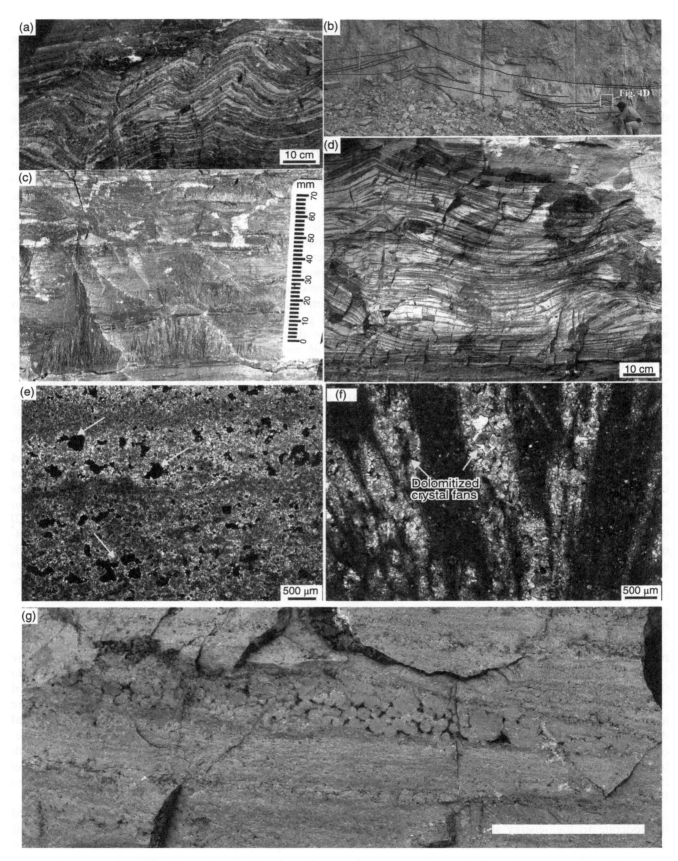

Figure 6.7 General aspects of cap carbonate in the Amazon Craton. (a) Microbialites. (b) Giant wave ripple. (c) Calcite crystal fans (pseudomorphs after aragonite) interbedded with undulated lamination. (d) Detail of (b), showing the climbing wave ripple cross-lamination and well-preserved crest of megaripple. (e) Fenestral porosity filled by bitumen. (f) Dolomitized crystal fans. (g) Macropeloids (scale bar = 2 cm). Bitumen impregnation detached all laminations of carbonates. *(See insert for color representation of the figure.)*

2005; *Hoffman*, 2011]. Despite the uncertain origin, the temporal restriction of tubestones associated with microbialites allows their use as a chronostratigraphic tool [*Romero et al.*, 2016].

The inference of a deepwater environment for the cap carbonate exposed in the Mirassol d'Oeste region using alone fenestrae and "keystone vugs" [*Font et al.*, 2006] is unwarranted. Keystone vugs or "beach bubble" is a typical irregular fenestra found in beach sands formed by water movement combined with air through the sediment [*Packard and Hills*, 2001]. In the absence of grain support provided by larger clasts, these vugs cannot be interpreted as keystone pores and do not appear to be a selective fabric. These features occur isolated and are not associated with other sedimentary structures indicative of exposition or deepwater environments.

In all occurrences of cap carbonate, the facies succession is retrogradational related to the installation of a shallow to moderately deep platform. The cap carbonate occurrence in the southwestern craton suggests a large flooding that reached the inland portion of the Amazon Craton with the development of an expressive dolomitic platform during the early Ediacaran. Carbonate precipitation in modern platforms is mainly aragonitic, and primary dolomicrite is restricted to lagoon and low-energy settings generally linked to the replacement of Ca by Mg in carbonate induced by sulfate-reducing bacteria in alkaline waters [*Vasconcelos et al.*, 1995]. Neoproterozoic dolomitic platforms are atypical settings with no modern analog.

6.5.2. Cap Limestone Cementstone

The 25 m thick cap limestone consists of fine-grained bituminous limestone with disseminated terrigenous grains interbedded with laminae of shales rich in pyrite (Fig. 6.6). In Tangará da Serra region, a laterally discontinuous marl/siltstone bed 20cm to 1.60m in thickness occurs at the base of the cap limestone, onlapping large-scale syncline and anticlines associated with subvertical fractures and faults, and chevron folds imprint at the top of cap limestone [cf., *Soares et al.*, 2013]. In this region, the marls are succeeded by terrigenous limestone with megaripple bedding displaying wavelengths between 1.10 and 1.3 m and amplitudes of up to 35 cm. Ripple marks occur locally associated with the top of the megaripple bedding. The most prominent neomorphic feature in these rocks is fibro-radial calcite crystals considered as aragonite pseudomorphs (Figs. 6.6 and 6.7), like those found in other Neoproterozoic cap carbonates worldwide [*Clough and Goldhammer*, 2000; *James et al.*, 2001; *Hoffman and Schrag*, 2002; *Corsetti et al.*, 2004; *Lorentz et al.*, 2004]. The crystals form columnar arrays, isolated or laterally connected by fine fibrous crystals. The crystal

fans reach 5cm in size and developed on planar laminae, while at the top of the crystals, the micritic laminae conform irregularly to the crystals surface, also detached by compaction. Ripple marks with interference pattern occur at the top of limestone beds with crystal fans (Fig. 6.7).

In the Mirassol d'Oeste and Tangará da Serra regions, deformational features consist of vertical to subvertical fractures and faults and intraformational conglomerate and breccia filling neptunian dykes, forming locally slump folds developed in deformed beds sandwiched by undeformed beds [*Soares et al.*, 2013]. The fine (micritic) limestone with crystal fans separated by shale laminae is the most common facies of the cap limestone and is distributed along the craton margin for hundreds of kilometers. The matured organic matter found in the form of bitumen occurs disseminated in the micrite and filling intercrystalline porosities and fractures (Fig. 6.6).

The fine-grained limestones, with no wave and tidal current structures, laterally continuous for hundreds of kilometers are compatible with a deeper water environment, below the storm wave base [*Pfeil and Read*, 1980; *Stow*, 1986]. The contact with the cap dolostone is marked by deposition of silt and lime muds interbedded with aragonite-rich layers (crystal fans). The well-preserved arborescent crystals suggest calm and deep waters just below the storm wave base. The abundance of crystals suggests that these facies were formed in a $CaCO_3$-supersaturated environment (high alkalinity; *Grotzinger and Knoll*, 1995; *Sumner and Grotzinger*, 1996; *Sumner*, 2002; *Corsetti et al.*, 2004; *Lorentz et al.*, 2004]. The presence of ripple marks at the top of the crystal fans indicates weak currents in a calm water environment. The main causes for aragonite precipitation would be changes in the ocean circulation and temperature and in atmospheric CO_2 concentration, which together would produce a rapid increase in aragonite saturation [*Sumner*, 2002].

The passive nature of lime mud deposition indicates that the contact is a flooding surface onlapping the giant wave ripple morphologies. The presence of siliciclastic grains disseminated in the limestone may be related to the first pulse of terrigenous continental influx into the platform. Turbidity currents may have generated the normal gradation found in the lime mudstone layers deposited above the storm wave base. The retrogradational tendency of the cap limestone is consistent with the deglaciation phase followed by further transgression, similar in other cap carbonates worldwide [cf., *Fairchild and Hambrey*, 1984; *Tucker*, 1986; *Kennedy*, 1996; *James et al.*, 2001; *Hoffman and Schrag*, 2002]. The megaripple bedding found in the cap limestone, at Tangará da Serra, was formed by migration of bedforms induced by oscillatory flow and, secondarily, currents in the offshore transition zone [cf., *De Raaf et al.*, 1977; *Aigner*, 1985].

The association of faults, slump structures, mass flow, neptunian dikes, and breccias indicates gravitational instability and rupture of partially lithified sediment compatible with deepwater sedimentation related to the slope of a ramp on the seafloor. The consistent orientation of observed brittle and ductile deformational structures is compatible with the regional tectonics of the Paraguay Belt and suggests earthquakes as the triggering mechanism for sediment deformation and large-scale mass failure [*Soares et al., 2013*].

The indigenous nature of the bitumen, primarily accumulated as hydrocarbon fluids migrated to high-porosity and high-permeability carbonates, further altered into bitumen or migrabitumen [*Sousa et al., 2016*]. These bitumen-rich carbonates are restricted to the Guia Formation, sandwiched by dolomitic units of the Araras Group with relative low porosity (Fig. 6.2). The primary precipitation of dolomite, crystal fans, and peloidal micrite is related to depositional processes [*Sansjofre et al., 2011; Bosak et al., 2013; Soares et al., 2013*]. Cap limestone is more diagenetically altered than cap dolostone despite both units being submitted to early and late diagenesis. Cap dolostones are mainly affected by early diagenetic processes such as dolomite neomorphism, dissolution (vug porosity), and cementation (Fig. 6.7e). Chemical compaction (stylolites and dissolution seams) is the only evidence of late diagenesis. In the cap limestone, the main early diagenetic processes were neomorphism of crystal fans and micrite and chemical compaction (Fig. 6.7f and g). Meteoric fluids could have favored the inversion of aragonite to calcite in crystal fans soon after burial. Dolomitization, fracturing, and precipitation of ferrous dolomite and calcite are observed locally (Fig. 6.7g). Dolomite fractures (Fig. 6.7h) can be associated only with syn-sedimentary deformation of these deposits [*Soares et al., 2013*), while limestone fractures are associated with burial (filled with calcite and ferrous dolomite and hydrocarbons) and more rarely to the deformation, which occurs near neptunian dykes [*Nogueira et al., 2003*].

6.6. PALEOBIOLOGY AND BIOSTRATIGRAPHY

Even with the abrupt environmental changes recorded at the end of the Cryogenian and beginning of the Ediacaran, the fossil record is consistent in the Marinoan cap carbonate. The Ediacaran fossil record in postglacial units includes microbialites (Fig. 6.8a–c), acritarchs and other organic-walled microfossils (Fig. 6.8d and e), and biomarkers (Fig. 6.8b), all of which were already recognized in the Marinoan cap carbonate [*Hidalgo, 2007; Brelaz, 2012; Romero, 2015; Sousa et al., 2016*]. The most abundant fossils are well-preserved microbialites, identified above the contact with the diamictites of the Puga

Formation, indicating, therefore, a rapid recovery of the benthic ecosystems at the end of the Marinoan glaciation [*Nogueira et al., 2003; Romero, 2015*]. On the other hand, biomarkers and organic-walled microfossils are less abundant, but still showing application for paleobiological considerations and putative biostratigraphical studies, as shown below.

The microbialites of the Mirassol d'Oeste Formation consist of simple microbial laminites, comprising stratiform, domic, and irregularly wavy stromatolites, the last ones being more common at the top of the postglacial sequence (Fig. 6.8a–c). They occur in ~65% of the deposits of the cap dolomite in its type locality at Terconi quarry [*Romero, 2015*]. These microbialites were formed by the constant alternation of thick and thin laminations composed mainly of peloids, interpreted as micritized colonies of coccoidal cyanobacteria. Besides, thick laminations present greater abundance of fenestrae in comparison to thinner laminae, interpreted as periods of higher biomass production and consequent gas release [*Romero, 2015*].

The abundance of microbialites indicates the development of microbial mats in the photic zone of a wide epicontinental sea, due to ocean transgression in the aftermath of the Marinoan event [*Bosak et al., 2013; Romero, 2015*]. The irregular microbial laminites are related to environmental changes due to mega-hurricanes that occurred during the postglacial phase [*Hoffman and Schrag, 2000; Romero, 2015*], which, in this case, resulted in environmental conditions unviable for the maintenance of benthic communities responsible for microbial growth.

The Marinoan cap carbonate was analyzed for Neoproterozoic biomarkers. Few studies were performed, but even so, they provided interesting data for the recovery of the biota following the Marinoan event. *Elie et al.* [2007] found short-chain acyclic isoprenoids which were interpreted as derived from bacterial chlorophyll. C_{27} steranes were also obtained through gas chromatography-mass spectrometry, pointing to the presence of rhodophyceous algae. Other identified biomarkers were 5,28,30-trisnorhopanes, the source of which was attributed to bacteria, and aryl isoprenoids, produced by green sulfur photosynthetic bacteria, although authors also cited a carotenoid origin.

The bitumen has low total organic carbon content, and gas chromatography trace analysis indicated the presence of the biomarkers n-C_{14} to n-C_{37} alkane. They suggest a major contribution of marine algae [*Sousa et al., 2016*]. The presence of C_{14}–C_{25} monomethyl alkanes, attest the occurrence of cyanobacteria, and the relative abundance of gammacerane indicates a well-stratified saline paleoenvironment during the deposition of the cap carbonate [*Sousa et al., 2016*]. The presence of mid-chain

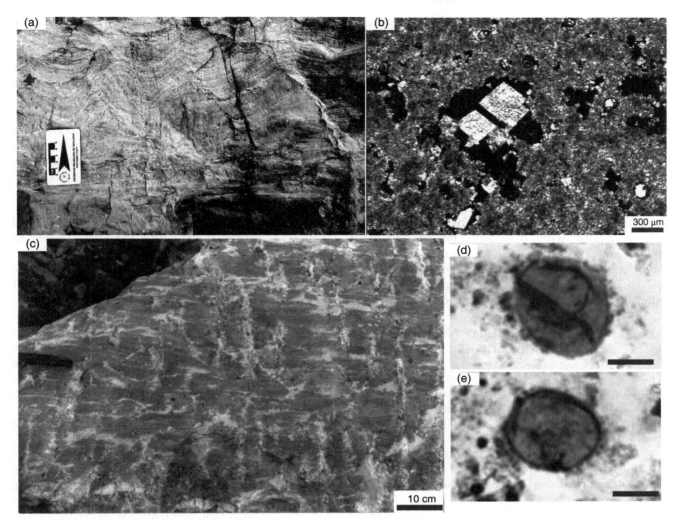

Figure 6.8 Representative fossils of the Marinoan cap carbonate in the Amazon Craton. (a) Typical microbial laminite. (b) Fenestral porosity filled by bitumen and euhedral dolomite. (c) Tubestone structures associated with microbial laminites. (d and e) *Leiosphaeridia* taxa from the intermediate part to top cap carbonate (scale bars = 20 µm). *(See insert for color representation of the figure.)*

monomethyl alkanes (C_{14}–C_{25}) is another biomarker commonly found in middle to late Proterozoic oils [*Elie et al.,* 2007; *Sansjofre et al.,* 2011; *Sousa et al.,* 2016].

An acritarch assembly was recovered in the upper dolomitic portion of the Marinoan cap carbonate, above the microbial laminites, and in the limestone part of the cap carbonate [*Hidalgo,* 2007; *Brelaz,* 2012]. Some specimens were also recovered from the Nobres Formation, the topmost unit of the Araras Group [*Hidalgo,* 2007].

The acritarchs in general are rare, spheroidal, and poorly preserved, although still subject to taxonomic classification [*Hidalgo,* 2007; *Brelaz,* 2012]. The cap carbonate assemblage is dominated by spheromorphic acritarchs, especially *Leiosphaeridia* (Figs. 6.8d and e), but presents subordinate prokaryotes, such as *Chlorogloeaopsis* and *Siphonophycus* [*Hidalgo,* 2007]. On the other hand,

the acritarch assembly reported for the Nobres Formation, the topmost unit of the Araras Group, comprises mainly acanthomorphic taxa, including *Cavaspina* sp., *Ericiasphaera* sp., *Appendisphaera barbata*, *Tanarium* sp., and *Micrhystridium pisinnum*, with cyanobacteria and other subordinate prokaryotes [*Hidalgo,* 2007].

Hidalgo [2007] suggested that the assemblage found in the Marinoan cap carbonate is typical of successions deposited after the Marinoan glaciation, correlating these assemblages to the *Ediacaran Leiosphere Palynoflora* (ELP) biozone of Australia, following the works of *Grey* [2005], indicating an early Ediacaran age for these deposits. However, there are two problems for the biostratigraphic associations on the Araras Group. The cap carbonate assemblage cannot be correlated with other proposed biozones for the early Ediacaran successions in

China, Siberia, and India. The post-Marinoan basal Ediacaran succession preserved in the Doushantuo Formation in China [*McFadden et al., 2009*], as well as acritarchs preserved in the Siberian platform [*Sergeev et al., 2011*], shows a record dominated by acanthomorphic forms (e.g., *Tianzhushania*), with only subordinate leiospheres. In Australia, the acanthomorphic forms succeed the ELP assemblage and coincide with the establishment of metazoans, but the lowermost *Tianzhushania* acanthomorph assemblage is probably missing [*Liu et al., 2013*].

Nonetheless, the lack of diagnostic images of acritarch specimens in the assemblages of the Araras Group weakens the interpretations of these biozones. In fact, the only acritarchs with diagnostic images (e.g., the *Leiosphaeridia* genus; Figure 6.8d and e) are not indicative of specific biozones in the Ediacaran, being preferable to infer the age of deposits through the set of sedimentologic, chemostratigraphic, and Pb-Pb dating.

6.7. CHEMOSTRATIGRAPHY AND GLOBAL CORRELATIONS

The $\delta^{13}C$ values of cap carbonates exposed in the Espigão d'Oeste and Chupinguaia regions are relatively higher than those found in the Mirassol d'Oeste and Tangará da Serra regions (Fig. 6.6). The limited diagenetic modification of the carbonates indicates that at least the carbon isotopic signal is primary. Despite the variations, the relative homogeneity of C isotopic values along entire succession suggests stability of seawater isotopic composition for a prolonged period. The cap carbonates in the southwestern Amazon Craton exhibit $\delta^{13}C$ values ranging between -2.34 and $-4.69‰$ VPDB and $\delta^{18}O$ values varying from -6.40 to $-8.72‰$ VPDB, with incipient covariance (Figs. 6.5 and 6.9). Basalts of the Anari Formation overlie locally the cap dolostone and promote recrystallization of carbonate and possible alteration of the $\delta^{13}C$ signal [*Gaia et al., 2017*]. At the classical occurrences of cap carbonate in the Amazon Craton, in the Mirassol d'Oeste and Tangará da Serra regions, the average $\delta^{13}C$ values of cap carbonates are around $-5‰$, with more negative values near $-9‰$ (Fig. 6.9). High values are coincident with stratigraphic surfaces also marked by high $\delta^{18}O$ values above $-6‰$ and high $^{87}Sr/^{86}Sr$ ratios up to 0.709 (Fig. 6.9). The homogeneous pattern of the $\delta^{13}C$ curve observed in the limestone succession (Guia Formation) with predominant values of $-2‰$ is observed for at least 300 m above the cap carbonate (Fig. 6.9). Upsection, the $\delta^{13}C$ values pass gradually to 0‰ persisting for more than a 100 m (Serra do Quilombo Formation) and become slightly negative ($-1‰$, Nobres Formation) near the unconformity with the Cambrian Raizama Formation (Fig. 6.9). $^{87}Sr/^{86}Sr$ values between 0.7074 and 0.7089 have been obtained for the limestones [*Nogueira et al., 2007; Alvarenga et al., 2008; Paula-Santos et al., 2017*]. However, *Romero et al.* [2013] obtained lower values of 0.7071–0.7073 that they interpreted as the primary isotopic signature of marine Sr. These authors attributed the previous more radiogenic values to a contribution from siliciclastic components in the carbonates. These Sr signatures are compatible with the lowest ones encountered elsewhere for the post-Marinoan carbonates [ca. 0.7072–0.7087; *Kaufman et al., 1993; Saylor et al., 1998; Jacobsen and Kaufman, 1999; Halverson et al., 2004, 2005, 2007, 2010, 2011*]. Finally, the integration of ichnologic and radiometric information, combined with finely calibrated curves of $\delta^{13}C$ and $^{87}Sr/^{86}Sr$, correlated mainly with the Otavi Group of Namibia, allowed to elaborate a robust chronostratigraphic framework for Neoproterozoic units in the southern Amazon Craton and northern Paraguay Belt (Fig. 6.2 and 6.9).

The typical late Ediacaran biota found in the Tamengo Formation in the southern Paraguay Belt [cf., *Becker-Kerber et al., 2017*] is not observed in the ~700 m of the Araras Group, and isotopic and geochronologic data indicate only lower Ediacaran strata in the northern Paraguay Belt. In the same way, ~100 m thick fine siliciclastics of the upper Serra Azul Formation, previously considered as a putative postglacial record and attributed to Gaskiers glaciation of ca. 580 Ma [*Alvarenga et al., 2007; Figueiredo et al., 2008*], are here reinterpreted as offshore deposits included at the base of the Cambrian Raizama Formation (Figs. 6.2 and 6.3). The mapping carried out in the western and northern portions of northern Paraguay Belt, which include the type section of the Serra Azul Formation, revealed that these rocks are tectonically deformed by brittle-to-brittle-ductile structures [*Santos, 2016*]. The array represents a set of normal oblique dextral faults trending east-west, north-south, northeast-southwest, and northwest-southeast and fracture cleavages and brittle-ductile asymmetrical synformal forced folds, with northeast plunging axes (Fig. 6.3). Normal oblique fault sets comprise graben and horst systems, determining an east-west trending tectonic contact between the Puga Formation and the fine siliciclastics of the basal Raizama Formation (Fig. 6.3). At some places, this tectonic contact places the Puga diamictites topographically on top of the Araras Group (Fig. 6.3). The apparent gradational contact between these two units comprises heavily weathered fine sediments, the glaciogenic diamictites (Puga Formation) and offshore shales (Raizama Formation), but is in fact an expressive unconformity that represents ~80 Ma (Figs. 6.2 and 6.3).

During the Cambrian, the offshore fine deposits of the Raizama Formation covered indiscriminately phyllites and metadiamictites of the Cuiabá Group, as well as the

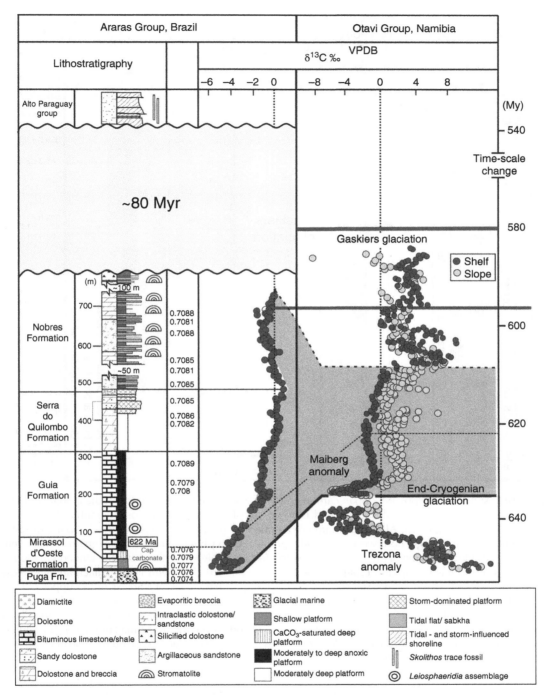

Figure 6.9 Carbon isotope and [87]Sr/[86]Sr chemostratigraphy and lithostratigraphy of the Araras Group in the southern Amazon Craton (Modified of *Nogueira et al.* [2007]) compared with composite $\delta^{13}C$ curve of the Otavi Group in Namibia [Halverson *et al.*, 2005; *Macdonald et al.*, 2009]. The ^{13}C isotopic trend of the Araras Group is correlated with the shelf values of Cryogenian Abeneb Subgroup. Approximately 500 m thick shelf deposits of Araras Group sections have been stretched to fit the Otavi Group section. The Maiberg anomaly is recorded in the base of Araras Group and marks the Cryogenian-Ediacaran boundary. In this context, the Araras Group does not transpass the lower Ediacaran. Middle to late Ediacaran is not recorded in the southern Amazon Craton. Reproduced with permission of Elsevier.

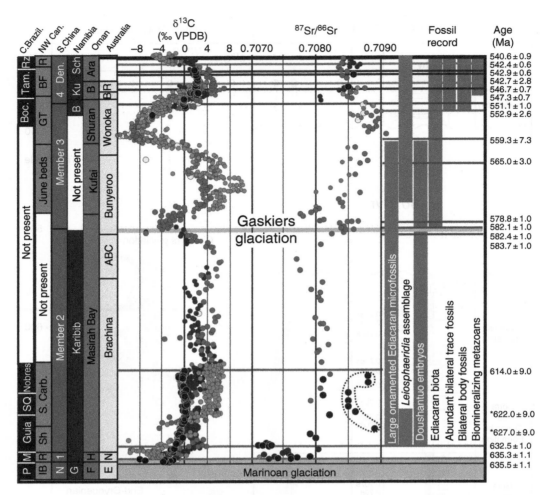

Figure 6.10 Insertion of the post-Marinoan carbonates of the southern Amazon Craton in the global correlations for the Ediacaran key locations with associated isotopic, geochronologic, and biostratigraphic data (Modified from *Macdonald et al.* [2013]). All the data are color coded for geographic location: Black, southern Amazon Craton, Central Brazil; orange, northwestern Canada; purple, South China; blue, Namibia; green, Oman; yellow, Australia; red, White Sea region of Russia; brown, Avalon Terrane of Newfoundland and the United Kingdom. The carbonates of the Araras Group and Espigão d'Oeste formations are close with lower Ediacaran age below of 614 Ma, corroborated by δ13C and 87Sr/86Sr values. Enriched positive values of 87Sr/86Sr in the upper portion of succession (dotted circle) represent samples probably contaminated with high content of siliciclastics. P, Puga Formation; M, Mirassol d'Oeste Formation; Guia, Guia Formation; SQ, Serra do Quilombo Formation; Nobres, Nobres Formation; Boc., Bocaina Formation; Tam., Tamengo Formation; Rz., Raizama Formation. For additional details and other abbreviation of units, see *Macdonald et al.* [2013]. Reproduced with permission of Elsevier. *(See insert for color representation of the figure.)*

Marinoan Puga diamictites and carbonate rocks of the Araras Group (Fig. 6.3). U-Pb dating of detrital zircons and Ar-Ar ages of detrital micas of diamictites exposed in the type area of the Serra Azul Formation indicated a maximum age of 640 ± 15 Ma, which is also compatible with a Marinoan depositional age [*McGee et al.*, 2014, 2015]. Thus, the diamictites or pebbly mudstone previously considered as recording the Gaskiers event (late Ediacaran) belongs, in fact, to the Marinoan Puga Formation. The apparent structural conformity between

the Puga Formation and offshore deposits of the Raizama Formation (Cambrian) observed in the type section of the Serra Azul Formation led to interpret a supposed stratigraphic continuity (Fig. 6.3). This bedding orientation, dipping to the north, represents the effect of the last tectonic tilting imprint in these rocks by implantation of Paleozoic Parecis and Paraná basins during Ordovician [*Santos*, 2016].

The δ13C curve of the Araras Group shows the same pattern as other Marinoan cap carbonate sequences worldwide, that is, northwestern Canada, Namibia, China, Oman, and Australia (Figs. 6.9 and 6.10). The increase in $^{87}Sr/^{86}Sr$ values during the Ediacaran is related to an increase in the continental input into the oceans, probably associated with the Pan-African-Brasiliano tectonics [*Jacobsen and Kaufman*, 1999].

A late Ediacaran age is indicated for the carbonate succession in the southern Paraguay Belt, recording an Ediacaran biota composed of acritarchs, vendotaenids, soft-bodied metazoans (*Corumbella*), and skeletal fossils (*Cloudina*) found in the Corumbá Group [cf., *Gaucher et al.*, 2003; *Boggiani et al.*, 2010; *Pacheco et al.*, 2015; *Sial et al.*, 2016; *Becker-Kerber et al.*, 2017]. The positive excursion of $\delta^{13}C$ with values of up to +5.5‰ and the U-Pb SHRIMP zircon age of an ash bed of 543 ± 3 Ma [*Babinski et al.*, 2008] confirm the proximity of the Ediacaran-Cambrian boundary. More recent data yielded U-Pb TIMS ages of 542 ± 0.3 Ma for ash beds in the upper Tamengo Formation and 555 ± 0.3 Ma for the upper Bocaina Formation, lower down in the succession [*Parry et al.*, 2017]. These carbonate rocks are thus younger than those found in the northern Paraguay Belt. The negative $\delta^{13}C$ values found in the lower Bocaina Formation are consistent with the Shuram-Wonoka anomaly, given the proximity of this unit to the Tamengo Formation [*Boggiani et al.*, 2010]. Thus, as previously commented, the Bocaina Formation probably does not represent a cap carbonate (Fig. 6.10).

6.8. THE TIME SCALE OF CAP CARBONATE DEPOSITION AT THE CRYOGENIAN-EDIACARAN BOUNDARY

The contact between Marinoan diamictites and the cap carbonate in the Amazon Craton is coincident with the Cryogenian-Ediacaran boundary as defined in the GSSP located in the Nuccaleena Formation, South Australia [*Knoll et al.*, 2004, 2006]. The contact between the cap dolostone and the diamictite has been described as erosional, indicative of a diastema and hiatus [*von der Borch et al.*, 1989; *Christie-Blick et al.*, 1990; *Fairchild*, 1993; *Dyson and von der Borch*, 1994; *Kennedy*, 1996]. This surface was considered conformable and gradational or a sharp contact without evidence of erosion [*Plummer*, 1978; *Alvarenga and Trompette*, 1992; *Kennedy*, 1993, 1996; *Hoffman et al.*, 1998a, 1998b; *Myrow and Kaufman*, 1999]. It shows the same stratigraphic relationships worldwide, and the presence of ice-rafted detritus in the last meters of diamictite suggests a gradational origin linked to the fast Cryogenian deglacial transgressions [*Hoffman and Macdonald*, 2010; *Hoffman*, 2011]. In the Amazon Craton, the contact between the cap dolostone and glacial diamictite is sharp and deformed plastically, formed by syn-sedimentary deformation related to post-

glacial isostatic rebound following the Marinoan glaciation [*Nogueira et al.*, 2003; *Soares et al.*, 2013; *Gaia et al.*, 2017]. Considering the Pb-Pb age of 622 Ma and the $\delta^{13}C$ negative values and low $^{87}Sr/^{86}Sr$ variations consistent with an early Ediacaran age, the contact surface between diamictites and cap carbonate unequivocally represents the Cryogenian-Ediacaran boundary.

The duration of cap dolostone deposition is a fundamental question to infer how long the transition between the extreme icehouse to greenhouse conditions lasted. A rapid deglaciation has been questioned by magnetic inversions measured in the cap dolostone at Terconi quarry, which suggested carbonate deposition over hundreds of thousands of years, based on the frequency of Cenozoic paleomagnetic inversions [*Trindade et al.*, 2003]. This actualistic point of view for Neoproterozoic paleomagnetic data has been questioned by new research about the inner-core inception and growth [*Davies*, 2015; *Hoffman et al.*, 2017]. Following these data, the solid inner core was small or absent at 635 Ma, and the geomagnetic reversals probably occur much faster in comparison with Cenozoic reversals, corroborating a fast carbonate precipitation. Other age estimates for cap carbonate deposition given by *Font et al.* [2010] considered the annual rate of accumulation of primary dolomite produced by recent microbial mats in Lagoa Vermelha (Rio de Janeiro), yielding an estimate of 10,000–400,000 years for the 20 m of the cap dolostone in Mirassol d'Oeste. This interpretation was based on the hypothesis that the entire cap dolostone resulted from the precipitation of dolomite linked to the metabolism of sulfate-reducing bacteria. Thus, the recurrence and internal composition of the microbial lamination would be similar to annual deposits. However, only a 10 m thick interval of the cap dolostone is microbial (see Fig. 6.6d). The first 2 m of cap dolostone (peloidal dolostone with flat and low-angle lamination) and the last 3 m (giant wave-rippled dolostone) at Terconi quarry were deposited under episodic effects of waves and storms [*Nogueira et al.*, 2003, 2007]. Thus, we consider that the Lagoa Vermelha is not a suitable modern analog, since that cap dolostone was not deposited in a lagoon environment but in a shallow dolomitic marine platform without modern equivalent.

A fast cap carbonate precipitation was favored by previous interpretations of postglacial stages and a smaller influence of ice-melt waters on carbonate fabric. In fact, dolomicrite precipitation with persistent C isotope patterns associated with seafloor cements corroborates fast precipitation of cap carbonate before the expressive postglacial detrital input [*Corsetti et al.*, 2004; *Lorentz et al.*, 2004; *Hoffman et al.*, 2017]. Paleohydraulic analysis of wave ripples and tidal laminae of the post-Marinoan cap carbonate in Australia indicated that water depths of

9–16 m remained nearly constant for ~100 years throughout 27 m of sediment accumulation [*Myrow et al.,* 2018]. This accumulation rate (0.2–0.27 m/yr), too great to evocate subsidence, indicates much more to an extraordinary rapid rate of sea-level rise favoring the rapid deglaciation models of snowball Earth hypothesis during the passage to greenhouse climate [*Myrow et al.,* 2018]. In addition, the deformed dolostone-diamictite contact, considered as a distinctive feature of Marinoan cap carbonates in the southern Amazon Craton, indicates rapid accumulation of dolomicrite over a still plastic and unlithified diamictite. This deformation was linked to a regional-scale event compatible with isostatic rebound or seismic shock [*Nogueira et al.,* 2003; *Soares et al.,* 2013]. This surface correlates the cap carbonate occurrences in the Amazon Craton for over 600 km and coincides with the Cryogenian-Ediacaran boundary.

6.9. FINAL REMARKS

The southern Amazon Craton is a highly relevant area for studying episodes related to the snowball Earth glaciations. Previous studies in this region led to the recognition of cap carbonate sequences that overlie the Marinoan diamictites, representing the best preserved Cryogenian-Ediacaran boundary record in South America. The main events observed are the rapid decay of Marinoan ice sheets and onset of anomalous cap carbonates accompanied by a distinctive pattern of secular changes in carbon and strontium isotopes. The cap carbonate in Amazonia was deposited in a subsiding coastal setting directly atop glaciogenic sediments influenced by isostatic rebound and was succeeded by long-term transgression with and implantation of a $CaCO_3$-supersaturated deep platform. Unlike other occurrences worldwide, the cap carbonate base exhibits soft-sediment deformation, indicating abrupt transition from icehouse to greenhouse conditions. This deformed contact is one of several sedimentological evidences that suggest rapid precipitation of cap carbonates, contrasting with previous paleomagnetic data that indicate a slower accumulation rate. The use of actualistic duration of paleomagnetic polarity reversals for the Cryogenian period is debatable and remains an important subject to better understand the amount of time involved in the Cryogenian-Ediacaran boundary.

The review of previous geological, geochemical, and geochronological data allows to reinterpret some points of the late Neoproterozoic sedimentary history of the southern Amazon Craton such as the following: (i) the Pb-Pb geochronology combined with $\delta^{13}C$ negative values and low $^{87}Sr/^{86}Sr$ ratios indicates an early Ediacaran age for the cap carbonate sequence, similar to those found in northwestern Canada, where the Gaskiers glaciation is not represented; (ii) the unconformity of ~80 Ma between lower Ediacaran and Lower Cambrian deposits represents an important stratigraphic surface for regional correlation; (iii) the Cryogenian, Ediacaran, and Cambrian rocks were deposited in shallow to moderately deep platformal settings related to an intracratonic basin that was tectonically inverted during the Ordovician; and (iv) the collisional event that formed the Paraguay Belt is recorded only in the metasedimentary rocks of the Cuiabá Group.

This work aims to stimulate discussion about the need of more robust stratigraphic data based on facies analysis and depositional systems, as well as a coherent interpretation of the key surfaces in the sedimentary succession. In addition, it is demonstrated that the fundamental role of structural geology is to provide a better stratigraphic reading of the deformed segments exposed in the southern Amazon Craton.

ACKNOWLEDGMENTS

The authors would like to thank the Instituto Nacional de Ciência e Tecnologia de Geociências da Amazônia (INCT/GEOCIAM), the Conselho Nacional de Desenvolvimento Científico e Tecnológico (CNPq), the Fundação de Amparo à Pesquisa do Estado de São Paulo (FAPESP), and the LABISE/UFPE for financial, logistic, and technical support during 20 years for the research group of Sedimentary Basin Analysis of Amazonia ("Grupo de Análise de Bacias Sedimentares da Amazônia" (GSED)) of the Universidade Federal do Pará. Thanks to Dr. Claudio Riccomini for inestimable scientific support and encouragement of GSED researches in the last years. The authors thank Dr. Alcides Nóbrega Sial and Dr. Kathryn Corcoran for the tireless guidance during manuscript submission process and the invaluable opportunity given to GSED to contribute with the Ediacaran Geology of Brazil. They also thank the anonymous reviewers and especially Dr. Claudio Gaucher for the constructive discussions about the geology of the southern Amazon Craton that significantly improved the manuscript.

REFERENCES

Ader, M., Macouin, M., Trindade, R. I. F., Hadrien, M.-H., Yang, Z., Sun, Z., & Besse, J. (2009). A multilayered water column in the Ediacaran Yangtze platform? Insights from carbonate and organic matter paired δ13C. Earth and Planetary Science Letters, *288*, 213–227.

Afonso, J. W. L., & Nogueira, A. C. R. (2018). Sedimentology and stratigraphy of Neoproterozoic-lower Paleozoic carbonate-siliciclastic succession in the Southwestern most

Amazon Craton, State of Rondônia, Brazil. Brazilian Journal of Geology, 48, 75–93.

Aigner, T. (1985). *Storm Depositional Systems: Dynamic Stratigraphy in Modern and Ancient Shallow-Marine Sequences.* Earth Sciences (Vol. 3). Berlín: Springer-Verlag, 17 p.

Alkmim, F. F., Marshak, S., & Fonseca, M. A. (2001). Assembly Western Gondwana in the Neoproterozoic: clues from the São Francisco craton region, Brazil. Geology, 29, 319–322.

Allen, P. A., & Hoffman, P. F. (2004). Extreme winds and waves in the aftermath of a Neoproterozoic glaciation. Nature, 433, 123–127.

Allen, P. A., & Hoffman, P. F. (2005). Formation of Precambrian sediment ripples. Nature, 433, 123–127.

Almeida, F. F. M. (1964). Geologia do Centro-Oeste Matogrossense. Boletim da Divisão de Geologia e Mineralogia, 219, 1–53. Rio de Janeiro, RJ: Ed. DNPM.

Almeida, F. F. M. (1984). Província Tocantins, setor sudoeste. In: F. F. M. Almeida, Y. Hasui, (Eds.), *O Pré-Cambriano do Brasil* (pp. 265–281). São Paulo, SP: Ed. Blücher Ltda.

Almeida, F. F. M., & Mantovani, M. S. M. (1975). Geologia e geocronologia do Granito São Vicente, Mato Grosso. Anais da Academia Brasileira de Ciências, 47, 451–458.

Alvarenga, C. J. S. (1988). Turbiditos e A Glaciação do Final do Proterozóico Superior No Cinturão Paraguai, Mato Grosso. Revista Brasileira de Geociências, 18, 323–327

Alvarenga, C. J. S., & Trompette, R. (1988). Upper Proterozoic glacial environment of the border of Amazonian Craton and its evolution towards the adjacent Paraguay Belt. (Mato Grosso, Brazil). Paper presented at Conference on Meeting Earth's Glacial Record-Proj. IGCP-UNESCO/UFMT, Cuiabá, Brazil.

Alvarenga, C. J. S., & Trompette, R. (1992). Glacially influenced sedimentation in the Later Proterozoico of the Paraguay Belt (Mato Grosso, Brazil). Palaeogeography, Palaeoclimatology, Palaeoecology, 92, 85–105.

Alvarenga, C. J. S., Santos, R. V., & Dantas, E. L. (2004). C–O–Sr isotopic stratigraphy of cap carbonates overlying Marinoan-age glacial diamictites in the Paraguay Belt, Brazil. Precambrian Research, 131, 1–21.

Alvarenga, C. J. S., Figueiredo, M. F., Babinski, M., & Pinho, F. E. C. (2007). Glacial diamictites of Serra Azul Formation (Ediacaran, Paraguay Belt): evidence of the Gaskiers glacial event in Brazil. Journal of South American Earth Science, 23, 236–241.

Alvarenga, C. J. S., Dardene, M. A., Santos, R. V., Brod, E. R., Gioia, S. M. C. L., Sial, A. N., Dantas, E. L., & Ferreira, V. (2008). Isotope stratigraphy of Neoproterozoic cap carbonates in the Araras Group, Brazil. Gondwana Research, 13, 469–479.

Alvarenga, C. J. S., Boggiani, P. C., Babinski, M., Dardenne, M. A., Figueiredo, M. F., Dantas, E. L., Uhlein, A., Santos, R. V., Sial, A. N., & Trompette, R. (2011). Glacially influenced sedimentation of Puga Formation, Cuiabá Group and Jacadigo Group, and associated carbonates of Araras and Corumbá groups, Paraguay Belt Brazil. Geological Society of London, 36, 487–497.

Arnaud, E., & Etienne, J. L. (2011). Recognition of glacial influence in Neoproterozoic sedimentary successions. In: E. Arnaud, G. P. Halverson, G. Shields-Zhou, (Eds.), *The Geological Record of Neoproterozoic Glaciations* (Vol. 36, pp. 39–50). London: Geological Society.

Arnott, R. W. C. (1993). Quasi-planar-laminated sandstone beds of the Lower Cretaceous Bootlegger Member, north-central Montana: evidence of combined-flow sedimentation. Journal of Sedimentary Research, 63, 488–494.

Arrouy, M. J., Warren, L. V., Quaglio, F., Poiré, D. G., Simões, M. G., Rosa, M. B., & Peral, L. E. G. (2016). Ediacaran discs from South America: probable soft-bodied macrofossils unlock the paleogeography of the Clymene Ocean. Scientific Reports, 6, 30590.

Babinski, M., Trindade, R. I. F., Alvarenga, J. C., Boggiani, P. C., Liu, D., & Santos, R. V. (2006). *Geochronological constraints on the Neoproterozoic glaciations in Brazil.* Snowball Earth 2006, Ascona Proceedings, Ascona, SZ (Vol. 1, pp. 19–20).

Babinski, M., Boggiani, P. C., Fanning, M., Simon, C. M., & Sial, A. N. (2008). *U-Pb shrimp geochronology and isotope chemostratigraphy (C, O, Sr) of the Tamengo Formation, Southern Paraguay belt, Brazil.* Paper presented at VI South American Symposium on Isotope Geology, San Carlos de Bariloche, AR.

Bandeira, J., McGee, B., Nogueira, A. C. R., Collins, A. S., & Trindade, R. (2012). Sedimentological and provenance response to Cambrian closure of the Clymene Ocean: the upper Alto Paraguai Group, Paraguay belt, Brazil. Gondwana Research, 21(2–3), 323–340.

Basei, M. A. S., & Brito Neves, B. B. (1992). Características geológicas da transição Proterozóico-Faneorozóico no Brasil. In: *Paleozóico Inferior de Iberoamérica* (pp. 331–342). Mérida: Universidad de Extremadura.

Becker-Kerber, B., Pacheco, M. L. A. F., Rudnitzki, I. D., Galante, D., Rodrigues, F., & Leme, J. M. (2017). Interações ecológicas em Cloudina do Ediacaran do Brasil: implicações para o aumento da biomineralização animal. Scientific Reports 7, 5482.

Boggiani, P. C. (1997). Análise estratigráfica da bacia Corumbá (Neoproterozóico) – Mato Grosso do Sul (Doctoral dissertation) Instituto de Geociências (183 p). Universidade de São Paulo.

Boggiani, P. C., Ferreira, V. P., Sial, A. N., Babinski, M., Trindade, R. I. F., Acenolaza, G., Toselli, A. J., & Parada, M. A. (2003). *The cap carbonate of the Puga Hill (Central South America) in the context of the post-Varanger Galciation.* Paper presented at IV South American Symposium on Isotope Geology, Salvador, Brasil.

Boggiani, P. C., Gaucher, C., Sial, A. N., Babinski, M., Simon, C. M., Riccomini, C., Ferreira, V. P., & Fairchild, T. R. (2010). Chemostratigraphy of the Tamengo Formation (Corumbá Group, Brazil): a contribution to the calibration of the Ediacaran carbon-isotope curve. Precambrian Research, 182, 382–401.

Bosak, T., Mariotti, G., Macdonald, F. A., Perron, J. T., & Pruss, S. B. (2013). Microbial sedimentology of stromatolites in Neoproterozoic cap carbonates. In: A. M. Bush, S. B. Pruss, J. L. Payne, (Eds.), *Ecosystem Paleobiology and Geobiology.* Paleontological Society (Vol. 19, pp. 51–77). Pittsburgh: The Paleontological Society.

Brelaz, L. C. (2012). Paleoambiente dos calcários e folhelhos betuminosos da Formação Guia, Neoproterozoico, sudoeste do Estado do Mato Grosso (Masters dissertation), 64 p. Instituto de Geociências, Universidade Federal do Pará.

Christie-Blick, N., Mountain, G. S., & Miller, K. G. (1990). *Seismic Stratigraphic Record of Sea-Level Change* (pp. 116–140). Washington: National Academy.

Clough, J.G. & Goldhammer, R.K. (2000). Evolution of the Neoproterozoic Katakturuk Dolomite ramp complex, northeastern Brooks Range, Alaska. Carbonate Sedimentation and Diagenesis in the Evolving Precambrian World, *67*, 209–241.

Cordani, U. G., Sato, K., Texeira, W., Tassinari, C. C. G., & Basei, M. A. S. (2000). Crustal evolution of the South American platform. In: U. G. Cordani, E. J. Milani, A. Thomaz Filho, D. A. Campos, (Eds.), *Tectonic Evolution of South America, International Geological* (Vol. *31*, pp. 19–40). Rio de Janeiro, RJ: 31º International Geological Congress.

Cordani, U. G., Pimentel M. M., Araújo C. E. G., Basei, M. A. S., Fuck, R. A., & Girardi, V. A. V. (2013). Was there an Ediacaran Clymene Ocean in central Brazil? American Journal of Sciences, *313*, 517–539.

Corsetti, F. A., & Grotzinger, J. P. (2005). Origin and significance of tube structures in Neoproterozoic Post-Glacial cap carbonates: example from noonday dolomite, Death Valley, United States. Palaios, *20*(4), 348–362.

Corsetti, F. A., Lorentz, N. J., & Pruss, S. B. (2004). Formerly-aragonite seafloor fans from Neoproterozoic strata, Death Valley and southeastern Idaho, United States: implications for cap carbonate formation and snowball Earth. In: G. Jenkins, M. McMenamin, L. Sohl, (Eds.), *The Extreme Proterozoic: Geology, Geochemistry, and Climate*. American Geophysical Union Geophysical Monograph Series (Vol. *146*, pp. 33–44). Washington, DC: American Geophysical Union.

Creveling, J. R., & Mitrovica, J. X. (2014). The sea-level fingerprint of a Snowball Earth deglaciation. Earth and Planetary Science Letters, *399*, 74–85.

D'Agrella-Filho, M. S., Babinski, M., Trindade, R. I. F., Van Schmus, W. R., & Ernesto, M. (2000). Simultaneous remagnetization and U-Pb isotope resetting in Neoproterozoic carbonates of the São Francisco Craton, Brazil. Precambrian Research, *99*, 179–196.

Davies, C. J. (2015). Cooling history of Earth's core with high thermal conductivity. Physics of the Earth and Planetary Interiors, *247*, 65–79.

De Raaf, J. D., Boersma, J. R., & Gelder, A. V. (1977). Wave-generated structures and sequences from a shallow marine succession, Lower Carboniferous, County Cork, Ireland. Sedimentology, *24*(4), 451–483.

Dumas, S. & Arnott, R. W. C. (2006). Origin of hummocky and swaley cross-stratification: the controlling influence of unidirectional current strength and aggradation rate. Geology, *34*(12), 1073–1076.

Dyson, I. A., & von der Borch, C. C. (1994). Sequence stratigraphy of an incised-valley fill: the Neoproterozoic Seacliff Sandstone, Adelaide Geosyncline, South Australia. In R. W. Dalrymple, R. Boyd, B. A. Zaitlin (Eds.), *Incised-Valley Systems*. Society for Sedimentary Geology, Special Publication (Vol. *51*, pp. 209–222). Tulsa, OK: SEPM.

Elie, M., Nogueira, A. C. R., Nedelec, A., Trindade, R. I. F., & Kenig, F. (2007). A red algal bloom in the aftermath of the Marinoan Snowball Earth. Terra Nova, *19*(5), 303–308.

Evans, D. A. D. (2006). Proterozoic low orbital obliquity and axial-dipolar geomagnetic field from evaporite palaeolatitudes. Nature, *444*, 51–55.

Fairchild, I. J. (1993). Balmy shores and icy wastes: the paradox of carbonates associated with glacial deposits in Neoproterozoic times. Sedimentology Review, *1*, 1–16.

Fairchild, I. J., & Hambrey, M. J. (1984). The Vendian succession of northeastern Spitsbergen: petrogenesis of a dolomite-tillite association. Precambrian Research, *26*(2), 111–167.

Figueiredo, M. F., & Babinski, M. (2008). *Sedimentary provenance of Serra Azul Formation (Ediacaran), Northern Paraguay Belt, Brazil*. Paper presented at VI South American Symposium on Isotope Geology, San Carlos de Bariloche, AR.

Figueiredo, M. F., Babinski, M., Alvarenga, C. J. S., & Pinho, F. E. C. (2008). Nova unidade litoestratigráfica registra glaciação ediacarana em Mato Grosso: Formação Serra Azul. Geologia USP, Série Científica, *8*(2), 65–74.

Font, E., Trindade, R. I. F., & Nedelec, A. (2005). Detrital remanent magnetization in haematite-bearing Neoproterozoic Puga cap dolostone, Amazon craton: a rock magnetic and SEM study. Geophysical Journal International, *163*, 491–500.

Font, E., Nédelec, A., Trindade, R. I. F., Macouin, M., & Charriere, A. (2006). Chemostratigraphy of the Neoproterozoic Mirassol D'Oeste cap dolostones (Mato Grosso, Brazil): an alternative model for Marinoan cap dolostone formation. Earth and Planetary Science Letters, *250*(1), 89–103.

Font, E., Nédélec, A., Trindade, R. I. F., & Moreau, C. (2010). Fast or slow melting of the Marinoan snowball Earth? The cap dolostone record. Palaeogeography, Palaeoclimatology, Palaeoecology, *295*, 215–225.

Frei, R., Døssing, L. N., Gaucher, C., Boggiani, P. C., Frei, K. M., Bech Árting, T., Crowe, S. A., Freitas, B. T. (2017). Extensive oxidative weathering in the aftermath of a late Neoproterozoic glaciation: evidence from trace element and chromium isotope records in the Urucum district (Jacadigo Group) and Puga iron formations (Mato Grosso do Sul, Brazil). Gondwana Research, *49*, 1–20.

Freitas, B. T., Warren, L. V., Boggiani, P. C., Almeida, R. P., & Piacentini, T. (2011). Tectono-sedimentary evolution of the Neoproterozoic BIF-bearing Jacadigo Group, SW-Brazil. Sedimentary Geology, *238*(1–2), 48–70.

Gaia, V. C. S., Nogueira, A. C. R., Domingos, F. H. G., Sansjofre, P., Bandeira, J., Oliveira, J. G. F., & Sial, A. N. (2017). The new occurrence of Marinoan cap carbonate in Brazil: the expansion of snowball Earth events to the southwesternmost Amazon Craton. Journal of South American Earth Sciences, *76*, 446–459.

Gaucher, C., Boggiani, P. C., Sprechmann, P., Sial, A. N., & Fairchild, T. R. (2003). Integrated correlation of the Vendian to Cambrian Arroyo del Soldado and Corumba Groups (Uruguay and Brazil): palaeogeographic, palaeoclimatic and palaeobiologic implications. Precambrian Research, *120*(3), 241–278.

Godoy, A. M., Pinho, F. E. C., Manzano, J. C., de Araújo, L. M. B., da Silva, J. A., & Figueiredo, M. (2010) Estudos isotópicos das rochas granitóides neoproterozóicas da Faixa de Dobramento Paraguai. Revista Brasileira de Geociências, *40*, 380–391.

Grey, K. (2005). Ediacaran palynology of Australia. Association of Australasian Palaeontologists, *31*, 439 p.

Grotzinger, J. P., & Knoll, A. H. (1995). Anomalous carbonate precipitates: is the Precambrian the key to the Permian? Palaios, *10*(6), 578–596.

Halverson, G. P., Maloof, A. C., & Hoffman, P. F. (2004). The Marinoan glaciation (Neoproterozoic) in northeast Svalbard. Basin Research, *16*, 297–324.

Halverson, G. P., Hoffman, P. F., Schrag, D. P., Maloof, A. C., & Rice, A. H. N. (2005). Toward a Neoproterozoic composite carbon-isotope record. GSA Bulletin, *117*, 1181–1207.

Halverson, G. P., Dudás, F. Ö., Maloof, A. C., & Bowring, S. A. (2007). Evolution of the 87Sr/86Sr composition of Neoproterozoic seawater. Palaeogeography, Palaeoclimatology, Palaeoecology, *256*(3–4), 103–129.

Halverson, G. P., Wade B. P., Hurtgen M. T., & Barovich, K. M. (2010). Neoproterozoic chemostratigraphy. Precambrian Research, *182*, 337–350.

Halverson, G. P., Poitrasson, F., Hoffman, P. F., Nédélec, A., Montel, J. M., & Kirby, J. (2011). Fe isotope and trace element geochemistry of the Neoproterozoic syn-glacial Rapitan iron formation. Earth and Planetary Science Letters, *309*(1–2), 100–112.

Harland, W. B. (1964a). Evidence of late Precambrian glaciation and its significance. In: A. E. M. Nairn, (Ed.), *Problems in Palaeoclimatology* (pp. 119–149). London: Interscience.

Harland, W. B. (1964b). Critical evidence for a great infra-Cambrian glaciation. Geologische Rundschau, *54*(1), 45–61.

Hidalgo, R. L. L. (2007). Vida após as glaciações globais neoproterozoicas: um estudo microfossilífero de capas carbonáticas dos crátons do São Francisco e Amazônico. PhD thesis Instituto de Geociências, Universidade de São Paulo, 197 p.

Hoffman, P. F. (2011). Strange bedfellows: glacial diamictite and cap carbonate from the Marinoan (635 Ma) glaciation in Namibia. Sedimentology, *58*(1), 57–119.

Hoffman, P. F., & Schrag, D. P. (2000). Snowball Earth. Scientific American, *282*, 68–75.

Hoffman, P. F., & Schrag, D. P. (2002). The Snowball Earth hypothesis: testing the limits of global changes. Terra Nova, *14*, 129–155.

Hoffman, P. F., & Macdonald, F. A. (2010). Sheet-crack cements and early regression in Marinoan (635 Ma) cap dolostones: regional benchmarks of vanishing ice-sheets? Earth and Planetary Science Letters, *300*(3–4), 374–384.

Hoffman, P. F., Kaufman, A. J., Halverson, G. P., & Schrag, D. P. (1998a). A Neoproterozoic snowball Earth. Science, *281*, 1342–1346.

Hoffman, P. F., Kaufman, A. J., & Halverson, G. P. (1998b). Comings and goings of global glaciations on a neoproterozoic tropical platform in Namibia. GSA Today, *8*(5), 1–9.

Hoffman, P. F., Halverson, G. P., Domack, E. W., Husson, J. M., Higgins, J. A., & Schrag, D. P. (2007). Are basal Ediacaran (635 Ma) post-glacial "cap dolostones" diachronous? Earth and Planetary Science Letters, *258*(1), 114–131.

Hoffman, P. F., Lamothe, K. G., LoBianco, S. J. C., Hodgskiss, M. S. W., Bellefroid, E. J., Johnson, B. W., Hodgin, E. B., &

Halverson, G. P. (2017). Sedimentary depocenters on Snowball Earth: case studies from the Sturtian Chuos Formation in Northern Namibia. Geosphere, *13*(3), 811–837.

Jacobsen, S. B., & Kaufman, A. J. (1999). The Sr, C and O isotopic evolution of neoproterozoic seawater. Chemical Geology, *161*, 37–57.

James, N. P., Narbonne, G. M., & Kyser, T. K. (2001). Late Neoproterozoic cap carbonates: Mackenzie Mountains, northwestern Canada: precipitation and global glacial meltdown. Canadian Journal of Earth Sciences, *38*, 1229–1262.

Kaufman, A. J., & Knoll, A. H. (1995). Neoproterozoic variations in the C isotopic composition of seawater: stratigraphic and biogeochemical implications. Precambrian Research, *73*, 27–49.

Kaufman, A. J., Hayes, J. M., Knoll, A. H., & Germs, G. J. B. (1991). Isotopic compositions of carbonates and organic carbon from upper Proterozoic successions in Namibia: stratigraphic variation and the effects of diagenesis and metamorphism. Precambrian Research, *49*(3), 301–327.

Kaufman, A. J., Jacobsen, S. B., Knoll, A. H. (1993). The Vendian record of Sr and C isotopic variations in seawater: implications for tectonics and paleoclimate. Earth and Planetary Science Letters, *120*(3–4), 409–430.

Kennedy, M. (1993). The Undoolya sequence: late Proterozoic salt influenced deposition, Amadeus basin, central Australia. Australian Journal of Earth Sciences, *40*(3), 217–228.

Kennedy, M. J. (1996). Stratigraphy, sedimentology, and isotopic geochemistry of Australian Neoproterozoic postglacial cap dolostones: deglaciation, $\delta^{13}C$ excursions, and carbonate precipitation. Journal of Sedimentary Research, *66*, 1050–1064.

Knoll, A. H., Walter, M. R., Narbonne, G. M., & Christie-Blick, N. (2004). A new period for the geologic time scale. Science, *305*(5684), 621–622.

Knoll, A. H., Javaux, E. J., Hewitt, D., & Cohen, P. (2006). Eukaryotic organisms in Proterozoic oceans. Philosophical Transactions of the Royal Society of London Biological Sciences, *361*(1470), 1023–1038.

Kump, L. R. (1991). Interpreting carbon-isotope excursions: Strangelove oceans. Geology *19*, 299–302.

Lamb, M. P., Fischer, W. W., Raub, T. D., Perron, J. T., & Myrow, P. M. (2012). Origin of giant wave ripples in snowball Earth cap carbonate. Geology, *40*(9), 827–830.

Liu, P., Yin, C., Chen, S., Tang, F., & Gao, L. (2013). The biostratigraphic succession of acanthomorphic acritarchs of the Ediacaran Doushantuo Formation in the Yangtze Gorges area, South China and its biostratigraphic correlation with Australia. Precambrian Research, *225*, 29–43.

Lorentz, N. J., Corsetti, F. A., & Link, P. K. (2004). Seafloor precipitates and C-isotope stratigraphy from the Neoproterozoic Scout Mountain Member of the Pocatello Formation, southeast Idaho: implications for Neoproterozoic earth system behavior. Precambrian Research, *130*(1), 57–70.

Macdonald, F. A., Jones, D. S., & Schrag, D. P. (2009). Stratigraphic and tectonic implications of a newly discovered glacial diamictite-cap carbonate couplet in southwestern Mongolia. Geology, *37*(2), 123–126.

Macdonald, F. A., Strauss, J. V., Sperling, E. A., Halverson, G. P., Narbonne, G. M., Johnston, D. T., Kunzmann, M., Schrag, D. P., & Higgins, J. A. (2013). The stratigraphic relationship between the Shuram carbon isotope excursion, the oxygenation of Neoproterozoic oceans, and the first appearance of the Ediacara biota and bilaterian trace fossils in northwestern Canada. Chemical Geology, *362*, 250–272.

Maciel, P. (1959). Tilito cambriano (?) no Estado de Mato Grosso. Boletim da Sociedade Brasileira de Geologia, *8*(1), 9–31.

McFadden, K. A., Xiao, S., Zhou, C., & Kowalewski, M. (2009). Quantitative evaluation of the biostratigraphic distribution of acanthomorphic acritarchs in the Ediacaran Doushantuo Formation in the Yangtze Gorges area, South China. Precambrian Research, *173*(1), 170–190.

McGee, B., Collins A. S., & Trindade R. I. F. (2012). G'day Gondwana – birth of a supercontinent: U/Pb ages for the post-orogenic São Vicente granite, Mato Grosso, Brazil. Gondwana Research, *21*, 316–322.

McGee, B., Collins, A. S., Trindade, R. I. F., & Jourdan, F. (2014). Investigating mid-Ediacaran glaciations and final Gondwana amalgamation using coupled sedimentology and ^{40}Ar/^{39}Ar detrital muscovite provenance from the Paraguay Belt, Brazil. Sedimentology, *62*(1), 130–154.

McGee, B., Collins, A. S., Trindade, R. I. F., & Payne, J. (2015). Age and provenance of the Cryogenian to Cambrian passive margin to foreland basin sequence of the Northern Paraguay Belt, Brazil. Geological Society of America Bulletin, *127*(1–2), 76–86.

Misi, A., & Veizer, J. (1998). Neoproterozoic carbonate sequences of the Una Group, Irecê Basin, Brazil: chemostratigraphy, age and correlations. Precambrian Research, *89*(1–2), 87–100.

Moura, C. A. V., & Gaudette, H. E. (1993). *Zircon ages of the basement orthogneisses of the Araguaia Belt, north-central Brazil*. Paper presented at 4o Congresso Brasileiro de Geoquimica, SBGq, Brasília, Brazil.

Myrow, P. M., & Kaufman, A. J. (1999). A newly discovered cap carbonate above Varanger-age glacial deposits in Newfoundland, Canada. Journal of Sedimentary Research, *69*(3), 784–793.

Myrow, P. M., Lamb, M. P., & Ewing, R. C. (2018). Rapid sea level rise in the aftermath of a Neoproterozoic snowball Earth. Science, *360*, 649–651, doi:10.1126/science.aap8612.

Nogueira, A. C. R. (2003). A plataforma carbonática Araras no sudoeste do Cráton Amazônico: estratigrafia, contexto paleoambiental e correlação com os eventos glaciais do Neoproterozóico. (Doctoral dissertation), Universidade de São Paulo, 173 p.

Nogueira, A. C. R., & Riccomini, C. (2006). O Grupo Araras (Neoproterozóico) na parte norte da Faixa Paraguai e Sul do Cráton Amazônico. Revista Brasileira de Geociências, *36*, 576–587.

Nogueira, A. C. R., Riccomini, C., & Sial, A. N. (2001). *Capa carbonática pós-Varanger no Sw do Cráton Amazônico, MT: evidência de glaciação global (Snowball Earth)*. Paper presented at Simpósio de Geologia da Amazônia 7o, Belém.

Nogueira, A. C. R., Riccomini, C., Sial, A. N., Moura, C. A. V., & Fairchild, T. R. (2003). Soft-sediment deformation at the base of the Neoproterozoic Puga cap carbonate (southwestern Amazon craton, Brazil): confirmation of rapid icehouse to greenhouse transition in snowball Earth. Geology, *31*, 613–616.

Nogueira, A. C. R., Riccomini, C., Sial, A. N., Moura, C. A. V., Trindade, R. I. F., & Fairchild, T. R. (2007). Carbon and strontium isotope fluctuations and paleoceanographic changes in the late Neoproterozoic Araras carbonate platform, Southern Amazon Craton, Brazil. Chemical Geology, *237*, 168–190.

Oliveira, R.S. 2010. Depósitos de rampa carbonática ediacarana do Grupo Corumbá, Região de Corumbá, Mato Grosso do Sul, (Master Thesis). Instituto de Geociências, Universidade Federal do Pará, 104 pp.

Pacheco, M. L. A. F., Galante, D., Rodrigues, F., Leme, J. M., Bidola, P., Hagadorn, W., Stockmar, M., Herzen, J., Rudnitzki, I. D., Pfeiffer, F., & Marques, A. C. (2015). Insights into the skeletonization, lifestyle, and affinity of the unusual Ediacaran fossil Corumbella. PLoS One, *10*(3), e0114219, doi:10.1371/journal.pone.0114219.

Packard, J. J., & Hills, D. (2001). *The Importance of Early (Penecontemporaneous) Meteoric Diagenesis in the Development of Limestone Porosity in the "Platform" of the Devonian Swan Hills Formation*. Paper presented at the CSPG Annual Convention.

Parry, L. A., Boggiani, P. C., Condon, D. J., Garwood, R. J., Leme, J. M., McIlroy, D., Brasier, M. D., Trindade, R., Campanha, G. A. C., Pacheco, M. L. A. F., Diniz, C. Q. C., & Liu, A. G. (2017). Ichnological evidence for meiofaunal bilaterians from the terminal Ediacaran and earliest Cambrian of Brazil. Nature Ecology and Evolution, *1*, 1455–1464.

Paula-Santos, G. M., Caetano-Filho, S., Babinski, M., Trindade, R. I. F., & Guacaneme, C. (2017). Tracking connection and restriction of West Gondwana São Francisco Basin through isotope chemostratigraphy. Gondwana Research, *42*, 280–305.

Pedreira, A. J., & Bahia, R. B. C. (2000). Sedimentary basins of Rondônia State, Brazil: response to the geotectonic evolution of the Amazonic craton. Revista Brasileira de Geociências, *30*(3), 477–480.

Pfeil, R. W., & Read, J. F. (1980). Cambrian carbonate platform margin facies, shady dolomite, southwestern Virginia, USA. Journal of Sedimentary. Petrology, *50*, 91–116.

Plummer, P. S. (1978) Note on the paleoenvironmental significance of the Nuccaleena Formation (Upper Precambrian), central Flinders Ranges, South Australia. Geological Society of Australia Journal, *25*, 395–402.

Riccomini, C., Nogueira, A. C., & Sial, A. N. (2007). Carbon and oxygen isotope geochemistry of Ediacaran outer platform carbonates, Paraguay Belt, central Brazil. Anais da Academia Brasileira de Ciências, *79*(3), 519–527.

Riding, R. (2000). Microbial carbonates: the geological record of calcified bacterial-algal mats and biofilms. Sedimentology, *47*, 179–214.

Romero, G. R. (2015). Geobiologia de microbialitos do Ediacarano da Faixa Paraguai e Sul do Craton Amazônico (MS e MT): Implicações paleoambientais, paleoecológicas e estratigráficas, (Doctoral dissertation). Universidade de São Paulo, http://www.teses.usp.br/teses/disponiveis/44/44141/tde-24022016-095246/pt-br.php.

Romero, J. A. S., Lafon, J. M., Nogueira, A. C. R., & Soares, J. L. (2013). Sr isotope geochemistry and Pb-Pb geochronology of the Neoproterozoic cap carbonates, Tangará da Serra, Brazil. International Geology Review, 55, 119.

Romero, G. R., Sanchez, E. A. M., Morais, L., Boggiani, P. C. & Fairchild, T. R. (2016). Tubestone microbialite association in the Ediacaran cap carbonates in the Southern Paraguay Fold Belt (SW Brazil): geobiological and stratigraphic implications for a Marinoan cap carbonate. Journal of South American Earth Sciences, 71, 172–181.

Rudnitzki, I. D., Romero, G. R., Hidalgo, R., & Nogueira, A. C. R. (2016). High frequency peritidal cycles of the upper Araras Group: implications for disappearance of the Neoproterozoic carbonate platform Amazon Craton. Journal of South American Earth Sciences, 65, 6778.

Sansjofre, P., Ader, M., Trindade, R. I. F., Elie, M., Lyons, J., Cartigny, P., & Nogueira, A. C. R. (2011). A carbon isotope challenge to the Snowball Earth. Nature, 478(7367), 93–97.

Santos, I. M. (2016). Revisão estratigráfica e tectônica dos Grupos Cuiabá e Araras no contexto da Faixa Paraguai (MT). (Masters dissertation). Universidade Federal do Pará, 128 p.

Santos, V. R., Alvarenga, C. J. S., Dardenne, M. A., Sial, A. N., & Ferreira, V. P. (2000). Carbon and oxygen isotope profiles across Meso-Neoproterozoic limestones from central Brazil: Bambuí and Paranoá groups. Precambrian Research, 104, 107–122

Santos, H. P., Mángano, M. G., Soares, J. L., Nogueira, A. C. R., Bandeira, J., & Rudnitzki, I. D. (2017). Ichnologic evidence of a Cambrian age in the Southern Amazon Craton: implications for the onset of the Western Gondwana history. Journal of South American Earth Sciences, 76, 482–488.

Saylor, B. Z., Kaufman, A. J., Grotzinger, J. P., & Urban, F. (1998). A composite reference section for terminal Proterozoic strata of Southern Namibia. Journal of Sedimentary Research, 68, 1223–1235.

Sergeev, V. N., Knoll, A. H., & Vorob'Eva, N. G. (2011). Ediacaran microfossils from the Ura Formation, Baikal-Patom Uplift, Siberia: taxonomy and biostratigraphic significance. Journal of Paleontology, 85(5), 987–1011.

Sial, A. N., Gaucher, C., Misi, A., Boggiani, P. C., Alvarenga, C. J. S. D., Ferreira, V. P., & Geraldes, M. (2016). Correlations of some Neoproterozoic carbonate-dominated successions in South America based on high-resolution chemostratigraphy. Brazilian Journal of Geology, 46(3), 439–488.

Silva, G. D., Paes, J. D. S., & Sais, G. S. (2015). Formação Serra do Caeté: a glaciação do sistema Puga, Neoproterozoico da Faixa Paraguai, Sudoeste de Mato Grosso. Contribuições a Geologia da Amazônia, 9, 157–1670.

Siqueira, L. P., & Teixeira L. B. (1993). Bacia dos Parecis: nova fronteira exploratória da Petrobrás. Paper presented at International Congress of the Brazilian Geophysics Society 3.

Soares, J. L., & Nogueira, A. C. R. (2008). Depósitos carbonáticos de Tangará da Serra (MT): uma nova ocorrência de capa carbonática neoproterozóica no sul do Cráton Amazônico. Revista Brasileira de Geociências, 38(4), 715–729.

Soares, J. L., Nogueira, A. C. R., Domingos, F., & Riccomini, C. (2013). Synsedimentary deformation and the paleoseismic record in Marinoan cap carbonate of the Southern Amazon Craton, Brazil. Journal of South American Earth Sciences, 48, 58–72.

Sousa, G. R., Jr., Nogueira, A. C. R., Santos, N., Eugênio, V., Moura, C. A. V., Araújo, B. Q., & Reis, F. A. M. (2016). Organic matter in the Neoproterozoic cap carbonate from the Amazonian Craton, Brazil. Journal of South American Earth Sciences, 72, 7–24.

Stow, D. A. V. (1986). Deep clastic seas. In: H. G. Reading, (Ed.), Sedimentary Environments: Processes, Facies and Stratigraphy (pp. 399–444). Oxford: Blackwell Scientific Public.

Sumner, D. Y. (2002). Decimeter-thick encrustations of calcite and aragonite on the sea-floor and implications for Neoarchaean and Neoproterozoic ocean chemistry. In: W. Altermann, P. Corcoran, (Eds.), Precambrian Sedimentary Environments: A Modern Approach to Ancient Depositional Systems. International Association of Sedimentologists Special Publication (Vol. 33, pp. 107–120). Oxford: Blackwell Science.

Sumner, D. Y. & Grotzinger, J. P. (1996). Were kinetics of Archean calcium carbonate precipitation related to oxygen concentration? Geology, 24, 119–122.

Torsvik, T. H., & Cocks, L. R. M. (2013). Gondwana from top to base in space and time. Gondwana Research, 24, 999–1030.

Trindade, R. I. F., Font, E., D'Agrella-filho, M. S. D., Nogueira, A. C. R., & Riccomini, C. (2003). Low-latitude and multiple geomagnetic reversals in the Neoproterozoic Puga cap carbonate, Amazon Craton. Terra Nova, 15, 441–446.

Trindade, R. I. F., D'Agrella Filho, M. S., Babinski, M., Font, E., & Neves, B. B. B. (2004). Paleomagnetism and geochronology of the Bebedouro cap carbonate: evidence for continental-scale Cambrian remagnetization in the São Francisco craton, Brazil. Precambrian Research, 128, 83–103.

Trivelli, G. G. B., Pierosan, R., & Ruiz, A. S. (2017). Geologia e petrologia do Granito São Vicente na região do Parque Estadual Águas Quentes, estado de Mato Grosso, Brasil. Geologia USP. Série Científica, 17(3), 29–48.

Trompette, R., Alvarenga, C. J. S., & Walde, D. (1998). Geological evolution of the Neoproterozoic Corumbá graben system (Brazil): depositional context of the stratified Fe and Mn ores of the Jacadigo Group. Journal of South America Earth Science, 11(6), 587–597.

Tucker, M. E. (1986). Formerly aragonitic limestones associated with tillites in the Late Proterozoic of Death Valley, California. Journal of Sedimentary Petrology, 56, 818–830.

Vasconcelos, C. O., McKenzie, J. A., Bernasconi, S., Grujic, D., & Tien, A. J. (1995). Microbial mediation as a possible mechanism for natural dolomite formation at low temperature. Nature, *337*, 220–222.

Von der Borch, C. C., Christie-Blick, N., & Grady, A. E. (1989) Depositional sequence analysis applied to Late Proterozoic Wilpena Group, Adelaide Geosyncline, South Australia. Australian Journal of Earth Science, *35*, 59–71.

Warren, L. V., Quaglio, F., Riccomini, C., Simões, M. G., Poiré, D. G., Strikis, N. M., Anelli, L. E., & Strikis, P. C. (2014). The puzzle assembled: Ediacaran guide fossil Cloudina reveals an old proto-Gondwana seaway. Geology *42*, 391394.

Xiao, S., Narbonne, G. M., Zhou, C., Laflamme, M., Grazhdankin, D. V., Moczydłowska-Vidal, M., & Cui, H. (2016). Toward an Ediacaran time scale: problems, protocols, and prospects. Episodes, *39*(4), 540

Zaine, M. F., Fairchild, T. R. (1992). *Considerações paleoambientais sobre a Formação Araras, Faixa Paraguai, estado do Mato Grosso*. Paper presented in 37º Brazilian Congress of Geology.

7

The Ediacaran-Cambrian Transition: A Resource-Based Hypothesis for the Rise and Fall of the Ediacara Biota

Alan J. Kaufman

ABSTRACT

Ediacaran oceans hosted a strange world of exotic soft-bodied forms that were a failed early evolutionary experiment in macroscopic life. These enigmatic organisms appear suddenly in sedimentary rocks as old as 573 Ma above a glacial diamictite in Newfoundland, and, in most successions, a profound and equally puzzling negative carbon cycle anomaly known as the Shuram excursion. The Ediacaran biota died out near the Ediacaran-Cambrian boundary (<541 Ma) coincident with a second strong negative $\delta^{13}C$ excursion. From Ediacaran ashes, the proliferation of complex feeding traces, soft-bodied arthropod tracks, and mineralized skeletons in the succeeding Fortunian stage led to a diverse landscape of modern phyla by the detonation of the Cambrian explosion around 529 Ma. This chapter provides a review of the profound changes in the carbon, sulfur, and strontium cycles across this critical transition in order to better understand the redox history of the oceans, as well as the tectonic, climatic, and biological events preserved in its sedimentary archive, and further proposes a novel resource-based hypothesis for the rise and fall of the Ediacaran biota.

7.1. A BIOGEOCHEMICAL PERSPECTIVE

From climatic, geochemical, and paleontological perspectives, the Ediacaran-Cambrian transition stands out as an interval of profound global change preserving evidence in its sedimentary archive for the coevolution of life and environment (Fig. 7.1). Across this critical transition, the initial driver of oceanographic and biological events is envisioned as tectonic in nature, related to widespread Pan-African orogeny as east and west Gondwana were sutured [*Squire et al., 2006*] in the aftermath of the ultimate Cryogenian ice age (aka snowball Earth event; *Kirschvink*, 1992; *Hoffman et al.*, 1998). Subsequent erosion of uplifted Ediacaran terrains delivered thick piles of sediments to the oceans, as well as bio-limiting nutrients, especially N, P, and Fe [*Derry et al., 1992; Kaufman et al., 1993*]. Silicate weathering of

the Transgondwanan mountains may have drawn enough CO_2 out of the atmosphere to contribute to episodic cooling of surface environments leading to the development of regional ice sheets and the deposition of glacial diamictites [cf., *Raymo et al.*, 1988], including the Gaskiers in Newfoundland, on at least eight paleocontinents [*Hoffman and Li*, 2009]. The weathering flux of nutrients to the oceans may have similarly contributed to global cooling by stimulating primary productivity, thereby increasing the flux of organic carbon to the deep oceans, as well as the buildup of oxygen in the atmosphere, which is a phenomenon broadly known as the Neoproterozoic Oxygenation Event (NOE) [*Derry et al.*, 1992; *Des Marais et al.*, 1992; *Kaufman et al.*, 1993, 1997; *Campbell and Allen*, 2008; *Hardisty et al.*, 2017; *Och and Shields-Zhou*, 2012].

In this review, the redox history of the oceans across the Ediacaran-Cambrian transition is considered in light of tumultuous changes in the exogenic carbon, sulfur, and strontium cycles; most notable are the profound negative $\delta^{13}C$ excursions that bracket the first and last

Department of Geology, Earth System Science Interdisciplinary Center, University of Maryland, College Park, MD, USA

Chemostratigraphy Across Major Chronological Boundaries, Geophysical Monograph 240, First Edition.
Edited by Alcides N. Sial, Claudio Gaucher, Muthuvairavasamy Ramkumar, and Valderez Pinto Ferreira.

Figure 7.1 The agronomic revolution or Cambrian substrate revolution depicted in this modified illustration by Peter Trusler reveals profound changes in bioturbation across the Ediacaran (E)-Cambrian (C) transition. The enigmatic Ediacaran biota (left) is rooted in ubiquitous microbial mats that carpeted the seafloor (Precambrian matgrounds) and largely sealed the anoxic sediments beneath from the free exchange of gases. The slow diffusion of sulfate into these mats provided an oxidant for microbial sulfate reduction, leading to the buildup of toxic H_2S, a proposed critical nutrient for the Ediacaran biota. The horizontal burrows of animals within or beneath the mats and the activities of the Ediacaran organisms on its surface, which appear very late in the Ediacaran game, had little effect on sedimentary layering or the release of gases to seawater. In contrast, the deep dive of early Cambrian (right) animals into the sediments in search of food and shelter disrupted the mats and allowed the free exchange of gases across the sediment-water interface, including O_2. Ventilation and mixing of the sediments is implicated in the demise of the Ediacaran biota, if H_2S was a critical physiological resource, as well as the buildup of sulfate in the oceans. Reproduced with permission of Peter Trusler. *(See insert for color representation of the figure.)*

appearance of the enigmatic Ediacaran biota in most successions. To some these large and complexly ornamented organisms appear as stem group animals [e.g., *Runnegar*, 1982; *Erwin et al.*, 2011; *Budd and Jensen*, 2017; *Droser and Gehling*, 2015], while to others these soft-bodied tubular and fractally constructed forms represent symbiotic lichens [*Retallack*, 2013] or as giant unicellular protists with no clear representatives in Cambrian and younger seas (i.e., the vendobiont hypothesis; *Seilacher*, 1989). Most modern workers, however, regard the Ediacaran biota as representative of a multitude of multicellular phylogenies united by their soft bodies and large size.

Ediacaran-type fossils (which have been reported at nearly 30 localities on 5 continents) are typically in the centimeter to decimeter range, with some giants ranging to more than a meter in length [*Narbonne*, 2005]. Understanding the position of these strange organisms, typically preserved as three-dimensional (3D) casts and molds in soft sands and silts or as carbonaceous compressions in thin bituminous limestones [*Wade*, 1968; *Grazhdankin*, 2004; *Narbonne*, 2005; *Fedonkin et al.*, 2007], in the tree of life is one of the greatest current challenges in paleobiology.

Rather than to take a morphological approach in placing ornaments of the Ediacaran biota on this tree, it seems worthwhile to apply a biogeochemical perspective and ask first whether these organisms might have had a common metabolic strategy. For example, osmotrophy has been considered viable in a dissolved organic carbon (DOC)-rich ocean [*Rothman et al.*, 2003; *Laflamme et al.*, 2009] given the presence of a large surface area presented by the tubular and fractal morphologies of the Ediacaran biota. Extended surface areas could allow for the simple diffusion of dissolved nonpolar organic molecules (and oxygen) through the thin walls of the organisms in order to fuel their metabolic activities, *assuming* that these were respiratory in nature. Osmotrophic feeding is an intriguing hypothesis, especially considering that many representative taxa lived below the photic zone [*Narbonne and Aitken*, 1995; *Dalrymple and Narbonne*, 1996; *MacNaughton et al.*, 2000; *Clapham et al.*, 2003; *Wood et al.*, 2003] and that there is scant evidence for any of these organisms having a mouth, gut, or anus. Notably, in modern ecosystems, only microscopic bacteria with large surface areas relative to total volume are able to survive by osmotrophy. If the organic membranes of the Ediacaran biota (including both tubular erniettamorphs and fractal rangeomorphs) were thin and their interiors were hollow or sand filled, then their surface area to volume would approach that of the osmotrophic bacteria [*Laflamme et al.*, 2009], allowing for this feeding strategy.

Since a large standing DOC pool was probably a long-standing feature of the Neoproterozoic oceans (in contrast to the dominant dissolved inorganic carbon (DIC) pool in modern seaways), the Ediacaran biota appear to have evolved largely associated with the rise of oxygen in surface environments [cf., *Nursall*, 1959; *Berkner and Marshall*, 1965] and may have seen their demise with fall of the breathing gas near the Ediacaran-Cambrian boundary [*Kimura and Watanabe*, 2001; *Amthor et al.*, 2003; *Schröder and Grotzinger*, 2007; *Wille et al.*, 2008; *Laflamme et al.*, 2013; *Zhang et al.*, 2018]. Following this logic, oxygen would have been one of the key metabolic resources for the Ediacaran biota, whether they were animals or not. Given the infaunal and epifaunal lifestyle of many of these large pneumatic organisms, I explore whether hydrogen sulfide (H_2S) produced below and within the ubiquitous microbial mats by sulfate-reducing microbes would have been the other critical metabolic resource and further that the earliest form of bioturbation was microbial growth into the sediments in order to access the corrosive, toxic, flammable, and smelly hydride gas. Rather than making the a priori assumption that these organisms were animals and somehow had to protect themselves from toxic H_2S in a symbiotic relationship [cf., *Dufor and McIlroy*, 2016], I hypothesize an alternative bacterial physiology [*Grazhdankin and Gerdes*, 2007] for the Ediacaran biota based on thioautotrophy (i.e., sulfide oxidation in the absence of light) that could explain their morphological disparity, as well as their spatial and temporal distribution in terminal Ediacaran seaways.

Bioturbation, or literally the burrowing and stirring of soft sediments by motile triploblastic (bilaterally symmetric with a true mouth, gut, and anus) animals in search of food and shelter, appears very late in the Ediacaran game; it has been suggested to play a role in the demise of the Ediacaran biota [e.g., *Laflamme et al.*, 2013; *Budd and Jensen*, 2017]. The oldest potential yet problematic burrows (attributed as *Archaeonassa*) are reported atop ~565 Ma deepwater turbidites in Newfoundland [*Liu et al.*, 2010, 2014], but simple, unbranched, and horizontal surface traces reliably attributed to bilaterians appear abundantly in sedimentary rocks around 550 Ma [*Jensen*, 2003]. These simple burrowing traces (e.g., *Helminthoidichnites*) are believed to represent the peristaltic movement of animals through the microbial mats at or very near to the sediment-water interface. Recently, more complex surface trackways of bilaterian animals with paired appendages have been reported alongside simple burrows in terminal Ediacaran strata of South China [*Chen et al.*, 2018]. Near the Ediacaran-Cambrian boundary, ichnofossils become more 3D with evidence of a deeper dive into sediments while probing or farming (agrichnia) for food [*Mángano et al.*, 2012; *Laing et al.*, 2018]. By piercing and ventilating Ediacaran matgrounds [*Seilacher and Pflüger*, 1994; *Bottjer et al.*, 2000; *Gougeon et al.*, 2018; *Hantsoo et al.*, 2018], Fortunian (the ratified basal stage of the Cambrian period; *Peng et al.*, 2012) bioturbation would have depleted the hydrogen sulfide resource required for the proposed Ediacaran metabolism and may have ultimately resulted in the Cambrian buildup of oceanic sulfate [*Canfield and Farquhar*, 2009].

The events described herein are viewed through a 50 million year window spanning the presently undefined terminal Ediacaran stage (TES) [*Xiao et al.*, 2016] and through the basal Cambrian Fortunian stage. One side is framed by the Shuram excursion (SE), an unprecedented and highly debated alkalinity and negative carbon cycle anomaly preserved in a variety of carbonate facies on multiple continents [*Knauth and Kennedy*, 2009; *Grotzinger et al.*, 2011; *Schrag et al.*, 2013; *Cui et al.*, 2017], while on the other side is the Cambrian explosion of modern animal phyla [*Rozanov et al.*, 1969]. Within this narrow temporal window, life got big [*Narbonne and Gehling*, 2003], smart [*Carbone and Narbonne*, 2014], and hard [*Germs*, 1972; *Grant*, 1990; *Grotzinger et al.*, 2000; *Porter*, 2007, 2010; *Zhuralev and Wood*, 2008].

7.2. CHEMOSTRATIGRAPHY

The profusion of chemostratigraphic studies of shallow marine sedimentary successions that accumulated across the Ediacaran-Cambrian transition is testament to the importance of this most critical transition in Earth's history, as well as the seminal isotopic research of *Bill Holser* [*Holser and Kaplan*, 1966; *Claypool et al.*, 1980], *Manfred Schidlowski et al.* [1975], and *Jan Veizer* [*Veizer and Compston*, 1976; *Veizer and Hoefs*, 1976]. These scientists were among the first to identify large swings in the carbon, sulfur, and strontium isotopic composition of seawater proxies preserved in Precambrian and Phanerozoic basins. The dynamic and expanding landscape of Ediacaran and Cambrian chemostratigraphic research over the past 30+ years owes much to these early isotopic explorers.

While current studies provide rigorous sequence stratigraphic, biostratigraphic, and radiometric architectures on which to hang chemostratigraphic data, most are still hindered by incompleteness of the sedimentary record, the absence of key lithologies (including interbedded volcanic ashes), and/or diagenetic insults. The large number of publications purporting composite reference sections is testament to the general acceptance of isotope stratigraphy as a critical tool in the Ediacaran and Cambrian periods. It is also a manifestation of the ambiguity inherent in projecting local systems onto the world stage [Corsetti and Kaufman, 2003]. Intrabasinal correlation remains a critical aspect of reconstructing historical events in deep time and is best served as an integrated dish. Technological advancements over the past 25 years have allowed for a proliferation of novel isotope measurements (e.g., Cr isotopes in carbonates; Frei et al., 2011) that purport great insight to environmental change, but these require further calibration, as well as stratigraphic and diagenetic tests [Hood et al., 2018] of whether they reflect local perturbations or global phenomenon.

The EARTHTIME initiative has accelerated advances in geochronology since 2003 by providing calibrated solutions to U-Pb zircon laboratories worldwide in order to compare standard results, to minimize bias, and to improve absolute calibration. Based on these efforts, sub-million year uncertainties on U-Pb zircon dates for both Ediacaran and Cambrian periods are more and more common, and in turn, we understand more about rates and the driving forces for biological and oceanographic change. Application of these ages, however, must still be tied to the stratigraphic record with the understanding that most successions represent missing time and that cryptic unconformities or miscorrelations can confound even the most precise dates. The development of laser ablation U-Pb zircon techniques is a tremendous advance allowing for rapid age characterization of both detrital and volcanic grains. However, if this technique is used to pre-characterize volcanic zircons, the lower precision age distribution of *all* grains determined by the rapid technique should be published alongside the higher precision ages of grains selected specifically for chemical abrasion single-crystal analyses.

7.2.1. Shuram Excursion

This critical transition chapter focuses solely on the TES and Fortunian intervals while acknowledging the coupled climatic and geochemical events of the Cryogenian

and earlier Ediacaran [Knoll et al., 2004] that set the stage for the origin of the Ediacaran biota and of animals. Foremost among the carbon isotope events in this tumultuous interval is the enigmatic Shuram excursion (henceforth abbreviated as SE). The origin of this biogeochemical anomaly and its relationship to the Gaskiers ice age is one of the greatest outstanding problems of Ediacaran Earth history [Xiao et al., 2016]. First recognized in middle Ediacaran strata from Oman [Burns and Matter, 1993] and South Australia [Pell et al., 1993], the deep stratigraphic divide was soon found on other continents preserving remarkably similar $\delta^{13}C$ trends [Grotzinger et al., 2011]. With nadir values dramatically lower than mantle inputs (ca. −5‰), its primary nature has long been questioned.

While the Gaskiers diamictite is remarkably well constrained by radiometric dates [Pu et al., 2016; see discussion below], the SE is not. In Ediacaran strata of Namibia, the biogeochemical anomaly is preserved in carbonate facies of the Kanies and Mara members of the Kuibis Subgroup [Wood et al., 2015], which sit hundreds of meters below an ~547 Ma U-Pb zircon age constraint [Grotzinger et al., 1995; recalculated in Bowring et al., 2007; Narbonne et al., 2012] but provide a minimum upper bound. Closer to the event, many researchers pin the top of the SE to the 551 Ma Doushantuo-Dengying boundary age in South China [Condon et al., 2005], but uncertainty as to whether the Miaohe volcanic ash that lies within the famed lagerstätte belongs to the Doushantuo or the Dengying is critical [Kaufman, 2005]. A recent study indicated that the ash lies above a notable sequence boundary [An et al., 2015], such that the uppermost interval of the Doushantuo (Member IV) could be much older than it currently appears in most chemostratigraphic compilations [see Xiao et al., 2016]. The onset of the Shuram anomaly, which is coincident with the global increase in seawater alkalinity as evidenced by the sudden appearance of thick-bedded carbonates in otherwise siliciclastic dominated successions, is equally uncertain. In the absence of dateable volcanic ashes, three non-zircon-based radiometric techniques have been attempted to constrain Doushantuo Member IV. Using Lu-Hf and Pb-Pb techniques on phosphorites some 20 m below the unit, Barfod et al. [2002] documented ages of 584±26 and 599.3±4.2 Ma, respectively. A similar age of 595±22 Ma was recently reported from Re-Os analysis of black shale from the base of Member IV [Zhu et al., 2013]; the significant uncertainty and high MSWD was suggested to result from temporal changes in the $^{187}Os/^{188}Os$ of seawater, so a subset of the analyses were used to estimate a depositional age of 591±3 Ma. Accepting any of these unconventional radiometric constraints, the beginning of the SE in South China could well be coincident with the end of the Gaskiers ice age or even be somewhat older.

With these age uncertainties in mind, there are a number of possible historical alternatives to consider. In the accompanying carbon isotope compilation (Fig. 7.2), the SE is shown to begin immediately above the 580 Ma Gaskiers diamictite (see correlation 2 in *Xiao et al.* (2016)), such that all of the Ediacaran biotas lie above the biogeochemical anomaly. Thus far, it is only in northwestern Canada, where a putative SE has been identified in carbonate strata assigned to the Gametrail Formation in the Wernecke and Ogilvie mountains [*Johnston et al.,* 2013; *Macdonald et al.,* 2013], that deepwater Ediacaran organism are interpreted to have lived before the inferred oxidation event. This view, based on proposed, but not unambiguous, interbasinal correlations, is complicated by the absence of an SE in the Mackenzie Mountains where a depauperate but recognizable assemblage of Ediacaran fossils (including complex and radially symmetrical discs and fronds, but no segmented forms) have previously been documented in upper Sheepbed strata below the Gametrail Formation. Previous chemostratigraphic research in the Mackenzie Mountains [*Kaufman et al.,* 1997] identified two strong positive $\delta^{13}C$ excursions in the Sheepbed Formation [the upper interval now characterized as the June beds, which contain the Ediacaran fossils; *Macdonald et al.,* 2013]. Although carbonate lithologies are lacking, it is plausible that the negative SE lies between these events, which would support the view entertained here that all of the soft-bodied life forms postdated the biogeochemical anomaly. To support the alternative conclusion, Ediacaran fossils would need to be discovered in the so-called June beds in the Wernecke and Ogilvie mountains, or a credible explanation for the absence of the SE in the thick Gametrail carbonates of the Mackenzie Mountains must be presented. Insofar as these remote successions in northern Canada may hold the key to our understanding of Ediacaran Earth history, further research is warranted.

Controversies surrounding the SE and its potential correlatives include (i) its origin: whether the event represents a long-lived disturbance of the global ocean DIC reservoir due to the oxidation of organic carbon or methane [*Rothman et al.,* 2003; *Fike et al.,* 2006; *Kaufman et al.,* 2007; *Bristow and Kennedy,* 2008; *McFadden et al.,* 2008; *Bjerrum and Canfield,* 2011; *Husson et al.,* 2015], (ii) reflects conditions conducive to authigenic carbonate precipitation [*Macdonald et al.,* 2013; *Schrag et al.,* 2013; *Cui et al.,* 2017], or (iii) results from meteoric or burial diagenesis [*Knauth and Kennedy,* 2009; *Derry,* 2010]. While we have documented clear petrographic and isotopic evidence for the presence of authigenic carbonate nodules derived from the anaerobic oxidation of methane associated with the SE in South China [*Cui et al.,* 2017] and Namibia, my personal bias is that this was a primary

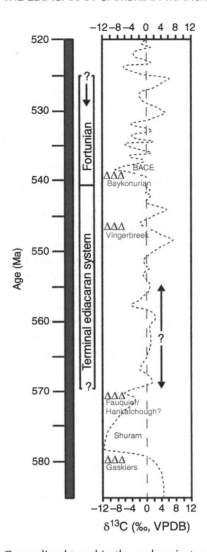

Figure 7.2 Generalized trend in the carbon isotope composition of marine carbonates through the TES and Fortunian stage of the Cambrian period. This compilation is based on data from a wide array of sources from successions in India [*Kaufman et al.,* 2006], Morocco [*Maloof et al.,* 2005], Namibia [*Kaufman et al.,* 1991; *Saylor et al.,* 1998; *Wood et al.,* 2015], South China [*McFadden et al.,* 2008; *Cui et al.,* 2015, 2016a, 2016b, 2017], Oman [*Fike et al.,* 2006], India [*Kaufman et al.,* 2006; *Tewari and Sial,* 2007], Siberia [*Knoll et al.,* 1995a; *Kaufman et al.,* 1996; *Maloof et al.,* 2010 and references therein; *Cui et al.,* 2016c], South Australia [*Husson et al.,* 2015], and the United States [*Corsetti and Kaufman,* 2003; *Hebert et al.,* 2010; *Verdel et al.,* 2011]. This compilation should be compared against those presented in *Xiao et al.* [2016] insofar as the upper reaches of the Shuram excursion are not pinned to the 551 Ma U-Pb age at the Doushantuo-Dengying contact (see text for explanation). Positions for ice ages are marked by ΔΔΔ. Note the uncertainties of the positions of the base of the TES and the end of the Fortunian, which relate to issues of correlating chemostratigraphic and biostratigraphic events between basins.

oceanographic phenomenon related to the progressive ventilation of the anoxic pre-Shuram oceans.

In this view, the sudden increase in ^{13}C-depleted alkalinity was a direct result of the delivery of nutrients and sulfate during intense weathering of uplifted Transgondwanan terrains. These continental fluxes would have stimulated anaerobic microbial sulfate reduction (MSR) throughout the oceans and the water column production of carbonate with strongly negative δ^{13}C compositions. The sulfate reducers would have dined on the abundant DOC (eventually depleting this resource) allowing for the oceans to become increasingly oxygenated [*McFadden et al., 2008; Och and Shields-Zhou*, 2012], thereby driving the anaerobes into the sediments, by the end of the biogeochemical anomaly. This scenario, including the progressive sense of ocean oxygenation, is supported by recent U and Mo isotope measurements of Shuram equivalent carbonates in the Doushantuo of South China [*Kendall et al.*, 2015], as well as from iodine abundance measurements of carbonates from the Shuram Formation of Oman [*Hardisty et al.*, 2017]. Evidence for intense weathering of the continents and the buildup of

oceanic sulfate is provided by time-series analyses of strontium (Fig. 7.3; indicating a significant increase in the delivery of radiogenic ^{87}Sr to the oceans) and sulfur isotope abundances [e.g., *Fike et al.*, 2006; *Kaufman et al.*, 2007; *McFadden et al.*, 2008] during the unprecedented carbon cycle perturbation. In fact, given the remarkable coupling of carbon, sulfur, and strontium isotope change during the event, either the SE represents a diagenetic conspiracy, or it is an indicator of truly global environmental change [*Halverson et al.*, 2007; *Kaufman et al.*, 2007; *Lee et al.*, 2015; *Cui et al.*, 2017].

Accepting this holistic scenario, by the end of the SE, shallow ocean water would have been oxygenated enough to stimulate the evolution and diversification of the Ediacaran biota. These organisms appear very near to the crossover from strongly negative-to-positive δ^{13}C compositions, with the earliest example perhaps being *Palaeopaschichnus*; most modern workers regard this Ediacaran form as a serially repeating body fossil [e.g., *Haines*, 2000; *Antcliffe et al.*, 2011], although it has alternatively been interpreted as the earliest evidence for bioturbation [*Rogov et al.*, 2012]. The fossil occurs in

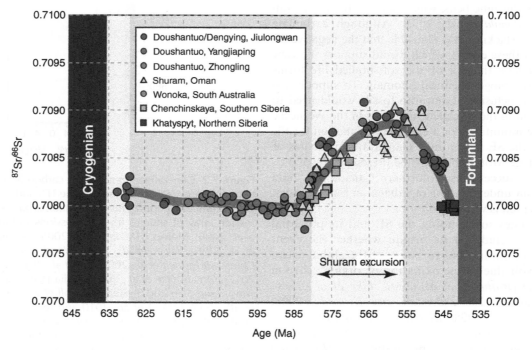

Figure 7.3 Generalized trend in the strontium isotope composition of well-preserved high-Sr marine limestones through the Ediacaran period modified from *Xiao et al.* [2016] with age constraints based on their "correlation 2" and assuming the Shuram excursion is pinned to the 551 Ma age for the Doushantuo-Dengying boundary (see discussion in text and Fig. 7.2 caption). This compilation is based on data from successions in South China [*Sawaki et al.*, 2010; *Cui et al.*, 2015], Oman [*Burns et al.*, 1994], South Australia [*Calver*, 2000], southern Siberia [*Melezhik et al.*, 2009], and northern Siberia [*Cui et al.*, 2016c]. The trend marked by the thick gray line indicates a plateau of ca. 0.7080 followed by a profound rise in ^{87}Sr/^{86}Sr values coincident with the Shuram excursion up to as high as 0.7090 and likely associated with intense weathering up uplifted terrains worldwide. The trend then declines back to ~0.7080 very near to the Ediacaran-Cambrian boundary. (*See insert for color representation of the figure.*)

Unit 8 of the Wonoka Formation of South Australia [Haines, 2000] just as the carbon isotope trend approaches the 0‰ value [Husson et al., 2015]. Similarly, in Namibia, the first of the soft-bodied Nama Assemblage appears in sandstone of the Kliphoek and Aar members [Hall et al., 2013], including Ernietta, Pteridinium, and Rangea, immediately above the SE preserved in the Kanies and Mara members.

7.2.2. Neoproterozoic Oxidation Event (NOE)

Due to the desert climate, the geological exposures in southern Namibia are spectacular, and the views are expansive. In terms of an origin story, it is here where time-series Ediacaran carbon and strontium isotope research was first hatched [Kaufman et al., 1991, 1993] on samples graciously donated by Gerard Germs, Wulf Hegenberger, and Andy Knoll, along with stratigraphic assistance from Charlie Hoffmann who has played a lasting role for a host of international investigators ever since. From the base of the Nama Group, one can see Ediacaran history play out as the carbonate lithologies become increasingly darker in color from the Mara to the Mooifontein reflecting their organic carbon contents [Saylor et al., 1998]. Associated with the tonal transition are dramatic increases in the ^{13}C and ^{87}Sr abundances of well-preserved limestones (Figs. 7.2 and 7.3) and the evolution of macroscopic life. The isotope changes preserved in carbonates are believed to reflect proportionally greater burial of ^{12}C-rich organic matter prompted by the coincidence of high primary productivity and enhanced rates of continental weathering. The Ediacaran Sr isotope record illustrated in Figure 7.3 is consistent with Pan-African uplift of Himalayan-scale mountains and the related erosional shock to the oceans [Asmerom et al., 1991; Derry and France-Lanord, 1996; Squire et al., 2006; Campbell and Allen, 2008]. Insofar as there is a linear relationship between sedimentation rate and organic carbon burial [Derry et al., 1992], atmospheric O_2 will increase during orogenic events but only if the buried organic mass (along with reduced iron and sulfur) exceeds that of the mass of organic carbon eroded [Kaufman et al., 1993; Campbell and Allen, 2008]. Geochemical models of secular trends in carbon, strontium, and neodymium isotope changes through the Ediacaran imply that most of the O_2 in the present atmosphere (ca. 21%) could have been generated [see also Des Marais et al., 1992] during the NOE [Och and Shields-Zhou, 2012].

An independent barometer for Ediacaran oxidation may come from elemental redox proxies, including Mo, V, and U [Scott et al., 2008; Och and Shields-Zhou, 2012; Partin et al., 2013] abundances in black shale, as well as Ce concentrations in carbonates [Wallace et al., 2017]. These elements have very minor concentrations in detrital minerals, so they are a more direct proxy of seawater conditions. Their relative enrichments in Ediacaran and Cambrian successions in comparison with older strata support the view of rising oxygen in surface environments, although none of these paleoproxies are particularly well calibrated. Iron speciation measurements (which are better calibrated in shale from modern oxic, ferruginous, and sulfidic environments; Poulton and Canfield, 2005) of fossiliferous Ediacaran sedimentary rocks in Newfoundland are consistent with the post-Gaskiers rise of deepwater oxygen that fostered the evolution and diversification of macroscopic life [Canfield et al., 2007, but see Sperling et al. [2015] for a contrary view]. While these analyses play a supporting role in our understanding of the NOE, the iron speciation proxy is considered by many to have only local significance; results from one basin to the next seem fairly clearly tied to the organic carbon and pyrite contents of the preserved shales in each basin.

Temporal changes in the sulfur cycle recorded in the isotopic compositions of sedimentary sulfates, carbonate-associated sulfate (CAS), and pyrite may also provide important constraints on Ediacaran and Cambrian redox conditions. Relative to measurements in older intervals, Ediacaran successions document a significant change in the magnitude of fractionation between reduced (pyrite) and oxidized (sulfate or CAS) phases [Canfield and Thamdrup, 1994; Canfield, 1998]. This redistribution of sulfur isotopes has been interpreted to reflect the onset of oxidative processes, including sulfur disproportionation. In disproportionation, sulfur is recycled via both reductive and oxidative pathways. On the reductive side, the magnitude of kinetic sulfur isotope fractionation correlates directly with extracellular sulfate concentrations. Experiments from pure cultures of sulfate reducers indicate maximal fractionation of 66‰ at sulfate concentrations similar to modern seawater at 28 mM [Sim et al., 2011], while fractionations may be suppressed at very low sulfate abundances (<200 μM) [Habicht et al., 2002]. On the oxidative side, the sulfide produced through MSR is typically reoxidized to elemental sulfur, which is subsequently disproportionated to sulfate and sulfide, by coupling with the reduction of O_2, NO_3^-, or iron and manganese compounds. Disproportionation reactions can thus augment the fractionations induced during MSR, resulting in isotopic contrasts between reactant sulfate and product sulfide of >70‰. Interpreting the environmental significance of stratigraphic variations in the magnitude of fractionation, however, requires the recognition that the δ^{34}S signatures of CAS and pyrite may have been inherited from different parts of the depositional basin [Cui et al., 2016a]. Sulfate incorporation into primary carbonate sediments would occur within the water column, whereas pyrite would form either in euxinic bottom waters or within sediments. Considering this spatial separation, local sulfate availability could dictate the sulfur isotopic difference between CAS and pyrite, particularly if pyrite is formed in non-bioturbated and microbially sealed sediments where the water-sediment interface represents a significant diffusion barrier [*Bottjer et al.,* 2000; *Fike et al.,* 2009].

Evidenced by the widespread deposition of marine evaporites near the end of the Ediacaran in Oman and elsewhere, some sulfur isotope studies conclude that the overall sulfate concentration of the oceans increased from the SE to the Ediacaran-Cambrian boundary [*Hurtgen et al.,* 2005; *Fike et al.,* 2006; *Halverson and Hurtgen,* 2007; *Wu et al.,* 2015]. This interpretation is consistent with the oxidation of pyrite in exposed continental rocks during Pan-African orogeny, but it could alternatively signify the release of sulfide from sealed mats (and its conversion to sulfate via microbial sulfide oxidation) associated with the onset of bioturbation [*Canfield and Farquhar,* 2009; *Wu et al.,* 2015; *Hantsoo et al.,* 2018]. Enhanced sulfide oxidation provides an alternative explanation for the large magnitude sulfur isotope fractionations recorded in Ediacaran strata that were previously attributed to disproportionation [*Wu et al.,* 2015] and is consistent with the physiological hypothesis for the Ediacaran biota presented here.

Complicating the view of a unidirectional rise in oxygen levels associated with the NOE is the recognition of rapid oscillations in the $\delta^{13}C$ compositions of carbonates (Fig. 7.2), which suggest an inherent instability in the carbon cycle continuing from the SE to the end of the Fortunian stage. Many recent studies have suggested that oceans through this transitional interval may have been stratified with deep ferruginous and episodically euxinic (containing free H_2S as in the Black Sea today) waters overlain by oxygenated conditions on shallow marine platforms [*Li et al.,* 2010; *Johnston et al.,* 2012; *Sperling et al.,* 2012, 2013a, 2013b; *Wood et al.,* 2015; *Bowyer et al.,* 2017]. Fluctuations in the chemocline could thus dictate the spatial distribution of the Ediacaran biota [*Tostevin et al.,* 2016], whether these large organisms were animals that would be affected by the absence of O_2 or the toxicity of H_2S or the osmotrophic sulfide-oxidizing organisms that would require access to both resources. Variably stratified conditions may help to explain the presence of "superheavy" pyrite (approaching or exceeding the $\delta^{34}S$ compositions of coeval sulfate preserved as trace phases in carbonate) in Ediacaran successions postdating the SE. The observation of spectacular ^{34}S enrichment in sedimentary sulfides has moved some researchers to otherwise suggest that the Ediacaran oceans had low (rather than high) sulfate concentrations [*Ries et al.,* 2009; *Shen et al.,* 2010; *Loyd et al.,* 2012,

2013]. The origin of the ^{34}S-enhanced pyrites largely remains a mystery but is consistent with rapid rates of MSR in the sediments (or in anoxic bottom waters) stimulated by the presence of abundant nutrients and organic substrates. More concerning, however, is the likelihood that previous researchers have underestimated the non-structurally bound sulfur components in these rocks by using incomplete leaching techniques [cf., *Marenco et al.,* 2008; *Tostevin et al.,* 2017], resulting in higher CAS abundances and lower $\delta^{34}S$ values for their seawater sulfate proxies.

7.2.3. Basal Cambrian Excursion (BACE)

Carbon isotope trends leading up to the Ediacaran-Cambrian boundary are characterized by a plateau of moderately positive $\delta^{13}C$ compositions that in some regions fall through the origin [*Pelechaty et al.,* 1996; *Smith et al.,* 2016] before a punctuated negative carbon isotope event known as the basal Cambrian excursion (BACE). This profound biogeochemical anomaly rivals the SE in its depth, if not its stratigraphic throw (Fig. 7.2). The negative excursion was first clearly seen in northwestern Canada and southwestern United States [*Narbonne et al.,* 1994; *Corsetti and Kaufman,* 1994] but soon recognized in carbonate-rich successions worldwide with nadir values near −8‰ or lower. It is closely associated with the first appearance datum (FAD) of *Treptichnus pedum* [*Corsetti and Hagadorn,* 2000] (the 3D ichnofabric presented as the poster child for the boundary) insofar as the initiation of widespread penetrative bioturbation is regarded as such a significant geobiological event that it is defined by the appearance of such diagnostic trace fossils [*Hantsoo et al.,* 2018].

Measurements of redox-sensitive trace elements and the widespread occurrence of black shale and phosphorites during the BACE in Iran [*Kimura and Watanabe,* 2001] and Oman [*Schröder and Grotzinger,* 2007] suggest widespread anoxia during this geochemical divide, representing a potential kill mechanism for the soft-bodied Ediacaran biota, as well as their shelly relatives *Cloudina* and *Namacalathus* [*Amthor et al.,* 2003]. New paleontological observations from southern Siberia, however, suggest that taxa attributed to Ediacaran and Fortunian skeletal biotas overlap without notable biotic turnover before the BACE [*Zhu et al.,* 2017]; in essence, the new data show that cloudinids and anabaritids had slightly longer ranges than previously understood with the former extending up through the BACE and the latter appearing below. If correct, this questions whether the biogeochemical anomaly was a significant factor in the mass extinction of the Ediacaran biota. While these shelly fossils are generally folded into the broad sweep of the Ediacaran organisms, their template-directed biomineralization sets them apart as true animals. On the other hand, the soft-bodied forms possessed a common physiological strategy as described in this thesis: they may be completely unrelated to animals. With only the possible exception of *Tirasiana* and *Swartpuntia* [*Hagadorn et al.*, 2000], the soft-bodied forms appear not to have made it across the BACE.

New evidence for widespread ocean anoxia in the prelude to the BACE is interpreted from U isotope measurements of carbonates from the Dengying Formation of South China [*Wei et al.*, 2018; *Zhang et al.*, 2018]. In these bituminous and fossiliferous sediments, *Cui et al.* [2016a] previously documented a significant positive $\delta^{13}C$ excursion and a dramatic shift in the $\delta^{34}S$ of pyrite coincident with the biological transition from *Conotubus* to *Cloudina*. These authors suggest that an increase in terrestrial weathering fluxes of nutrients, sulfate, and alkalinity stimulated primary productivity, biomineralization, and the spread of anoxic subtidal and basinal environments. The new $\delta^{238}U$ results similarly suggest the spread of oceanic anoxia, which may have played an important role in the demise of the soft-bodied Ediacaran biota and further may have stimulated animal motility [*Zhang et al.*, 2018]. An alternative way of looking at the biological arms race that would be consistent with the resource-based hypothesis is that the onset of bioturbation by motile animals could have released H_2S from the microbial mats into the water column and therefore allowed for the spread of anoxia.

The driving factor(s) for the spread of ocean anoxia in the lead-up to the Ediacaran-Cambrian boundary remains a mystery, but interrogation of the emerging Sr isotope record of the terminal Ediacaran may shed some light on the problem. Recent $^{87}Sr/^{86}Sr$ results from the Khatyspyt Formation in the Olenek uplift of Arctic Siberia indicate values around 0.7080 (Fig. 7.3), which indicate a sharp decline from values that characterize the SE and the TES elsewhere in the world [cf., *Kaufman et al.*, 1993; *Narbonne et al.*, 1994; *Halverson et al.*, 2007]. Insofar as these bituminous and richly fossiliferous sedimentary rocks (the Ediacaran biota here are preserved as carbonaceous compressions in carbonate and shale rather than as casts and molds in sandstone; *Grazhdankin et al.*, 2008) were deposited at the doorstep of the Ediacaran-Cambrian boundary, the sharp decline in $^{87}Sr/^{86}Sr$ of these exceptionally well preserved limestones may record a significant hydrothermal event. If correct, the venting of reduced hydrothermal fluids into the world oceans may well have resulted in the consumption of dissolved oxygen and the spread of anoxic (and perhaps euxinic) bottom waters. Support for the end-Ediacaran rifting hypothesis notably comes from Sr and Cr isotope measurements of TES carbonates associated with *Cloudina* in Uruguay [*Frei et al.*, 2011].

In sedimentary rocks deposited during the BACE, uranium isotope compositions [*Wei et al.*, 2018; *Zhang et al.*, 2018] notably reflect more oxidized conditions, but this may be analogous to the SE where ventilation of the ocean resulted first in the depletion of reduced components until free oxygen could build up in the water column. Molybdenum isotope measurements in basal Cambrian sedimentary rocks in South China similarly suggest oxidizing conditions by the end of the BACE, which could have then triggered the evolution and diversification of modern animal phyla [*Wen et al.*, 2011].

7.2.4. The Cambrian Explosion

The global sedimentary record of carbon isotope variations reveals a prolonged interval of carbon cycle instability in carbonate-dominated Fortunian strata, with rapid oscillations between negative and positive extremes [*Knoll et al.*, 1995a; *Kaufman et al.*, 1996; *Maloof et al.*, 2010], and a rise in $^{87}Sr/^{86}Sr$ compositions [*Kaufman et al.*, 1996]. It is likely that the redox landscape of Fortunian ocean water was similar to that suggested from geochemical studies of the TES with oscillations in the chemocline being particularly important in some basins, while not in others. Analyses of sedimentary pyrite and CAS in transitional Ediacaran to Cambrian successions record a significant sulfur isotope shift [*Fike and Grotzinger*, 2008; *Cui et al.*, 2016a, 2016c; *Hantsoo et al.*, 2018]. Geochemical box modeling suggests that sulfur remobilization through bioturbation would have enhanced oxidative processes (including sulfide oxidation) and increased oceanic sulfate concentration [*Canfield and Farquhar*, 2009]. With the rise of oxygen at the base of the Fortunian, the short fuse to the Cambrian explosion was ignited, leading to detonation near to a positive carbon isotope anomaly known as the ZHUCE (Fig. 7.2) event some 10–15 Ma later [*Kaufman et al.*, 2012].

The continuous record of invertebrate FADs with no turnover by extinction in the midst of Fortunian carbon cycle instability is unusual and consistent with the idea that Fortunian ecosystems may have been more tolerant of new genetic variants [*Knoll et al.*, 1995a] and oscillating redox environments. This in part may be due to the novelty of bioturbation. Some of the earliest Fortunian animals were able to penetrate the widespread microbial mats that dominated the Ediacaran period [*McIlroy and Logan*, 1999; *Buatois et al.*, 2014; *Carbone and Narbonne*, 2014; *Gougeon et al.*, 2018]. By piercing and ventilating Ediacaran matground [*Seilacher and Pflüger*, 1994; *Bottjer et al.*, 2000], bioturbation may have provided greater opportunities for early animals by expanding habitable ecosystems [*McIlroy and Logan*, 1999; *Knoll and Bambach*, 2000]. Ecosystem expansion via bioturbation might explain the macroevolutionary lag between the divergence of major animal clades during the Cryogenian and early Ediacaran periods and their diversification during the Cambrian explosion [*Erwin et al.*, 2011].

Numerous studies worldwide have documented an increase in the size, diversity, and architectural complexity of animal burrows from the late Ediacaran to the early Cambrian [see *Gougeon et al.*, 2018 for a recent compilation]; however, there is less agreement about the impact of these changes on seafloor bioturbation. Some local and global ichnological studies [*Droser and Bottjer*, 1988; *McIlroy and Logan*, 1999; *Mángano and Buatois*, 2014] document a stepwise increase in the abundance and depth of bioturbation through the Cambrian, while others conclude that sediment mixing was not significant until the late Cambrian [*Tarhan and Droser*, 2014] or even the late Silurian [*Tarhan et al.*, 2015].

To assess the geochemical effects of penetrative bioturbation, high-resolution organic carbon and pyrite sulfur analyses were recently conducted of the siliciclastic dominated basal Cambrian Global Boundary Stratotype Section and Point (GSSP) (Chapel Island Formation, Newfoundland, Canada). Along these wave-washed exposures at Fortune Head (giving a name to the basal Cambrian stage), a positive $\delta^{13}C$ excursion in organic matter was noted to start at *exactly* the Ediacaran-Cambrian boundary and returns to stably ^{13}C-depleted values hundreds of meters higher coincident with enrichments of ^{13}C in carbonate carbon toward seawater values [*Hantsoo et al.*, 2018]. Pyrite in these sediments notably underwent significant ^{34}S depletion at the Ediacaran-Cambrian boundary that continued upsection toward the end of the Fortunian. This observation is suggestive of the growth of the oceanic sulfate reservoir spurred by progressive ventilation and oxygenation of shallow sediments as a direct consequence of penetrative bioturbation, which further stimulated the oxidative sulfur cycle. These authors suggest that sediment ventilation in the basal Fortunian spurred a temporary increase in MSR and benthic sulfur cycling under low-oxygen conditions. A decline in bottom water pO_2 could have resulted from more efficient biological pumping, and MSR may have generated the organic carbon isotope excursion that terminates in the upper Fortunian strata. The end of the organic carbon isotope excursion implies stabilization of carbon and oxygen cycling in the shallow substrate, potentially related to a balance between the production, export, and remineralization of organic matter and/or to a stabilization of water column pO_2. Overall, these data from a siliciclastic dominated succession shed new light on the Cambrian explosion. They attest to the geochemical significance of the initiation of sediment ventilation by animals at the dawn of the Phanerozoic, with oxygenation of the shallow substrate soon followed by the appearance of biomineralized small shelly fossils (SSFs).

7.3. CLIMATE

Insofar as the global Cryogenian glaciations, which ended around 635 Ma [*Hoffmann et al.*, 2004; *Condon et al.*, 2005], have been considered as evolutionary bottlenecks [*Hoffman and Schrag*, 2002], it is important to understand the temporal and spatial distribution of Ediacaran ice ages relative to that of the Ediacaran biota, as well as the wild swings in the carbon and sulfur cycles that characterize the period. Evidence for icehouse conditions through the upper reaches of the Ediacaran period comes from the preservation of diamictites (unsorted or poorly sorted sedimentary rocks) of confirmed or presumed glacial origin, some with postglacial faux cap carbonates, as well as deeply incised paleo-valleys interpreted as related to glacio-eustacy.

Foremost among these is the ca. 580 Ma Gaskiers diamictite in Newfoundland (and its sister on the Bonavista Peninsula, the Trinity diamictite, which is constrained to have lasted <340,000 years; *Pu et al.*, 2016). The Gaskiers (Fig. 7.4a) is the only Ediacaran diamictite to have thus far been directly dated by U-Pb techniques. Notably, the short duration of the Gaskiers ice age is inconsistent with predictive durations for snowball Earth events, although a low-latitude ($19.1 \pm 11.1°$) paleomagnetic determination [*Pisarevsky et al.*, 2011] for the Avalon terrain in which it sits and a thin yet discontinuous cap carbonate (see inset) with a negative-to-positive $\delta^{13}C$ trend [*Myrow and Kaufman*, 1999] are consistent with other aspects of Cryogenian global glaciation.

Many researchers have attempted to correlate the Gaskiers glaciation with the SE as recorded in middle Ediacaran carbonates worldwide [see review in *Xiao et al.*, 2016], but the potential for repetition [*McFadden et al.*, 2008] or diachroneity of the biogeochemical anomaly and the lack of post-Cryogenian glacial diamictites in critical successions have severely hindered our ability to demonstrably connect climatic and geochemical events. Based on their lithological characteristics and indirect radiometric constraints, other potential Gaskiers equivalents include the Mortensnes [*Halverson et al.*, 2005] and Moelv diamictites [*Bingen et al.*, 2005] in northern and southern Norway; the Hankalchough [*Xiao et al.*, 2004] and Hongtiegou [*Shen et al.*, 2010] diamictites in northwestern China; the Egan diamictite in Australia [*Corkeron and George*, 2001; *Corkeron*, 2007] and the Croles Hill diamictite in nearby Tasmania [*Calver et al.*, 2004]; the Serra Azul diamictite in Brazil [*Alvarenga et al.*, 2007]; and the Squantum diamictite in the Boston Basin, United States [*Thompson and Bowring*, 2000]. Ediacaran successions in these areas are typically dominated by siliciclastics [*Hoffman and Li*, 2009], which complicate their correlation with carbon cycle anomalies typically recorded in carbonate facies. Nonetheless,

Figure 7.4 (a) White to red colored Gaskiers diamictite overlain by a 50 cm thick white carbonate (left of the field assistant) on the shore of Conception Bay at Harbour Main, Newfoundland. Inset shows the brecciated and potentially karstified upper surface of the thin carbonate filled with green mudstone of the overlying Drook Formation. (b) Green meta-basalt of the basal Catoctin Formation with diapirs (flame structures) of Fauquier Formation carbonate injected between the chilled pillow margins on Goose Creek near Aldie, Virginia, United States. Inset illustrates a hyaloclastic texture at contact between the Catoctin meta-basalt and marble of the Fauquier cap carbonate, indicating that the sediments were water-saturated during emplacement of the volcanic rocks. Reproduced with permission of Elsevier. (c) A 20 m thick diamictite along the Tas-Yuryakh River lying unconformably above Turkut Formation dolomites in the Olenek uplift, Arctic Siberia, Russia. The freshly exposed diamictite has a green-gray sandy to clayey calcareous matrix with abundant cobble- to boulder-sized clasts in a weakly stratified pile. The randomly oriented clasts (see inset) are primarily derived from the Turkut and underlying Khatyspyt formations, but they also consist of occasional green igneous rocks and metamorphic rocks of exotic origin. Subrounded clasts plucked from the surrounding carbonate-rich matrix are notably faceted. *(See insert for color representation of the figure.)*

strongly to moderately negative $\delta^{13}C$ compositions are noted in postglacial carbonates above the Hankalchough, Egan, and Serra Azul diamictites, and these have been provisionally correlated to the SE. Paleomagnetic data notably support a low-latitude position for both the Egan and Hankalchough diamictites [*Hoffman and Li*, 2009], which, assuming synchronicity of the presumed Gaskiers equivalents, would blur the lines between Cryogenian and middle Ediacaran ice ages.

Radiometric, stratigraphic, and paleontological data from Laurentia, Africa, China, and Siberia further suggest the possibility of post-Gaskiers Ediacaran ice ages, although there is scarce evidence for striated clasts, till pellets, or dropstones in most of these deposits. For example, in northern Virginia, United States, diamictite of the Fauquier Formation is stratigraphically overlain by a thick red bed sandstone and a 20 m thick carbonate horizon that preserves a strong negative-to-positive $\delta^{13}C$ trend [*Hebert et al.*, 2010]. The carbonate has a conformable soft-sediment contact with pillow basalts of the overlying Catoctin Formation (Fig. 7.4b), which is constrained by a U-Pb zircon age of around 571 Ma. Application of the Catoctin age to the Fauquier diamictite supports the view of a regional glaciation nine million years younger than the Gaskiers [see also *Linnemann et al.*, 2018 for a broadly equivalent post-Gaskiers glacial diamictite in peri-Gondwanan West Africa] and within uncertainty of the earliest fossiliferous Ediacaran strata in Avalonia. An even younger Ediacaran ice age (the Vingerbreek glaciation) is suggested by the presence of deep and laterally extensive paleo-valleys interpreted to have formed during glacio-eustacy and filled with a heterogeneous mix of limestone-clast conglomerate, graywacke, quartzite, shale, carbonate, and tillite (with abundant faceted pebbles). Radiometric constraints [*Grotzinger et al.*, 1995] support a ca. 547 Ma age for the Vingerbreek event in Namibia and its potential correlative in South Africa [*Schwellnus*, 1941; *Germs*, 1972, 1983; *Germs and Gresse*, 1991; *Praekelt et al.*, 2008; *Germs and Gaucher*, 2012]. If the drawdown surface is correctly tied to the regional buildup of ice sheets (as consistent with a moderate negative shift in carbonate $\delta^{13}C$ values at the base of the member; *Saylor et al.*, 1998), the Ediacaran biota would have felt its chilling effects.

Finally, glacial deposits or deep paleo-valleys believed to be related to glacio-eustacy near the Ediacaran-Cambrian boundary have been described from throughout central Asia [see review in *Chumakov*, 2009], West Africa [*Bertrand-Sarfati et al.*, 1995], and Namibia [*Germs*, 1972, 1995]. These ice age deposits are best known from Kazakhstan and Kyrgyzstan and are collectively part of the Baykonurian glaciation. Some of these glacial deposits occur above sedimentary rocks containing Ediacaran fossils (in one case from the East Sayan

Mountains of southern Siberia *Cloudina* have been found in diamictite matrix) and are interbedded or lie below those containing elements of the basal Fortunian *Anabarites trisulcatus* Zone.

Supporting the view of a boundary ice age, a 20 m thick diamictite [*Kaufman et al.*, 2009] composed of subrounded to angular cobble- to boulder-sized clasts (many faceted) of underlying lithologies in a gray-green sandy to clayey calcareous matrix was recently recognized along the Tas-Yuryakh River in the Olenek uplift of Arctic Siberia (Fig. 7.4c). The Tas-Yuryakh diamictite was deposited unconformably above carbonates of the Turkut Formation (similarly containing elements of the *A. trisulcatus* Zone) at the same level where a volcanic tuff was identified and radiometrically dated at 543 ± 1 Ma [*Bowring et al.*, 1993]. This weakly stratified pile includes seven distinct levels interpreted as lodgment till [*Arnaud and Eyles*, 2002; *Arnaud*, 2008] that accumulated at the interface of a grounded ice sheet in a shallow marine environment. Lacking evidence for tectonic uplift and erosion of the platform, it is likely that incision of underlying lithologies resulted from glacio-eustatic lowering of sea level by 150 m or more. The paleokarst is a prominent hiatal surface that shows systematic variations in the amount of erosion throughout the Olenek uplift [*Pokrovsky and Venagradov*, 1991; *Pokrovsky and Missarzhevsky*, 1993; *Knoll et al.*, 1995b; *Pelechaty et al.*, 1996] with progressive downcutting of the Turkut and equivalent strata across the basin. This unconformity partially erases the profound negative $\delta^{13}C$ excursion that characterizes the Ediacaran-Cambrian boundary (Fig. 7.2) in the Olenek uplift and is expressed throughout the eastern Siberian platform [*Khomentovsky*, 1990], suggesting glacio-eustacy as a potential driver. However, a diatreme that cuts through underlying strata is recognized along the Khorbusuonka and Mattaia rivers, which is the likely source of the dated volcanic tuffs, and undoubtedly contributed to the glacially remobilized volcano-sedimentary succession. The Tas-Yuryakh diamictite is immediately overlain by mixed carbonate and siliciclastic rocks of the early Cambrian Mattaia Formation, which preserves traces of animal activity as well as diagnostic SSFs and reef-building archeocyathids representing the most rapid phase of animal diversification in Earth's history [*Rozanov et al.*, 1969].

The observations of a widespread Baykonurian ice age suggest a glacial divide between the enigmatic Ediacaran biota and the succeeding Cambrian explosion. Insofar as oxygen was a critical resource for the Ediacaran biota, colder TES temperatures associated with icehouse conditions (starting with the Gaskiers event and continuing episodically toward the Fortunian) would have further saturated seawater with the breathing gas. While increased O_2 concentrations would be consistent with the "cold cradle" hypothesis [*Vickers-Rich*, 2007] for the biota, the breathing gas is 28× less soluble than CO_2 and 83× less soluble than H_2S at 20°C, which would have had a significant effect on the pH and toxicity (to animals) of the water column and pore fluids. Notably, a report of Mo isotope variations across the Ediacaran-Cambrian boundary in Oman and South China [*Wille et al.*, 2008] suggests that H_2S may have spread across shallow depositional environments, hence providing a potential kill mechanism for the Ediacaran biota, assuming these organisms were animals. Alternatively, the coup de grâce may have been simply the drawdown of oxygen as ocean anoxia spread in the TES [see *Kimura and Watanabe*, 2001; *Zhang et al.*, 2018], especially if the biota was dependent on sulfide oxidation as a common metabolic strategy. Recognition of a boundary glaciation provides a physical mechanism for sea-level drawdown and the enhanced upwelling of H_2S as well as ^{13}C-depleted HCO_3^- produced through sulfate reduction in deep anoxic waters, which contributed to the strong negative $\delta^{13}C$ excursion preserved in marine carbonates. In turn, the ecospace vacated by the Ediacaran biota may have been repopulated with Cambrian faunas in the glacial aftermath, in likely response to rising oxygen levels in the ocean and atmosphere.

7.4. ERNIETTAVILLE: EARLY BIOTURBATION?

The resource-based hypothesis for the rise and fall of the Ediacaran biota was formulated in the midst of chemostratigraphic and paleontological research of lower Nama Group (Kuibis Subgroup) sedimentary rocks in southern Namibia just east of Aus on Farm Aar and the surrounding region. Observations of the shallow marine Nama Assemblage in southern Africa were augmented by several excursions to Mistaken Point in Newfoundland to study the deepwater habitats of the long-ranging Avalon Assemblage and to South Australia to view and describe the cosmopolitan White Sea Assemblage.

In all but one debatable basin [*Macdonald et al.*, 2013: see discussion above], the Ediacaran biota appear in the stratigraphic record after the SE, which on several continents preserves coupled changes in the carbon, sulfur, and strontium isotope composition of carbonates [*Cui et al.*, 2015 and references therein] that together suggest the rise of oxidants in seawater and of oxygen in the atmosphere. These changes were most likely driven by Pan-African orogeny and silicate weathering that delivered sediments, nutrients, and oxidants into the oceans, promoting primary productivity and organic carbon burial [*Kaufman et al.*, 1993; *Squire et al.*, 2006; *Och and Shields-Zhou*, 2012; *Planavsky*, 2018], as well as the rise of sulfate concentrations [*Fike et al.*, 2006; *Kaufman et al.*, 2007]. In light of these coupled biogeochemical changes, the evolution of the Ediacaran biota is an origins story based on the rise of oxygen in seawater. On one hand, the oxygen could have promoted eukaryotic metabolisms including the formation of sterols [*Runnegar*, 1991; *Catling et al.*, 2005; *Budd*, 2008] and of collagen [*Towe*, 1981; *Saul*, 2009], but on the other the rise of O_2 could have stimulated autotrophic sulfide oxidation that led to the evolution and diversification of mat-like organisms connected to a common metabolic strategy.

My initial interest in the preservation of *Rangea*, *Pteridinium*, and *Ernietta* stemmed from the observation of yellow coatings composed of jarosite, a product of the oxidative weathering of early diagenetic pyrite [*Darroch et al.*, 2012; *Hall et al.*, 2013], on freshly exposed fossiliferous surfaces (Fig. 7.5a; see also Fig 7.6d as presented in color in *Vickers-Rich et al.*, 2013). Astrobiological interest in jarosite stems from its presence on the surface of Mars, which suggests wet, acid, and sulfate-rich conditions early in planetary history [*Squyres et al.*, 2004]. In association with these Ediacaran fossils, the presence of jarosite is arguably related to the growth of pyrite "death masks" [*Gehling*, 1999; *Anderson et al.*, 2011], which help to explain their unusual preservation as exquisitely detailed impressions in fine sands and silts [but see *Newman et al.*, 2016; *Tarhan et al.*, 2016 for an alternative explanation for mat and fossil preservation based on silica saturation of the Ediacaran ocean).

While the death mask hypothesis is particularly attractive, there is no a priori reason that the living organic membranes were not at least partially pyritized while the organisms were alive. Given how rapidly the organic sheets would decay, there must have been available H_2S and Fe^{2+} in pore fluids to support mineralization in life. Given the remarkably low solubility of both hematite ($K_{sp} \sim 10^{-45}$) and pyrite ($K_{sp} \sim 10^{-19}$), an even partially oxidized water column would have been lacking in these reduced aqueous species. This view is consistent with Fe speciation constraints that suggest the Ediacaran biota lived below waters that were at least episodically oxic and low in iron [*Wood et al.*, 2015]. The sediments below the ubiquitous mats, however, would necessarily have been anoxic in order to support anaerobic MSR and its output of hydrogen sulfide. Under reducing conditions, Fe^{3+} preserved as oxide coatings on detrital minerals could have been reduced and mobilized within pore fluids, which could then react with hydrogen sulfide to form pyrite on the fractal interior walls of the organism. Interior pyritization could have conferred an advantage to the Ediacaran biota by providing structural support for the organisms or to regulate the free diffusion of O_2 to the interior of the organism. With a predominantly endobenthic lifestyle [*Crimes and Fedonkin*, 1996; *Grazhdankin and Seilacher*, 2002; *Seilacher and Gishlick*, 2014], it is the buried portion of the organisms that would be initially pyritized.

Figure 7.5 (a) Freshly exposed *Pteridinium* surfaces coated with yellow colored jarosite, a hydrous sulfate of potassium and iron that forms as a product of pyrite weathering, at Aarhauser on Farm Aar, near Aus, Namibia [*Hall et al.,* 2013]. Inset shows a typical iron oxidize patina on an exposed and weathered surface of *Pteridinium* at the same locality. Reproduced with permission of Elsevier. (b) "Mud chip" breccia associated with *Ernietta*-bearing sandstone from Ernietta Hill on Farm Aar. The mud chips have the iron oxide patina indicated above, and a partially exposed erniettid is exposed on the surface (see red arrow). This suggests that the chips represent the surface connection between infaunal *Ernietta* bases and their epibenthic fronds. (c) Sock-shaped sandstone concretions with flat upper surfaces likely to represent *Ernietta* specimens that have lost their tubular covering through exposure and weathering on Windy Peak, Farm Aar. These specimens are the same size and general shape to *Ernietta* preserved in situ with tubular structures (d) (yellow scale bar 1 cm). *(See insert for color representation of the figure.)*

Figure 7.6 Exceptionally preserved *Ernietta* and *Rangea* specimens discovered in sandy gutter casts on Farm Aar, southern Namibia. (a) The most complete *Ernietta* specimen known to date [*Ivantsov et al.,* 2016], with white arrow pointing to the position of the connection between infaunal base and epibenthic frond (see Fig. 7.5b: coin is 2.26 cm in diameter). Reproduced with permission of Wiley. (b) Illustrated reconstruction of the *Ernietta* specimen depicted in (a) with tubular bilayer shown. Reproduced with permission of Peter Trusler. (c) Illustrated reconstruction of *Rangea* with a tubular core and sixfold symmetry of vanes including at least three orders of fractal folding. Reproduced with permission of Peter Trusler. (d) The most complete *Rangea* specimen known to date [*Vickers-Rich et al.,* 2013] showing sand-filled base of the organism (image of same specimen in *Vickers-Rich et al.* [2013] shows patina of yellow jarosite, a weathering product of pyrite).

Other saclike, bulbous, baggy, and pimpled holdfasts might then reflect early pyritization and the requirement of the mat-like organism to migrate within the sediment in order to harvest additional sulfide. The migration of the organic walls would be far more realistic if the interior (rather than the exterior) of the organism were preferentially pyritized. Pyritization in life could explain why *Ernietta* bases are so common and why the Ediacaran fossil record, in general, is replete with holdfasts but proportionally fewer fronds. Supporting this argument, close examination of specimens preserving both holdfast and frond reveals torsion in the latter, while the former appears cemented in place [*Peter Trusler*, pers. Comm., 2018; see Fig. 7.7e). While many of the Ediacaran taxa clearly lived above the sea bottom and were buried and preserved during storm events [*Vickers-Rich et al.*, 2013; *Ivantsov et al.*, 2016], they nonetheless had holdfasts that connected them to the substrate, allowing for a direct connection to the H$_2$S resource within and beneath the mats.

Evidence for the pyrite molds, however, weathers quickly away once fossils are exposed, often leaving behind iron-rich surface veneers (Fig. 7.5a inset). In the case of *Ernietta*, the veneer is usually missing, leaving behind bulbous, lumpy, and sock-shaped sandstone balls, each with a single round to elliptical flat surface. These specimens (Fig. 7.5c), which litter the grounds where well-preserved isolated and in situ specimens of the same size and shape, but with tubular external morphology (Fig. 7.5d), are found, are interpreted to represent the interior sandstone fills of the organic-walled organisms. Recall that thin walls and inert interiors would allow the surface area/volume of the Ediacaran biotas to approach that of osmotrophic bacteria. Furthermore, fields of sub-mat sandstone surfaces associated with individual *Ernietta* specimens are similarly likely to be the preserved expression of continuous undulating sheets of organic membranes, with the top surfaces typified by ovoid and oxidized "mud chip" breccias (Fig. 7.5b). Given traces of tubular impressions on the surface of the chips and the co-occurrence of partial *Ernietta* specimens emanating from them (see arrow), the mud chips are alternatively interpreted as the surface expression of connections between individual *Ernietta* bases and their fronds *or of* fields of fronds emanating from a single expansive sheet of under-mat organic structures.

On Farm Aar, complete specimens of *Ernietta* and *Rangea* were transported during storms and exquisitely preserved in sandy channels and gutter casts [*Vickers-Rich et al.*, 2013; *Elliott et al.*, 2016; *Ivantsov et al.*, 2016]. The discovery of these unique lagerstätte allowed for the most detailed reconstruction of the organisms to date (Fig. 7.6). *Ernietta* has a sac-shaped body with walls constructed of two parallel layers of vertically arranged tubes that form a palisade-like structure, which form a zigzag suture at the base of the sand-filled anchor. Extending beyond the bag-like base, the parallel tubes continue into two facing fanlike structures (Fig. 7.6a and b). The connection of the two parts (see white arrow in Fig. 7.6a) would represent the position of the mud chips discussed above. Other representatives of the Nama Assemblage, including *Pteridinium* and *Namalia*, are strikingly similar to *Ernietta* being composed of palisades of tubes, but in larger and more complex boat and bell shapes [*Grazhdankin and Seilacher*, 2002]. In contrast, *Rangea* consist of an axial bulb and stalk that extends into a cylindrical cone; the axial structure forms the foundation for six vanes arranged radially around the axis, with each vane consisting of a bilaminar sheet composed of repetitive tubular elements exhibiting at least three orders of self-similar branching (Fig. 7.6c and d). The common tubular constructional elements of the Nama Assemblage, including *Swartpuntia* [*Narbonne et al.*, 1997], suggest an adaptation to an infaunal lifestyle in order to maximize gas exchange and nutrition [*Grazhdankin and Seilacher*, 2002, 2005]. In this sense, the growth of tubular biofilms from the sediments (siphoning microbially produced H$_2$S into the fronds where O$_2$ would diffuse in from the water column through osmotrophy) provides a mechanism to gather the two critical resources necessary to sustain thioautotrophy, thus potentially representing the earliest form of bioturbation (*contra Dzik*, 1999). The double layers of parallel tubes in *Ernietta* and its relatives might play physiological roles with one delivering H$_2$S to the fronds and the other O$_2$ to the holdfast.

One of the fundamental properties of biological membranes is that they are barriers to the permeation of polar molecules resulting from the fact that the paraffinic interior of bilayer lipid membranes is hydrophobic [*Widomska et al.*, 2007]. Although H$_2$S is slightly polar and has some hydrogen-bonding capability, it too is hydrophobic, which lends to its toxicity as it will readily partition into cells [*Riahi and Rowley*, 2014]. The permeability of H$_2$S through biological membranes is four orders of magnitude more than water but is less than that of nonpolar O$_2$ [*Subczynski et al.*, 1989] by 0.5 to 2 orders of magnitude based on experimental [*Mathai et al.*, 2014] and model calculations [*Riahi and Rowley*, 2014]. The bottom line is that if the thin walls of the Ediacaran fronds were similar to lipid membranes, they would diffuse in O$_2$ faster than the H$_2$S could escape out, although the rates would undoubtedly be dependent on the concentration gradients inside and outside of the organisms and their effective surface areas [*Laflamme et al.*, 2009].

The irregularities of microbial mats reflected in the varied expression of microbially induced sedimentary structures (MISS) [*Noffke et al.*, 2001] suggest the potential

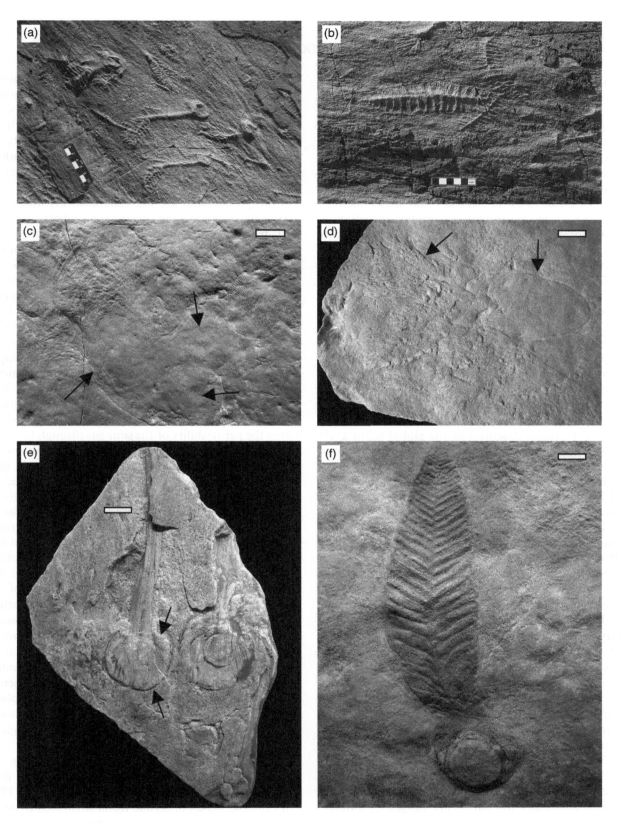

Figure 7.7 (a) Conception-style preservation [*Narbonne*, 2005] of *Charniodiscus spinosus and Cychrus procerus* in positive relief on a surface of the Mistaken Point Formation of Newfoundland. The fossil surface is overlain by a thin bed of volcanic ash (visible as dark layer on the upper right). (b) Spindle-shaped *Fractofusus misrai* with primary and secondary branches on the same surface in Newfoundland with visible volcanic ash. Fossils in (a) and (b) are members of the Avalon Assemblage. Images (c–f) are from the Ediacaran member in South Australia and archived at the South Australia Museum (white scale bars are 5 cm in length). These are representatives of the White Sea Assemblage. (c) *Dickinsonia costata* moving (see arrows) and resting traces. (d) *Kimberella quadrata* (right arrow) and associated trace fossil *Kimberichnus teruzzi* (left arrow). (e) *Aspidella* with holdfast and stalk, but no frond. (f) *Arborea arborea*. *Source:* Government of South Australia. *(See insert for color representation of the figure.)*

that tubes could form through wrinkling and then propagate into sediments. While microbial mats are normally composed of horizontally stratified communities, they can exhibit sedimentary structures defined by dynamic physiochemical gradients, as well as the diversity and physiology of its denizens [*Krumbein et al.,* 1994]. Oxygenic photoautotrophs usually occupy the highest level in the mats, while anoxygenic photoautotrophs and chemolithoautotrophs lie beneath, and these microbes utilize H_2S as an electron donor in the reduction of carbon dioxide to organic matter. Of particular concern here are the non-photosynthetic sulfide oxidizers that use molecular oxygen from the water column and sulfide produced lower in the mat through MSR [*Zhelezinskaia et al.,* 2014] to sustain their metabolic activities. While photosynthetic sulfide oxidizers are also possible, any common metabolic strategy for the Ediacaran biota would have to accommodate the deepwater forms living far below the photic zone, especially representatives of the Avalon Assemblage (Fig. 7.7a and b).

According to *Friedrich et al.* [2001] and references therein, the biological oxidation of hydrogen sulfide to sulfate by a phylogenetically diverse array of prokaryotes is one of the major reactions of the global sulfur cycle. Sulfur oxidation in the Archaea is restricted to some members of the order *Sulfolobales,* while bacterial oxidation of sulfur is mediated by both anaerobic phototrophs and aerobic lithotrophs. The latter is the metabolic strategy considered here given the likelihood of a terminal Neoproterozoic buildup of oxygen in the atmosphere and oceans, although other oxidants are known to play a role in this metabolism [*Ghosh and Dam,* 2009]. Aerobic sulfur-oxidizing bacteria belong to nearly 15 known genera, including *Acidithiobacillus, Beggiatoa,* and *Pseudomonas.* These autotrophic bacteria fix carbon dioxide to create food either via the reductive pentose phosphate cycle or via the reductive tricarboxylic acid cycle. With the ability to harness energy and produce their own food through sulfide oxidation, the earliest sediment miners arguably developed a wide variety of bauplans to accommodate environmental gradients. For example, the *rangeomorphs* from Avalonia might have had to significantly increase their fractal surface areas and/or their height above the sediment-water interface (Fig. 7.7a and b) in order to deal with lower O_2 abundances in deep bottom waters [*Mills et al.,* 2014]. If correct, ecological tiering of the Avalonian life forms would be a strategy for harvesting oxygen at different levels above the substrate rather than organic molecules [*Clapham and Narbonne,* 2002; *Ghisalberti et al.,* 2014]. In support for the dominance of prokaryotic sulfide-oxidizing communities in the TES, biomarker studies in Baltica indicate unusually high hopane/sterane ratios [*Pehr et al.,* 2018]. This finding suggests that these large

and fractally constructed organisms flourished in oligotrophic and bacterially dominated marine environments. The most parsimonious interpretation then would be that the Ediacaran biota themselves were either bacterial colonies or, more likely given their complexity and multicellularity, that they symbiotically hosted sulfide-oxidizing bacteria within their bodies [cf., *Burzynski et al.,* 2017].

More problematic are the Ediacaran organisms that appear untethered to the substrate, have been characterized as bilaterian, and/or are associated with feeding traces, including dickinsonids, sprigginids, parvancorinids, and the monospecific genus of *Kimberella* [*Gehling et al.,* 2014; *Droser and Gehling,* 2015]. Notwithstanding arguments suggesting that specimens lacking bilateral symmetry are taphonomic variants, none of these forms have a demonstrable mouth, gut, or anus. The movement traces of *Dickinsonia* (Fig. 7.7c) and its relatives have been interpreted as evidence for osmotrophic feeding on the mats [*Laflamme et al.,* 2009], but these organisms might alternatively move between fresh surfaces in an effort to quickly absorb H_2S into their systems in order to recharge the resource for the sulfide oxidizers within. Given the permeability of the sour gas through biological membranes, this process might better explain the preservation of both movement and resting traces. On the other hand, *Kimberella* [see Fig. 7.7d] has been interpreted as a molluskan-grade organism [*Fedonkin and Waggoner,* 1997] and is documented in close association with fans of scratches (see arrow) interpreted as traces of mat grazing [*Fedonkin et al.,* 2007]. Here it is the absence of an anus and/or of Ediacaran coprolites that poses a problem, which could be solved if the proposed siphon that scratched the mat was designed to evacuate it of H_2S rather than organic matter. It is difficult to shoehorn these complex and apparently motile forms with those of sessile infaunal and fractal grades [*Droser and Gehling,* 2015], but their bilateral symmetry is debatable as are issues related to their physiology. If there was a common metabolism shared by all of the Ediacaran biota, then the unorthodox speculations entertained here could explain a lot about their morphological disparity (i.e., resource gathering in different sedimentary and basinal regimes) as well as their demise.

It should not be overlooked that sulfide-oxidizing symbionts are harbored within rare modern animals living under unusual circumstances [*Nelson and Fisher,* 1995]. For example, tube worms like *Riftia pachyptila* (from the polychaete family *Siboglinidae*) living near hydrothermal vents [*Bright and Lallier,* 2010], gutless phallodriline oligochaete worms living in oxygen-deficient regions of the ocean [*Blazejak et al.,* 2005], and free-living nematodes within the family *Desmodoridae* [*Bergin et al.,* 2018] all benefit from symbioses with sulfide-oxidizing microbes.

While the symbionts within the tube worms at deep ocean ridges utilize H_2S emanating from black smokers and O_2 directly from seawater, those in marginal marine settings benefit from the ability of the host animals to migrate between sulfide- and oxygen-rich sedimentary environments. Other modern animals (e.g., *Nereis*) are known to tolerate high sulfide environments, and these might be important modern analogs for true Ediacaran metazoans. It has been suggested that sulfur-based metabolisms helped shape initial symbiotic events, leading to the evolution of both unicellular and multicellular eukaryotes [*Overmann and van Gemerden*, 2000; *Theissen et al.*, 2003; *Mentel and Martin*, 2008].

7.5. METAZOANS TAKE THE STAGE

While biomarker and possible fossil evidence suggests a pre-Ediacaran origin of sponges [*Love et al.*, 2009; *Maloof et al.*, 2010; *Brocks et al.*, 2016], Ediacaran evidence for the earliest verifiable metazoans comes in the form of Doushantuo embryos preserving a range of cleavage stages, differentiated multicellularity, and the absence of a cell wall, as well as the presence of diapause egg cysts [*Xiao et al.*, 1998, 2014; *Yin et al.*, 2007]. The animal affinity of these fossils has nonetheless been challenged by observations along the present-day Namibian coast of giant sulfide-oxidizing bacteria of the genus *Thiomargarita* [*Bailey et al.*, 2007], which are quite unlikely to be fossilized [*Cunningham et al.*, 2012]. The sudden appearance of template-directed biomineralization in the cloudinids, which appear worldwide in carbonate facies some time after the SE, may be further evidence for the earliest Ediacaran metazoans. Otherwise, clear evidence of animal activity comes from wormlike organisms that crawled through the mats using peristaltic motions that left increasingly complex trace fossil behind [*Carbone and Narbonne*, 2014; *Meyer et al.*, 2014; *Schiffbauer et al.*, 2016; *Budd and Jensen*, 2017]. These animals apparently mined the mats for nutrients and O_2, but they would also have to have contended with abundant H_2S if they ventured too deep [*Chen et al.*, 2013]. In this case, these early metazoans would have to be at least episodically tolerant of the toxic gas, unless they too harbored sulfide-oxidizing chemosymbionts.

The dramatic end of the Ediacaran biota near the Ediacaran-Cambrian boundary may represent the first great mass extinction, the biotic replacement of soft-bodied metazoans by Cambrian animals, or their gradual disappearance in the fossil record as a result of the elimination of the matgrounds in which they thrived [*Laflamme et al.*, 2013]. The latter is countered by the preservation of Ediacaran-style mats and environments in basal Fortunian strata [*Buatois et al.*, 2014], but evaluating the other possibilities requires an approach that integrates geochemical, ecological, and physiological information. In their 2013 review, Laflamme and colleagues conclude that behavioral innovations associated with predation and ecosystem-wide changes, reflected in the matground to mixground transition (Fig. 7.1), likely spelled the death knell for the Ediacaran biota. In this case, the first vertical perforations of the matgrounds near the end of the TES would have dramatically increased the flux of reduced gases to seawater, including both H_2S and CH_4, that would have combined with available O_2 and hastened the spread of deep ocean anoxia. Evidence for anoxia around the boundary comes from trace element, as well as Mo and U isotope studies [*Kimura and Watanabe*, 2001; *Schröder and Grotzinger*, 2007; *Wille et al.*, 2008; *Wei et al.*, 2018; *Zhang et al.*, 2018 and references therein], although the incomplete preservation of the stratigraphic record in both South China and Oman complicates the timing of events as recorded in widely separated basins. If sulfide oxidation was the common metabolic strategy, the decline of oxygen at the boundary would have wiped out the Ediacaran biota, at the same time promoting MSR and the production of authigenic carbonate (either in the water column or in sediments) strongly depleted in ^{13}C [*Schrag et al.*, 2013; *Cui et al.*, 2017], which could explain the negative Ediacaran-Cambrian boundary $\delta^{13}C$ excursion noted in most successions worldwide.

7.6. CONCLUSIONS

In order to explain the stratigraphic and spatial distribution of the Ediacaran biotas, one must place them in the environmental context in which they were found. From a chemostratigraphic point of view, the large and fractally constructed organisms, most of which were rooted in the substrate among the ubiquitous microbial mats, are tightly framed between two of the most profound negative carbon cycle anomalies in Earth's history: the older SE and a younger event near the Ediacaran-Cambrian boundary known as the BACE. Assuming these bookend geochemical events had environmental drivers and are not simply globally distributed diagenetic artifacts, the spatial and temporal distribution of the Ediacaran organisms should be understood in terms of changes in global ocean chemistry, emerging ecological opportunities, and evolutionary innovations.

Researchers often seek out specific morphological characteristics of the Ediacaran biota in order to fit these enigmatic organisms into preferred interpretations as animal, plant, lichen, or microbial clades, with Adolf Seilacher suggesting that the biota fits into a phylum distinct from any modern representative. Given the similarity of primary tubular and fractally folded units in the construction of these organisms, one might consider whether they all had a unifying metabolic strategy to explain their morphological complexity rather than trying to fit strange morphologies into recognizable categories.

The resource-based hypothesis entertained here stems from geochemical clues for both the rise of O_2 in the oceans and atmosphere in middle Ediacaran time and the continued production of H_2S in the sediments. Expanding upon the vendobiont hypothesis with the likelihood of multicellularity and symbiosis, the geochemical clues as well as the morphological construction and disparity of the Ediacaran biota are consistent with the utilization of these two critical resources required for thioautotrophy. These organisms are envisioned to have absorbed oxygen from the water column and pumped hydrogen sulfide from the sediments in order to sustain their unique physiology and explain their morphological peculiarities. If true, the Ediacaran biotas were among the earliest ecosystem engineers. Their search for resources both in the sediments and in the water column would represent the earliest form of bioturbation, which would have propagated a shift in the chemical and physical properties of the sediments [*Schiffbauer et al.*, 2016]. Furthermore, insofar as their occupation of the subsurface environment delivered autotrophic organic matter to the sediments, their activities would have connected the pelagic to benthic realms for the first time in Earth's history [*Erwin et al.,* 2011].

Whether a heterotrophic animal, eukaryotic symbiont, or bizarre bacterial life form is dependent on sulfide oxidation to make a living, the decline of oxygen and spread of hydrogen sulfide into the water column near to the Ediacaran-Cambrian boundary would spell trouble for any organism that could not move away from toxic conditions [*Chen et al.*, 2018; *Zhang et al.,* 2018]. With the assumption that these bizarre life forms were not animals, the biotic replacement hypothesis would not apply, leaving mass extinction of the Ediacaran experiment the most likely explanation for the biological crisis at the Ediacaran-Cambrian boundary. Given my predilection for geochemistry, the spread of anoxia and loss of the O_2 resource near the end of the Ediacaran period remain my preferred interpretation; however, there could be an alternative (or supplementary) cause for the mass extinction, assuming these tubular and fractal pneumatic organisms were unusual bacterial colonies or complex eukaryotic bags that housed sulfide-oxidizing bacteria. In either case, predation by the emerging army of motile and carnivorous metazoans would have punctured the organic walls of the Ediacaran biota, thereby deflating the H_2S-filled balloons and leading to their ultimate demise.

ACKNOWLEDGMENTS

The author wishes to thank the editors and publishers of this important volume, especially Alcides Sial and Ritu Bose, for their insight and patience, as well as to a laundry list of card-carrying paleontologists who over the past 30+ years have entertained my questions and listened to my often heretical ideas about the life across the Ediacaran-Cambrian transition. Foremost among these are Andy Knoll, Gerard Germs, Guy Narbonne, David Bottjer, Frank Corsetti, Steve Rowland, Shuhai Xiao, Zhou Chuanming, Nick Butterfield, Dima Grazhdankin, Patricia Vickers-Rich, Tom Rich, Peter Trusler, Les Kriesfeld, Doug Erwin, Maoyan Zhu, Jim Gehling, James Schiffbauer, Gregory Retallack, Tom Holtz, John Merck, Simon Darroch, and Marc Laflamme. The ideas entertained in this chapter are my own and are not intended to throw shade on any of my esteemed colleagues who no doubt could effectively counter many of my arguments. Thanks also to the three external reviewers and to a wide range of international field geologists, stratigraphers, geochronologists, and paleomagnetists with whom I have worked (and argued with) over the decades as we focus on the peculiarities of the Ediacaran and Cambrian world.

REFERENCES

Alvarenga, C.J.S., Figueiredo, M.F., Babinski, M., and Pinho, F.E.C., 2007. Glacial diamictites of Serra Azul Formation (Ediacaran, Paraguay belt): evidence of the Gaskiers glacial event in Brazil. Journal of South American Earth Sciences *23*: 236–241.

Amthor, J.E., Grotzinger, J.P., Schroder, S., Bowring, S.A., and Ramezani, J., 2003. Extinction of *Cloudina* and *Namacalathus* at the Precambrian-Cambrian boundary in Oman. Geology *31*:431–34.

An, Z., Jiang, G., Tong, J., Tian, L., Ye, Q., Song, H., and Song, H., 2015. Stratigraphic position of the Ediacaran Miaohe biota and its constraints on the age of the upper Doushantuo $\delta^{13}C$ anomaly in the Yangtze Gorges area, South China. Precambrian Research *271*: 243–253.

Anderson, E.P., Schiffbauer, J.D., and Xiao, S., 2011. Taphonomic study of Ediacaran organic-walled fossils confirms the importance of clay minerals and pyrite in Burgess Shale-type preservation. Geology *39*: 643–646.

Antcliffe, J.B., Gooday, A.J., and Brasier, M.D., 2011. Testing the protozoan hypothesis for Ediacaran fossils: a developmental analysis of *Palaeopascichnus*. Palaeontology *54*: 1157–1175.

Arnaud, E., 2008. Deformation in the Neoproterozoic Smalfjord Formation, northern Norway: an indicator of glacial depositional conditions? Sedimentology *55*: 335–356.

Arnaud, E. and Eyles, C.N., 2002. Glacial influence on Neoproterozoic sedimentation: the Smalfjord Formation, northern Norway. Sedimentology *49*: 765–788.

Asmerom, Y., Jacobsen, S.B., Butterfield, N.J., and Knoll, A.H., 1991. Sr isotope variations in Late Proterozoic seawater: implications for crustal evolution. Geochimica et Cosmochimica Acta *55*: 2883–2894.

Bailey, J.V., Joye, S.B., Kalanetra, K.M., Flood, B.E., and Corsetti, F.A., 2007. Evidence of giant sulphur bacteria in Neoproterozoic phosphorites. Nature *445*: 198–201.

Barfod, G.H., Albarède, F., Knoll, A.H., Xiao, S., Télouk, P., Frei, R., and Baker, J., 2002. New Lu-Hf and Pb-Pb age constraints on the earliest animal fossils. Earth and Planetary Science Letters *201*: 203–212.

Bergin, C. Wentrup, C., Brewig, N., Blazejak, A., Erséus, C., Giere, O., Schmid, M., De Wit, P., and Dubilier, N., 2018. Acquisition of a Novel Sulfur-Oxidizing Symbiont in the Gutless Marine Worm *Inanidrilus exumae*. Applied and Environmental Microbiology *84*: e02267-17.

Berkner, L.V. and Marshall, L.C., 1965. History of major atmospheric components. Proceedings of the National Academy of Sciences of the USA *53*: 1215–1226.

Bertrand-Sarfati, J., Moussine-Pouchkine, A., Amard, B., and Ahmed, A.A.K., 1995. First Ediacaran fauna found in Western Africa and evidence for an Early Cambrian glaciation. Geology *23*: 133–136.

Bingen, B., Griffin, W.L., Torsvik, T.H., and Saeed, A., 2005. Timing of Late Neoproterozoic glaciation on Baltica constrained by detrital zircon geochronology in the Hedmark Group, south-east Norway. Terra Nova *17*: 250–258.

Bjerrum, C.J. and Canfield, D.E., 2011. Towards a quantitative understanding of the late Neoproterozoic carbon cycle. Proceedings of the National Academy of Sciences of the USA *108*: 5542–5547.

Blazejak, A., Erséus, C., Amann, R., and Dubilier, D., 2005. Coexistence of bacterial sulfide oxidizers, sulfate reducers, and spirochetes in a gutless worm (*Oligochaeta*) from the Peru Margin. Applied and Environmental Microbiology *71*: 1553–1561.

Bottjer, D.J., Hagadorn, J.W., and Dornbos, S.Q., 2000. The Cambrian substrate revolution. GSA Today *10*: 1–9.

Bowring, S.A., Grotzinger, J.P., Isachsen, C.E., Knoll, A.H., Pelechaty, S., and Kolosov, P., 1993. Calibrating rates of Early Cambrian evolution. Science *261*: 1293–1298.

Bowring, S.A., Grotzinger, J.P., Condon, D.J., Ramezani, J., Newall, M.J., and Allen, P.A., 2007. Geochronologic constraints on the chronostratigraphic framework of the Neoproterozoic Huqf Supergroup, Sultanate of Oman. American Journal of Science *307*: 1097–1145.

Bowyer, F., Wood, R.A., and Poulton, S.W., 2017. Controls on the evolution of Ediacaran metazoan ecosystems: a redox perspective. Geobiology *15*: 516–551.

Bright, M. and Lallier, F.H., 2010. The biology of Vestimentiferan tubeworms. Oceanography and Marine Biology *48*: 213–266.

Bristow, T.F. and Kennedy, M.J., 2008. Carbon isotope excursions and the oxidant budget of the Ediacaran atmosphere and ocean. Geology *36*: 863–866.

Brocks, J.J., Jarrett, A.J.M., Sirantoine, E., Kenig, F., Moczydłowska, M., Porter, S., and Hope, J., 2016. Early sponges and toxic protists: possible sources of cryostane, an age diagnostic biomarker antedating Sturtian Snowball Earth. Geobiology *14*: 129–149.

Buatois, L.A., Narbonne, G.M., Mángano, M.G., Carmona, N.B., and Myrow, P., 2014. Ediacaran matground ecology persisted into the earliest Cambrian. Nature Communications *5*: 3544.

Budd, G.E., 2008. The earliest fossil record of the animals and its significance. Philosophical Transactions of the Royal Society B: Biological Sciences *363*: 1425–1434.

Budd, G.E. and Jensen, S., 2017. The origin of the animals and a 'Savannah' hypothesis for early bilaterian evolution. Biological Reviews of the Cambridge Philosophical Society *92*: 446–473.

Burns, S.J. and Matter, A., 1993. Carbon isotopic record of the latest Proterozoic from Oman. Eclogae Geologicae Helvetiae *86*: 595–607.

Burns, S.J., Haudenschild, U., and Matter, A., 1994. The strontium isotopic composition of carbonates from the late Precambrian (~560-540 Ma) Huqf Group of Oman. Chemical Geology *111*: 269–282.

Burzynski, G.R., Narbonne, G.M., Dececchi, A.T., and Dalrymple, R.W., 2017. The ins and outs of Ediacaran discs. Precambrian Research *300*: 246–260.

Calver, C.R., 2000. Isotope stratigraphy of the Ediacarian (Neoproterozoic III) of the Adelaide Rift Complex, Australia, and the overprint of water column stratification. Precambrian Research *100*: 121–150.

Calver, C.R., Black, L.P., Everard, J.L., and Seymour, D.B., 2004. U–Pb zircon age constraints on Late Neoproterozoic glaciation in Tasmania. Geology *32*: 893–896.

Campbell, I.H. and Allen, C.M., 2008. Formation of supercontinents linked to increases in atmospheric oxygen. Nature Geoscience *1*: 554–558.

Canfield, D.E., 1998. A new model for Proterozoic ocean chemistry. Nature *396*: 450–453.

Canfield, D.E. and Thamdrup, B., 1994. The production of ^{34}S-depleted sulfide during bacterial disproportionation of elemental sulfur. Science *266*: 1973–1975.

Canfield, D. E. and Farquhar, J., 2009. Animal evolution, bioturbation, and the sulfate concentration of the oceans. Proceedings of the National Academy of Sciences of the USA *106*: 8123–8127.

Canfield, D.E., Poulton, S.W., and Narbonne, G.M., 2007. Late-Neoproterozoic deep-ocean oxygenation and the rise of animal life. Science *315*: 92–95.

Carbone, C. and Narbonne, G.M., 2014. When life got smart: the evolution of behavioral complexity through the Ediacaran and early Cambrian of NW Canada. Journal of Paleontology *88*: 309–330.

Catling, D.C., Glein, C.R., Zahnle, K.J., and McKay, C.P., 2005. Why O_2 is required by complex life on habitable planets and the concept of planetary "oxygenation time". Astrobiology *5*: 415–438.

Chen, Z., Zhou, C., Meyer, M., Xiang, K., Schiffbauer, J.D., Yuan, X., and Xiao, S., 2013. Trace fossil evidence for Ediacaran bilaterian animals with complex behaviors. Precambrian Research *224*: 690–701.

Chen, Z., Chen, X., Zhou, C., Yuan, X., and Xiao, S., 2018. Late Ediacaran trackways produced by bilaterian animals with paired appendages. Science Advances *4*: 1–8.

Chumakov, N.M., 2009. The Baykonurian glaciohorizon of the Late Vendian. Stratigraphy and Geological Correlation *17*: 373–381.

Clapham, M.E. and Narbonne, G.M., 2002. Ediacaran epifaunal tiering. Geology *30*: 627–630.

Clapham, M.E., Narbonne, G.M., and Gehling, J.G., 2003. Paleoecology of the oldest known animal communities: Ediacaran assemblages at Mistaken Point, Newfoundland. Paleobiology *29*: 527–544.

Claypool, G.E., Holser, W.T., Kaplan, I.R., Sakai, H., and Zak, I., 1980. The age curves of sulfur and oxygen isotopes in marine sulfates and their mutual interpretation. Chemical Geology 28: 199–260.

Condon, D., Zhu, M., Bowring, S., Wang, W., Yang, A., and Jin, Y., 2005. U–Pb ages from the Neoproterozoic Doushantuo Formation, China. Science 308: 95–98.

Corkeron, M.L., 2007. 'Cap carbonates' and Neoproterozoic glacigenic successions from the Kimberley region, north-west Australia. Sedimentology 54: 871–903.

Corkeron, M.L. and George, A.D., 2001. Glacial incursion on a Neoproterozoic carbonate platform in the Kimberley region, Australia. Geological Society of America Bulletin 113: 1121–1132.

Corsetti, F.A. and Kaufman, A.J., 1994. Chemostratigraphy of Neoproterozoic-Cambrian units, White-Inyo Region, eastern California and western Nevada: implications for global correlation and faunal distribution. Palaios 9: 211–219.

Corsetti, F.A. and Hagadorn, J.W., 2000. Precambrian-Cambrian transition: Death Valley, United States. Geology 28: 299–302.

Corsetti, F.A. and Kaufman, A.J., 2003. Stratigraphic investigations of carbon isotope anomalies and Neoproterozoic ice ages in Death Valley, California. Geological Society of America Bulletin 115: 619–632.

Crimes, T.P. and Fedonkin, M.A. 1996. Biotic changes in platform communities across the Precambrian-Phanerozoic boundary. Rivista Italiana di Paleontologia e Stratigrafia 102: 317–331.

Cui, H., Kaufman, A.J., Xiao, S., Zhu, M., Zhou, C., and Lui, X.-M., 2015. Redox architecture of an Ediacaran ocean margin: integrated chemostratigraphic ($\delta^{13}C$–$\delta^{34}S$–$^{87}Sr/^{86}Sr$–Ce/Ce*) correlation of the Doushantuo Formation, South China. Chemical Geology 405: 48–62.

Cui, H., Kaufman, A.J., Xiao, S., Peek, S., Cao, H., Min, X., Cai, Y., Siegel, Z., Liu, X.-M., Schiffbauer, J.D., and Martin, A., 2016a. Environmental context for the terminal Ediacaran biomineralization of animals. Geobiology 14: 344–363.

Cui, H., Xiao, S., Zhou, C., Peng, Y., Kaufman, A.J., and Plummer, R.E., 2016b. Phosphogenesis associated with the Shuram Excursion: Petrographic and geochemical observations from the Ediacaran Doushantuo Formation of South China. Sedimentary Geology 341: 134–146.

Cui, H., Grazhdankin, D.V., Xiao, S., Peek, S., Rogov, V.I., Bykova, N.V., Sievers, N.E., Liu, X.-M., and Kaufman, A.J., 2016c. Redox-dependent distribution of early macroorganisms: evidence from the terminal Ediacaran Khatyspyt Formation in Arctic Siberia. Palaeogeography, Palaeoclimatology, Palaeoecology 461: 122–139.

Cui, H., Kaufman, A.J., Xiao, S., and Zhou, C., 2017. Methane-derived authigenic carbonates from the uppermost Doushantuo Formation in South China: was the Ediacaran Shuram Excursion a globally synchronized early diagenetic event? Chemical Geology 450: 59–80.

Cunningham, J.A., Thomas, C.-W., Bengtson, S., Marone, F., Stampanoni, M., Turner, F.R., Bailey, J.V., Raff, R.A., Raff, E.C., and Donoghue, P.C.J., 2012. Experimental taphonomy of giant sulphur bacteria: Implications for the interpretation of the embryo-like Ediacaran Doushantuo fossils. Proceedings of the Royal Society B (Biological Sciences) 279: 1857–1864.

Dalrymple, R.W. and Narbonne, G.M., 1996. Continental slope sedimentation in the Sheepbed Formation (Neoproterozoic; Windermere Supergroup), Mackenzie Mountains, N.W.T. Canadian Journal of Earth Sciences 33: 848–862.

Darroch, S.A.F., Laflamme, M., Schiffbauer, J.D., and Briggs, D.E.G., 2012. Experimental formation of a microbial death mask. Palaios 27: 293–303.

Derry, L.A., 2010. A burial diagenesis origin for the Ediacaran Shuram–Wonoka carbon isotope anomaly. Earth and Planetary Science Letters 294: 152–162.

Derry, L.A. and France-Lanord, C., 1996. Neogene growth of the sedimentary organic carbon reservoir. Paleoceanography 11: 267–275.

Derry, L.A., Kaufman, A.J., and Jacobsen, S.B., 1992. Sedimentary cycling and environmental change in the Late Proterozoic: evidence from stable and radiogenic isotopes. Geochimica et Cosmochimica Acta 56: 1317–1329.

Des Marais, D.J., Strauss, H., Summons, R.E., and Hayes, J.M., 1992. Carbon isotope evidence for the stepwise oxidation of the Proterozoic environment. Nature 359: 605–609.

Droser, M.L. and Bottjer, D.J., 1988. Trends in depth and extent of bioturbation in Cambrian carbonate marine environments, western United States. Geology 16: 233–236.

Droser, M.L. and Gehling, J.G., 2015. The advent of animals: the view from the Ediacaran. Proceedings of the National Academy of Sciences of the USA 112: 4865–4870.

Dufor, S.C. and McIlroy, D., 2016. Ediacaran pre-placozoan diploblasts in the Avalonian biota: the role of chemosynthesis in the evolution of early animal life, in Brasier, A.T., McIlroy, D., and McLoughlin, N. (eds) Earth System Evolution and Early Life: A Celebration of the Work of Martin Brasier. Geological Society, London, Special Publications 448, pp. 211–219, Geological Society of London, London.

Dzik, J., 1999. Organic membranous skeleton of the Precambrian metazoans from Namibia. Geology 27: 519–522.

Elliott, D.A., Trusler, P.W., Narbonne, G.M., Vickers-Rich, P., Morton, N., Hall, M., Hoffmann, K.H., and Schneider, G.I.C., 2016. Ernietta from the Late Ediacaran Nama Group, Namibia. Journal of Paleontology 90: 1017–1026.

Erwin, D.H., Laflamme, M., Tweedt, S.M., Sperling, E.A., Pisani, D., and Peterson, K.J., 2011. The Cambrian conundrum: early divergence and later ecological success in the early history of animals. Science 334: 1091–1097.

Fedonkin, M.A. and Waggoner, B.M., 1997. The Late Precambrian fossil Kimberella is a mollusc-like bilaterian organism. Nature 388: 868–871.

Fedonkin, M.A., Simonetta, A., and Ivantzov, A.Y., 2007. New data on Kimberella, the Vendian mollusk-like organism (White Sea region, Russia): Palaeoecological and evolutionary implications. Geological Society of London Special Publication 286: 157–179.

Fike, D.A. and Grotzinger, J.P., 2008. A paired sulfate–pyrite $\delta^{34}S$ approach to understanding the evolution of the Ediacaran–Cambrian sulfur cycle. Geochimica et Cosmochimica Acta 72: 2636–2648.

Fike, D.A., Grotzinger, J.P., Pratt, L.M., and Summons, R.E., 2006. Oxidation of the Ediacaran ocean. Nature 444: 744–747.

Fike, D.A., Finke, N., Zha, J., Blake, G., Hoehler, T.M., and Orphan, V.J., 2009. The effect of sulfate concentration on (sub) millimeter scale sulfide δ³⁴S in hypersaline cyanobacterial mats over the diurnal cycle. Geochimica et Cosmochimica Acta 73, 6187–6204.

Frei, R., Gaucher, C., Dossing, L.N., and Sial, A.N., 2011. Chromium isotopes in carbonates: a tracer for climate change and for reconstructing the redox state of ancient seawater. Earth and Planetary Science Letters 312: 114–125.

Friedrich, C.G., Rother, D., Bardischewsky, F., Quentmeier, A., and Fischer, J., 2001. Oxidation of reduced inorganic sulfur compounds by bacteria: emergence of a common mechanism? Applied and Environmental Microbiology 67: 2873–2882.

Gehling, J.G., 1999. Microbial mats in terminal Proterozoic siliciclastics: Ediacaran death masks. Palaios 14: 40–57.

Gehling, J.G., Runnegar, B.N., and Droser, M.L., 2014. Scratch traces of large Ediacara bilaterian animals. Journal of Paleontology 88: 284–298.

Germs, G.J.B., 1972. The stratigraphy and paleontology of the lower Nama Group, South West Africa. Precambrian Research Unit, University of Cape Town, South Africa, Bulletin 12: 1–250.

Germs, G.J.B., 1983. Implications of sedimentary and depositional environmental analysis of the Nama Group in South West Africa/Namibia. Geological Society of South Africa, Special Publications 11: 89–114.

Germs, G.J.B., 1995. The Neoproterozoic of southwestern Africa, with emphasis on platform stratigraphy and paleontology. Precambrian Research 73: 137–151.

Germs, G.J.B. and Gresse, P.G., 1991. The foreland basin of the Damara and Gariep orogens in Namaqualand and southern Namibia: correlations and basin dynamics. South African Journal of Geology 94: 159–169.

Germs, G.J.B. and Gaucher, C., 2012. Nature and extent of a late Ediacaran (ca. 547 Ma) glaciogenic erosion surface in southern Africa. South African Journal of Geology 115: 91–102.

Ghisalberti, M., Gold, D.A., Laflamme, M., Clapham, M.E., Narbonne, G.M., Summons, R.E., Johnston, D.T., and Jacobs, D.K., 2014. Canopy flow analysis reveals the advantage of size in the oldest communities of multicellular eukaryotes. Current Biology 24: 305–309.

Ghosh, W. and Dam, B., 2009. Biochemistry and molecular biology of lithotrophic sulfur oxidation by taxonomically and ecologically diverse bacteria and archaea. FEMS Microbiology Reviews 33: 999–1043.

Gougeon, R.C., Mángano, M.G., Buatois, L.A., Narbonne, G.M., and Laing, B.A., 2018. Early Cambrian origin of the shelf sediment mixed layer. Nature Communications 9: 1909–1916.

Grant, S.W.F., 1990. Shell structure and distribution of Cloudina, a potential index fossil for the terminal Proterozoic. American Journal of Science 290A: 261–294.

Grazhdankin, D., 2004. Patterns of distribution in the Ediacaran biotas: facies versus biogeography and evolution. Paleobiology 30: 203–221.

Grazhdankin, D. and Seilacher, A., 2002. Underground Vendobionta from Namibia. Palaeontology 45: 57–78.

Grazhdankin, D. and Seilacher, A., 2005. A re-examination of the Nama-type Vendian organism Rangea schneiderhoehni. Geological Magazine 142: 571–582.

Grazhdankin, D. and Gerdes, H.Y., 2007. Ediacaran microbial colonies. Lethaia 40: 201–210.

Grazhdankin, D.V., Balthasar, U, Nagovitsin, K.E., and Kochnev, B.B., 2008. Carbonate-hosted Avalon-type fossils in arctic Siberia. Geology 36: 803–806.

Grotzinger, J.P., Bowring, S.A., Saylor, B.Z., and Kaufman, A.J., 1995. Biostratigraphic and geochronologic constraints on early animal evolution. Science 270: 598–604.

Grotzinger, J.P., Watters, W.A., and Knoll, A.H., 2000. Calcified metazoans in thrombolite-stromatolite reefs of the terminal Proterozoic Nama Group, Namibia. Paleobiology 26: 334–359.

Grotzinger, J.P., Fike, D.A., and Fischer, W.W., 2011. Enigmatic origin of the largest-known carbon isotope excursion in Earth's history. Nature Geoscience 4: 285–292.

Habicht, K.S., Gade, M., Thamdrup, B., Berg, P., and Canfield, D.E., 2002. Calibration of sulfate levels in the Archean ocean. Science 298: 2372–2374.

Hagadorn, J.W., Fedo, C.M., and Waggoner, B.M., 2000. Early Cambrian Ediacaran-type fossils from California. Journal of Paleontology 74: 731–740.

Haines, P.W., 2000. Problematic fossils in the late Neoproterozoic Wonoka Formation, South Australia. Precambrian Research 100: 97–108.

Hall, M., Kaufman, A.J., Vickers-Rich, P., Ivantsov, A., Trusler, P., Linnemann, U., Hofmann, M., Elliott, D., Cui, H., Fedonkin, M., Hoffmann, K.-H., Wilson, S.A., Schneider, G., and Smith, J., 2013. Stratigraphy, palaeontology and geochemistry of the late Neoproterozoic Aar Member, southwest Namibia: reflecting environmental controls on Ediacara fossil preservation during the terminal Proterozoic in African Gondwana. Precambrian Research 238: 214–232.

Halverson, G.P. and Hurtgen, M.T., 2007. Ediacaran growth of the marine sulfate reservoir. Earth and Planetary Science Letters 263: 32–44.

Halverson, G.P., Hoffman, P.F., Schrag, D.P., Maloof, A.C., and Rice, A.H.N., 2005. Toward a Neoproterozoic composite carbon-isotope record. Bulletin of the Geological Society of America 117: 1181–1207.

Halverson, G.P., Dudás, F.Ö., Maloof, A.C., and Bowring, S.A., 2007. Evolution of the ⁸⁷Sr/⁸⁶Sr composition of Neoproterozoic seawater. Palaeogeography, Palaeoclimatology, Palaeoecology 256: 103–129.

Hantsoo, K.G., Kaufman, A.J., Cui, H., Plummer, R.E., and Narbonne, G.M., 2018. Effects of bioturbation on carbon and sulfur cycling across the Ediacaran–Cambrian transition at the GSSP in Newfoundland, Canada. Canadian Journal of Earth Science. doi:https://doi.org/10.1139/cjes-2017-0274.

Hardisty, D.S., Lu, Z., Bekker, A., Diamond, C.W., Gill, B.C., Jiang, G., Kah, L.C., Knoll, A.H., Loyd, S.J., Osburn, M.R., Planavsky, N.J., Wang, C., Zhou, C., and Lyons, T.W., 2017. Perspectives on Proterozoic surface ocean redox from iodine contents in ancient and recent carbonate. Earth and Planetary Science Letters 463: 159–170.

Hebert, C.L., Kaufman, A.J., Penniston-Dorland, S.C., and Martin, A.J., 2010. Radiometric and stratigraphic constraints on terminal Ediacaran (post-Gaskiers) glaciation and metazoan evolution. Precambrian Research 182: 402–412.

Hoffman, P.F. and Schrag, D.P., 2002. The snowball Earth hypothesis: testing the limits of global change. Terra Nova *14*: 129–155.

Hoffman, P.F. and Li, Z.-X., 2009. A palaeogeographic context for Neoproterozoic glaciation. Palaeogeography Palaeoclimatology Palaeoecology *277*: 158–172.

Hoffman, P.F., Kaufman, A.J., Halverson, G.P., and Schrag, D.P., 1998. A Neoproterozoic snowball Earth. Science *281*: 1342–1346.

Hoffmann, K.-H., Condon, D.J., Bowring, S.A., and Crowley, J.L., 2004. U-Pb zircon date from the Neoproterozoic Ghaub Formation, Namibia: constraints on Marinoan glaciation. Geology *32*: 817–820.

Holser, W.T. and Kaplan, I.R., 1966. Isotope geochemistry of sedimentary sulfates. Chemical Geology *1*: 93–135.

Hood, A.v.S., Planavsky, N.J., Wallace, M.W., and Wang, X., 2018. The effects of diagenesis on geochemical paleoredox proxies in sedimentary carbonates. Geochimica et Cosmochimica Acta *232*: 265–287.

Hurtgen, M.T., Arthur, M.A., and Halverson, G.P., 2005. Neoproterozoic sulfur isotopes, the evolution of microbial sulfur species, and the burial efficiency of sulfide as sedimentary sulfide. Geology *33*: 41–44.

Husson, J.D., Higgins, J.A., Maloof, A.C., and Schoene, B., 2015. Ca and Mg isotope constraints on the origin of Earth's deepest $\delta^{13}C$ excursion. Geochimica et Cosmochimica Acta *160*: 243–266.

Ivantsov, A.Y., Narbonne, G.M., Trusler, P.W., Greentree, C., and Vickers-Rich, P., 2016. Elucidating Ernietta: new insights from exceptional specimens in the Ediacaran of Namibia. Lethaia *49*: 540–554.

Jensen, S., 2003. The Proterozoic and earliest Cambrian trace fossil record; patterns, problems, and perspectives. Integrative and Comparative Biology *43*: 219–228.

Johnston, D.T., Poulton, S.W., Goldberg, T., Sergeev, V.N., Podkovyrov, V., Vorob'eva, V.G., Bekker, A., and Knoll, A.H., 2012. Late Ediacaran redox stability and metazoan evolution. Earth and Planetary Science Letters *335–336*: 25–35.

Johnston, D.T., Poulton, S.W., Tosca, N.J., O'Brien, T., Halverson, G.P., Schrag, D.P., and Macdonald, F.A., 2013. Searching for an oxygenation event in the fossiliferous Ediacaran of northwestern Canada. Chemical Geology *362*: 273–286.

Kaufman, A.J., 2005. The calibration of Ediacaran time. Science *308*: 59–60.

Kaufman, A.J., Hayes, J.M., Knoll, A.H., and Germs, G.J.B., 1991. Isotopic compositions of carbonates and organic carbon from upper Proterozoic successions in Namibia: stratigraphic variation and the effects of diagenesis and metamorphism. Precambrian Research *49*: 301–327.

Kaufman, A.J., Jacobsen, S.B., and Knoll, A.H., 1993. The Vendian record of C- and Sr-isotopic variations: implications for tectonics and paleoclimate. Earth and Planetary Science Letters *120*: 409–430.

Kaufman, A.J., Knoll, A.H., Semikhatov, M., Grotzinger, J.P., Jacobsen, S.B., and Adams, W.R., 1996. Isotopic chemostratigraphy of Precambrian-Cambrian boundary beds in the Western Anabar Region, Northern Siberia. Geological Magazine *133*: 509–533.

Kaufman, A.J., Narbonne, G.M., and Knoll, A.H., 1997. Isotopes, ice ages, and terminal Proterozoic earth history. Proceedings of the National Academy of Sciences of the USA *94*: 6600–6605.

Kaufman, A.J., Jiang, G., Christie-Blick, N., Banerjee, D., and Rai, V., 2006. Stable isotope record of the terminal Neoproterozoic Krol platform in the Lesser Himalayas of northern India. Precambrian Research *147*: 156–185.

Kaufman, A.J., Corsetti, F.A., and Varni, M.A., 2007. The effect of rising atmospheric oxygen on carbon and sulfur isotope anomalies in the Neoproterozoic Johnnie Formation, Death Valley, USA. Chemical Geology *237*: 47–63.

Kaufman, A.J., Grazhdankin, D., Rogov, V., Peek, S., Kochnev, B., Nagovitsin, K., Bykova, N., and Xiao, S., 2009. A glacial divide between Ediacaran extinction and the Cambrian explosion of life. Geological Society of America Abstracts with Programs *41*: 395.

Kaufman, A.J., Peek, S., Aaron, J., Cui, H., Grazhdankin, D., Rogov, V., Xiao, S., Buchwaldt, R., and Bowring, S. 2012. A Shorter Fuse for the Cambrian Explosion? 2012 GSA Annual Meeting. The Geological Society of America, Charlotte, NC.

Kendall, B., Komiya, T., Lyons, T.W., Bates, S.M., Gordon, G.W., Romaniello, S.J., Jiang, G., Creaser, R.A., Xiao, S., McFadden, K., Sawaki, Y., Tahata, M., Shu, D., Han, J., Chu, X., and Anbar, A.D., 2015. Uranium and molybdenum isotope evidence for an episode of widespread ocean oxygenation during the late Ediacaran Period. Geochimica et Cosmochimica Acta *156*: 173–193.

Khomentovsky, V.V., 1990. Vendian of the Siberian Platform, in Sokolov, B.S. and Fedonkin, M.A. (eds) *Regional Geology: The Vendian System*, pp. 103–183, Berlin, Springer-Verlag, v. 2.

Kimura, H. and Watanabe, Y., 2001. Oceanic anoxia at the Precambrian-Cambrian boundary. Geology *29*: 995–998.

Kirschvink, J.L., 1992. Late Proterozoic low-latitude global glaciation: the snowball Earth, in Schopf, J.W. and Klein, C. (eds) *The Proterozoic Biosphere: A Multidisciplinary Study*, pp. 51–58, Cambridge University Press, Cambridge. 1348 pp.

Knauth, L.P. and Kennedy M.J., 2009. The late Precambrian greening of the Earth. Nature *460*: 728–732.

Knoll, A.H. and Bambach, R.K., 2000. Directionality in the history of life: diffusion from the left wall or repeated scaling of the right? Paleobiology *26*: 1–14.

Knoll, A.H., Kaufman, A.J., Semikhatov, M.A., Grotzinger, J.P., and Adams, W.R., 1995a. Sizing up the sub-Tommotian unconformity in Siberia. Geology *23*: 1139–1143.

Knoll, A.H., Grotzinger, J.P., Kaufman, A.J., and Kolosov, P., 1995b. Integrated chronostratigraphy of the terminal Proterozoic successions of the Olenek Uplift, northern Siberia. Precambrian Research *73*: 251–270.

Knoll, A.H., Walter, M.R., Narbonne, G.M., and Christie-Blick, N., 2004. A new period for the Geologic Time Scale. Science *305*: 621–622.

Krumbein, W.E., Paterson, D.M., and Stal, L.J. 1994. *Biostabilization of Sediments*, BIS-Verlag, Oldenburg, 526p.

Laflamme, M., Xiao, S., and Kowalewski, M., 2009. Osmotrophy in modular Ediacara organisms. Proceedings of the National Academy of Sciences of the USA *106*: 14438–14443.

Laflamme, M., Darroch, S.A.F., Tweedt, S.M., Peterson, K.J., and Erwin, D.H., 2013. The end of the Ediacara biota: extinction, biotic replacement, or Cheshire Cat? Gondwana Research 23: 558–573.

Laing, B.A., Buatois, L.A., Mángano, M.G., Narbonne, G.M., and Gougeon, R.C., 2018. Gyrolithes from the Ediacaran-Cambrian boundary section in Fortune Head, Newfoundland, Canada: exploring the onset of complex burrowing. Palaeogeography, Palaeoclimatology, Palaeoecology 495: 171–185.

Lee, C., Love, G.D., Fischer, W.W., Grotzinger, J.P., and Halverson, G.P., 2015. Marine organic matter cycling during the Ediacaran Shuram excursion. Geology 43: 1103–1106.

Li, C., Love, G.D., Lyons, T.W., Fike, D.A., Sessions, A.L., and Chu, X., 2010. A stratified redox model for the Ediacaran ocean. Science 328: 80–83.

Linnemann, U., Pidal, A.P., Hofmann, M., Drost, K., Quesada, C., Gerdes, A., Marko, L., Gärtner, A., Zieger, J., Ulrich, J., Krause, R., Vickers-Rich, P., and Horak, J., 2018. A ~565 Ma old glaciation in the Ediacaran of peri-Gondwanan West Africa. International Journal of Earth Sciences 107: 885–911.

Liu, A.G., McIlroy, D., Brasier, M.D., 2010. First evidence for locomotion in the Ediacara biota from the 565 Ma Mistaken Point Formation, Newfoundland. Geology 38: 123–126.

Liu, A.G., McIlroy, D., Matthews, J.J., and Brasier, M.D., 2014. Confirming the metazoan character of a 565 Ma trace-fossil assemblage from Mistaken Point, Newfoundland. Palaios 29: 420–430.

Love, G.D., Grosjean, E., Stalvies, C., Fike, D.A., Grotzinger, J.P., Bradley, A.S., Kelly, A.E., Bhatia, M., Meredith, W., Snape, C.E., Bowring, S.A., Condon, D.J., and Summons, R.E., 2009. Fossil steroids record the appearance of Demospongiae during the Cryogenian Period. Nature 457: 718–721.

Loyd, S.J., Marenco, P.J., Hagadorn, J.W., Lyons, T.W., Kaufman, A.J., Sour-Tovar, F., and Corsetti, F.J., 2012. Sustained low sulfate concentration in the Neoproterozoic to Cambrian ocean: insights from carbonates of northwestern Mexico and eastern California. Earth and Planetary Science Letters 339–340: 79–94.

Loyd, S.J., Marenco, P.J., Hagadorn, J.W., Lyons, T.W., Kaufman, A.J., Sour-Tovar, F., and Corsetti, F.A., 2013. Local $\delta^{34}S$ variability in ~580Ma carbonates of northwestern Mexico and the Neoproterozoic marine sulfate reservoir. Precambrian Research 224: 551–569.

Macdonald, F.A., Strauss, J.V., Sperling, E.A., Halverson, G.P., Narbonne, G.M., Johnston, D.T., Kunzmann, M., Schrag, D.P., and Higgins, J.A., 2013. The stratigraphic relationship between the Shuram carbon isotope excursion, the oxygenation of Neoproterozoic oceans, and the first appearance of the Ediacara biota and bilaterian trace fossils in northwestern Canada. Chemical Geology 362: 250–272.

MacNaughton, R.B., Narbonne, G.M., and Dalrymple, R.W., 2000. Neoproterozoic slope deposits, Mackenzie Mountains, northwestern Canada: implications for passive margin development and Ediacaran faunal ecology. Canadian Journal of Earth Sciences 37: 997–1020.

Maloof, A.C., Schrag, D.P., Crowley, J.L., and Bowring, S.A., 2005. An expanded record of Early Cambrian carbon cycling from the Anti-Atlas Margin, Morocco. Canadian Journal of Earth Sciences 42: 2195–2216.

Maloof, A.C., Porter, S.M., Moore, J.L., Dudás, F.Ö., Bowring, S.A., Higgins, J.A., Fike, D.A., and Eddy, M.P., 2010. The earliest Cambrian record of animals and ocean geochemical change. Geological Society of America Bulletin 122: 1731–1774.

Mángano, M.G. and Buatois, L.A., 2014. Decoupling of body-plan diversification and ecological structuring during the Ediacaran–Cambrian transition: evolutionary and geobiological feedbacks. Proceedings of the Royal Society of London B: Biological Sciences 281: 20140038.

Mángano, M.G., Bromley, R.G., Harper, D.A.T., Nielsen, A.T., Paul Smith, M., and Vinther, J., 2012. Nonbiomineralized carapaces in Cambrian seafloor landscapes (Sirius Passet, Greenland): opening a new window into early Phanerozoic benthic ecology. Geology 40: 519–522.

Marenco, P.J., Corsetti, F.A., Hammond, D.E., Kaufman, A.J., and Bottjer, D.J., 2008. Oxidation of pyrite during extraction of carbonate associated sulfate. Chemical Geology 247: 124–132.

Mathai, J.C., Missnerb, A., Kügler, P., Saparovb, S.M., Zeidela, M.L., Lee, J.K., and Pohlb, P., 2014. No facilitator required for membrane transport of hydrogen sulfide. Proceedings of the National Academy of Sciences of the USA 106: 16633–16638.

McFadden, K.A., Huang, J., Chu, X., Jiang, G., Kaufman, A.J., Zhou, C., Yuan, X., and Xiao, S., 2008. Redox instability and biological evolution in the Ediacaran Doushantuo Formation. Proceedings of the National Academy of Sciences of the USA 105: 3197–3202.

McIlroy, D. and Logan, G.A., 1999. The impact of bioturbation on infaunal ecology and evolution during the Proterozoic-Cambrian transition. Palaios 14: 58–72.

Melezhik, V.A., Pokrovsky, B.G., Fallick, A.E., Kuznetsov, A.B., and Bujakaite, M.I., 2009. Constraints on $^{87}Sr/^{86}Sr$ of Late Ediacaran seawater: insight from Siberian high-Sr limestones. Journal of the Geological Society 166: 183–191.

Mentel, M. and Martin, W., 2008. Energy metabolism among eukaryotic anaerobes in light of Proterozoic ocean chemistry. Philosophical Transactions of the Royal Society of London B Biological Sciences 363: 2717–2729.

Meyer, M., Xiao, S., Gill, B.C., Schiffbauer, J.D., Chen, Z., Zhou, C., and Yuan, X., 2014. Interactions between Ediacaran animals and microbial mats: insights from Lamonte trevallis, a new trace fossil from the Dengying Formation of South China. Palaeogeography, Palaeoclimatology, Palaeoecology 396: 62–74.

Mills, D.B., Ward, L.M., Jones, C., Sweeten, B., Forth, M., Treusch, A.H., and Canfield, D.E., 2014. Oxygen requirements of the earliest animals. Proceedings of the National Academy of Sciences of the USA 111: 4168–4172.

Myrow, P.M. and Kaufman, A.J., 1999. A newly discovered cap carbonate above Varanger-age glacial deposits in Newfoundland, Canada. Journal of Sedimentary Research 69: 784–793.

Narbonne, G.M., 2005. The Ediacara biota: Neoproterozoic origin of animals and their ecosystems. Annual Review in Earth and Planetary Sciences 33: 421–442.

Narbonne, G.M. and Aitken, J.D., 1995. Neoproterozoic of the Mackenzie Mountains, northwestern Canada. Precambrian Research 73: 101–121.

Narbonne, G.M. and Gehling, J.G., 2003. Life after Snowball: the oldest complex Ediacaran fossils. Geology 31: 27–30.

Narbonne, G.M., Kaufman, A.J., and Knoll, A.H., 1994. Integrated carbon isotope and biostratigraphy of the upper Windermere Group, MacKenzie Mountains, N.W. Territories, Canada. The Bulletin of the Geological Society of America 106: 1281–1292.

Narbonne, G.M., Saylor, B.Z., and Grotzinger, J.P., 1997. The youngest Ediacaran fossils from southern Africa. Journal of Paleontology 71: 953–967.

Narbonne, G.M., Xiao, S., and Shields, G., 2012. The Ediacaran Period, in Gradstein, F., Ogg, J., and Ogg, G. (eds.) Geologic Timescale 2012, pp. 427–449, Elsevier, Boston, MA.

Nelson, D.C. and Fisher, C.R., 1995. Chemoautotrophic and methanotrophic endosymbiotic bacteria at deep-sea vents and seeps, in Karl, D.M. (ed.) The Microbiology of Deep-Sea Hydrothermal Vents, pp. 125–167, CRC Press, Boca Raton, FL.

Newman, S.A., Mariotti, G., Pruss, S., and Bosak, T., 2016. Insights into cyanobacterial fossilization in Ediacaran siliciclastic environments. Geology 44: 579–582.

Noffke, N., Gerdes, G., Klenke, T., and Krumbein, W.E., 2001. Microbially induced sedimentary structures: a new category within the classification of primary sedimentary structures. Journal of Sedimentary Research 71: 649–656.

Nursall, J.R., 1959. Oxygen as a prerequisite to the origin of the Metazoa. Nature 183: 1170–1172.

Och, L. and Shields-Zhou, G.A., 2012. The Neoproterozoic oxygenation event: environmental perturbations and biogeochemical cycling. Earth-Science Reviews 110: 26–57.

Overmann, J. and van Gemerden, H., 2000. Microbial interactions involving sulfur bacteria: implications for the ecology and evolution of bacterial communities. FEMS Microbiology Reviews 24: 591–599.

Partin, C.A., Bekker, A., Planavsky, N.J., Scott, C.T., Gill, B.C., Li, C., Podkovyrov, V., Maslov, A., Konhauser, K.O., Lalonde, S.V., Love, G.D., Poulton, S.W., and Lyons, T.W., 2013. Large-scale fluctuations in Precambrian atmospheric and oceanic oxygen levels from the record of U in shales. Earth and Planetary Science Letters 369–370: 284–293.

Pehr, K., Love, G.D., Kuznetsov, A., Podkovyrov, V., Junium, C.K., Shumlyanskyy, L., Sokur, T., and Bekker, A., 2018. Ediacara biota flourished in oligotrophic and bacterially dominated marine environments across Baltica. Nature Communications 9: 1807.

Pelechaty, S., Kaufman, A.J., and Grotzinger, J.P., 1996. Evaluation of $\delta^{13}C$ isotope stratigraphy for intrabasinal correlation: data from Vendian strata of the Olenek uplift and Kharaulakh Mountains, Siberian platform, Russia. The Bulletin of the Geological Society of America 108: 992–1003.

Pell, S.D., McKirdy, D.M., Jansyn, J., and Jenkins, R.J.F., 1993. Ediacaran carbon isotope stratigraphy of South Australia: an initial study. Transactions of the Royal Society of South Australia 117: 153–161.

Peng, S., Babcock, L.E., and Cooper, R.A., 2012. The Cambrian Period, in Gradstein, F. (ed) The Geological Time Scale, pp. 437–488, Elsevier, Amsterdam.

Pisarevsky, S.A., McCausland, P.J., Hodych, J.P., O'Brien, S.J., Tait, J.A., Murphy, J.B., and Colpron, M., 2011. Paleomagnetic study of the late Neoproterozoic Bull Arm and Crown Hill formations (Musgravetown Group) of eastern Newfoundland: implications for Avalonia and West Gondwana paleogeography. Canadian Journal of Earth Sciences 49: 308–327.

Planavsky, N.J., 2018. From orogenies to oxygen. Nature Geoscience 11: 9–11.

Pokrovsky, B.G. and Venagradov, V.E., 1991. Isotopic composition of strontium, oxygen and carbon in upper Precambrian carbonates of the western area of the Anabar uplift (Kotyikan River). Akademiya Nauk SSSR, Doklady 320: 1245–1250 (in Russian).

Pokrovsky, B.G. and Missarzhevsky, V., 1993. Isotope correlation of Precambrian and Cambrian of the Siberian platform. Akademiya Nauk SSSR Doklady 329: 768–771 (in Russian).

Porter, S.M., 2007. Seawater chemistry and early carbonate biomineralization Science 316: 1302.

Porter, S.M., 2010. Calcite and aragonite seas and the de novo acquisition of carbonate skeletons. Geobiology 8: 256–277.

Poulton, S.W. and Canfield, D.E., 2005. Development of a sequential extraction procedure for iron: implications for iron partitioning in continentally derived particulates. Chemical Geology 214: 209–221.

Praekelt, H.E., Germs, G.J.B., and Kennedy, J.H. 2008. A distinct unconformity in the Cango Caves Group of the Neoproterozoic to early Paleozoic Saldania Belt in South Africa: its regional significance. South African Journal of Geology 111: 357–368.

Pu, J.P., Bowring, S.A., Ramezani, J., Myrow, P., Raub, T.D., Landing, E., Mills, A., Hodgin, E., and Macdonald, F.A., 2016. Dodging snowballs: geochronology of the Gaskiers glaciation and the first appearance of the Ediacaran biota. Geology 44: 955–958.

Raymo, M.E., Ruddiman, W.F., and Froelich, P.N., 1988. Influence of late Cenozoic mountain building on ocean geochemical cycles. Geology 16: 649–653.

Retallack, G.J., 2013. Ediacaran life on land. Nature 432: 89–92.

Riahi, S. and Rowley, C.N., 2014. Why can hydrogen sulfide permeate cell membranes? Journal of the American Chemical Society 136: 15111–15113.

Ries, J.B., Fike, D.A., Pratt, L.M., Lyons, T.W., and Grotzinger, J.P., 2009. Superheavy pyrite ($\delta^{34}S_{pyr} > \delta^{34}S_{CAS}$) in the terminal Proterozoic Nama Group, southern Namibia: a consequence of low seawater sulfate at the dawn of animal life. Geology 37: 743–746.

Rogov, V., Marusin, V., Bykova, N., Goy, Y., Nagovitsin, K., Kochnev, B., Karlova, G., and Grazhdankin, D., 2012. The oldest evidence of bioturbation on Earth. Geology 40: 395–398.

Rothman, D.H., Hayes, J.M., Summons, R.E., 2003. Dynamics of the Neoproterozoic carbon cycle. Proceedings of the National Academy of Sciences of the USA 100: 8124–8129.

Rozanov, A.Y., Volkova, N.A., Voronova, L.C., Krylov, I.N., Keller, B.M., Korolyuk, I.K., Lendzion, K., Michniak, R., Pykhova, N.G., and Sidorov, A.D., 1969. The Tommotian Stage and the Cambrian Lower Boundary Problem, Amerind, New Delhi (1981 translation), 359 p.

Runnegar, B., 1982. Oxygen requirements, biology and phylogenetic significance of the late Precambrian worm *Dickinsonia*, and the evolution of the burrowing habit. Alcheringa 6: 223–239.

Runnegar, B., 1991. Precambrian oxygen levels estimated from the biochemistry and physiology of early eukaryotes. Global and Planetary Change 5: 97–111.

Saul, J.M., 2009. Did detoxification processes cause complex life to emerge? Lethaia 42: 179–184.

Sawaki, Y., Ohno, T., Tahata, M., Komiya, T., Hirata, T., Maruyama, S., Windley, B.F., Han, J., Shu, D., and Li, Y., 2010. The Ediacaran radiogenic Sr isotope excursion in the Doushantuo Formation in the Three Gorges area, South China. Precambrian Research 176: 46–64.

Saylor, B.Z., Kaufman, A.J., Grotzinger, J.P., and Urban, F., 1998. A composite reference section for terminal Proterozoic strata of southern Namibia. Journal of Sedimentary Research 68: 1223–1235.

Schidlowski, M., Eichmann, R., and Junge, C.E., 1975. Precambrian sedimentary carbonates: carbon and oxygen isotope geochemistry and implications for the terrestrial oxygen budget. Precambrian Research 2: 1–69.

Schiffbauer, J.D., Huntley, J.W., O'Neil, G.R., Darroch, S.A.F., Laflamme, M., and Cai, Y., 2016. The latest Ediacaran Wormworld fauna: setting the ecological stage for the Cambrian Explosion. GSA Today 26: 4–11.

Schrag, D.P., Higgins, J.A., Macdonald, F.A., and Johnston, D.T., 2013. Authigenic carbonate and the history of the global carbon cycle. Science 339: 540–543.

Schröder, S. and Grotzinger, J.P., 2007. Evidence for anoxia at the Ediacaran–Cambrian boundary: the record of redox-sensitive trace elements and rare earth elements in Oman. Journal of the Geological Society of London 164: 175–187.

Schwellnus, C.M., 1941. The Nama tillite in the Klein Karas Mountains, South West Africa. Transactions of the Geological Society of South Africa 44: 19–33.

Scott, C., Lyons, T.W., Bekker, A., Shen, Y., Poulton, S.W., Chu, X., and Anbar, A.D., 2008. Tracing the stepwise oxygenation of the Proterozoic Ocean. Nature 452: 456–459.

Seilacher A., 1989. Vendozoa: organismic construction in the Proterozoic biosphere. Lethaia 22: 229–39.

Seilacher, A. and Pflüger, F., 1994, From biomats to benthic agriculture: a biohistoric revolution, in Krumbein, W.E. (eds.) *Biostabilization of Sediments*, pp. 97–105, Bibliotheks und Informationssystem der Carl von Ossietzky Universität Oldenburg (BIS), Oldenburg, Germany.

Seilacher, A. and Gishlick, A.D., 2014. Vendobionts: lost life forms of Ediacaran times, in Seilacher, A. and Gishlick, A.D. (eds) *Morphodynamics*, pp. 133–148, CRC Press, Taylor & Francis Group, Boca Raton, FL.

Shen, B., Xiao, S., Zhou, C., Kaufman, A.J., and Yuan, X., 2010. Carbon and sulfur isotope chemostratigraphy of the Neoproterozoic Quanji Group of the Chaidam Basin, NW China: basin stratification in the aftermath of an Ediacaran glaciation postdating the Shuram event? Precambrian Research 177: 241–252.

Sim, M.S., Bosak, T., and Ono, S., 2011. Large sulfur isotope fractionation does not require disproportionation. Science 333, 74–77.

Smith, E.F., Nelson, L.L., Strange, M.A., Eyster, A.E., Rowland, S.M., Schrag, D.P., and Macdonald, F.A., 2016. The end of the Ediacaran: two new exceptionally preserved body fossil assemblages from Mount Dunfee, Nevada, USA. Geology 44: 911–914.

Sperling, A., Halverson, G.P., Knoll, A.H., Macdonald, F.A., and Johnston, D.T., 2012. A basin redox transect at the dawn of animal life. Earth and Planetary Science Letters 371–372: 143–155.

Sperling, E.A., Frieder, C.A., Raman, A.V., Girguis, P.R., Levin, L.A., and Knoll, A.H., 2013a. Oxygen, ecology, and the Cambrian radiation of animals. Proceedings of the National Academy of Sciences of the United States of America 110: 13,446–13,451.

Sperling, E.A., Halverson, G.P., Knoll, A.H., Macdonald, F.A., and Johnston, D.T., 2013b. A basin redox transect at the dawn of animal life. Earth and Planetary Science Letters 371–372: 143–155.

Sperling, E.A., Wolock, C.J., Morgan, A.S., Gill, B.C., Kunzmann, M., Halverson, G.P., Macdonald, F.A., Knoll, A.H., and Johnston, D.T., 2015, Statistical analysis of iron geochemical data suggests limited late Proterozoic oxygenation. Nature 523: 451–454.

Squire, R.J., Campbell, I.H., Allen, C.M., and Wilson, C.J.L., 2006. Did the Transgondwanan Supermountain trigger the explosive radiation of animals on Earth? Earth and Planetary Science Letters 250: 116–133.

Squyres, S.W., Grotzinger, J.P., Arvidson, R.E., Bell, J.F., III, Calvin, W., Christensen, P.R., Clark, B.C., Crisp, J.A., Farrand, W.H., Herkenhoff, K.E., Johnson, J.R., Klingelhofer, G., Knoll, A.H., McLennan, S.M., McSween, H.Y., Jr., Morris, R.V., Rice, J.W., Jr., Rieder, R., and Soderblom, L.A., 2004. *In situ* evidence for an ancient aqueous environment at Meridiani Planum, Mars. Science 306: 1709–1714.

Subczynski, W.K., Hyde, J.S., and Kusumi, A., 1989. Oxygen permeability of phosphatidylcholine-cholesterol membranes. Proceedings of the National Academy of Sciences of the USA 86: 4474–4478.

Tarhan, L.G. and Droser, M.L., 2014. Widespread delayed mixing in early to middle Cambrian marine shelfal settings. Palaeogeography, Palaeoclimatology, Palaeoecology 399: 310–322.

Tarhan, L.G., Droser, M.L., Planavsky, N.J., and Johnston, D.T., 2015. Protracted development of bioturbation through the early Palaeozoic Era. Nature Geoscience 8: 865–869.

Tarhan, L.G., Hood, A.v.S., Droser, M.L., Gehling, J.G., and Briggs, D.E.G., 2016. Exceptional preservation of soft-bodied Ediacara Biota promoted by silica-rich oceans. Geology 44: 951–954.

Tewari, V.C. and Sial, A.N., 2007. Neoproterozoic – Early Cambrian isotopic variation and chemostratigraphy of the Lesser Himalaya, India, East Gondwana. Chemical Geology 237: 64–88.

Theissen, U., Hoffmeister, M., Grieshaber, M., and Martin, W., 2003. Single eubacterial origin of eukaryotic sulfide:quinone oxidoreductase, a mitochondrial enzyme conserved from the early evolution of eukaryotes during anoxic and sulfidic times. Molecular Biology and Evolution 20, 1564–1574.

Thompson, M.D., and Bowring, S.A., 2000. Age of the Squantum "Tillite," Boston Basin, Massachusetts: U-Pb zircon constraints on terminal Neoproterozoic glaciation. American Journal of Science 300: 630–655.

Tostevin, R., Wood, R.A., Shields, G.A., Poulton, S.W., Guilbaud, R., Bowyer, F., Penny, A.M., He, T. A. Curtis, K. H. Hoffmann, and M. O. Clarkson, 2016. Low-oxygen waters limited habitable space for early animals. Nature Communications 7: 12818.

Tostevin, R., He, T., Turchyn, A.V., Wood, R.A., Penny, A.M., Bowyer, F., Antler, G., and Shields, G.A., 2017. Constraints on the late Ediacaran sulfur cycle from carbonate associated sulfate. Precambrian Research 290: 113–125.

Towe, K.M., 1981. Biochemical keys to the emergence of complex life, in Billingham, J. (ed.) Life in the Universe, pp. 297–305, MIT Press, Cambridge, MA.

Veizer, J. and Compston, W., 1976. $^{87}Sr/^{86}Sr$ in Precambrian carbonates as an index of crustal evolution. Geochimica et Cosmochimica Acta 40: 905–914.

Veizer, J. and Hoefs, J., 1976. The nature of $^{18}O/^{16}O$ and $^{13}C/^{12}C$ secular trends in sedimentary carbonate rocks. Geochimica et Cosmochimica Acta 40: 1387–1395.

Verdel, C., Werneke, B.P., and Bowring, S.A., 2011. The Shuram and subsequent Ediacaran carbon isotope excursions from southwest Laurentia, and implications for environmental stability during metazoan radiation. Geological Society of America Bulletin 123: 1539–1559.

Vickers-Rich, P., 2007. Saline giants, cold cradles and global playgrounds of Neoproterozoic Earth: the origin of the Animalia. Geological Society London Special Publications 286: 447–448.

Vickers-Rich, P., Ivantsov, A.Y., Trusler, P., Narbonne, G.M., Hall, M., Wilson, S.A., Greentree, C., Fedonkin, M.A., Elliott, D.A., Hoffmann, K.H., and Schneider, G.I.C., 2013. Reconstructing Rangea: new discoveries from the Ediacaran of southern Namibia. Journal of Paleontology 87: 1–15.

Wade, M., 1968. Preservation of soft-bodied animals in Precambrian sandstones at Ediacara, South Australia. Lethaia 1: 238–267.

Wallace, M.W., Hood, A.vS., Shuster, A., Greig, A., Planavsky, N.J., and Reed, C.P., 2017. Oxygenation history of the Neoproterozoic to early Phanerozoic and the rise of land plants. Earth and Planetary Science Letters 466: 12–19.

Wei, G.-Y., Planavsky, N.J., Tarhan, L.G., Chen, X., Wei, W., Li, D., and Ling, H.-F., 2018. Marine redox fluctuation as a potential trigger for the Cambrian Explosion. Geology 46: 587–590.

Wen, H., Carignan, J., Zhang, Y., Fan, H., Cloquet, C., and Liu, S., 2011. Molybdenum isotopic records across the Precambrian-Cambrian boundary. Geology 39: 775–778.

Widomska, J., Raguz, M., and Subczynski, W.K., 2007. Oxygen permeability of the lipid bilayer membrane made of calf lens lipids. Biochimica et Biophysica Acta 1768: 2635–2645.

Wille, M., Nagler, T.F., Lehmann, B., Schröder, S., and Kramers, J.D., 2008. Hydrogen sulphide release to surface waters at the Precambrian/Cambrian boundary. Nature 453: 767–769.

Wood, D.A., Dalrymple, R.W., Narbonne, G.M., Gehling, J.G., and Clapham, M.E., 2003. Paleoenvironmental analysis of the late Neoproterozoic Mistaken Point and Trepassey formations, southeastern Newfoundland. Canadian Journal of Earth Science 40: 1375–1391.

Wood, R.A., Poulton, S.W., Prave, A.R., Hoffmann, K.-H., Clarkson, M.O., Guilbaud, R., Lyne, J.W., Tostevin, R., Bowyer, F., Penny, A.M., Curtis, A., and Kasemann, S.A., 2015. Dynamic redox conditions control late Ediacaran metazoan ecosystems in the Nama Group, Namibia. Precambrian Research 261: 252–271.

Wu, N., Farquhar, J., and Fike, D.A., 2015. Ediacaran sulfur cycle: insights from sulfur isotope measurements ($\Delta^{33}S$ and $\delta^{34}S$) on paired sulfate–pyrite in the Huqf Supergroup of Oman. Geochimica et Cosmochimica Acta 164: 352–364.

Xiao, S., Yun, Z., and Knoll, A.H., 1998. Three-dimensional preservation of algae and animal embryos in a Neoproterozoic phosphorites. Nature 391: 553–558.

Xiao, S., Bao, H., Wang, H., Kaufman, A.J., Zhou, C., Li, G., Yuan, X., and Ling, H., 2004. The Neoproterozoic Quruqtagh Group in eastern Chinese Tianshan: evidence for a post-Marinoan glaciation. Precambrian Research 130: 1–26.

Xiao, S., Muscente, A.D., Chen, L., Zhou, C., Schiffbauer, J.D., Wood, A.D., Polys, N.F., and Yuan, X., 2014. The Weng'an biota and the Ediacaran radiation of multicellular eukaryotes: National Science Review 1: 498–520.

Xiao, S., Narbonne, G.M., Zhou, C., Laflamme, M., Grazhdankin, D.V., Moczydłowska-Vidal, M., and Cui, H., 2016. Towards an Ediacaran time scale: problems, protocols, and prospects. Episodes 39: 540–555.

Yin, L., Zhu, M., Knoll, A.H., Yuan, X., Zhang, J., and Hu, J., 2007. Doushantuo embryos preserved inside diapause egg cysts. Nature 446: 661–663.

Zhang, F., Xiao, S., Kendall, B., Romaniello, S.J., Cui, H., Meyer, M., Gilleaudeau, G.J., Kaufman, A.J., and Anbar, A.D., 2018. Extensive marine anoxia during the terminal Ediacaran Period. Science Advances 4: eaan8983.

Zhelezinskaia, I., Kaufman, A.J., Farquhar, J., and Cliff, J., 2014. Large sulfur isotope fractionations associated with Neoarchean microbial sulfate reduction. Science 346: 742–744.

Zhu, B., Becker, H., Jiang, S.-Y., Pi, D.-H., Fischer-Gödde, M., and Yang, J.-H., 2013. Re–Os geochronology of black shales from the Neoproterozoic Doushantuo Formation, Yangtze platform, South China. Precambrian Research 225: 67–76.

Zhu, M., Zhuravlev, A.Yu., Wood, R.A., Zhao, F., and Sukhov, S.S., 2017. A deep root for the Cambrian explosion: Implications of new bio- and chemostratigraphy from the Siberian Platform. Geology 45: 459–462.

Zhuravlev, A. Yu., and Wood, R.A., 2008. Eve of biomineralization: Controls on skeletal mineralogy. Geology 36: 923–926.

Part III
Paleozoic

8

$\delta^{13}C$ Chemostratigraphy of the Ordovician-Silurian Boundary Interval

Stig M. Bergström[1] and Daniel Goldman[2]

ABSTRACT

A review of the $\delta^{13}C$ chemostratigraphy of five stratigraphically apparently continuous Ordovician-Silurian boundary sections in northern Europe, North America, and Asia suggests that the level of the systemic boundary falls in an interval with relatively uniform carbon isotope values. Hence, the systemic boundary cannot be defined in terms of $\delta^{13}C$ chemostratigraphy. However, comparison between biostratigraphy and chemostratigraphy indicates that the graptolite-defined base of the Silurian is located at a stratigraphic level only a little higher than the end of the Hirnantian carbon isotopic excursion (HICE), which is the largest $\delta^{13}C$ excursion known in the Ordovician.

In the latest Ordovician, high-latitude Gondwana glaciations resulted in some very significant sea-level changes. One such regressive sea-level event occurred just below the HICE interval, and another during the time of falling post-peak HICE values. There seems to be a close correlation between the main glaciations, which tend to be marked by significant stratigraphic gaps at low to mid-latitudes, and major faunal extinction horizons. These extinctions, which are among the largest known in the Earth's Phanerozoic history, resulted in striking differences between the Late Ordovician and Early Silurian marine faunas.

8.1. INTRODUCTION

The Ordovician-Silurian boundary interval represents a remarkable episode in the Earth's history. One of the greatest Phanerozoic extinction events occurred during the latest Ordovician, and this event was accompanied by catastrophic global climate fluctuations and sea-level changes. In fact, based on timing and scope of these environmental changes, scientists have suggested that the Late Ordovician and Early Silurian can be used as a deep time analog for Plio-Pleistocene orbitally forced glacial-interglacial cyclicity [e.g., *Elrick et al.*, 2013; *Melchin et al.*, 2013]. Precise correlation between Ordovician-

Silurian boundary successions is thus critical to understanding of the relations between the Earth's physical and biological systems during this key interval of our planet's history. As a part in a recent revolution in sedimentary geology, geochemical data collected from sedimentary successions spanning the Ordovician-Silurian boundary provide a new tool useful for both making global correlations and interpreting paleoenvironmental changes.

During the past two decades, a very large amount of chemostratigraphic research has been carried out in lower Paleozoic sedimentary successions around the world that has led to very significant improvements in our understanding of stratigraphic relations at both local and regional scales. Most of the investigations in the Ordovician-Silurian boundary interval have been based on $\delta^{13}C_{carb}$ and $\delta^{13}C_{org}$ [e.g., *Underwood et al.*, 1997; *Melchin and Holmden*, 2006; *Gorjan et al.*, 2012], but recent pioneer studies using $\delta^{34}S_{pyr}$ [e.g., *Young et al.*,

[1] *School of Earth Sciences, Division of Earth History, The Ohio State University, Columbus, OH, USA*

[2] *Department of Geology, University of Dayton, Dayton, OH, USA*

Chemostratigraphy Across Major Chronological Boundaries, Geophysical Monograph 240, First Edition.
Edited by Alcides N. Sial, Claudio Gaucher, Muthuvairavasamy Ramkumar, and Valderez Pinto Ferreira.
© 2019 the American Geophysical Union. Published 2019 by John Wiley & Sons, Inc.

2016] and $^{87}Sr/^{86}Sr$ [*Saltzman et al.*, 2014] suggest that also these isotopes have great potential as tools for both correlation and understanding of paleoenvironmental changes. Because very little has been published on the use of the two latter isotopes in the Ordovician-Silurian boundary successions, the present review is focused on $\delta^{13}C$ chemostratigraphy.

A considerable amount of $\delta^{13}C$ chemostratigraphic work has in recent years been carried out in uppermost Ordovician successions, especially in Baltoscandia and North America, but fewer such investigations have been conducted in the lowermost Silurian. A possible reason for this may be that especially in many regions and particularly in nontropical successions, much of the lower Llandovery is developed in shaly noncalcareous facies that tends to be unsuitable for $\delta^{13}C_{carb}$ analysis. However, recent studies [e.g., *Bergström et al.,* 2014] have shown that when using $\delta^{13}C_{org}$, many shales produce isotope curves that are similar to those based on $\delta^{13}C_{carb}$ and hence are useful for chemostratigraphic correlations. Depending on the prevailing lithology in the study successions, in the present con-tribution, we deal with both $\delta^{13}C_{carb}$ and $\delta^{13}C_{org}$ for our chemostratigraphic discussions.

Locally and regionally, the interval of the Ordovician-Silurian boundary, which herein is taken to include the uppermost Ordovician Hirnantian Stage and the lower-most Silurian Rhuddanian Stage, has a varied lithology ranging from sandstone to shale to limestone and dolostone. Furthermore, as is the case in most of the North American Midcontinent and the Siberian platform, the systemic boundary is in many regions marked by an unconformity of variable magnitude with uppermost Ordovician and/or lowermost Silurian strata being absent (Fig. 8.1). Although this more or less prominent stratigraphic gap makes it easy to locate the systemic boundary, it obviously makes it difficult to analyze the precise timing and nature of, for instance, depositional and biological events across the boundary interval. In the present contribution, we therefore focus on the stratigraphically most complete successions and specifically on those with good biostratigraphic control in order to tie, as far as possible, the $\delta^{13}C$ curves to stan-dard graptolite, conodont, and chitinozoan zones. For

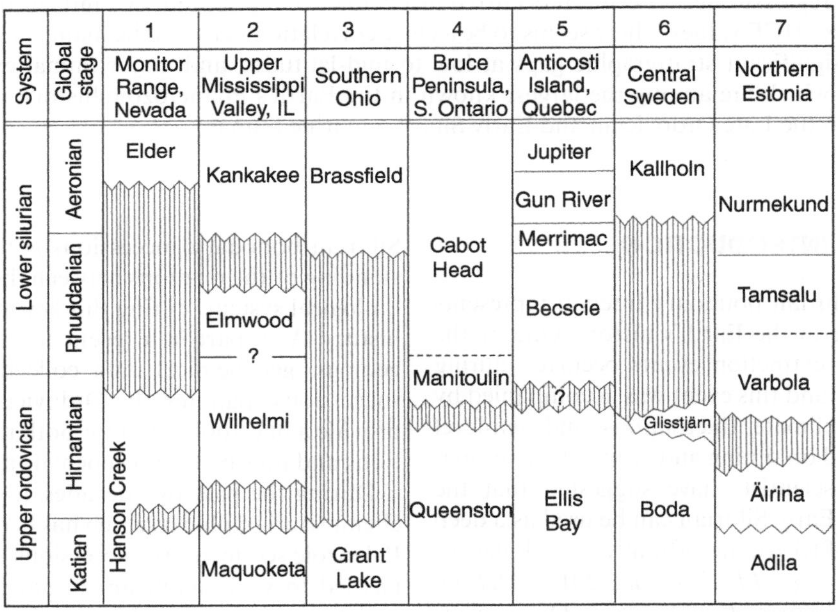

System	Global stage	1 Monitor Range, Nevada	2 Upper Mississippi Valley, IL	3 Southern Ohio	4 Bruce Peninsula, S. Ontario	5 Anticosti Island, Quebec	6 Central Sweden	7 Northern Estonia
Lower silurian	Aeronian	Elder	Kankakee	Brassfield	Cabot Head	Jupiter / Gun River / Merrimac	Kallholn	Nurmekund
Lower silurian	Rhuddanian		Elmwood / ?		Cabot Head / Manitoulin	Becscie		Tamsalu / Varbola
Upper ordovician	Hirnantian	Hanson Creek	Wilhelmi	Grant Lake	Queenston	? / Ellis Bay	Glisstjärn / Boda	Äirina
Upper ordovician	Katian	Hanson Creek	Maquoketa	Grant Lake	Queenston	Ellis Bay	Boda	Adila

Figure 8.1 Diagram showing the common occurrence of stratigraphic gaps in the Ordovician-Silurian boundary interval as illustrated by seven important successions in North America and Baltoscandia. All of these sequences have full or partial $\delta^{13}C$ chemostratigraphic control. (1) The Monitor Range, Nevada, succession (Based on chemostratigraphy after *LaPorte et al.* [2009] and *Bergström et al.* [2014]). (2) The Upper Mississippi Valley succession in Illinois (Based on chemostratigraphy by *Bergström et al.* [2011]). (3) The succession in Adams County, southern Ohio (After unpublished chemostratigraphy by Bergström and Kleffner). (4) The Bruce Peninsula succession in southern Ontario, Canada (based on chemostratigraphy after *Bergström et al.* [2011]). (5) The western Anticosti Island succession, Quebec, Canada (Based on chemostratigraphy after *Young et al.* [2010] and *Jones et al.* [2011]). (6) The succession in the Siljan region, south-central Sweden (Based on chemostratigraphy after *Schmitz and Bergström* [2007]. Reproduced with permission of Taylor & Francis Ltd.). (7) The northern Estonia succession (Based on chemostratigraphy after *Ainsaar et al.,* [2015]. Reproduced with permission of Elsevier).

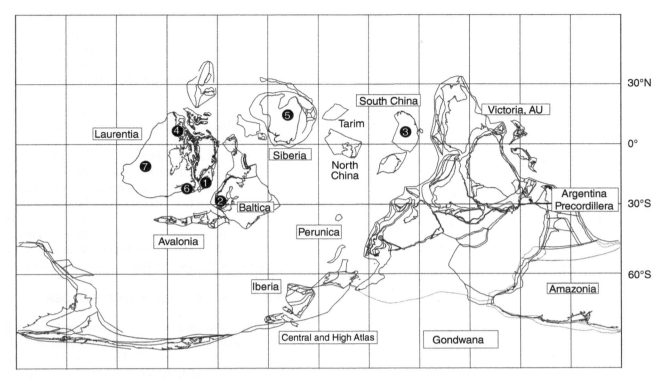

Figure 8.2 Inferred Late Ordovician paleogeographic position of some important sections dealt with herein. Map generated from BugPlates software [*Torsvik*, 2009]. Figured localities are as follows. (1) Dob's Linn, Scotland. (2) Röstånga, S. Sweden. (3) Wangjiwan, Yangtze platform, China. (4) Truro Island, Canadian Arctic. (5) Mirny Creek, eastern Siberia. (6) Anticosti Island. (7) Canada; G, Monitor Range, Nevada.

definition of stages, stage slices, and graptolite zones in the Ordovician part of the study interval, see *Bergström et al.* [2009]. Corresponding data for the Silurian are summarized by *Cramer et al.* [2011].

We will first review the chemostratigraphy of the systemic boundary interval at Dob's Linn, southern Scotland, which is the officially ratified Global Stratotype Section and Point (GSSP) of the Ordovician-Silurian boundary. Then we will deal with the isotope stratigraphy in four other especially important boundary sections around the world (Fig. 8.2). Finally, we will briefly discuss the relations between the chemostratigraphy and biologic, glacial, and eustatic events during the latest Ordovician and earliest Silurian. This is appropriate in view of the fact that the Ordovician-Silurian boundary interval is of special paleontological and environmental interest. It is the time of the second largest among the five major mass extinctions in the Phanerozoic with a loss of >90% of the marine fauna during the Hirnantian Stage. It was also a time interval with major Gondwana glaciations that had a profound impact on the sea level globally. The exact relationship between these events and isotope geology remains a matter of widespread discussion.

8.2. DOB'S LINN, SOUTHERN SCOTLAND

This locality (Fig. 8.2:1), which has doubtless become one of the most famous lower Paleozoic localities in the world since it was described in detail almost 150 years ago [*Lapworth*, 1878], serves as the global reference locality (GSSP) for the base of the Silurian System. For summary descriptions of this locality, see, for instance, *Williams* [1988], *Melchin and Williams* [2000], and *Rong et al.* [2008]. The succession is developed in graptolite-bearing shales. The Ordovician-Silurian boundary, which is currently defined as the first appearance level (FAL) of the graptolite *Akidograptus ascensus*, is 1.6 m above the base of the Birkhill Shale (Fig. 8.3). The Birkhill Shale is dark gray to black shale with virtually no other fossils than graptolites, a fact that has seriously hampered the recognition of the level of the systemic boundary in non-graptolitic successions.

In a pioneer study, *Underwood et al.* [1997] presented the first δ¹³C_org curve through the systemic boundary interval at Dob's Linn. Recent biostratigraphic research has led to some adjustment of the biostratigraphy in the boundary interval, but as illustrated in Figure 8.3, the base of the Silurian is at about the level where baseline

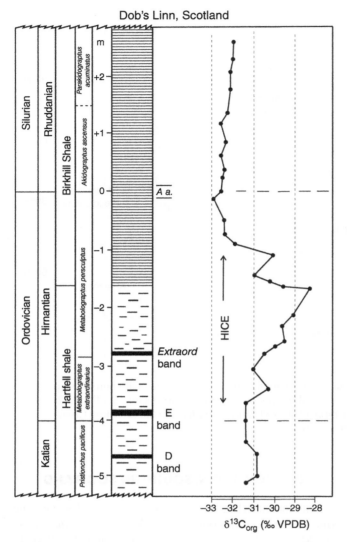

Figure 8.3 $\delta^{13}C_{org}$ curve through the Ordovician-Silurian boundary interval at the GSSP of the base of the Silurian at Dob's Linn, Scotland (After *Underwood et al.*, [1997]. Reproduced with permission of Geological Society of London). Note the range of the HICE through the *M. extraordinarius* and *M. persculptus* zones. Also note that the lower portion of the HICE exhibits a rather gradual, rather than abrupt, increase in $\delta^{13}C_{org}$ values.

$\delta^{13}C_{org}$ isotope values are reached after the comparatively conspicuous Hirnantian isotope excursion (HICE). This positive excursion ranges across the boundary between the Hartfell and Birkhill shales up to ~0.5 m above the base of the Birkhill Shale. The carbon isotope values through the basal 3 m of the Silurian, which correspond to the *A. ascensus* and *Parakidograptus acuminatus* zones, are quite uniform. This interval and that just below the systemic boundary appear to represent continuous deposition, there being no conglomerate and other lithologic

evidence of a sedimentary break. It should also be noted that in this section, the beginning of the HICE is not as abrupt as in many other sections and it appears to correspond rather closely with the base of the *Metabolograptus extraordinarius* Zone. Unfortunately, in the upper portion of the Hartfell Shale, graptolites are restricted to a few thin beds of dark shale in the gray mudstone, which makes it impossible to recognize the precise level of some graptolite zone boundaries.

In summary, in Dob's Linn succession, the $\delta^{13}C_{org}$ isotope curve is well tied into the graptolite biostratigraphy, but other biostratigraphic control is unfortunately lacking. In this key section, the systemic boundary level appears conformable, and it is not marked by any notable perturbation in the isotope curve. The HICE is clearly recognizable, and post-HICE baseline isotope values are reached just below the level of the systemic boundary.

8.3. RÖSTÅNGA, SOUTHERN SWEDEN

The $\delta^{13}C$ chemostratigraphy has been described in several Ordovician-Silurian boundary sections in Sweden and the East Baltic, but virtually all of these successions lack adequate graptolite biostratigraphic control, and/or the systemic boundary interval has a hiatus. For instance, recent chemostratigraphic investigations suggest that in several important Estonian sections, the Ordovician-Silurian boundary is located within an interval traditionally referred to the lower Llandovery [e.g., *Ainsaar et al.*, 2015].

However, there is at least one exception to the biostratigraphically incompletely controlled, or stratigraphically incomplete, Baltoscandian successions, namely, the Röstånga succession in the province of Skåne (Scania) in southernmost Sweden (Fig. 8.2:2). Although virtually all rock exposures are of only very limited size in the small outcrop area of lower Paleozoic rocks near the village of Röstånga, the relatively common presence of fossils, especially graptolites and trilobites, in the Ordovician-Lower Silurian Röstånga succession has attracted much paleontologic and biostratigraphic work all the way back to the nineteenth century (for pertinent references, see *Bergström et al.* [2014, 2016]). The dominantly shaly, relatively thin but stratigraphically virtually complete succession was deposited in an outer shelf to upper slope, moderately deep environment near the margin of the Baltic platform. Although only parts of the succession are currently exposed, recent core drillings have provided a wealth of new information from unexposed stratigraphic intervals.

Hence, the Ordovician-Silurian boundary interval in this area [*Troedsson*, 1918], although no longer accessible in outcrops, is well represented in the Röstånga-1 core. This core was drilled in 1997 [*Bergström et al.,* 1999] and subsequently has been subjected to a variety of geologic

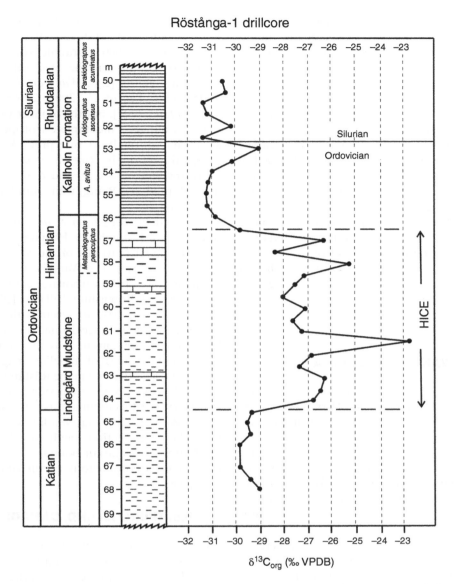

Figure 8.4 δ¹³C$_{org}$ curve through the Ordovician-Silurian boundary interval in the Röstånga-1 drill core, southern Sweden (Slightly modified after *Bergström et al.,* [2014]. Reproduced with permission of Taylor & Francis Ltd.). Note the abrupt increase in isotopic values at the base of the Hirnantian and the return to baseline values in the *M. persculptus* Zone in the topmost Hirnantian.

investigations [e.g., *Pålsson,* 2002; *Koren' et al.,* 2003; *Badawy et al.,* 2013; *Bergström et al.,* 2014, 2016; *Maletz and Ahlberg,* 2016; *Young et al.,* 2016]. Based on graptolites [*Koren' et al.,* 2003], the Ordovician-Silurian boundary, as defined by the appearance level of *A. ascensus,* is at a drill core depth of 52.7 m, which is ~3.2 m above the base of the Kallholn Formation (Fig. 8.4). The latter formation is a dark gray to black, richly graptolite-bearing shale. The underlying Katian-early Hirnantian Lindegård Mudstone is a gray, rather sparsely fossiliferous unit which, in addition to the mudstone, includes occasional beds of limestone, sandstone, and conglom-

erate. The Lindegård-Kallholn contact is gradational, and there is no lithologic evidence of the presence of a stratigraphic gap at the level of the Ordovician-Silurian boundary.

As seen in Figure 8.4, the HICE is prominently developed in the upper 8 m of the Lindegård Mudstone. From baseline values of ~−30‰, the excursion reaches peak value of >−23‰. As is the case in Dob's Linn succession, the end of the HICE is a short interval below the Ordovician-Silurian boundary. The boundary itself is in an interval of baseline values between −30 and −32‰. Right below the systemic boundary is a minor excursion with $\delta^{13}C_{org}$ values up to ~−29‰, but this is controlled by only two data points in the Röstånga succession, and its significance is questionable. However, a similar minor excursion has been recorded in the Hirnantian GSSP section at the Wangjiwan river section in China [cf., *Bergström et al.*, 2014; Fig. 15].

Although comparable in important respects to the Dob's Linn $\delta^{13}C_{org}$ isotope curve, the Röstånga $\delta^{13}C_{org}$ curve is also similar in general shape to $\delta^{13}C_{carb}$ curves, such as that from western Anticosti Island, Quebec [see *Bergström et al.*, 2014; Fig. 11]. However, in the latter succession, the precise level of the Ordovician-Silurian boundary remains uncertain in the absence of decisive biostratigraphic evidence from zonal graptolites. Although having no formal status as a global reference section, the chemostratigraphically and biostratigraphically well controlled, and apparently stratigraphically continuous, Röstånga succession is one of the best Ordovician-Silurian boundary sequences known anywhere in the world.

8.4. WANGJIWAN, CHINA

In view of the fact that it is the global reference section (GSSP) of the latest Ordovician Hirnantian Stage, the Wangjiwan section is also of particular importance for a study of the chemostratigraphy of the Ordovician-Silurian boundary. Detailed recent biostratigraphic investigations indicate that the Wangjiwan succession, although quite condensed, appears to be without obvious gaps at the systemic boundary, which makes it of special interest for the present chemostratigraphic review.

At the small town of Wangjiwan (Fig. 8.2:3), there are two outcrops that complement each other, the well-exposed but weathered highway section known as Wangjiwan North, which has yielded most of the fossils, and the less weathered but less well exposed outcrop along the river banks about 100–200 m south of the highway section, which has been referred to as the Wangjiwan Riverside section [*Chen et al.*, 2006].

The lithologic succession (Fig. 8.5) is closely similar at these outcrops. The lower Hirnantian consists of dark shale of the topmost portion of the Wufeng Formation. This interval is overlain by a ~0.3 m thick bed of impure limestone referred to as the Kuanyichiao bed (or formation) that contains a variety of shelly fossils but no graptolites. The shelly fossils include elements of the geographically widespread *Hirnantia* fauna [*Rong*, 1984]. The shaly Lungmashi Formation is present above this unit, and its lowermost part yields graptolites of the youngest Hirnantian *Metabolograptus persculptus* Zone. Silurian graptolites of the *A. ascensus* Zone appear as low

as ~0.2 m above the base of the Lungmashi Formation, a level taken to be the systemic boundary.

The $\delta^{13}C$ chemostratigraphic work at Wangjiwan has been carried out on the less weathered Wangjiwan Riverside section [e.g., *Wang et al.*, 1997; *Chen et al.*, 2005, 2006; *Yan et al.*, 2008, 2009; *Fan et al.*, 2009; *Zheng et al.*, 2009], but the pioneer work by these geologists resulted in incomplete isotope curves that were somewhat difficult to interpret. A more completely sampled section by *Gorjan et al.* [2012] resulted in more distinctive $\delta^{13}C_{carb}$ and $\delta^{13}C_{org}$ curves, and these are adopted as Wangjiwan standard curves herein (Fig. 8.5).

As noted by *Bergström et al.* [2014], the Wangjiwan $\delta^{13}C_{org}$ curve through the Hirnantian succession shows obvious similarity to that of the coeval interval at Dob's Linn (cf., Fig. 8.3). The Wangjiwan $\delta^{13}C_{carb}$ curve closely resembles those from carbonate successions, such as that of the Swedish Borenshult drill core [cf., *Bergström et al.*, 2012]. As is the case in several other stratigraphically complete boundary successions, the Ordovician-Silurian boundary at Wangjiwan falls in a post-HICE interval with isotope values of baseline magnitude. As noted above, and as is the case in the Röstånga succession, there is a small positive $\delta^{13}C_{org}$ excursion just below the systemic boundary, but this excursion is not visible in the $\delta^{13}C_{carb}$ curve. We conclude that similar to the carbon curves from the other sections dealt with herein, the level of the Ordovician-Silurian boundary at Wangjiwan is not marked by a conspicuous perturbation in the carbon isotope curves.

8.5. TRURO ISLAND, CANADIAN ARCTIC

Chemostratigraphic investigations using $\delta^{13}C$ through the Ordovician-Silurian boundary interval have been carried out at several localities in the United States and Canada (for references, see *Bergström et al.* [2014]). Unfortunately, at most of these, as is the case across the wide Midcontinent, the systemic boundary is marked by a stratigraphic gap of variable magnitude [cf., *Bergström and Boucot*, 1988], and adequate graptolite biostratigraphy is missing. Even in some sections with adequate graptolite control, such as those in the Monitor Range in the Great Basin of Nevada (Fig. 8.2:7; e.g., *LaPorte et al.*, 2009), there is a substantial stratigraphic gap at the systemic boundary level (Fig. 8.1) that includes much of the lower Llandovery [*Bergström et al.*, 2014].

The stratigraphically most complete Ordovician-Silurian boundary successions are best known from graptolite-bearing dolomite-shale successions, especially in the Cornwallis Island region in the Canadian Arctic (Fig. 8.2:4). The most informative succession with isotope

Wangjiwan (river), China

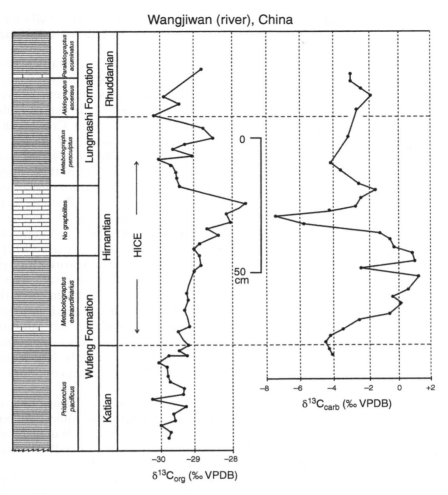

Figure 8.5 δ¹³C$_{org}$ and δ¹³C$_{carb}$ curves through the Ordovician-Silurian boundary interval at the Wangjiwan (Riverside) section, China, which is very near the GSSP of the global Hirnantian stage (Slightly modified after *Gorjan et al.* [2012]. Reproduced with permission of Elsevier). This is a stratigraphically very condensed succession, the total thickness of the Hirnantian stage being slightly less than one meter. However, the Hirnantian δ¹³C$_{carb}$ curve has a rather typical HICE appearance with a rapid increase in isotope values at the base of the Hirnantian; the δ¹³C$_{org}$ curve is more similar to the Dob's Linn HICE curve with its essentially background-size isotopic values in the lower Hirnantian.

chemostratigraphy may be that on Truro Island, where both graptolite biostratigraphy and δ¹³C$_{org}$ chemostratigraphy have been investigated in some detail [e.g., *Melchin*, 1987; *Melchin et al.*, 1991, 2003; *Melchin and Holmden*, 2006]. Although it has an extremely remote location, and is partly covered, the Truro Island succession is highly significant in providing a biostratigraphically well controlled δ¹³C$_{org}$ curve (Fig. 8.6) in what appears to be a stratigraphically continuous succession across the Ordovician-Silurian boundary. A very thin interval at the contact between the *M. persculptus* and the *A. ascensus* zones is covered, but the isotope curve through the Hirnantian and basalmost Silurian [*Melchin and*

Holmden, 2006] has a characteristic shape with a prominent HICE followed by baseline values across the systemic boundary (Fig. 8.6). As noted by *Bergström et al.* [2014], the Truro Island δ¹³C$_{org}$ curve exhibits a close similarity to the coeval Swedish Röstånga curve (Fig. 8.4), which, in combination with the graptolite biostratigraphy, allows for a very close correlation of especially the Hirnantian interval between these geographically widely separated localities. As is the case in our other study localities, the Ordovician-Silurian boundary is not marked by any very obvious perturbation in the isotope curve which is characterized by baseline values from the end of the HICE into the Rhuddanian Stage.

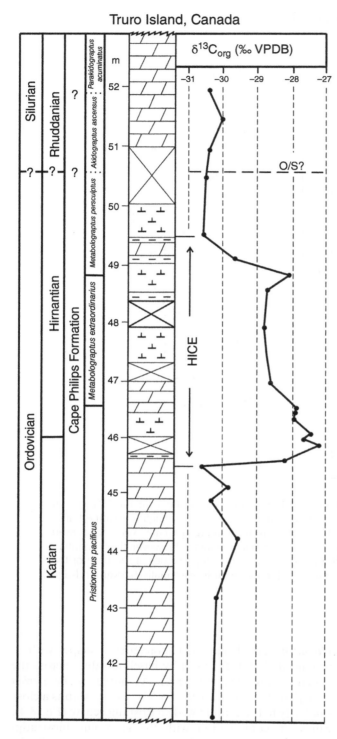

Figure 8.6 $\delta^{13}C_{org}$ curve through the Ordovician-Silurian boundary interval on Truro Island, Canadian Arctic (Slightly modified after *Melchin and Holmden* [2006]. Reproduced with permission of Elsevier). Note the characteristic shape of the HICE, which shows close similarity to the Röstånga curve (cf., Fig. 8.4 herein). As in the other successions discussed herein, the Ordovician-Silurian boundary level is not marked by a prominent isotope excursion.

8.6. MIRNY CREEK, OMULEV MOUNTAINS, EASTERN SIBERIA

One of the most important, stratigraphically apparently complete, and biostratigraphically well controlled Katian-Rhuddanian successions in Asia, if not in the entire world, is exposed along the extremely remotely located Mirny Creek (Fig. 8.2:5; e.g., *Koren' et al.*, 1988; *Zhang and Barnes*, 2007; *Koren' and Sobolevskaya*, 2008; *Kaljo and Martna*, 2011; *Kaljo et al.*, 2012]. The >160 m thick upper Katian to lower Rhuddanian succession consists of alternating beds of carbonate, shale, siltstone, marlstone, and conglomerate-breccia (Fig. 8.7). The presence of some biostratigraphic index species of graptolites and conodonts [cf., *Kaljo et al.*, 2012] provides a useful biostratigraphic framework through the Ordovician-Silurian boundary interval. The position of this systemic boundary is taken to be the level of appearance of *A. ascensus*, which is located just below the first occurrence of an Early Silurian conodont fauna, which includes *Kockelella* cf. *manitoulinensis* and stratigraphically slightly higher in the *A. ascensus* Zone, the basal Silurian conodont zone index species *Distomodus kentuckyensis*.

The $\delta^{13}C_{carb}$ curve published by *Kaljo et al.* [2012] has good biostratigraphic calibration, but its shape through the Hirnantian interval is unusual although the HICE can be distinguished broadly (Fig. 8.7). An increase in isotope values near the base of the *M. extraordinarius* Zone is followed by an interval of rather uniform isotope values of ~+2‰ up to the middle of the upper Hirnantian *M. persculptus* Zone, where there is a brief interval of peak values of ~+5‰ similar to those of the HICE in many successions, such as that of the Viki drill core, Estonia [*Hints et al.*, 2014; Fig. 2]. This peak value interval is followed by decreasing values of ~+1‰ up to the base of the Silurian, where there is a distinct negative excursion in the *A. ascensus* Zone. This pronounced drop in isotope values has not been recognized in any of the other successions dealt with herein. However, a somewhat similar negative excursion around the systemic boundary has been recorded from the Czech Republic by *Mitchell et al.* [2011; Fig. 1]. Although being somewhat "noisy," as a whole the Mirny Creek isotope curve exhibits some broad similarity to that of Truro Island (cf., Fig. 8.6), but its HICE values tend to be relatively lower compared with its pre-HICE and post-HICE baseline values. It should be noted that a rather limited number of isotope samples have so far been available from the comparatively thick Mirny Creek succession and more samples are clearly needed to clarify details in its isotope curve.

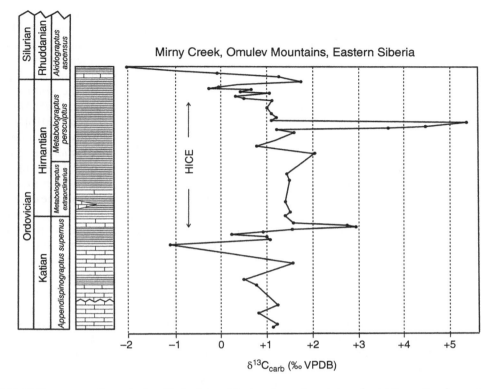

Figure 8.7 $\delta^{13}C_{carb}$ curve through the Ordovician-Silurian boundary interval at Mirny Creek, eastern Siberia (Slightly modified after *Kaljo et al.* [2012]. http://www.kirj.ee/21624/?tpl=1061&c_tpl=1064. Licensed under CC BY 4.0). The isotope curve through this important succession has good graptolite [e.g., *Koren' et al.,* 1988] and conodont [*Zhang and Barnes,* 2007] biostratigraphic control through the upper Katian-lower Rhuddanian interval. The Hirnantian segment of the isotope curve, which exhibits some similarity to the Truro Island and Röstanga-1 curves, has little more than baseline values except for raised values at the beginning of the Hirnantian and a brief interval in middle portion of the *M. persculptus* Zone. There is also a notable negative curve trend in the lowermost Rhuddanian that is not obvious in the other successions dealt with herein, but the significance of this feature remains unclear.

8.7. POSSIBLE RELATIONS BETWEEN δ¹³C CHEMOSTRATIGRAPHY, EUSTACY, AND EXTINCTION EVENTS

In the Earth's Phanerozoic history, the Ordovician-Silurian boundary interval was a time period characterized by conspicuous local and global sea-level changes, among which the major ones are interpreted to reflect significant glaciations on Gondwana. For informative reviews, see *Ghienne* [2011] and *Melchin et al.* [2013]. Furthermore, there were at least two profound extinction events that together rank as the second largest turnover in the marine faunas during the Phanerozoic. Despite a large amount of recent research, the possible relations between these events and the isotope chemostratigraphy are still partly poorly understood. A detailed discussion of this is outside the scope of the present chapter, but we will provide a few brief comments on these important and interesting matters.

8.7.1. Relations Between Isotope Chemostratigraphy and Eustacy

As discussed above, the δ¹³C curve through the systemic boundary interval may be subdivided into to broad divisions, the lower-middle Hirnantian distinctive positive excursion (HICE) and the upper Hirnantian-lowermost Silurian (lower Rhuddanian) interval with rather uniform isotope values of essentially baseline magnitude. The global stratigraphic significance of the end of the HICE is demonstrated by the fact that this level was taken as the boundary of the Hirnantian Stage Slices Hi1 and Hi2 of *Bergström et al.* [2009]. These two isotope curve subdivisions can be recognized virtually globally in $\delta^{13}C_{carb}$ and $\delta^{13}C_{org}$ curves from both carbonate and fine-clastic marine successions although there are some local variations in the magnitude of isotope values. The beginning of the HICE is very near the base of the Hirnantian Stage in the Hirnantian GSSP at Wangjiwan, China (Fig. 8.5). In apparently more or less stratigraphically continuous successions, such as that at Dob's Linn, the isotope curve in the beginning of the HICE increases gradually, but at many other localities, the isotope curve rises rather abruptly. This may reflect the presence of a stratigraphic gap at the Katian-Hirnantian stage boundary or perhaps local geological influence of seawater and sediment geochemistry [see, e.g., *Melchin et al., 2013*]. In the former case, the common widespread presence of such an unconformity suggests a eustatic sea-level fall caused by a major glaciation on Gondwana. A striking illustration of this glaciation is provided by the Don Braulio Formation of the Precordillera of Argentina (Fig. 8.2), where shales and mudstones containing shelly fossils of the Hirnantian *Hirnantia* fauna directly overlie a 15–20 m thick diamictite unit [cf., *Peralta, 2003*].

The presence of diverse shelly faunas and reefs in eastern Canada, the East Baltic, and China [cf., *Copper, 2001*] and bahamitic limestones at localities well away from the equatorial zone, such as in Sweden [*Bergström et al., 2006, 2014*], suggests that the mid-Hirnantian rocks in these regions were deposited in relatively warm water. Some authors have interpreted the entire Hirnantian as a period of peak glaciation, but this idea is not in agreement with the faunal and lithologic evidence, and we prefer to interpret most of the HICE interval as representing an interglacial sedimentation period with a slightly raised sea level.

In an interval broadly corresponding to a part of the falling values of the HICE, there is a prominent stratigraphic gap in many successions around the world (Fig. 8.1). In the past, for instance, in the East Baltic, this post-peak HICE unconformity has generally been considered to be the Ordovician-Silurian boundary [cf., *Ainsaar et al., 2015*]. This interpretation was also influenced by the fact that some Silurian-type shelly fossils appeared just above this unconformity. It seems very likely that this gap reflects another period of major glaciation on Gondwana. As noted above, the level of the Ordovician-Silurian boundary is now formally defined by graptolite biostratigraphy, and this horizon is well above both the HICE and the upper Hirnantian unconformity. Hence, $\delta^{13}C$ chemostratigraphy cannot be used to very precisely locate the level of the systemic boundary as it occurs in an interval dominated by approximately baseline values.

8.7.2. Relations Between Extinction Events and Isotope Chemostratigraphy

Past investigations have shown that there are at least two major extinction intervals in the uppermost Ordovician (Fig. 8.8). The oldest, which was referred to as "the first phase of extinction" by *Brenchley et al.*

[2003], took place right before the beginning of the HICE and hence appears to coincide with the latest Katian regressive event and the major glaciation referred to above. This extinction tremendously affected the shelly, graptolite, and chitinozoan faunas. It is particularly well documented in the case of the graptolites [cf., *Melchin and Mitchell*, 1991; *Chen et al.*, 2005; *Melchin et al.*, 2011; *Cooper et al.*, 2014; *Sheets et al.*, 2016]. It is of interest to note that in the apparently continuous Dob's Linn succession, the pre-HICE *Pristionchus pacificus* Zone has a relatively diverse graptolite fauna (about 15 species), but only three taxa, two of which are not identified to species, have been recorded from the overlying *M. extraordinarius* Zone [cf., *Williams*, 1988]. Hence, there appears to have been a conspicuous pre-HICE graptolite extinction event also in this important succession. There is a second but less significant level of extinction in graptolites [cf., *Chen et al.*, 2005; Fig. 3] that occurs in the *M. persculptus* Zone. This level appears to correspond to a position near the end of the HICE and be at least broadly coeval with the late Hirnantian regression. It is especially well documented from China, where it is followed by a conspicuous recovery of species diversity in the earliest Rhuddanian *A. ascensus* and *P. acuminatus* zones [*Chen et al.*, 2005].

The pre-HICE extinction event is less obvious in the conodont record [e.g., *Barnes and Bergström*, 1988]. During the Hirnantian, there is a rather gradual disappearance of characteristic Ordovician conodont taxa, such as *Aphelognathus*, *Amorphognathus*, *Belodina*, *Hamarodus*, *Phragmodus*, *Plectodina*, *Plegagnathus*, and *Strachanognathus*. The late Katian conodont faunas are particularly diverse in North America [cf., e.g., *Sweet*, 1979a, 1979b], whereas the coeval Baltoscandic faunas do not contain nearly as many species [e.g., *Männik and Viira*, 2012; *Goldman et al.*, 2014]. The lowermost Rhuddanian is characterized by conodont faunas having a strikingly different aspect compared with the Late Ordovician ones. These include representatives of *Decoriconus*, *Distomodus*, *Icriodella*, *Kockelella* (?), *Oulodus*, and *Rexroadus*, but as a whole, the earliest Silurian conodont faunas are less diverse than those from the Late Ordovician. Virtually all conodont collections from the Ordovician-Silurian boundary interval come from successions without precise graptolite control, and it is currently unclear if the conodont fauna change is in the uppermost *M. persculptus* Zone or at the base of the *A. ascensus* Zone. Regardless, this conodont turnover event appears to be stratigraphically higher than most of the HICE.

In summary, there appear to be two important extinction events in the study interval, one just before the HICE and one near the end of the HICE. Neither of these seems to be directly related to this highly significant

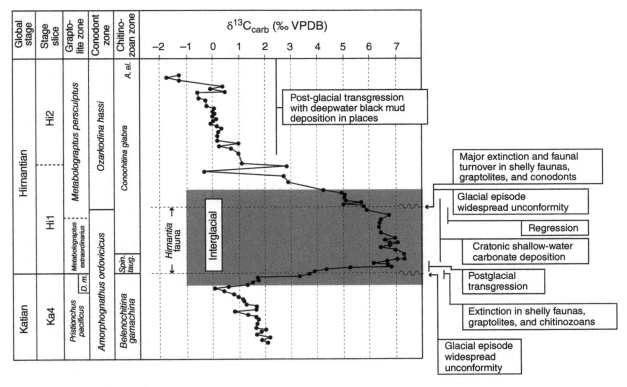

Figure 8.8 Model of the inferred relations between chemostratigraphy, eustatic events, and major faunal extinction intervals in the Ordovician-Silurian boundary interval. This model, here applied to the stratigraphically apparently continuous succession in the Röstånga-1 drill core from southern Sweden, is similar to one proposed by *Bergström et al.* [2014; Fig. 19] for the Monitor Range succession in Nevada, which has a major stratigraphic gap at the systemic boundary. Although not identified in the Röstånga-1 drill core, in many other successions worldwide, there are widespread, more or less prominent stratigraphic gaps at the base of the Hirnantian and in the middle-upper part of Stage Slice Hi1 that are interpreted to reflect glaciations on Gondwana. These appear to correlate with significant global extinction events. As noted by *Bergström et al.* [2014], these extinction events are not at the end of the Ordovician as is commonly erroneously stated in the literature but in the early-middle Hirnantian. Reproduced with permission of Taylor & Francis Ltd.

chemostratigraphic excursion event. Rather, they may be associated with regressive eustatic events and/or climatic events caused by glaciations on Gondwana.

8.8. CONCLUSIONS

A review of the δ¹³C chemostratigraphy of five stratigraphically apparently continuous Ordovician-Silurian boundary sections in northern Europe, North America, and Asia suggests that the level of the systemic boundary falls in an interval with relatively uniform δ¹³C$_{carb}$ and δ¹³C$_{org}$ values. Hence, because the boundary level is not marked by a significant perturbation in the isotope curve, the precise level of the systemic boundary cannot be defined in terms of δ¹³C chemostratigraphy. However, comparison between biostratigraphy and chemostratigraphy indicates that the graptolite-based base of the Silurian is located at a stratigraphic level only a little

higher than the end of the HICE, which is the largest δ¹³C excursion known in the Ordovician.

In the latest Ordovician, there were some very significant sea-level changes which are likely to correspond to Gondwana glaciations at high latitudes. One such event occurred just below the HICE, and another during the time of falling post-peak HICE values. These regression events are in several regions separated by locally richly fossiliferous marine carbonates that at subtropical latitudes include warm-water limestones with locally developed reefs that appear to have been deposited during an interglacial. Because the HICE began at the beginning of this interglacial and seems to have ended after the end of this interglacial, there does not appear to be a close correlation between the range of this excursion and the glacial episodes. On the other hand, there seems to be a close correlation between the main glaciations, which tend to be marked by significant stratigraphic gaps at low to mid-latitudes, and major faunal extinction horizons. These extinctions, which are among the very largest known in the Earth's Phanerozoic history, resulted in striking differences between the Late Ordovician and Early Silurian marine faunas. Unfortunately, the evidence at hand is insufficient to completely understand the complex casual relationships between the sedimentary geochemistry, the Late Ordovician/Early Silurian climate history, and these important faunal turnovers.

REFERENCES

Ainsaar, L., Truumees, J. & Meidla, T., 2015. The position of the Ordovician-Silurian boundary in Estonia tested by high-resolution $\delta^{13}C$ chemostratigraphic correlation. *In*: M. Ramkumar (ed.): Chemostratigraphy: Concepts, Techniques and Applications. Amsterdam: Elsevier, 395–412.

Badawy, A.S., Ahlberg, P., Calner, M., Mellqvist, K. & Valda, V., 2013. Palynology and sedimentology of the Upper Ordovician-lowermost Llandovery in the Fågelsång-1 core, southern Sweden. *In*: A. Lindskog & K. Mellqvist (eds.): Proceedings of the 3rd ICCP 591 Annual Meeting, Lund, Sweden, Lund University, 9–19 June 2013, 39–41.

Barnes, C.R. & Bergström, S.M., 1988. Conodont biostratigraphy of the uppermost Ordovician and lowermost Silurian. *In*: I.R.M. Cocks & R.B. Rickards (eds.): A Global Analysis of the Ordovician-Silurian Boundary. Bulletin of the British Museum of Natural History (Geology), *41*. London: British Museum (Natural History), 325–343.

Bergström, S.M. & Boucot, A., 1988. The Ordovician-Silurian boundary in the United States. *In*: I.R.M. Cocks & R.B. Rickards (eds.): A Global Analysis of the Ordovician-Silurian Boundary. Bulletin of the British Museum of Natural History (Geology), *41*. London: British Museum (Natural History), 273–284.

Bergström, S.M., Huff, W.D., Koren', T., Larsson, K., Ahlberg, P. & Kolata, D.E., 1999. The 1997 core drilling through Ordovician and Silurian strata at Röstånga, S. Sweden: preliminary stratigraphic assessment and regional comparison. GFF, *121*, 127–135.

Bergström, S.M., Saltzman, M. & Schmitz, B., 2006. First record of the Hirnantian (Upper Ordovician) $\delta^{13}C$ excursion in the North American Midcontinent and its regional implications. Geological Magazine, *143*, 657–678.

Bergström, S.M., Chen, X., Gutiérrez-Marco, J.-C. & Dronov, A., 2009. The new chronostratigraphic classification of the Ordovician System and its relations to regional series and stages and to $\delta^{13}C$ chemostratigraphy. Lethaia, *42*, 97–107.

Bergström, S.M., Kleffner, M., Schmitz, B. & Cramer, B.D., 2011. Revision of the position of the Ordovician–Silurian boundary in southern Ontario: regional chronostratigraphic implications of the $\delta^{13}C$ chemostratigraphy of the Manitoulin Formation and associated strata. Canadian Journal of Earth Sciences, *48*, 1447–1470.

Bergström, S.M., Lehnert, O., Calner, M. & Joachimski, M.M., 2012. A new upper Middle Ordovician-Lower Silurian drill-core standard succession from Borenshult in Östergötland,

southern Sweden. 2. Significance of $\delta^{13}C$ chemostratigraphy. GFF, *134*, 39–63.

Bergström, S.M., Eriksson, M.E., Young, S.A., Ahlberg, P. & Schmitz, B., 2014. Hirnantian (latest Ordovician) $\delta^{13}C$ chemostratigraphy in southern Sweden and globally: a refined integration with the graptolite and conodont zone successions. GFF, *136*, 355–386.

Bergström, S.M., Eriksson, M.E., Schmitz, B., Young, S.A. & Ahlberg, P., 2016. Upper Ordovician $\delta^{13}C$ chemostratigraphy, K-bentonite stratigraphy, and biostratigraphy in southern Scandinavia: a reappraisal. Palaeogeography, Palaeoclimatology, Palaeoecology, *454*, 175–188.

Brenchley, P.J., Carden, G.A., Hints, L., Kaljo, D., Marshall, J.D., Martma, T., Meidla, T. & Nõlvak, J., 2003. High-resolution stable isotope stratigraphy of Upper Ordovician sequences: constraints on the timing of bioevents and environmental changes associated with mass extinction and glaciation. Bulletin of the Geological Society of America, *115*, 89–104.

Chen, X., Fan, J.-X., Melchin, M.J. & Mitchell, C.E., 2005. Hirnantian (latest Ordovician) graptolites from the Upper Yangtze region, China. Palaeontology, *48*, 235–280.

Chen, X., Rong, J.-Y., Fan, J.-X., Zhan, R.B., Mitchell, C.E., Harper, D A.T., Melchin, M.J., Peng, P., Finney, S.C. & Wang, X.-F., 2006. The global boundary stratotype section and point (GSSP) for the base of the Hirnantian Stage (the uppermost of the Ordovician System). Episodes, *29*, 183–196.

Cooper, R.A., Safler, P.M., Munnecke, A. & Crampton, J.S., 2014. Graptoloid evolutionary rates track Ordovician-Silurian global climate change. Geological Magazine, *151*, 249–364.

Copper, P., 2001. Reefs during multiple crises towards the Ordovician-Silurian boundary, Anticosti Island, eastern Canada, and world-wide. Canadian Journal of Earth Sciences, *38*, 153–171.

Cramer, B.D., Brett, C.E., Melchin, M.J., Männik, P., Kleffner, M.A., McLaughlin, P.I., Loydell, D.K., Munnecke, A., Jeppsson, L., Corradini, C., Brunton, F.R. & Saltzman, M.R., 2011. Revised correlation of Silurian provincial series of North America and global and regional chronostratigraphic units with $\delta^{13}C_{carb}$ chemostratigraphy. Lethaia, *44*, 185–203.

Elrick, M., Reardon, D., Labor, W., Marin, J., Desrochers, A. & Pope, M., 2013. Orbital-scale climate change and glacioeustasy during the early Late Ordovician (pre-Hirnantian) determined from $\delta^{18}O$ values in marine apatite. Geology, *41*, 775–778.

Fan, J.-X., Peng, P. & Melchin, M.J., 2009. Carbon isotopes and event stratigraphy near the Ordovician-Silurian boundary, Yichang, South China. Palaeogeography, Palaeoclimatology, Palaeoecology, *276*, 160–169.

Ghienne, J.-F., 2011. The Late Ordovician glacial record: state of the Art. *In*: J.-C. Gutiérrez-Marco, I. Rábano & D. García-Bellido (eds.): Ordovician of the World. Cuadernos del Museo Geominero, *14*. Madrid: Instituto Geológico y Minero de España, 13–19.

Goldman, D., Bergström, S.M., Sheets, D. & Pantle, C., 2014. A CONOP9 composite taxon range chart for Ordovician conodonts from Baltoscandia and a framework for biostratigraphic correlation and maximum-likelihood biodiversity analyses. GFF, *36*, 342–354.

Gorjan, P., Kaiho, K. Fike, D.A. & Chen, X., 2012. Carbon and sulphur-isotope geochemistry of the Hirnantian (Late Ordovician) Wangjiwan (Riverside) section, South China: global correlation and environmental event interpretation. Palaeogeography, Palaeoclimatology, Palaeoecology, *337–338*, 14–22.

Hints, O., Martma, T., Männik, P., Nolvák, J., Põldvere, A., Shen, Y. & Viira, V., 2014. New data on Ordovician stable isotope record and conodont biostratigraphy from the Viki reference drill core, Saaremaa Island, western Estonia. GFF, *136*, 100–104.

Jones, D.S., Fike, D.A., Finnegan, S., Fisher, W.W., Schrag, O.P. & McKay, D., 2011, Terminal Ordovician carbon isotope stratigraphy and glacioeustatic sea-level change across Anticosti Island (Quebec, Canada). Geological Society of America Bulletin, *123*, 1645–1664.

Kaljo, D. & Martna, T., 2011. Carbon isotope trend in the Mirny Creek area, NE Russia, its specific features and possible implications of the uppermost Ordovician stratigraphy. *In*: J.C. Gutiérrez-Marco, I. Rábano, & D. Garcia-Bellido (eds.): Ordovician of the World. Cuadernos del Museo Geominero, *14*. Madrid: Instituto Geológica y Minero de España, 267–273.

Kaljo, D., Männik, P., Martma, T. & Nõlvak, J., 2012. More about the Ordovician-Silurian transition beds at Mirny Creek, Omulev Mountains, NE Russia: carbon isotopes and conodonts. Estonian Journal of Earth Sciences, *61*, 277–294.

Koren', T. & Sobolevskaya, R.F., 2008, The regional stratotype section and point for the base of the Hirnantian Stage (the uppermost Ordovician) at Mirno Creek, Omulev Mountains, northeast Russia. Estonian Journal of Earth Sciences, *57*, 1–10.

Koren', T., Oradovskaya, M.M. & Sobolevskaya, R.F., 1988. The Ordovician-Silurian boundary beds of the north-east USSR. Bulletin British Museum Natural History (Geology), *43*, 133–138.

Koren', T., Ahlberg, P. & Nielsen, A.T., 2003. The post-*persculptus* and pre-*ascensus* graptolite fauna in Scania, southwestern Sweden: Ordovician or Silurian? *In*: G. Ortega & G.L. Albanesi (eds.): Proceedings of the 7th International Graptolite Conference and Field Meeting of the Subcommission of Silurian Stratigraphy. Serio Correlación Geológica, *18*, 133–138. Tucuman, Argentina: INSUGEO.

LaPorte, D.F., Holmden, C., Patterson, W.P., Loxton, J.D., Melchin, M.L., Mitchell, C.E., Finney, S.C. & Sheets, H.D., 2009. Local and global perspectives on carbon and nitrogen cycling during the Hirnantian glaciation. Palaeogeography, Palaeoclimatology, Palaeoecology, *276*, 185–195.

Lapworth, C., 1878. The Moffat Series. Quarterly Journal of the Geological Society of London, *34*, 240–346.

Maletz, J. & Ahlberg, P. 2016. The Ordovician succession of the Fågelsång-2 drill core, Scania, southern Sweden. *87th Annual Conference of the Paläontologischen Gesellschaft, Dresden. Abstracts*, p. 104–105.

Männik, P. & Viira, V., 2012. Ordovician conodont diversity in the northern Baltic. Estonian Journal of Earth Sciences, *61*, 1–14.

Melchin, M.L., 1987. The Ordovician graptolites from the Cape Phillips Formation, Canadian Arctic Islands. Bulletin of the Geological Society of Denmark, *35*, 191–202.

Melchin M.J. & Mitchell C.E., 1991. Late Ordovician extinction in the Graptoloidea. *In*: C.R. Barnes & S.H. Williams (eds.), Advances in Ordovician Geology. Geological Survey of Canada Papers, *90-9*. Ottawa: Geological Survey of Canada, 143–156.

Melchin, M.L. & Williams, S.H., 2000. A restudy of the akidograptine from Dob's Linn and a proposed redefined zonation of the Silurian stratotype. *In*: P. Cockle, G.A. Wilson, G.A. Brock, M.J. Engerbretsen & A. Simpson (eds.): Palaeontology Down-Under 2000, 61–63. Orange, NSW: Geological Society of Australia.

Melchin, M.L. & Holmden, C., 2006. Carbon isotope chemostratigraphy in Arctic Canada: sea-level forcing of carbonate platform weathering and implication for Hirnantian global correlation. Palaeogeography, Palaeoclimatology, Palaeoecology, *234*, 186–200.

Melchin, M.L., McCracken, A.D. & Oliff, F.J., 1991. The Ordovician-Silurian boundary on Cornwallis and Truro Islands, Arctic Canada: preliminary data. Canadian Journal of Earth Sciences, *28*, 1854–1862.

Melchin, M.L., Holmden, C. & Williams, S.H., 2003. Correlation of graptolite biozones, chitinozoan biozones, and carbon isotope curves through the Hirnantian. *In:* G.L. Albanesi, M.S. Beresi & S.H. Peralta (eds.): Ordovician from the Andes. Serie Correlación Geológica, *17*. Tucuman, Argentina: INSUGEO, 101–104.

Melchin, M.J., Mitchell, C.E., Naczk-Cameron, A., Fan, J. & Loxton, J., 2011. Phylogeny and adaptive radiation of the Neograpta (Graptoloidea) during the Hirnantian mass extinction and Silurian recovery. Proceedings of the Yorkshire Geology Society, *58*, 281–309.

Melchin, M.L., Mitchell, C.E., Holmden, C., & Storch, P., 2013. Environmental changes in the Late Ordovician–Early Silurian: review and new insights from black shales and nitrogen isotopes. Geological Society of America Bulletin, *125*, 1635–1670.

Mitchell, C.R., Storch, P., Holmden, C., Melchin, M.J. & Gutiérrez-Marco, J.-C., 2011. New stable isotope data and fossils from the Hirnantian Stage in Bohemia and Spain: implications for correlation and paleoclimate. *In*: J.-C. Gutiérrez-Marco, I. Rábano & D. Garcia-Bellido (eds.): Ordovician of the World. Cuadernos del Museum Geominero, *14*. Madrid: Instituto Geológico y Minero de España, 371–378.

Pålsson, C., 2002. Upper Ordovician graptolites and biostratigraphy of the Röstånga 1 core, Scania, S. Sweden. Bulletin of the Geological Society of Denmark, *49*, 9–23.

Peralta, S.H., 2003. Don Braulio Creek, Villicum Range and Rinconada Area, Chica de Zonda Range, eastern Precordillera. *In*: S.H. Peralta, G.L. Albanesi & G. Ortega (eds.): Ordovician and Silurian of the Precordillera, San Juan Province. 9th International symposium on the Ordovician System, 7th International Graptolite Conference & Field

Meeting of the Subcommission on Silurian Stratigraphy. Field Trip Guide. Miscelánea, *10*. Tucuman: Instituto Superior de Correlación Geológica (INSUGEO), 23–43.

Rong, J.-Y., 1984. Brachiopods of the latest Ordovician in the Yichang district. *In*: Nanjing Institute of Geology and Palaeontology, Academia Sinica (ed.): Stratigraphy and Paleontology of Systemic Boundaries in China. Ordovician-Silurian Boundary, *1*. Hefei: Anhui Science and Technology Publishing House, 111–178.

Rong, J., Melchin, M.L., Williams, S.H., Koren', T.N. & Verniers, J., 2008. Report of the restudy of the defined global stratotype of the base of the Silurian System. Episodes, *31*, 315–318.

Saltzman, M.R., Edwards, C.T., Leslie, S.A., Dwyer, G.S., Bauer, J.A., Repetski, J.E., Harris, A.G. & Bergström, S.M., 2014. Calibration of a conodont apatite-based Ordovician Sr-87/Sr-86 curve to biostratigraphy and geochronology: implications for stratigraphic resolution. Geological Society of America Bulletin, *126*, 1551–1568.

Schmitz, B. & Bergström, S.M., 2007, Chemostratigraphy in the Swedish Upper Ordovician: regional significance of the Hirnantian δ¹³C excursion (HICE) in the Boda Limestone of the Siljan region. GFF, *129*, 133–140.

Sheets, H.D., Mitchell, C.E., Melchin, M.J., Loxton, J., Storch, P., Carlucci, K.L. & Hawkins, A.D., 2016. Graptolite community responses to global climate change and the Late Ordovician mass extinction. Proceedings of the National Academy of Sciences, *113*, 8380–8385. doi:10.1073/pnas.1602102113.

Sweet, W.C., 1979a. Late Ordovician conodonts and biostratigraphy of the Western Midcontinent Province. Brigham Young University Geology Studies, *26*, 45–85.

Sweet, W.C., 1979b. Conodonts and Conodont Biostratigraphy of Post-Tyrone Ordovician Rocks of the Cincinnati Region. *U.S. Geological Survey Professional Paper*, *1066*. Washington, DC: U.S. Government Printing Office, G1–G26.

Torsvik, T.H., 2009. *BugPlates: Linking biogeography and palaeogeography* (software). IGCP Project 503: Early Palaeozoic biogeography and paleogeography. StatoilHydro.

Troedsson, G., 1918. Om Skånes Brachiopodskiffer. Lunds Universitets Årsskrift, NF, 2, *15*(3), 1–110.

Underwood, C.J., Cowley, S., Marshall, J.D. & Brenchley, P.J. 1997. High-resolution carbon isotope stratigraphy of the basal Silurian stratotype (Dob's Linn, Scotland) and its global correlation. Journal of the Geological Society London, *154*, 709–718.

Wang, K., Chatterton, R.D.E. & Wang, Y., 1997. An organic carbon isotope record of Late Ordovician to Early Silurian marine sedimentary rocks, Yangtze Sea, South China: implication for CO_2 changes during the Hirnantian glaciation. Palaeogeography, Palaeoclimatology, Palaeoecology, *132*, 147–158.

Williams, S.H., 1988. Dob's Linn: the Ordovician-Silurian boundary stratotype. Bulletin of the British Museum of Natural History (Geology), *43*, 17–30.

Yan, D., Chen, D., Wang, Q., Wang, J., and Chu, Y., 2008. Environmental redox changes of the ancient sea in the Yangtze area during the Ordovician-Silurian transition. Acta Geologica Sinica (English Edition), *82*, 677–695.

Yan, Q., Chen, D., Wang, Q., Wang, J. & Chu, Y., 2009. Environmental redox changes of the Yangtze Sea during the Ordovician-Silurian transition. Acta Geologica Sinica (English Edition) *87*, 679–689.

Young, S.A., Saltzman, M.R., Ausich, W.I., Derochers, A. & Kaljo, D., 2010. Did changes in atmospheric CO_a coincide with latest Ordovician glacial-interglacial cycles? Palaeogeography, Palaeoclimatology, Palaeoecology, *296*, 376–388.

Young, S.A., Gill, B.C., Edwards, C.T., Saltzman, M.R. & Leslie, S.A., 2016. Middle-Late Ordovician (Darriwilian-Sandbian) decoupling of global sulfur and carbon cycles: isotopic evidence from eastern and southern Laurentia. Palaeogeography, Palaeoclimatology, Palaeoecology, *458*, 118–132.

Zhang, S. & Barnes, C.R., 2007. Late Ordovician to Early Silurian conodont faunas from the Koluma Terrane, Omulev Mountains, northeastern Russia, and their paleobiogeographic affinity. Journal of Paleontology, *81*, 490–512.

Zheng, T., Shen, Y., Zhan, R., Shen, S. & Chen, C., 2009. Large perturbation of the carbon and sulphur cycles associated with the Late Ordovician mass extinction in South China. Geology, *37*, 299–302.

9

Chemostratigraphy Across the Permian-Triassic Boundary: The Effect of Sampling Strategies on Carbonate Carbon Isotope Stratigraphic Markers

Martin Schobben[1,2], Franziska Heuer[2], Melanie Tietje[2], Abbas Ghaderi[3], Dieter Korn[2], Christoph Korte[4], and Paul B. Wignall[1]

ABSTRACT

A major extinction pulse occurred just below the conodont-defined Permian-Triassic boundary. Global-scale compilations of increasingly larger paleontological, sedimentological, and geochemical datasets further amplify our understanding of this event by unraveling temporospatial patterns. Robust stratigraphic frameworks are an integral part of these worldwide compilations. Bulk carbonate rock carbon isotope records are a widely used, and easy-to-generate, stratigraphic tool; it can substitute for biostratigraphy. However, inconsistencies in the amplitude and shape of stratigraphic carbon isotope patterns have also hampered the successful linkages of different geographic locations. In this study, we focus on the multicomponent nature of various limestone facies. We show how sampling strategies can be adapted in order to retrieve material from this multicomponent system that will most likely represent secular patterns in marine dissolved inorganic carbon $\delta^{13}C$. By obtaining multiple and randomly chosen bed-internal samples, we reveal that the magnitude of bed-internal $\delta^{13}C$ variations can differ between distinct lithologies. However, we also note that the largest within-bed variations (in excess of 0.5‰) do not correspond to obvious textural variations for this specific rock. Bed-internal variations do not always trace marine dissolved inorganic carbon $\delta^{13}C$ and might cause identification of ambiguous isotope stratigraphic markers that do not allude to globally correlative signals.

9.1. INTRODUCTION

9.1.1. Chemostratigraphy Across the Permian-Triassic Boundary

The abrupt change in fossil assemblages across the Permian-Triassic (P-Tr) transition has been documented already since the first half of the 19th century. These early paleontological documentations have played a central role in our attempts to subdivide Earth history, the discipline of stratigraphy. The documented change in P-Tr rock sequences was attributed to a significant faunal turnover, but it was not until more recent that it was recognized as a catastrophic event with elevated extinction rates [*Schindewolf*, 1954; *Newell*, 1962]. The P-Tr boundary is nowadays regarded as the most severe event of species richness loss during post-Cambrian time [*Raup and Sepkoski*, 1982; *Bambach et al.*, 2004]. The causes for this great depletion of biodiversity have been dominantly sought in extrinsic Earth-bound drivers [*Erwin*, 1993, 1994].

The main contending drivers are a rapid global temperature increase [*Holser et al.*, 1989; *Joachimski et al.*, 2012], ocean chemistry changes [*Wignall and Twitchett*, 1996; *Isozaki*, 1997], and the emplacement of the Siberian

[1] *School of Earth and Environment, University of Leeds, Leeds, United Kingdom*

[2] *Museum für Naturkunde - Leibniz Institute for Evolution and Biodiversity Science, Berlin, Germany*

[3] *Department of Geology, Faculty of Sciences, Ferdowsi University of Mashhad, Mashhad, Iran*

[4] *Department of Geosciences and Natural Resource Management, University of Copenhagen, Copenhagen, Denmark*

Chemostratigraphy Across Major Chronological Boundaries, Geophysical Monograph 240, First Edition.
Edited by Alcides N. Sial, Claudio Gaucher, Muthuvairavasamy Ramkumar, and Valderez Pinto Ferreira.
© 2019 the American Geophysical Union. Published 2019 by John Wiley & Sons, Inc.

Traps basalts [*Holser and Magaritz*, 1987; *Svensen et al.*, 2009; *Black et al.*, 2012]. Enhanced volcanic activity is often regarded as a proximal driver that can set a series of events in motion (i.e., feedbacks from the Earth system) which act as a synergistic killing agent [*Wignall*, 2007]. The antagonists and synergistic effects of such scenarios can be reconciled with the bulk of the sedimentological and paleontological data. For example, the global increased occurrence of laminated sediments and small-sized framboidal pyrite has been attributed to widespread anoxia [*Wignall and Twitchett*, 1996; *Wignall and Newton*, 2003]. On the other hand, changing geographical distributions of species and selective extinctions have been tied to physiological factors corresponding to the combined effects of thermal stress, hypercapnia, carbonate undersaturation, and marine oxygen depletion [*Knoll et al.*, 2007; *Clapham and Payne*, 2011; *Sun et al.*, 2012]. The isotope chemical composition of sediments and individual fossils, and their ability to be a proxy for chemical and physical parameters of the ancient environment, has yielded a robust testimony of the previous outlined environmental deterioration. Notably, temporal fluctuations in carbon, sulfur, oxygen, boron, and calcium isotope composition hallmark the P-Tr boundary beds and are suggestive of drastic environmental changes, such as seawater warming, perturbations of the carbon and sulfur biogeochemical cycles, and deteriorating marine chemical conditions hazardous to (most) life [*Holser and Magaritz*, 1987; *Holser et al.*, 1989; *Baud et al.*, 1989; *Newton et al.*, 2004; *Korte and Kozur*, 2010; *Luo et al.*, 2010; *Payne et al.*, 2010; *Joachimski et al.*, 2012; *Clarkson et al.*, 2016].

Although significant progress has been made in reconstructing the timeline of environmental change and (corresponding) biological responses [*Shen et al.*, 2011; *Burgess et al.*, 2014], controversy still exists with respect to the duration and timing of events as well as their spatial patterns. For example, the actual mode of biodiversity depletion is still matter of controversy, with hypotheses favoring one abrupt single pulsed extinction [*Jin et al.*, 2000; *Wang et al.*, 2014] and multiple extinction peaks [*Song et al.*, 2012] and/or encompassing regionally diachronous events [*Algeo et al.*, 2012]. Hence, attempts to refine the P-Tr stratigraphic framework are not only important to better gauge the rate of evolutionary and environmental processes but also help merge paleontological and environmental data from different regions in the world. Again, isotope geochemical records do play a pivotal role in refining the global P-Tr stratigraphic framework, with a potentially unrivaled temporal resolving power (i.e., bed-by-bed sampling). In addition, isotope geochemical records are more easily accessible comparative to biostratigraphic data, where often only a select group of specialists can identify stratigraphic

marker species. Especially carbon isotope records, based on bulk micritic carbonate rock, have been a key player in such global correlative frameworks, where these records almost uniformly reveal a trend to more ^{13}C-depleted values over the P-Tr transition (Fig. 9.1; for a review, see *Korte and Kozur*, 2010). At a more stratigraphic resolved scale, some authors have identified carbon isotope excursions superimposed on this long-term negative δ^{13}C trend. These second-order features can potentially yield high-resolution stratigraphic frameworks [*Cao et al.*, 2009; *Richoz et al.*, 2010], but their application in global compilations is not always apparent [*Schobben et al.*, 2017]. Important in this respect is that carbonate carbon isotope composition might also reflect changes in the predominant carbonate polymorphs [*Brand et al.*, 2012a; *Heydari et al.*, 2013; *Li and Jones*, 2017] or diagenetic artifacts [*Heydari et al.*, 2001; *Cao et al.*, 2010; *Schobben et al.*, 2016; *Li and Jones*, 2017] and, as such, does not always trace a global signal. Hence, the nature of these second-order carbon isotope signals as well as their recorder, bulk micritic rock, should be better understood in order to better grasp their value for P-Tr chemostratigraphic frameworks.

9.1.2. Bulk Rock Carbon Isotope Composition and Its Diagenetic Potential

The application of carbonate isotope systems in reconstructing past environments became apparent soon after the theoretical foundations behind isotope fractionations had been laid; isotope mass spectrometry was developed, and the first measurements on natural substances had been performed [*Dole and Slobod*, 1940; *Nier*, 1947; *Urey*, 1947]. Since then, the usage of temporal trends and variations of carbonate isotope composition to trace climatic and oceanic conditions through time has rooted itself firmly in geological studies [*Emiliani*, 1955; *Shackleton and Turner*, 1967; *Shackleton and Kennett*, 1976; *Scholle and Arthur*, 1980; *Veizer et al.*, 1999; *Zachos et al.*, 2001].

Nonetheless, the possible problems with this one-to-one translation of carbonate isotope signals to environmental parameters were soon recognized [*Urey*, 1947]. In particular, the adverse effect of sediment burial, subsequent subjection of rock to immense pressure and temperatures as well as percolating fluids, has been considered to reset a large portion of the carbonate archive [*Scholle*, 1977; *Brand and Veizer*, 1981; *Frank and Bernet*, 2000]. However, the preservation of unstable carbonate deposits as old as the Paleoarchean [*Allwood et al.*, 2006; *Duda et al.*, 2016] the preservation of primary textures from rock of different ages [*Lowenstam*, 1961; *Al-Aasm and Veizer*, 1986; *Munnecke and Samtleben*, 1996; *Ullmann and Korte*, 2015] and the differential diagenetic potential of individual carbonate rock constituents

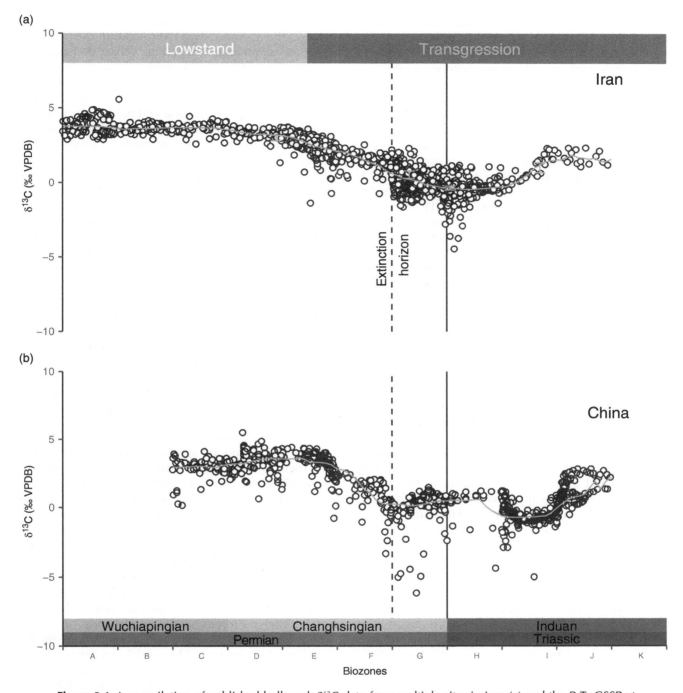

Figure 9.1 A compilation of published bulk rock δ¹³C data from multiple sites in Iran (a) and the P-Tr GSSP at Meishan in South China (b) (Modified from *Schobben et al.* [2017]. https://www.clim-past.net/13/1635/2017/. Licensed under a CC BY 3.0.). The data points are placed on a dimensionless timeline to ensure a better comparison of trends in the δ¹³C data of both regions. The dimensionless timeline is based on conodont zones. (The biostratigraphic scheme and sources regarding sea-level change and the extinction horizon are given in Supporting Information Text S1.) The trendline is based on a subsampling routine as outlined in *Schobben et al.* [2017]. The yellow error bars represent the value ranges of newly generated δ¹³C data for this study. *(See insert for color representation of the figure.)*

[*Brand and Veizer*, 1980, 1981; *Marshall*, 1992] alludes to the potential of, at least, a partial retention of primary isotopic signatures. And, indeed, long-term trajectories of carbonate carbon and oxygen isotope composition, based on database compilations, seem to point to an overarching temporal trend controlled by environmental parameters [*Veizer et al.*, 1999; *Prokoph et al.*, 2008; *Veizer and Prokoph*, 2015].

When zooming in on specific time windows, such as the P-Tr boundary interval, these trends can become less clear, and correlations can become difficult (Section 9.1.1). Although calcite fossils are used for high-resolution carbon and oxygen studies [*Samtleben et al.*, 2001], single-component studies (i.e., brachiopod shell calcite) have yielded only incomplete temporal sequences in the P-Tr boundary beds [*Gruszczynski et al.*, 1989; *Kearsey et al.*, 2009; *Brand et al.*, 2012a; *Schobben et al.*, 2014]. This notable absence of continuous shelly records coincides with elevated extinction rates among many marine species during the end-Permian biotic crisis [*Jin et al.*, 2000].

Bulk rock carbonate samples form an alternative for probing the carbon isotope composition of ancient seawater dissolved inorganic carbon (DIC) and are widely applied for chemostratigraphic studies of the P-Tr boundary interval [e.g., *Korte and Kozur*, 2010]. However, a limestone sample is often composed of multiple components, with some primary constituents (e.g., biogenic material) and other textures being added during or after deposition (e.g., pore-occluding cements). On top of that, each component has its specific reactivity and can dissolve or be replaced after initial deposition at the seabed. This alteration follows thermodynamic laws prescribed by the original mineralogy and associated lattice defects [*Busenberg and Plummer*, 1989], as well as porosity, water to rock ratios, and many other chemical and physical aspects of the host rock and diagenetic environment [*Scholle*, 1977; *Brand and Veizer*, 1980, 1981; *Munnecke and Samtleben*, 1996; *Frank and Bernet*, 2000]. The majority of ancient carbonate rock constituents are therefore diagenetically altered (i.e., stabilized to less reactive counterparts) by either interaction with meteoric water [*Brand and Veizer*, 1980, 1981; *Knauth and Kennedy*, 2009], marine-derived fluids [*Marshall*, 1992; *Munnecke and Samtleben*, 1996; *Schobben et al.*, 2016], or fluids deep in Earth's interior [*Scholle*, 1977]. This stabilization results in many cases in an altered oxygen isotope composition of bulk rock samples and to a lesser degree in the alteration of bulk rock carbon isotope composition [*Brand et al.*, 2012a, 2012b; *Schobben et al.*, 2016]. The isotope-specific diagenetic potential has been related to the higher availability of oxygen in the percolating fluids [*Brand and Veizer*, 1981; *Banner and Hanson*, 1990; *Marshall*, 1992].

Nonetheless, bulk rock isotope records also exhibit some advantages over the single-component equivalent.

Notably, a well-mixed pre-lithified carbonate mud would average the isotope signal throughout the temporal sequence, eliminating anomalous values that might be unrepresentative for long-term operation of the biogeochemical carbon cycle [*Holser*, 1997]. Such smearing would make geochemical studies more compatible with other time-averaged geological and paleontological data, resulting from sedimentary processes and analytical artifacts. For example, time averaging is introduced by conventional geological and paleontological field analyses, which often rely on bed-to-bed sampling [*Flessa*, 2001; *Flügel*, 2010]. However, the bed-internal homogeneity, including $\delta^{13}C$, might vary as a function of bioturbation intensity [*Schobben et al.*, 2017], and a priori knowledge of the rock-internal composition would be needed to confidently assess the extent of this smearing effect. In addition, the chemistry of some calcite fossils is potentially influenced by biologically induced isotope fractionations, for example, shells specific to certain brachiopod species [*Auclair et al.*, 2003; *Brand et al.*, 2003; *Parkinson et al.*, 2005] and tests of photosymbiont-bearing foraminifera [*Spero and Williams*, 1988], thereby further underlining the importance of bulk rock isotope investigations.

Contradictory to bulk rock analysis, sampling of brachiopod shell and foraminifera test calcite for carbon and oxygen isotope analysis has received much attention and follows strict protocols [*Samtleben et al.*, 2001; *Korte et al.*, 2005; *Löwemark et al.*, 2005; *Brand et al.*, 2012b]. Due to the compound-specific diagenetic potential of bulk rock, a standardized sampling strategy would also be beneficial for bulk carbonate isotope analysis. This is of further relevance as, with the advancement of technology, bulk rock carbonate carbon isotope measurements have become routine. Furthermore, stratigraphic trends, wiggles, and rhythms within bulk rock based carbon isotope records are often used as an easy tool to date newly explored carbonate sequences. Henceforward, the aim of this work is to review and, where possible, revise current sampling strategies for bulk rock carbon isotope analysis to ensure the construction of reliable carbon isotope chemostratigraphic frameworks for the P-Tr boundary interval. The results can potentially be extrapolated to other time intervals as well as applied to other geochemical proxies (cf., spatial variability in carbonate-associated sulfate $\delta^{34}S$; *Present et al.*, 2015).

9.1.3. Sampling Strategies and Carbon Isotope Trends

To diminish the effect of sampling obvious diagenetically altered components, two widely applied methodologies are adopted in bulk rock based carbonate carbon isotope studies: (i) drilling of fresh cut surfaces, thereby directly targeting the best preserved sections, and (ii) trimming rock pieces with a saw, discarding weathered surfaces, cracks, veins, and other odd textures, after which the remainder is crushed and ground to a fine powder. Both methods rely on visual examination, often without further optical aids. Although drilling samples seems more precise, one might wonder how well such often sub-mg-sized drilled samples do represent the whole limestone bed, as it is possible that the targeted area is not representative of the whole specimen. Such effects could, furthermore, become increasingly important when the lithology was originally less well mixed, creating a higher order of rock-internal spatial variation of the isotope signal [*Schobben et al.*, 2017]. For example, the P-Tr transition is marked by a reduction of burrowing organisms, and an increase of internal heterogeneity in deposits could be expected for that time period.

To further elucidate the potential significance of distortion of temporal carbon isotope trends by within-bed isotope variability, we conducted a small survey of published P-Tr chemostratigraphic studies (Fig. 9.2). Note that the applied method could introduce anomalous $\delta^{13}C$ values by drilling of sections of the matrix that contains calcite which is not representative for oceanic DIC (i.e., a

relative high content of diagenetically altered components). In the survey, we focused on previously assigned isotope stratigraphic markers in published $\delta^{13}C$ records, and we divided these markers based on shape according to the following definitions:

1. The appointed isotope stratigraphic marker consists of an excursion of less than three data points, forming a distinct excursion superimposed on the background values (Fig. 9.2a).

2. The described isotope stratigraphic marker is a monotonic trend of five consecutive data points, which is further defined as a trend where the individual values do not deviate by more than 0.3‰ from the monotonic trend (Fig. 9.2b).

3. The described isotope stratigraphic marker is an isotope shift, where a set of data points can be divided into two distinct plateaus, while in stratigraphic order. In turn, variability within the plateau does not deviate by more than 0.3‰ from the plateau-specific baseline (Fig. 9.2c).

Isotope stratigraphic markers classified under definition 1 can be regarded as second-order $\delta^{13}C$ excursions [cf., *Schobben et al.*, 2016]. These second-order excursions

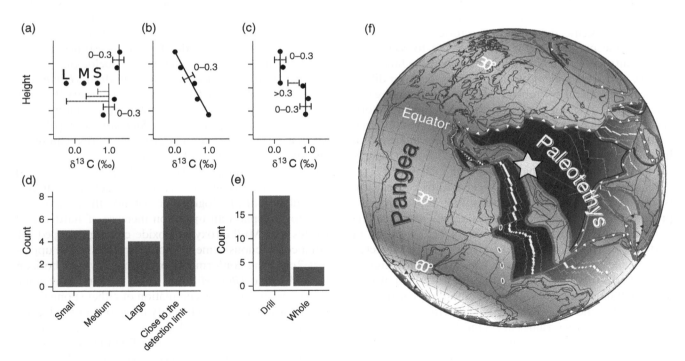

Figure 9.2 A survey of isotope stratigraphic markers as defined in previous studies on P-Tr carbonate rock exposed in Iran (see the Supporting Information Text S1 and Table S1 for data and sources). The survey evaluates the methodology of material retrieval and classifies the isotope events. The isotope events are defined according to the following set of definitions. (a) Second-order carbon isotope excursions: *S* is small (0.3–0.5‰), *M* is medium (0.5–1‰), and *L* is large (>1‰). If the amplitude is smaller than 0.3‰, we labeled the excursion as "close to the detection limit," as the value is close to the analytical bias. (b) Monotonic carbon isotope trends. (c) Carbon isotope shifts of more than 0.3‰. (d) Bar plot that counts the magnitude of the identified second-order excursions. (e) Bar plot that counts the material retrieval method for isotope analysis, cq: drilling or grounding of the whole rock. (f) Paleogeographic situation (Adapted from *Stampfli and Borel* [2002]. Reproduced with permission of Elsevier) of the sampling sites covered by the survey and the investigatory object of the current study.

form the main target of this survey, as it can easily be perceived that these features result from the inclusion of a diagenetically altered sample. A positively identified second-order excursion is subsequently classified according to its magnitude (small = 0.3–0.5‰, medium = 0.5–1.0‰, and large > 1.0‰) by taking the distance of the peak to the average of the nearest set of bracketing values (Fig. 9.2a, d). As a last step, we tallied how many of these studies rely on drilling or processing of whole rock samples to retrieve material for isotope analysis (Fig. 9.2e). One could envision that the sampling resolution, availability of samples, and sedimentological hiatuses are equally important factors in shaping patterns in carbon isotope records (besides the sampling strategy). Hence, this exercise should only be viewed as a rough guidance in quantifying published isotope stratigraphic markers.

When evaluating the results of this survey, we can conclude that chemostratigraphic studies in the past often relied on small second-order excursions (<0.3‰) to define an isotope stratigraphic marker. By inference, these stratigraphic markers under the previous definition would have to equate to global carbon cycle perturbations (Fig. 9.2d). However, given the bulk rock multicomponent nature and component-specific diagenetic trajectories, such isotope stratigraphic markers might equally relate to secondary signals entrained in an anomalous section of the limestone bed (i.e., elevated content of altered constituents or differences in the mineral assemblage). This survey therefore justifies a more in-depth investigation of within-bed isotope variations. Since the inclusion of anomalous bed-internal carbon isotope values might alter patterns in temporal bulk carbonate based carbon isotope sequences, thereby hampering the stratigraphic potential of such records.

As such, in this study, we target within-bed carbon isotope variations by a blind sampling approach on a predefined grid overlain on individual rock samples. The samples obtained from P-Tr boundary beds exposed in northwestern Iran (Zal section) are, furthermore, subjected to a detailed petrological study. This study will therefore help (i) identify C isotope variations entrained within individual limestone beds, with or without a clear relation to textural elements; (ii) clarify whether the inclusion of anomalous C isotope values in chemostratigraphic studies can compromise isotope stratigraphic markers, which are based on second-order carbon isotope excursions; and (iii) assess the ideal method to obtain material representative for secular changes in marine DIC-δ^{13}C.

9.2. GEOLOGICAL SETTING

Material is collected from the Zal section (38°73.30′N, 45°58.00′E) located in the Julfa region of northwestern Iran. This site was situated on the northwestern Iranian terrane as part of an amalgamation of smaller terranes,

known as the Cimmerian microcontinent, bordered by the Neotethys on the southern side and the Paleotethys on its northern fringes [Şengör, 1990; Ruban et al., 2007; Muttoni et al., 2009]. The terranes comprised a breakaway from Gondwana during a rift event, which commenced in the Late Carboniferous to Early Permian, and a northward drift moved the plates close to the equator during the Late Permian (Fig. 9.2; Stampfli and Borel, 2002; Muttoni et al., 2009). The extensional tectonics created a half-graben system, where the tilted fault blocks formed the accommodation space for carbonate accumulation [Horacek et al., 2007].

The P-Tr boundary beds exposed in northwestern Iran are stratigraphically complete carbonate-dominated facies with rich fossil assemblages [Stepanov et al., 1969; Aghai et al., 2009; Richoz et al., 2010; Ghaderi et al., 2014; Leda et al., 2014]. The lithological units bracketing the P-Tr boundary can be divided into three distinct units: (i) the 5.1 m thick Late Permian Paratirolites Limestone, consisting of red nodular argillaceous limestones, intercalated by some shaly horizons; (ii) a latest Permian 0.6 m thick calcareous claystone, from which the base is used as a reference horizon to demarcate the end-Permian mass extinction [Kozur, 2007]; and (iii) an expanded lithological unit of gray thick-bedded platy limestones bearing mass accumulations of paper pectins belonging to the genus Claraia and carrying the conodont [Hindeodus parvus; Yin et al., 2001] defined system boundary [Ghaderi et al., 2014]. Where the latter two units comprise the Elikah Formation, as defined by Stepanov et al. [1969], a unit that is traceable over a large area and that can reach over hundreds of meters of thickness.

A microfacies study by Leda et al. [2014] has revealed that behind this rough division lies a much more diverse accumulation of lithologies. For instance, the Paratirolites Limestone displays an upsection increase of hardground features and Mn/Fe-(oxyhydr)oxide coatings, topped off with a conspicuous acme in sponge spicules in the uppermost 2 cm. The conformable overlying clay-rich horizon [Aras Member; Ghaderi et al., 2014] is marked by some horizons with massive accumulation of either sponge spicules or ostracod valves. The platy limestone of the Elikah Formation harbors several distinct lithologies, among which are oncoid-bearing limestone, finely laminated bindstone with alternating couplets of coarser dark micritic and light sparitic layers, and peloid-bearing lithologies. These lithologies typified by a coarse crystalline matrix are common among these post-extinction strata, which have been assigned to a predominant microbial carbonate factory for the Elikah Formation, opposed to a skeletal origin for the youngest Permian carbonate-bearing unit [Richoz et al., 2010; Leda et al., 2014].

Large-amplitude sea-level fluctuations (e.g., shallowing-upward sequences) or signs of subaerial exposure (e.g., paleokarsts and/or paleosols) have not been detected in the studied sequence. Instead, deposition of these units continued in a deep marine environment underneath of the storm wave base [*Richoz et al.*, 2010; *Leda et al.*, 2014]. A shallowing of facies only occurs roughly 300 m higher up in the sequence, although deposition continues in a marine setting [*Horacek et al.*, 2007].

9.3. MATERIALS AND METHODS

9.3.1. Sample Selection

In order to probe for isotopic variations within limestone beds, four distinct bedded rock units have been selected for this study: a Late Permian red nodular limestone (Zl 11), a sponge wackestone (Zl 18a) 67 cm above the latest Permian extinction horizon [*Kozur*, 2005; *Ghaderi et al.*, 2014], a laminated bindstone (Zl 23) containing the administrative conodont-defined P-Tr boundary [*Schobben et al.*, 2015], and an Early Triassic oncoid wackestone (Zl 35). The dimensions of the rock samples, roughly 5 × 5 cm, formed another selection criterion; they enabled the preparation of thin sections and polished rock slabs for a subsequent petrographic study and systematic spatial carbonate carbon and oxygen isotope study, respectively. By selecting rock samples with distinctive textural (e.g., spar cement infill) and sedimentary features (e.g., lamination), we strive to re-create the circumstances under which a majority of chemostratigraphic studies have been carried out. In particular, it re-creates the somewhat subjective selection of areas for micro drilling and subsequent bulk rock carbon isotope analyses on the obtained material. The re-creation of these circumstances is critical as it allows the identification of isotopic variability introduced not only by drilling of the most visually obvious textures and structures but also by sampling of features that might escape detection when relying on observations made with the naked eye.

9.3.2. Petrographic Analysis

Thin sections have been prepared for detailed studies of sedimentary structures and textures contained within the limestone samples that were selected for this isotopic analysis. A polarizing microscope (Zeiss Axioskop 40) fitted with a camera (AxioCam MRc5) and the software AxioVision for image processing was used to inspect and document sedimentary and textural features. The remaining rock fragments after thin section preparation were ground and mounted in an epoxy resin to create a flat surface. Thereafter, the mounted sample is further ground and polished up to a roughness of 1 μm. The sampling surfaces were scanned with a commercial scanner. In addition, backscattered electron (BSE) images were taken

with a JEOL JSM-6610LV SEM at the Museum für Naturkunde, Berlin, with a 15.0 kV beam acceleration as the operating condition. Generated BSE images give a general overview of mineralogies and porosity contained in the polished rock slab. This enabled petrographic analyses in terms of abundance of matrix and components (bioclasts and nonskeletal clasts), matrix composition, and postdepositional features (e.g., cracks, veins, and stylolites). Facies assignment is largely based on the carbonate ramp model [cf., facies distribution controlled by energy levels; *Flügel*, 2010, p. 664–668], as the studied region does not seem to harbor the characteristics of a rimmed platform (i.e., reefs or sand shoals) but rather comprised isolated carbonate platforms. The extent of bioturbation was assessed by a description of ichnofabrics in terms of burrow density and size, burrow overlap, the sharpness of the fabric boundaries, and the existence of internal bedding, thereby identifying gradations of bioturbation ranging from "no bioturbation" to "complete bioturbation" [*Taylor and Goldring*, 1993]. Regarding the recognition of diagenetic components, we particularly focused on their chronological sequences of appearance after deposition of the initial carbonate mud.

9.3.3. Sampling Strategy and Carbonate Carbon and Oxygen Isotope Analysis

The polished rock fragments remaining after thin section preparation were cleaned and polished with 1 μm abrasive to remove all remains of the carbon coating used for SEM imaging. A grid with equally spaced intervals of 10 mm was used for targeting sample areas for isotope analysis (i.e., blind sampling), yielding 25 samples of ~2 mg per site. Sampling was performed with a Dremel 8100® and a spherical-shaped diamond-impregnated drill bit ($\varnothing$ = 2 mm). To avoid sampling of the epoxy resin, sampling sites deviated slightly from the prescribed grid, in some instances (especially notable for sample Zl 18a). The initially obtained material was discarded to account for contamination of the sample surface and contamination on the drill bit. Drilled powders were transferred to glass reaction vessels (LABCO®) and flushed with helium. The carbonate was left to react with 50 μl of anhydrous phosphoric acid (±102%) for at least 1.5 h. An IsoPrime triple collector isotope ratio mass spectrometer (Department of Geosciences and Natural Resource Management, University of Copenhagen) in continuous flow setup was used to measure carbon and oxygen isotope composition of CO_2 generated by acid digestion. IAEA standards (NBS-18 and NBS-19) were used to calibrate pure CO_2 (99.995%), subsequently used as a reference gas for the carbonate carbon and oxygen isotope measurements. Results were corrected for weight-dependent isotope ratio bias using multiple measurements of an in-house standard LEO (Carrara marble, $\delta^{13}C = 1.96‰$ and $\delta^{18}O = -1.93‰$), covering the entire range of signal intensities encountered in the samples. External reproducibility was monitored by replicate analysis of the in-house standard LEO. Long-term accuracy was better than 0.1‰ (2 standard deviations) for $\delta^{13}C$ and better than 0.2‰ (2 sd) for $\delta^{18}O$. All isotope values are reported in ‰ relative to VPDB and in standard δ-notation.

9.4. CARBONATE MICROFACIES

Argillaceous red, nodular, bioturbated wackestone (Zl 11) contains abundant bioclasts: ammonoids, ostracod valves, echinoderms, and shell fragments of different sizes (Figs. 9.3a, 9.6a–d, and S1). Ostracod valves are either filled by micritic or sparitic cements (Fig. 9.7a). The matrix is distinct by an absence of internal bedding and well-defined boundaries on the interface between burrow and matrix (Fig. 9.7c), suggesting high bioturbation. The combined sedimentary features and biogen assemblage are suggestive of deposition under fully marine and oxygenated conditions, below the storm wave base, in an outer ramp-basinal setting. Although the nodular fabric of the unit might suggest exhumation of the early lithified carbonate mud by weak currents under episodically sediment-starved conditions, followed by encrustation and coating with Mn/Fe-(oxyhydr)oxide precipitates [*Leda et al.*, 2014]. Parts of the matrix contain microspar and inequigranular dolomite with a morphology ranging from anhedral to subhedral (Fig. 9.7d). Diagenetic textures include thin microcracks, spar-filled cavities, and calcite veins which traverse the matrix. The calcite veins crosscut each other, suggesting temporally distinct generations of postdepositional carbonate addition.

Gray/yellow nodular sponge wackestone (Zl 18a) is conspicuous because of the unusual high accumulation of sponge spicules comparative to other intervals of the studied sequence (Figs. 9.4a, 9.7e, f, and S2). Besides sponge spicules, the sample contains gastropod remains (bellerophontids), spar-filled ostracod tests, and unidentifiable shell fragments. The sponge remains do not occur in life position, suggesting some transport or reworking. The amorphous silica of the sponge spicules appears to have been replaced by calcite [*Leda et al.*, 2014]. It cannot conclusively be determined whether the observed sediment reworking is related to animal activity, although if connected it could be classified as low bioturbation, based on an absence of burrows. The indices are in agreement with an outer ramp-basinal setting, often inferred for these mass occurrences of sponge spicules [*Flügel*, 2010, p. 496–497]. A conspicuous area at the upper part of the lower bed is marked by a distinctive yellowish color, whereas the remainder of the matrix is gray. The contact of the color transition is, however,

poorly defined, and individual clots of yellowish matrix can also be discerned at lower stratigraphic positions within the same sample (Fig. 9.4a). The matrix partly constitutes microspar and inequigranular dolomite crystals with an anhedral and subhedral morphology. The yellowish areas are preferentially associated with a fine crystalline matrix and abundant dolomite precipitates and, as such, could suggest a connection with the previous outlined color variation of the rock (Fig. 9.7f). Alternatively, higher Fe-(oxyhydr)oxide contents (e.g., goethite) have been connected with yellow color variations of carbonate rock [*Flügel*, 2010, p. 54]. The matrix is crossed by irregular anastomosing sets of stylolites and contains some voids filled with large sparite crystals.

Laminated bindstone (Zl 23) created by the accumulation of alternating layers of differential-sized grains or crystals and small discontinuity surfaces (Figs. 9.5a and 9.7h). The lamination bounds have a somewhat wrinkly structure, and individual laminae occasionally taper off in a lateral direction or branch. SEM imaging reveals that the larger-sized particles have irregular-shaped boundaries and therefore likely represent secondary mineral growths (Fig. S3). The concordance of these filled vugs to distinct laminae leads us to suggest that these structures comprise occluded fenestral fabrics, possibly associated with microbial mats. The fine-grained planar lamination is likely deposited in a protected inner ramp or outer ramp-basinal setting, below the storm wave base. The near absence of sediment reworking is attributed to the lack of (metazoan) organisms. The matrix seems to exist for a large extent of a microcrystalline spar, with veins cutting through laminae (Fig. 9.8a). Some of the laminae contacts are associated with stylolites. Stylolites, in turn, traverse some of the few observable calcite veins, confirming their chronological superposition in terms of the diagenetic alteration sequence.

Oncoid wackestone/floatstone (Zl 35) has a matrix consisting of grayish clotted micrite traversed by abundant clay seams, and entrained with smaller (10–100 µm) sparite aggregates, relative to the oncoids (500–1000 µm) (Figs. 9.6a and 9.8c, d). The irregular rounded cortex gives the oncoids a cloud-like appearance. The laminae of the cortex are randomly distributed, discontinuous, and overlapping [mode R; *Logan et al.*, 1964]. The recrystallized nucleus takes up more volume than the cortex and consists of large sparite crystals and dolomite with relative large pore spaces (well visible on the BSE image; Fig. S4). The laminae are composed of micrite, which commonly have been associated with the (altered) products of microbial-mediated calcite nucleation (porostromata) or sediment trapping and binding by microbial biofilm (spongiostromata) [type 1; *Flügel*, 2010, p. 127–128]. This particular microbial mode of oncoid formation has been linked with a deepwater setting and low background

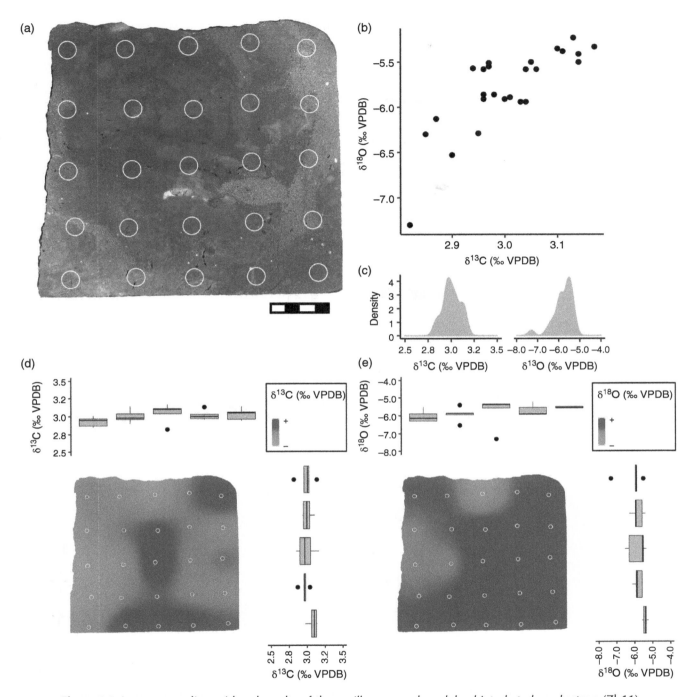

Figure 9.3 Isotope sampling grid and results of the *argillaceous red, nodular, bioturbated wackestone* (Zl 11) (digits within the scale bars represent 2.5 mm). (a) Polished rock slab with sampling grid (circumference of individual drill sites is representative of actual dimensions). (b) Crossplot of δ¹³C versus δ¹⁸O. (c) Probability density distributions for δ¹³C and δ¹⁸O. (d) Spatial grid of δ¹³C and box plots (boxes depict the IQR, and bars within the boxes the median) for cumulative variation across the horizontal and vertical axis. (e) Spatial grid of δ¹⁸O and box plots for cumulative variation across the horizontal and vertical axis. (*See electronic version for color representation of the figure.*)

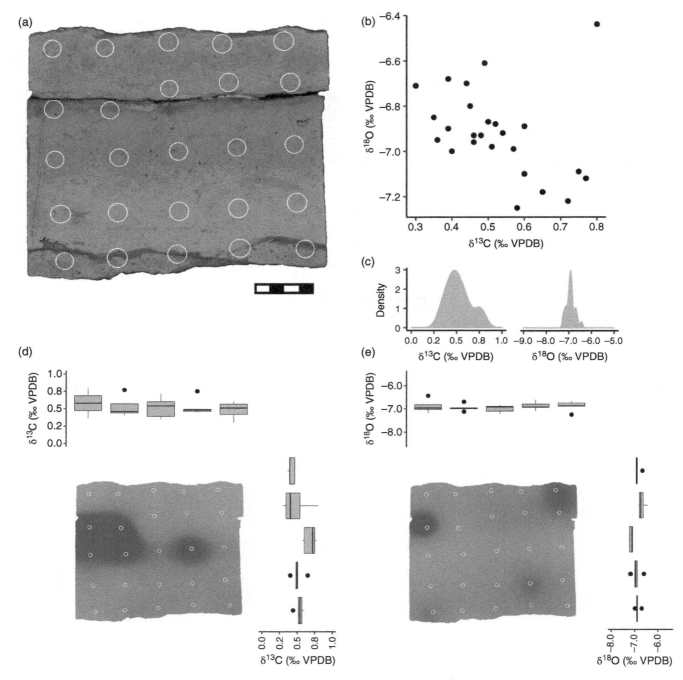

Figure 9.4 Isotope sampling grid and results of the *gray/yellow nodular sponge wackestone* (Zl 18a). See caption and legends of Figure 9.3 for explanations for panels a–e. (*See electronic version for color representation of the figure.*)

sedimentation rates (i.e., the oncoid does not move across the seabed by currents). This aligns with the irregular shape of the structures, corroborating with a low-energy environment, such as a protected inner ramp or an outer ramp-basinal environment [*Peryt*, 1981; *Ratcliffe*, 1988;

Flügel, 2010]. There is a notable absence of metazoan benthic organisms and traces of bioturbation. Hence, the undisturbed substrate possibly enabled the growth of these microbial structures. However, locally hydrographic, ecological, and chemical changes could equally have

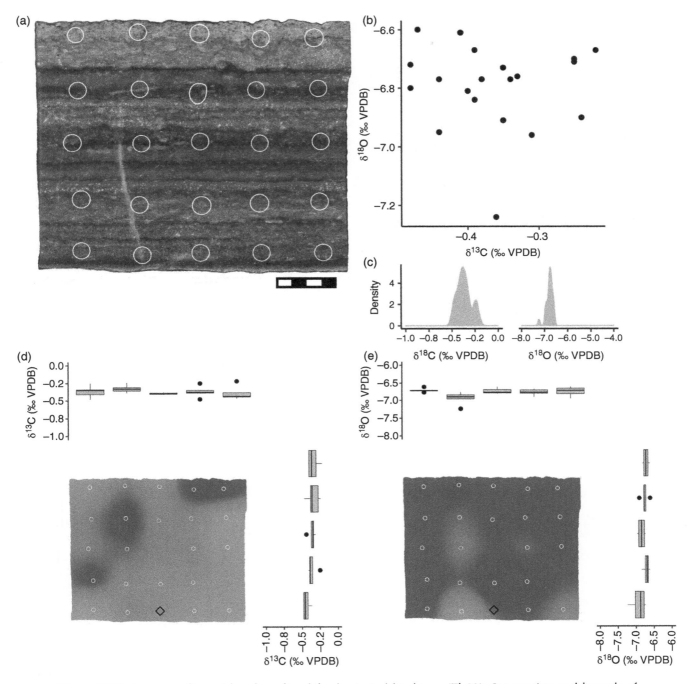

Figure 9.5 Isotope sampling grid and results of the *laminated bindstone* (Zl 23). See caption and legends of Figure 9.3 for explanations for panels a–e. The diamond in panel (d) and (e) demarcates an anomalous isotope value (see text).

stimulated the formation of these microbial seafloor fabrics. A distinct load cast, enveloping an oncoid, can be observed in the polished rock fragment, suggesting soft-sediment deformation by burial before complete lithification of the matrix (Fig. 9.6a). Multiple cracks,

occasionally filled with sparite, run across the sample specimen cutting across the matrix and oncoid cortex; this brittle deformation seems to concentrate in the area underneath of the load cast (Fig. 9.6a). A solid recrystallized oncoid nucleus prior to compaction and deformation

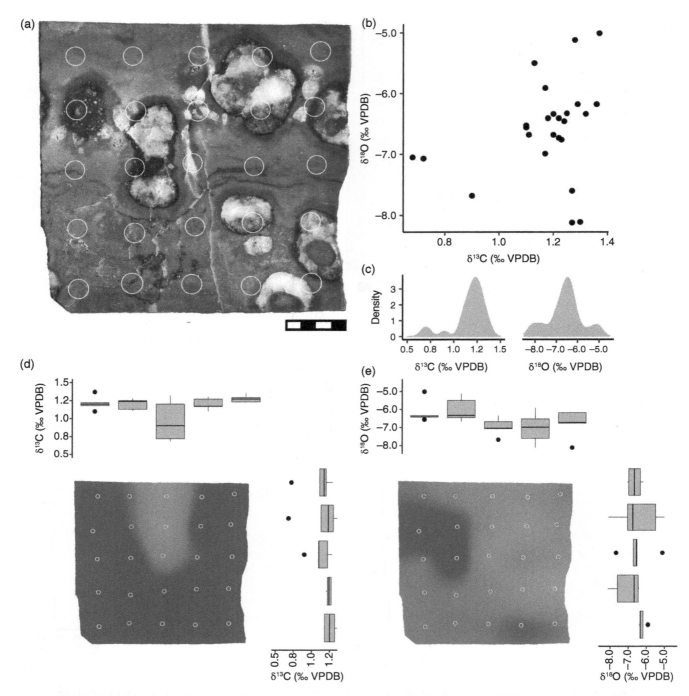

Figure 9.6 Isotope sampling grid and results of the *oncoid wackestone-floatstone* (Zl 35). See caption and legends of Figure 9.3 for explanations for panels a–e.

seems necessitated to explain the combined ductile and brittle deformation of the matrix. Stylolites crosscut and offset the calcite veins (Fig. 9.8b). Combined, these diagenetic features enable the determination of the sequence of events after deposition, with oncoid lithification, nuclei replacement by sparite and dolomite, subsequent intrusion of calcite veins and cracking, and finally stylolitization by pressure solution.

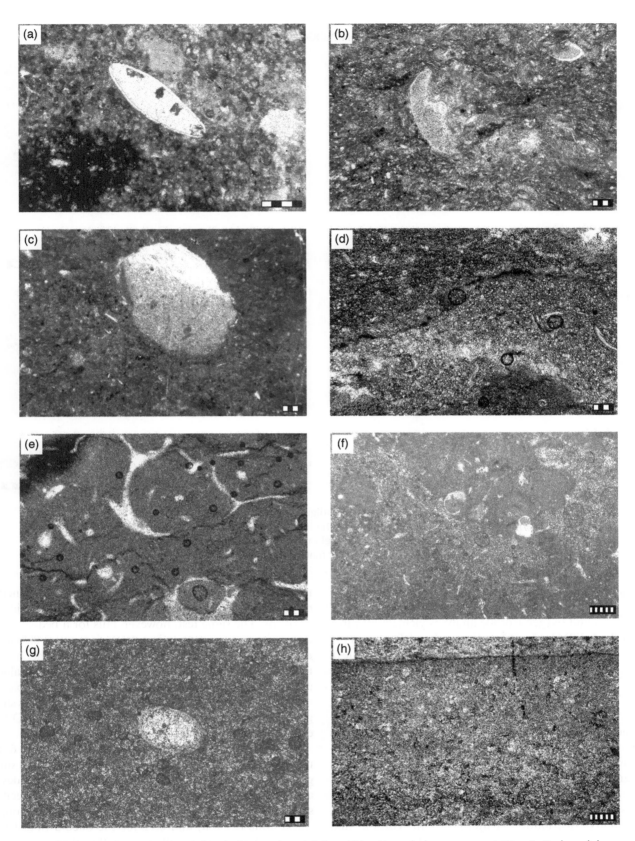

Figure 9.7 Microscopy images of rock thin sections (digits within the scale bars represent 50 μm). *Red, nodular, bioturbated wackestone* (Zl 11) with (a) ostracod test filled with sparite, (b) echinoderm fragment, (c) burrow with a geopetal infill, and (d) a dolomitized section of the matrix. *Gray/yellow nodular sponge wackestone* (Zl 18a) with (e) sponge spicules and (f) a transition from a fine micritic matrix to a more crystalline matrix, corresponding to the color transition from gray to yellow upsection (Fig. 9.4). *Laminated bindstone* (Zl 23) with (g) an ostracod test with a fine crystalline infill and (h) laminae formed by grain size and mineralogy differences and small discontinuity surfaces. (*See electronic version for color representation of the figure.*)

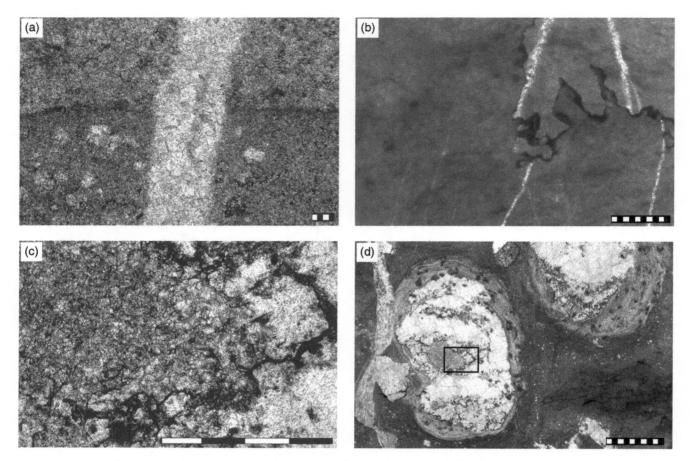

Figure 9.8 Microscopy images of rock thin sections (digits within the scale bars represent 50 μm). *Laminated bindstone* (Zl 23) with (a) a calcite vein crosscutting the internal lamination. *Oncoid floatstone* with (b) calcite veins interrupted and offset by stylolitization, (c) nuclei of oncoid partly consisting of dolomite and large-sized sparite crystals, and (d) oncoids within clotted micritic matrix (the inset corresponds to panel c).

9.5. CARBON AND OXYGEN ISOTOPE RESULTS

Carbon and oxygen isotope analyses yielded 98 values from the 4 different stratigraphic horizons (25 values per rock sample, except for Zl 35 which yielded 23 values). In this description, we exclude one data point of both $\delta^{13}C$ and $\delta^{18}O$ dataset (Zl 23), which stands out with 2.8 and 5.8‰ ^{13}C- and ^{18}O-enriched value, respectively, relative to bed-specific median value.

The pre-extinction nodular limestone (Zl 11) displays the highest $\delta^{13}C$ and $\delta^{18}O$ values, whereas the post-extinction carbonates are all comparatively depleted in ^{13}C and ^{18}O (Fig. 9.1 and Table 9.1 for summary statistics and the online Dataset S1 for the unprocessed data). The bed-specific sample sets do not produce, in all instances, probability distribution patterns (both carbon and oxygen isotope composition) that conform with normality distribution, based on Shapiro-Wilk test ($p < 0.05$) (Figs. 9.3c–9.6c, S5, and S6). Nonetheless, a comparison of the variance in the sample space (Fligner-Killeen test)

throughout the studied sequence suggests homogeneity of the carbon isotope variability among the individual horizons. The same test, however, fails for a bed-specific comparison of the oxygen isotope variance ($p = 0.23$ and $p < 0.05$ for $\delta^{13}C$ and $\delta^{18}O$, respectively). On the other hand, the sample sets for both carbon and oxygen isotope values are statistically distinguishable, based on a rank-based nonparametric Kruskal-Wallis H test ($p < 0.05$). This suggests that both the median $\delta^{13}C$ and the median $\delta^{18}O$ for each limestone bed comprise a distinguishable value.

In agreement with the relative larger bed-internal $\delta^{13}C$ ranges (Table 9.1), a bootstrap sample test (Fig. 9.9) suggests that the probability of obtaining a $\delta^{13}C$, which deviates more than 0.3‰ from the median of the bed-specific sample set, equates to a probability of 0.08 for the sponge wackestone (Zl 18a) and 0.12 for the oncoid floatstone (Zl 35). This test essentially mimics the effect of drilling just one site for every sampled limestone bed and suggests that ~1 out of 10 samples would deviate significantly from the bed-specific median. These previously defined

Table 9.1 Summary Statistics of Carbon and Oxygen Isotope Composition.

Age	Sample ID		Median	IQR (50%)	IPR (95%)	Range
Late Permian	Zl 11	$\delta^{13}C$	3.0	0.1	0.3	0.4
		$\delta^{18}O$	−5.6	0.4	1.5	2.1
Post-extinction latest Permian	Zl 18a	$\delta^{13}C$	0.5	0.2	0.5	0.5
		$\delta^{18}O$	−6.9	0.2	0.7	0.8
Permian-Triassic boundary	Zl 23	$\delta^{13}C$	−0.4	0.1	0.2 (1.7)	0.2 (3.4)
		$\delta^{18}O$	−6.8	0.1	0.5 (3.0)	0.6 (6.3)
Early Triassic	Zl 35	$\delta^{13}C$	1.2	0.1	0.7	0.7
		$\delta^{18}O$	−6.5	0.7	3.0	3.0

Values in brackets demarcate deviating results of summary statistics upon inclusion of a single anomalous data point.

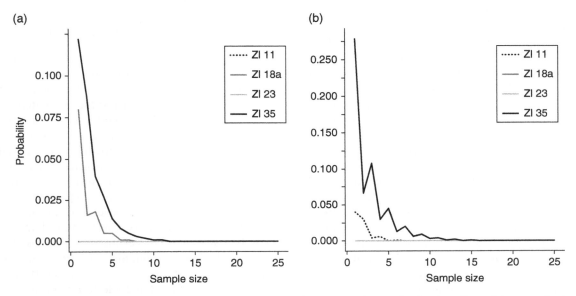

Figure 9.9 Bootstrap sampling exercise comprising multiple routines that probe for the effect of the sample size (i.e., the number of samples analyzed per limestone bed) on the median within-bed isotope value. In other words, this approach mimics the effect of drilling individual subsamples from limestone beds (i.e., 1, 2, ..., 25 drilled samples per stratigraphic horizon) on the obtained median carbon (a) and oxygen (b) isotope value. The cutoff values are based on whether a carbon isotope value deviates 0.3‰ from the bed-specific median and estimates the probability that a randomly chosen within-bed isotope value is therefore not representative for the whole rock value. The cutoff value is chosen to approximate the smallest wiggles used in carbon isotope chemostratigraphic studies (Section 9.1.3). A cutoff value of 1‰ was chosen for bed-internal $\delta^{18}O$, or a 4 °C difference of the fluid, based on the equilibrium fractionation.

deviations are relevant, as some of the isotope stratigraphic markers defined by independent chemostratigraphic studies have the same range (Fig. 9.2d). On the other hand, increasing the sample size of the averaged value (i.e., virtually increasing the number of drilled samples per bed) diminishes the probability of obtaining such anomalous $\delta^{13}C$ values, and such deviations are eliminated with samples sizes larger than 5 and 10 for Zl 18a and Zl 35, respectively (Fig. 9.9). However, the small $\delta^{13}C$ ranges for the nodular limestone (Zl 11) and the laminated bindstone (Zl 23) do not exhibit such deviations from the bed-internal median, as their value range is too small.

The bed-specific oxygen isotope composition is marked by a high interbed variability comparative to the $\delta^{13}C$ of sampled limestone beds (Table 9.1). Furthermore, a similar constructed bootstrap sample test suggests that the probability of sampling a $\delta^{18}O$ that deviates more than 1.0‰ [or a 4 °C difference of the solution when taken as a face value; *Kim and Neil*, 1997] from the median of the specific limestone bed is 0.33 and 0.08 for the nodular limestone (Zl 11) and the oncoid floatstone (Zl 35), respectively. Smaller value ranges of the sponge wackestone (Zl 18a) and the oncoid floatstone (Zl 35) (Fig. 9.9) cannot create such deviations from the median bed-internal $\delta^{18}O$. Spearman's rank correlation further reveals a strong correlation between $\delta^{13}C$ and $\delta^{18}O$ for Zl 11 ($\rho = 0.78$), but only weak correlations exist for the remainder of the samples ($-0.50 \leq \rho \leq 0.50$) (Figs. 9.3b–9.6b).

9.6. DISCUSSION

9.6.1. Controls on Limestone Formation: Local Versus Allochthonous Sources

The trajectories of carbon and oxygen isotope modification under progressive carbonate stabilization have been studied extensively [*Brand and Veizer*, 1980, 1981; *Marshall*, 1992]. Even the formation of micritic limestones, and some of their crucial building blocks (micrite and microspar), is still not fully understood [*Munnecke and Samtleben*, 1996; *Melim et al.*, 2002; *Lucia*, 2017]. These components will, however, ultimately determine the measured bulk rock C and O isotope composition. This is based on a straightforward observation; the large drill site compared to most of the observed biogenic and diagenetic textures ensures predominant excavation of the matrix material (Figs. 9.3a–9.6a). The oncoid wacke/floatstone with a more heterogeneous spatial distribution of different textures seems to be the exception to this generalized observation (Fig. 9.6a). The formation of micrite crystals was originally thought to be the product of a two-stepped process of early diagenetic cementation (with allochthonous dissolved carbonate sources) and aggrading neomorphism in the burial diagenetic realm [*Folk*, 1965]. However, more recent studies suggest the contrary and view micrite and microspar as similar components that only differ in size. In this view, a temporally discrete dissolution-crystallization process results in early diagenetic cementation with carbonate sourced from selective (albeit partial) dissolution of unstable argonite and high-Mg calcite as well as the smallest size fraction of the precursor components [*Lucia*, 2017].

Following the previously discussed controversies, many of the proposed mechanisms explaining the chemical and textural alteration of ancient limestone have been skewed toward a lithification mode with allochthonous dissolved carbonate sources, notably meteoric water [*Bathurst*, 1975; *Brand and Veizer*, 1980, 1981; *Bathurst*, 1993]. This was presumably fueled by numerous studies on Quaternary carbonate platforms (e.g., the Bahamas) which were subjected to rapid eustatic sea-level changes caused by glacial-interglacial cycles [*Bathurst*, 1975, 1993]. This might not always be representative for the ancient situation [*Munnecke and Samtleben*, 1996; *Melim et al.*, 2002]. Proposed closed systems of carbonate stabilization agree with multicomponent studies (micritic rock vs. brachiopod calcite/conodont apatite) on the Permian and Triassic strata, which suggest a potential preservation of

relative trends in bulk rock carbon and oxygen isotope composition [*Kearsey et al.*, 2009; *Brand et al.*, 2012a; *Schobben et al.*, 2016]. In line with these findings, current samples are congruent with a deepwater facies without petrological indices of meteoric influenced alteration. Stratigraphically more expanded petrological studies on the same site found also no indices for meteoric vadose or marine vadose diagenesis [*Horacek et al.*, 2007; *Richoz et al.*, 2010; *Leda et al.*, 2014]. The lack of those diagnostic features, together with an overall sea-level rise during the latest Permian to Early Triassic [Figure 1 and *Hallam and Wignall*, 1999], lends further support to a diagenetic model that does *not* rely on freshwater sources.

9.6.2. Within-Bed Carbon Isotope Variations

Within-bed carbon isotope variability can roughly divide the rock samples in groups with either a narrow or a broad $\delta^{13}C$ distribution (Fig. 9.9). The pre-extinction nodular wackestone (Zl 11) and the laminated bindstone (Zl 23) exhibit the smallest bed-internal carbon isotope variation. However, their petrographic features suggest widely different modes of deposition, that is, well-mixed versus laminated, as well as lithification to a micritic compared to a crystalline matrix, respectively. The original lime mud of the nodular limestone is presumably predominantly detrital, derived of skeleton disintegration by an active infaunal community. On the other hand, the laminated bindstone contains elements that invoke formation by a microbial mat community. This lime mud could then have been derived from detrital sources (trapping and bindings by a biofilm) but could be an authigenic product (metabolism-induced calcification) [*Riding*, 2000; *Dupraz and Visscher*, 2005]. After exclusion of three anomalous $\delta^{13}C$ values introduced by a calcite vein (visible in the spatial grid of Fig. 9.6e), the oncoid floatstone (Zl 35) exhibits a limited range of $\delta^{13}C$ values, again in conjunction with a texture signified by microbial contributions. The recrystallized nucleus of the oncoids does not seem to be a source for differences in bed-internal $\delta^{13}C$, even though these structures have contributed significantly to the retrieved volume of some of the samples (Fig. 9.6a). This result is, however, in agreement with evidence for a temporally discrete event of early dissolution and subsequential cementation of the oncoid core based on the petrological investigation (Section 9.4). Hence, the surrounding matrix provided carbonate for early cementation, which therefore presumably explains the invariant within-bed $\delta^{13}C$.

The post-extinction Permian sponge wackestone bed (Zl 18a) contains the broadest within-bed $\delta^{13}C$ distribution (entirely contributed to the matrix), where the yellow coarse-grained matrix aligns with enriched $\delta^{13}C$ values, comparative to the remainder of the sample surface (Figs. 9.5c, d, and 9.7f). As the lateral $\delta^{13}C$ anomaly is bedding conform, it appears that the original rock-internal $\delta^{13}C$ variation is preserved and can possibly be attributed to the less bioturbated matrix. In addition, a notable accumulation of dolomite (possibly connected with the yellowish coloration) occurs at the same stratigraphic height and might point to a polymorph-controlled carbon isotope signature (Fig. 9.5a). The anhedral to subhedral morphology of these dolomite precipitates might allude to precursor particles overgrown by a secondary cement, without a temporally discrete dissolution event [*Lucia*, 2017]. This connects dolomitization with an early diagenetic process, where it precipitates directly underneath of the sediment-water interface in the vicinity of sulfate-reducing microbes [*Vasconcelos et al.*, 1995; *Wacey et al.*, 2007; *Bontognali et al.*, 2014]. The carbon isotope composition of dolomite precipitated in equilibrium is heavier than equivalent calcite [*Sheppard and Schwarcz*, 1970], whereas microbial-mediated dolomite displays often varying degrees of ^{13}C depletion and enrichment, controlled by DIC sourced from degraded organic matter [*Meister et al.*, 2007; *Wacey et al.*, 2007].

9.6.3. Within-Bed Oxygen Isotope Variations

In the absence of meteoric water, the recorded low $\delta^{18}O$ (Table 9.1) must correspond to either ^{18}O-depleted sources, for instance, the interaction of pore fluids with clay minerals [*Lawrence et al.*, 1977], or precipitation under elevated (burial) temperatures [*Irwin et al.*, 1977; *Scholle*, 1977]. As a quantitative significant external source is absent (e.g., a predominance of carbonate sediments), a thermal origin is the more likely mechanism of bulk rock ^{18}O depletion. This suggests that diagenesis continues after initial stabilization, but whereas carbon isotope alteration seizes, oxygen isotope alteration is increased during these later stages of burial diagenesis, and owes to the large temperature dependence of the equilibrium O isotope fractionation [*Kim and Neil*, 1997], as opposed to the slight to virtually independent $\delta^{13}C$ variations with changing temperatures inferred for the equilibrium C isotope fractionation [*Emrich et al.*, 1970; *Romanek et al.*, 1992].

The bed-internal oxygen isotope content of different samples exhibits opposing systematics; the nodular limestone (Zl 11) records high oxygen isotope variability that correlates with the carbon isotope variations. The latter contrasts with the narrower $\delta^{13}C$ range and the absence of a systematic C-O isotope relation as exhibited by the laminated bindstone and the sponge wackestone. However, these patterns of bed-internal O isotope variability for different sampled bedding horizons likely equate to other sources than cited for the overall ^{18}O depletion. For example, the effect of polymorphism and polymorph-specific equilibrium isotope fractionations, such as the calcite-aragonite O isotope offset which has a value of 0.8–4.5‰ (at 25 °C) [*Tarutani et al.*, 1969] and mineralogy and mineral-specific equilibrium isotope fractionations, such as the dolomite-calcite O isotope offset which has a value of 0.7–6.8‰ (at 25 °C) [*O'Neil and Epstein*, 1966; *Zheng*, 1999]. Nonetheless, in natural environments, dolomite-calcite $\delta^{18}O$ offsets range up to several per mil [*Montanez and Read*, 1992; *Wacey et al.*, 2007; *Brand et al.*, 2012b] and could therefore explain the minimal but pronounced ^{18}O-depleted band at the dolomite-enriched layer of the sponge wackestone (Zl 18a). The diagenetically altered components can inherit the original mineral- and polymorph-specific isotope value [*Zheng*, 1999; *Brand et al.*, 2012a, 2012b], so large bed-internal variation and textural-specific O isotope signatures of the oncoid wackestone (Zl 35) can allude to precursor minerals. Furthermore, initial lithification does only occlude a fraction of the initial porosity [*Frank and Bernet*, 2000; *Lucia*, 2017]. Hence, ^{18}O-depleted carbonate can be added to the carbonate rock as intraparticle and intergranular burial cement derived from pressure solution under increasing overburden of the sediment pile [*Frank and Bernet*, 2000]. Temporally distinct additions of carbonate cement (as noted for Zl 11, with superimposed generations of calcite veins; Section 9.4) under increasing burial temperatures could, in such scenario, be an additional source of $\delta^{18}O$ variance. The latter observation invokes that the spatial distribution of porosity might be a driving factor behind bed-internal oxygen isotope variability. Essentially, we can conclude with some certainty that differences in bed-internal oxygen isotope variance result from the spatial distribution of mineralogical assemblages and/or physical properties of the partially lithified precursor sediment.

9.6.4. A Model for P-Tr Carbonate Diagenesis and Isotope Signal Retention

Our observations can be combined in a parsimonious scenario in which C-O systematics of the pre-extinction nodular limestone (Zl 11) corresponds with a mosaic of carbonate polymorphs produced by a range of mainly obligate calcifiers. Phase-specific porosity, crystal size, and the original polymorphs constrain early chemical cementation, resulting in roughly linear trajectories of C isotope imprinting. Whether the average C isotope composition of micrite compounds are translatable on a one-to-one basis to marine DIC conditions is debated, and component-specific isotope offsets (bulk rock vs. brachiopod shell) are interpreted as a function of the predominant polymorph and related specific isotope shifts [*Brand et al.*, 2012a] or weak diagenetic overprinting steered by the original organic carbon content of the sediment [*Schobben et al.*, 2016]. In conjunction with these notions, bed-internal variations in carbon isotope composition would in such scenario either correspond to the spatial distributions of the original polymorph assemblages and their respective C isotope signal or to carbonate component-specific porosity and diagenetic potential. Upon entering the burial diagenetic domain, both the precursor polymorph assemblage distribution and the spatial differential distribution of porosity (as voids created by burrowers, partial exhumation, or texture-specific internal structures) could have defined spatial pattern of oxygen isotope composition. The former spatial differentiation would then be a function of the differential retention of the original polymorph-specific O isotope signal or the relative volume to accommodate ^{18}O-depleted burial cements. We deem a scenario of lime mud compositional control on burial diagenesis likely, as it could as well explain the strong correlation between rock-internal C and O isotope variability (Fig. 9.3b), where ^{13}C-depleted sections comprise patches of a distinct textures with an elevated diagenetic potential.

In the same scenario, a less diverse mineralogic- and polymorphic assemblage of carbonate mud grains by non-obligate calcifying microbes yields rock with a more homogenized diagenetic potential and developmental trajectories. Microbial-induced carbonate production lacks a strong control on the calcification process, and local water chemistry controls the composition of the final precipitate, presumably aragonite for the Late Permian to Early Triassic [*Lowenstein et al.*, 2001; *Brand et al.*, 2012a]. This translates into homogeneous crystalline textures and suppressed $\delta^{18}O$ variation. These lithologies and corresponding internal C-O systematics fit in a post-extinction seafloor ecosystem where many metazoan organisms had been removed, creating a niche for microbial communities [*Baud et al.*, 1997; *Woods et al.*, 1999; *Pruss and Bottjer*, 2004; *Baud et al.*, 2007; *Woods*, 2009; *Kershaw et al.*, 2012].

9.6.5. The Effect of Sampling Strategies on P-Tr Chemostratigraphy

The combined spatial isotope sampling approach and petrological investigation revealed two distinct within-bed carbon isotope anomalies (Zl 18a and Zl 35), which have no association with water column DIC isotope composition. When accepting the overall high retention capability of carbon isotope composition in the here-studied limestone, these samples would have the potential to introduce a bias of more than 0.3‰ relative to the overall bed-internal median in 1 out of 10 drilled samples (Fig. 9.9). Such deviations have been classified as isotope stratigraphic markers when included in a temporal carbon isotope sequence (Section 9.1.3 and Fig. 9.2). This observation proves that we should be more critical in evaluating our C isotope data when constructing chemostratigraphic frameworks. Nonetheless, one can argue that some of the here-presented $\delta^{13}C$ variability could be easily circumvented under normal sampling conditions. For example, the ^{13}C-depleted calcite vein of the oncoid wackestone (Zl 35) would have been recognized, even without a polished rock surface, and likely omitted from any C isotope chemostratigraphic study. Such obvious exclusion criteria become less clear with the sponge wackestone (Zl 18a), where the yellow horizon does not conform to an easily recognizable diagenetic artifact. The ^{13}C-enriched bed-internal band of this horizon appears to be a clear example of polymorph (or perhaps microbial) controlled C isotope deviations and might not correspond to an oceanic DIC-C isotope value [*Brand et al.*, 2012a, 2012b]. Also striking is the absence of the internal variation in some of the more obvious diagenetic textural indices, that is, the large crystals inhabiting the core of the oncoids (Figs. 9.6a and 9.8c, d). Such structures would presumably be avoided in chemostratigraphic studies, but these textures do not yield anomalous $\delta^{13}C$ values. The conclusion is that even with a careful petrological examination, it is not always possible to conclude whether textures yield globally representative DIC-C isotope values. And inclusions of such unrepresentative C isotope values in past chemostratigraphic studies might have skewed temporal trends in bulk carbonate based C isotope record.

9.6.6. Recommendations for Future Sampling Strategies

As a more positive note, the current study does provide some means to at least reduce the chance of incorporating unrepresentative C isotope values in chemostratigraphic studies. The probability estimates of Figure 9.9 suggest that the chance of obtaining DIC-unrepresentative values can be reduced by additional sampling (i.e., drilling) of each individual limestone bed. Hence, an increased number of within-bed subsamples could make relative variations in the temporal $\delta^{13}C$ records a more reliable tracer of marine DIC-$\delta^{13}C$ variations. Nonetheless, the odds of sampling less variable C isotope values would have decreased if sampling of calcite veins would have been avoided (Zl 35) and proves the value of a careful visual examination before sample processing for isotope analysis, at least, to some degree. Based on this small study, we can make some further tentative suggestions for future sampling strategies:

1. A chemostratigraphic study would benefit from analyzing the C isotope composition of at least 5, but even better, more than 10 drilled samples per bedding horizon, thereby reducing the chance of sampling marine DIC-unrepresentative $\delta^{13}C$ values (Fig. 9.9).

2. Alternatively, the ground material obtained from multiple drill sites (on one limestone bed) could be homogenized and subsequently subsampled before performing the isotope analysis. This method would be less costly and labor intensive than the former proposed method.

3. Lastly, a representative volume of the matrix material could be derived from trimming pieces of larger rock fragments with a saw. After this, the material is ground and homogenized before the isotope measurement. However, this method appears less accurate in avoiding obvious diagenetic textures.

Studies that rely on bed-internal stratigraphic trends would benefit from sampling bedding-conform lateral transects to decrease the chance of sampling DIC-unrepresentative C isotope values. The value of such detailed chemostratigraphic studies varies on the lithology, as a well-mixed carbonate mud (Zl 11) seems to produce a within-bed homogenized $\delta^{13}C$ value. On the other hand, the current investigation confirms the existence of within-bed stratigraphic $\delta^{13}C$ rhythms, which can be created by an early diagenetic mode of carbonate mineral formation. Hence, both the inclusion of carbonate rock with signs of intense mixing and authigenic (microbial-steered) carbonate addition should be viewed with caution and warrant a petrological investigation prior to high-resolution carbon isotope chemostratigraphic studies.

The carbon and oxygen isotope date and supporting information can be found on the Zenodo repository (DOI: 10.5281/zenodo.1470697).

ACKNOWLEDGMENTS

The authors like to thank Harri Wyn Williams (University of Leeds) for his help in preparing the mounted and polished rock slabs, Hans-Rudolf Knöfler (Museum für Naturkunde, Berlin) for making thin sections, Kirsten Born (Museum für Naturkunde, Berlin) for performing the SEM analyses, and Bo Petersen (University of Copenhagen) for conducting the isotope measurements. The study benefited from support from the Deutsche Forschungsgemeinschaft (DFG) (projects KO1829/12-1, KO1829/12-2, KO2011/8-1, KO1829/18-1, and FOR 2332); MS is currently funded by a DFG Research Fellowship (SCHO 1689/1-1).

REFERENCES

Aghai, P. M., D. Vachard, and K. Krainer (2009), Transported foraminifera in Palaeozoic deep red nodular limestones exemplified by latest Permian, Revista Espanola de Micropaleontologia, 41(1–2), 197–213.

Al-Aasm, I. S., and J. Veizer (1986), Diagenetic stabilization of aragonite and low-Mg calcite, II. Stable isotopes in rudists. Journal of Sedimentary Research, 56(6), 763–770.

Algeo, T., C. M. Henderson, B. Ellwood, H. Rowe, E. Elswick, S. Bates, T. Lyons, J. C. Hower, C. Smith, B. Maynard, L. E. Hays, R. E. Summons, J. Fulton, and K. H. Freeman (2012), Evidence for a diachronous Late Permian marine crisis from the Canadian Arctic region, Geological Society of America Bulletin, 124(9–10), 1424–1448, doi:10.1130/B30505.1.

Allwood, A. C., M. R. Walter, B. S. Kamber, C. P. Marshall, and I. W. Burch (2006), Stromatolite reef from the Early Archaean era of Australia, Nature, 441(7094), 714–718, doi:10.1038/nature04764.

Auclair, A. C., M. M. Joachimski, and C. Lécuyer (2003), Deciphering kinetic, metabolic and environmental controls on stable isotope fractionations between seawater and the shell of Terebratalia transversa (Brachiopoda), Chemical Geology, 202(1–2), 59–78, doi:10.1016/S0009-2541(03)00233-X.

Bambach, R. K., A. H. Knoll, and S. C. Wang (2004), Origination, extinction, and mass depletions of marine diversity. Paleobiology, 30(4), 522–542, doi:10.1666/0094-8373(2004)030⟨0522:OEAMDO⟩2.0.CO;2.

Banner, J. L., and G. N. Hanson (1990), Calculation of simultaneous isotopic and trace element variations during water-rock interaction with applications to carbonate diagenesis, Geochimica et Cosmochimica Acta, 54(11), 3123–3137, doi:10.1016/0016-7037(90)90128-8.

Bathurst, R. G. C. (1975), Cementation, in Developments in Sedimentology 12: Carbonate Sediments and Their Diagenesis, chap. 10, pp. 415–457, Elsevier, Amsterdam, London, New York, Tokyo, doi:10.1016/S0070-4571(08)70904-X.

Bathurst, R. G. C. (1993), Microfabrics in carbonate diagenesis: A critical look at forty years in research, in Carbonate Microfabrics, chap. 1, edited by R. Rezak and D. L. Lavoie, pp. 3–14, Springer-Verlag, New York.

Baud, A., M. Magaritz, and W. T. Holser (1989), Permian-Triassic of the Tethys: Carbon isotope studies, Geologische Rundschau, 78(2), 649–677.

Baud, A., S. Cirilli, and J. Marcoux (1997), Biotic response to mass extinction: The lowermost Triassic microbialites, Facies, 36, 238–242.

Baud, A., S. Pruss, and S. Richoz (2007), The lower Triassic anachronistic carbonate facies in space and time, Global and Planetary Change, 55(1–3), 81–89.

Black, B. A., L. T. Elkins-Tanton, M. C. Rowe, and I. U. Peate (2012), Magnitude and consequences of volatile release from the Siberian Traps, Earth and Planetary Science Letters, 317–318, 363–373, doi:10.1016/j.epsl.2011.12. 001.

Bontognali, T. R. R., J. A. Mckenzie, R. J. Warthmann, and C. Vasconcelos (2014), Microbially influenced formation of Mg-calcite and Ca-dolomite in the presence of exopolymeric substances produced by sulphate-reducing bacteria, Terra Nova, 26(1), 72–77, doi:10.1111/ter.12072.

Brand, U., and J. Veizer (1980), Chemical diagenesis of a multicomponent carbonate system-1: Trace elements, Journal of Sedimentary Research, 50(4), 1219–1236, doi:10.1306/212F7BB7-2B24-11D7-8648000102C1865D.

Brand, U., and J. Veizer (1981), Chemical diagenesis of a multicomponent carbonate system-Z: Stable isotopes, Journal of Sedimentary Petrology, 51(3), 987–997, doi:10.1306/212F7DF6-2B24-11D7-8648000102C1865D.

Brand, U., A. Logan, N. Hiller, and J. Richardson (2003), Geochemistry of modern brachiopods: Applications and implications for oceanography and paleoceanography, Chemical Geology, *198*(3–4), 305–334, doi:10.1016/ S0009-2541(03)00032-9.

Brand, U., R. Posenato, R. Came, H. Affek, L. Angiolini, K. Azmy, and E. Farabegoli (2012a), The end-Permian mass extinction: A rapid volcanic CO_2 and CH_4 climatic catastrophe, Chemical Geology, *322*, 121–144, doi:10.1016/ j. chemgeo.2012.06.015.

Brand, U., G. Jiang, K. Azmy, J. Bishop, and I. P. Montañez (2012b), Diagenetic evaluation of a Pennsylvanian carbonate succession (Bird Spring Formation, Arrow Canyon, Nevada, U.S.A.)–1: Brachiopod and whole rock comparison, Chemical Geology, *308–309*, 26–39, doi:10.1016/j.chemgeo. 2012. 03.017.

Burgess, S. D., S. Bowring, and S. Shen (2014), High-precision timeline for Earth's most severe extinction, Proceedings of the National Academy of Sciences, *111*(9), 3316–3321, doi:10.1073/pnas.1317692111.

Busenberg, E., and L. N. Plummer (1989), Thermodynamics of magnesian calcite solid-solutions at 25°C and 1 atm total pressure, Geochimica et Cosmochimica Acta, *53*(6), 1189–1208, doi:10.1016/0016-7037(89)90056-2.

Cao, C., G. D. Love, L. E. Hays, W. Wang, S. Shen, and R. E. Summons (2009), Biogeochemical evidence for euxinic oceans and ecological disturbance presaging the end-Permian mass extinction event, Earth and Planetary Science Letters, *281*(3–4), 188–201, doi:10.1016/j.epsl.2009.02.012.

Cao, C., S. Shen, R. E. Summons, and C. L. K. Colonero (2010), Pattern of $\delta^{13}C_{carb}$ and implications for geological events during the Permian-Triassic transition in South China, Journal of Earth Science, *21*(S1), 118–120, doi:10.1002/gj.1220.

Clapham, M. E., and J. L. Payne (2011), Acidification, anoxia, and extinction: A multiple logistic regression analysis of extinction selectivity during the Middle and Late Permian, Geology, *39*(11), 1059–1062, doi:10.1130/G32230. 1.

Clarkson, M. O., R. A. Wood, S. W. Poulton, S. Richoz, R. Newton, and S. A. Kasemann (2016), Dynamic anoxic-ferruginous conditions during the end-Permian mass extinction and recovery, Nature Communications, *7*, 1–9, doi:10.1038/ ncomms12236.

Dole, M., and R. L. Slobod (1940), Isotopic composition of oxygen in carbonate rocks and iron oxide ores, Journal of the American Chemical Society, *62*(3), 471–479.

Duda, J.-P., M. J. Van Kranendonk, V. Thiel, D. Ionescu, H. Strauss, N. Schafer, and J. Reitner (2016), A rare glimpse of paleoarchean life: Geobiology of an exceptionally preserved microbial mat facies from the 3.4 Ga strelley pool formation, Western Australia, PLoS ONE, *11*(1), e0147629, doi:10.1371/ journal.pone.0147629.

Dupraz, C., and P. T. Visscher (2005), Microbial lithification in marine stromatolites and hypersaline mats, Trends in Microbiology, *13*(9), 429–438, doi:10.1016/j.tim.2005.07.008.

Emiliani, C. (1955), Pleistocene temperatures, The Journal of Geology, *63*(6), 538–578, doi:10.1086/626295.

Emrich, K., D. H. Ehhalt, and J. C. Vogel (1970), Carbon isotope fractionation during the precipitation of calcium carbonate, Earth and Planetary Science Letters, *8*(5), 363–371, doi:10.1016/0012-821X(70)90109-3.

Erwin, D. H. (1993), The Great Paleozoic Crisis. Life and Death in the Permian, 327 pp., Columbia University Press, New York.

Erwin, D. H. (1994), The Permo-Triassic extinction, Nature, *367*(6460), 231–236.

Flessa, K. (2001), Time-averaging, in Palaeobiology II, chap. 3, edited by D. E. Briggs and P. R. Crowther, pp. 292–296, Blackwell Publishing, Malden, Oxford, Victoria, and Berlin.

Flügel, E. (2010), Microfacies of Carbonate Rocks, second ed., 984 pp., Springer-Verlag, Berlin and Heidelberg, doi:10.1007/978-3-662-08726-8.

Folk, R. L. (1965), Some aspects of recrystallization in ancient limestones, in Dolomitization and Limestone Diagenesis, chap. 3, edited by C. P. LLoyd and C. Murray, pp. 14–48, SEPM Special Publication Volume *13*, Society for Sedimentary Geology (SEPM), Broken Arrow, OK, doi:10.2110/ pec.65.07.0014.

Frank, T. D., and K. Bernet (2000), Isotopic signature of burial diagenesis and primary lithological contrasts in periplatform carbonates (Miocene, Great Bahama Bank), Sedimentology, *47*(6), 1119–1134, doi:10.1046/j.1365-3091. 2000.00344.x.

Ghaderi, A., L. Leda, M. Schobben, D. Korn, and A. R. Ashouri (2014), High-resolution stratigraphy of the Changhsingian (Late Permian) successions of NW Iran and the Transcaucasus based on lithological features, conodonts and ammonoids, Fossil Record, *17*(1), 41–57, doi:10.5194/ fr-17-41-2014.

Gruszczynski, M., S. Halas, A. Hoffman, and K. Makowski (1989), A brachiopod calcite record of the oceanic carbon and oxygen isotope shifts at the Permian/Triassic transition, Nature, *337*, 64–68, doi:10.1038/337064a0.

Hallam, A., and P. B. Wignall (1999), Mass extinctions and sea-level changes, Earth Science Reviews, *48*, 217–250, doi:10.1016/S0012-8252(99)00055-0.

Heydari, E., W. J. Wade, and J. Hassanzadeh (2001), Diagenetic origin of carbon and oxygen isotope compositions of Permian-Triassic boundary strata, Sedimentary Geology, *143*, 191–197, doi:10.1016/S0037-0738(01)00095-1.

Heydari, E., N. Arzani, M. Safaei, and J. Hassanzadeh (2013), Ocean's response to a changing climate: Clues from variations in carbonate mineralogy across the Permian-Triassic boundary of the Shareza Section, Iran, Global and Planetary Change, *105*, 79–90, doi:10.1016/j.gloplacha.2012.12.013.

Holser, W. T. (1997), Geochemical events documented in inorganic carbon isotopes, Palaeogeography Palaeoclimatology Palaeoecology, *132*, 173–182.

Holser, W. T., and M. Magaritz (1987), Events near the Permian-Triassic boundary, Modern Geology, *11*, 155–180.

Holser, W. T., H. P. Schönlaub, M. J. Attrep, K. Boeckelmann, P. Klein, M. Magaritz, H. Mauritsch, E. Pak, J.-M. Schramm, K. Stattegger, and R. Schmöller (1989), A unique geochemical record at the Permian/Triassic boundary, Nature, *337*, 39–44.

Horacek, M., S. Richoz, R. Brandner, L. Krystyn, and C. Spötl (2007), Evidence for recurrent changes in Lower Triassic oceanic circulation of the Tethys: The $\delta^{13}C$ record from marine sections in Iran, Palaeogeography, Palaeoclimatology,

Palaeoecology, *252*(1–2), 355–369, doi:10.1016/j.palaeo. 2006.11.052.

Irwin, H., C. Curtis, and M. Coleman (1977), Isotopic evidence for source of diagenetic carbonates formed during burial of organic-rich sediments, Nature, *269*(5625), 209–213, doi:10.1038/269209a0.

Isozaki, Y. (1997), Permo-Triassic boundary superanoxia and stratified superocean: Records from lost deep sea, Science, *276*(5310), 235–238, doi:10.1126/science.276.5310.235.

Jin, Y., Y. Wang, W. Wang, Q. Shang, C. Cao, and D. Erwin (2000), Pattern of marine mass extinction near the Permian-Triassic boundary in South China, Science, *289*(5478), 432–436, doi:10.1126/science.289.5478.432.

Joachimski, M. M., X. Lai, S. Shen, H. Jiang, G. Luo, B. Chen, J. Chen, and Y. Sun (2012), Climate warming in the latest Permian and the Permian-Triassic mass extinction, Geology, *40*(3), 195–198, doi:10.1130/ G32707.1.

Kearsey, T., R. J. Twitchett, G. D. Price, and S. T. Grimes (2009), Isotope excursions and palaeotemperature estimates from the Permian/Triassic boundary in the Southern Alps (Italy), Palaeogeography, Palaeoclimatology, Palaeoecology, *279*(1–2), 29–40, doi:10.1016/j.palaeo.2009.04.015.

Kershaw, S., S. Crasquin, Y. Li, P.-Y. Collin, M.-B. Forel, X. Mu, A. Baud, Y. Wang, S. Xie, F. Maurer, and L. Guo (2012), Microbialites and global environmental change across the Permian-Triassic boundary: A synthesis, Geobiology, *10*(1), 25–47, doi:10.1111/j.1472-4669.2011.00302.x.

Kim, S.-T., and J. R. O. Neil (1997), Equilibrium and nonequilibrium oxygen isotope effects in synthetic carbonates, Geochimica et Cosmochimica Acta *61*(16), 3461–3475.

Knauth, L. P., and M. J. Kennedy (2009), The late Precambrian greening of the Earth, Nature, *460*(7256), 728–32, doi:10.1038/nature08213.

Knoll, A. H., R. K. Bambach, J. L. Payne, S. Pruss, and W. W. Fischer (2007), Paleophysiology and end-Permian mass extinction, Earth and Planetary Science Letters, *256*(3–4), 295–313, doi:10.1016/j.epsl.2007.02.018.

Korte, C., and H. W. Kozur (2010), Carbon-isotope stratigraphy across the Permian-Triassic boundary: A review, Journal of Asian Earth Sciences, *39*(4), 215–235, doi:10.1016/j. jseaes.2010.01.005.

Korte, C., T. Jasper, H. W. Kozur, and J. Veizer (2005), $\delta^{18}O$ and $\delta^{13}C$ of Permian brachiopods: A record of seawater evolution and continental glaciation, Palaeogeography, Palaeoclimatology, Palaeoecology, *224*(4), 333–351, doi:10.1016/ j.palaeo.2005.03.015.

Kozur, H. (2005), Pelagic uppermost Permian and the Permian-Triassic boundary conodonts of Iran. Part II: Investigated sections and evaluation of the conodont faunas, Hallesches Jahrbuch Geowissenschaften, Reihe B, Beiheft, *19*, 49–86.

Kozur, H. (2007), Biostratigraphy and event stratigraphy in Iran around the Permian-Triassic Boundary (PTB): Implications for the causes of the PTB biotic crisis, Global and Planetary Change, *55*(1–3), 155–176, doi:10.1016/j. gloplacha.2006.06.011.

Lawrence, J., J. Drever, T. Anderson, and H. Brueckner (1977), Importance of alteration of volcanic material in the sediments of Deep Sea, Drilling Site 323: Chemistry, $^{18}O/^{16}O$, $^{87}Sr/^{86}Sr$, Geochimica et Cosmochimica Acta, *43*, 573–588.

Leda, L., D. Korn, A. Ghaderi, V. Hairapetian, U. Struck, and W. U. Reimold (2014), Lithostratigraphy and carbonate microfacies across the Permian-Triassic boundary near Julfa (NW Iran) and in the Baghuk Mountains (Central Iran), Facies, *60*(1), 295–325, doi:10.1007/ s10347-013-0366-0.

Li, R., and B. Jones (2017), Diagenetic overprint on negative $\delta^{13}C$ excursions across the Permian/Triassic boundary: A case study from Meishan section, China, Palaeogeography, Palaeoclimatology, Palaeoecology, *468*, 18–33, doi:10.1016/ j.palaeo.2016.11.044.

Logan, B. W., R. Rezak, and R. N. Ginsburg (1964), Classification and environmental significance of Algal stromatolites, The Journal of Geology, *72*(1), 68–83.

Löwemark, L., W.-L. Hong, T.-F. Yui, and G.-W. Hung (2005), A test of different factors influencing the isotopic signal of planktonic foraminifera in surface sediments from the northern South China Sea, Marine Micropaleontology, *55*, 49–62, doi:10.1016/j.marmicro.2005.02.004.

Lowenstam, H. A. (1961), Mineralogy, $^{18}O/^{16}O$ ratios, and strontium and magnesium contents of recent and fossil brachiopods and their bearing on the history of the oceans, The Journal of Geology, *69*(3), 241–260, doi:10.1086/626740.

Lowenstein, T. K., M. N. Timofeeff, S. T. Brennan, L. A. Hardie, and R. V. Demicco (2001), Oscillations in Phanerozoic seawater chemistry: Evidence from fluid inclusions, Science, *294*(5544), 1086–1088, doi:10.1126/science. 1064280.

Lucia, F. J. (2017), Observations on the origin of micrite crystals, Marine and Petroleum Geology, *86*, 823–833, doi:10.1016/j.marpetgeo.2017.06.039.

Luo, G., L. R. Kump, Y. Wang, J. Tong, M. A. Arthur, H. Yang, J. Huang, H. Yin, and S. Xie (2010), Isotopic evidence for an anomalously low oceanic sulfate concentration following end-Permian mass extinction, Earth and Planetary Science Letters, *300*(1–2), 101–111, doi:10.1016/j.epsl.2010.09.041.

Marshall, J. D. (1992), Climatic and oceanographic isotopic signals from the carbonate rock record and their preservation, Geology Magazine, *129*(2), 143–160, doi:10.1017/ S0016756800008244.

Meister, P., J. A. McKenzie, C. Vasconcelos, S. Bernasconi, M. Frank, M. Gutjahr, and D. P. Schrag (2007), Dolomite formation in the dynamic deep biosphere: Results from the Peru Margin, Sedimentology, *54*(5), 1007–1031, doi:10.1111/ j.1365-3091.2007.00870.x.

Melim, L., P. Swart, A. Munnecke, G. Eberli, and H. Westphal (2002), Questioning carbonate diagenetic paradigms: Evidence from the Neogene of the Bahamas, Marine Geology, *185*(1–2), 27–53.

Montanez, I. P., and J. F. Read (1992), Fluid-rock interaction history during stabilization of early dolomites, Upper Knox Group (Lower Ordovician), U.S. Appalachians, Journal of Sedimentary Petrology, *62*(5), 753–778, doi:10.1306/ D42679D3-2B26-11D7-8648000102C1865D.

Munnecke, A., and C. Samtleben (1996), The formation of micritic limestones and the development of limestone-marl alternations in the Silurian of Gotland, Sweden, Facies, *34*, 159–176.

Muttoni, G., A. Zanchi, F. Berra, M. Balini, M. Gaetani, and M. Mattei (2009), The drift history of Iran from the Ordovician to the Triassic, Geological Society, London, Special Publications, *312*(1), 7–29, doi:10.1144/SP312.2.

Newell, N. D. (1962), Paleontological gaps and geochronology, Journal of Paleontology, *36*(3), 592–610.

Newton, R., E. Pevitt, P. Wignall, and S. Bottrell (2004), Large shifts in the isotopic composition of seawater sulphate across the Permo-Triassic boundary in northern Italy, Earth and Planetary Science Letters, *218*(3–4), 331–345, doi:10.1016/S0012-821X(03)00676-9.

Nier, A. O. (1947), A mass spectrometer for isotope and gas analysis, Review of Scientific Instruments, *18*(6), 398–411, doi:10.1063/1.1740961.

O'Neil, J. R., and S. Epstein (1966), Oxygen isotope fractionation in the system dolomite-calcite-carbon dioxide, Science, *152*(3719), 198–201.

Parkinson, D., G. B. Curry, M. Cusack, and A. E. Fallick (2005), Shell structure, patterns and trends of oxygen and carbon stable isotopes in modern brachiopod shells, Chemical Geology, *219*(1–4), 193–235, doi:10.1016/j.chemgeo.2005.02.002.

Payne, J. L., A. V. Turchyn, A. Paytan, D. J. Depaolo, D. J. Lehrmann, M. Yu, and J. Wei (2010), Calcium isotope constraints on the end-Permian mass extinction, Proceedings of the National Academy of Sciences of the United States of America, *107*(19), 8543–8548, doi:10.1073/pnas.0914065107.

Peryt, T. M. (1981), Phanerozoic oncoids: An overview, Facies, *4*, 197–214.

Present, T. M., G. Paris, A. Burke, W. W. Fischer, and J. F. Adkins (2015), Large carbonate associated sulfate isotopic variability between brachiopods, micrite, and other sedimentary components in Late Ordovician strata, Earth and Planetary Science Letters, *432*, 187–198, doi:10.1016/j.epsl.2015.10.005.

Prokoph, A., G. A. Shields, and J. Veizer (2008), Compilation and time-series analysis of a marine carbonate $\delta^{18}O$, $\delta^{13}C$, $^{87}Sr/^{86}Sr$ and $\delta^{34}S$ database through Earth history, Earth-Science Reviews, *87*(3–4), 113–133, doi:10.1016/j.earscirev.2007.12.003.

Pruss, S. B., and D. J. Bottjer (2004), Late Early Triassic microbial reefs of the western United States: A description and model for their deposition in the aftermath of the end-Permian mass extinction, Palaeogeography, Palaeoclimatology, Palaeoecology, *211*(1–2), 127–137, doi:10.1016/j.palaeo.2004.05.002.

Ratcliffe, K. T. (1988), Oncoids as environmental indicators in the Much Wenlock Limestone Formation of the English Midlands, Journal of the Geological Society, *145*(1), 117–124, doi:10.1144/gsjgs.145.1.0117.

Raup, D. M., and J. J. J. Sepkoski (1982), Mass extinctions in the marine fossil record, Science, *215*, 10–12.

Richoz, S., L. Krystyn, A. Baud, R. Brandner, M. Horacek, and P. Mohtat-Aghai (2010), Permian-Triassic boundary interval in the Middle East (Iran and N. Oman): Progressive environmental change from detailed carbonate carbon isotope marine curve and sedimentary evolution, Journal of Asian Earth Sciences, *39*(4), 236–253, doi:10.1016/j.jseaes.2009.12.014.

Riding, R. (2000), Microbial carbonates: The geological record of calcified bacterial-algal mats and biofilms, Sedimentology, *47*(1), 179–214.

Romanek, C. S., E. L. Grossman, and J. W. Morse (1992), Carbon isotopic fractionation in synthetic aragonite and calcite: Effects of temperature and precipitation rate, Geochimica et Cosmochimica Acta, *56*(1), 419–430, doi:10.1016/0016-7037(92)90142-6.

Ruban, D. A., H. Zerfass, and W. Yang (2007), A new hypothesis on the position of the Greater Caucasus Terrane in the Late Palaeozoic-Early Mesozoic based on palaeontologic and lithologic data, Trabajos de Geologia, *27*, 19–27.

Samtleben, C., A. Munnecke, T. Bickert, and J. Pätzold (2001), Shell succession, assemblage and species dependent effects on the C/O-isotopic composition of brachiopods: Examples from the Silurian of Gotland, Chemical Geology, *175*(1–2), 61–107, doi:10.1016/S0009-2541(00)00364-8.

Schindewolf, O. (1954), Über die möglichen Ursachen der grossen erdgeschichtlichen Faunenschnitte, Neues Jahrbuch fuer Geologie und Palaeontologie Monatshefte, *10*, 457–465.

Schobben, M., M. M. Joachimski, D. Korn, L. Leda, and C. Korte (2014), Palaeotethys seawater temperature rise and an intensified hydrological cycle following the end-Permian mass extinction, Gondwana Research, *26*, 675–683, doi:10.1016/j.gr.2013.07.019.

Schobben, M., A. Stebbins, A. Ghaderi, H. Strauss, D. Korn, and C. Korte (2015), Flourishing ocean drives the end-Permian marine mass extinction, Proceedings of the National Academy of Sciences, *112*(33), 10,298–10,303, doi:10.1073/pnas.1503755112.

Schobben, M., C. V. Ullmann, L. Leda, D. Korn, U. Struck, W. U. Reimold, A. Ghaderi, T. J. Algeo, and C. Korte (2016), Discerning primary versus diagenetic signals in carbonate carbon and oxygen isotope records: An example from the Permian-Triassic boundary of Iran, Chemical Geology, *422*, 94–107, doi:10.1016/j.chemgeo.2015.12.013.

Schobben, M., S. van de Velde, J. Gliwa, L. Leda, D. Korn, U. Struck, C. V. Ullmann, V. Hairapetian, A. Ghaderi, C. Korte, R. J. Newton, S. W. Poulton, and P. B. Wignall (2017), Latest Permian carbonate-carbon isotope variability traces heterogeneous organic carbon accumulation and authigenic carbonate formation, Climate of the Past, *13*, 1635–1659, doi:10.5194/cp-13-1635-2017.

Scholle, P. A. (1977), Chalk diagenesis and its relation to petroleum exploration: Oil from chalks, a modern miracle?, American Association of Petroleum Geologists Bulletin, *61*(7), 982–1009.

Scholle, P. A., and M. A. Arthur (1980), Carbon isotope fluctuations in Cretaceous pelagic limestones: Potential stratigraphic and petroleum exploration tool, American Association of Petroleum Geologists Bulletin, *64*(1), 67–87.

Şengör, A. M. C. (1990), A new model for the late Palaeozoic–Mesozoic tectonic evolution of Iran and implications for Oman, Geological Society, London, Special Publications, *49*(1), 797–831, doi:10.1144/GSL.SP.1992.049. 01.49.

Shackleton, N. J., and C. Turner (1967), Correlation between marine and terrestrial pleistocene successions, Nature, *216*, 1079–1082.

Shackleton, N. J., and J. P. Kennett (1976), Paleotemperature history of the Cenozoic and the initiation of Antarctic glaciation; Oxygen and carbon isotope analyses in DSDP sites

277, 279 and 281, Initial Reports of the Deep Sea Drilling Project, *29*, 743–755, doi:10.2973/dsdp.proc.37.1977.

Shen, S.-Z., J. L. Crowley, Y. Wang, S. A. Bowring, D. H. Erwin, P. M. Sadler, C.-Q. Cao, D. H. Rothman, C. M. Henderson, J. Ramezani, H. Zhang, Y. Shen, X.-D. Wang, W. Wang, L. Mu, W.-Z. Li, Y.-G. Tang, X.-L. Liu, L.-J. Liu, Y. Zeng, Y.-F. Jiang, and Y.-G. Jin (2011), Calibrating the end-Permian mass extinction, Science, *334*(6061), 1367–72, doi:10.1126/science.1213454.

Sheppard, S. M. F., and H. P. Schwarcz (1970), Fractionation of carbon and oxygen isotopes and magnesium between coexisting metamorphic calcite and dolomite, Contributions to Mineralogy and Petrology, *26*(3), 161–198.

Song, H., P. B. Wignall, J. Tong, and H. Yin (2012), Two pulses of extinction during the Permian-Triassic crisis, Nature Geoscience, *6*(1), 52–56, doi:10.1038/ngeo1649.

Spero, H., and D. Williams (1988), Extracting environmental information from planktonic foraminiferal ^{13}C data, Nature, *335*, 717–719, doi:10.1038/335717a0.

Stampfli, G. M., and G. Borel (2002), A plate tectonic model for the Paleozoic and Mesozoic constrained by dynamic plate boundaries and restored synthetic oceanic isochrons, Earth and Planetary Science Letters, *196*(1), 17–33.

Stepanov, D., F. Golshani, and J. Stöcklin (1969), Upper Permian and Permian-Triassic boundary in North Iran, Geological Survey of Iran, Report, *12*, 1–73.

Sun, Y., M. M. Joachimski, P. B. Wignall, C. Yan, Y. Chen, H. Jiang, L. Wang, and X. Lai (2012), Lethally hot temperatures during the early Triassic Greenhouse, Science, *338*(6105), 366–370, doi:10.1126/science.1224126.

Svensen, H., S. Planke, A. G. Polozov, N. Schmidbauer, F. Corfu, Y. Y. Podladchikov, and B. Jamtveit (2009), Siberian gas venting and the end-Permian environmental crisis, Earth and Planetary Science Letters, *277*(3–4), 490–500, doi:10.1016/j.epsl.2008.11.015.

Tarutani, T., R. N. Clayton, and T. K. Mayeda (1969), The effect of polymorphism and magnesium substitution on oxygen isotope fractionation between calcium carbonate and water, Geochimica et Cosmochimica Acta, *33*(8), 987–996, doi:10.1016/0016-7037(69)90108-2.

Taylor, A. M., and R. Goldring (1993), Description and analysis of bioturbation and ichnofabric, Journal of the Geological Society, *150*(1), 141–148, doi:10.1144/gsjgs.150.1.0141.

Ullmann, C. V., and C. Korte (2015), Diagenetic alteration in low-Mg calcite from macrofossils: A review, Geological Quarterly, *59*(1), 3–20, doi:10.7306/ gq.1217.

Urey, H. C. (1947), The thermodynamic properties of isotopic substances, Journal of the Chemical Society (Resumed), *(582)*, 562–581.

Vasconcelos, C., J. McKenzie, S. Bernasconi, D. Grujic, and A. J. Tien (1995), Microbial mediation as a possible mechanism for natural dolomite formation at low temperature, Nature, *377*, 220–222.

Veizer, J., and A. Prokoph (2015), Temperatures and oxygen isotopic composition of Phanerozoic oceans, Earth-Science Reviews, *146*, 92–104, doi:10.1016/j.earscirev.2015.03.008.

Veizer, J., D. Ala, K. Azmy, P. Bruckschen, D. Buhl, F. Bruhn, G. A. F. Carden, A. Diener, S. Ebneth, Y. Godderis, T. Jasper, C. Korte, F. Pawellek, O. G. Podlaha, and H. Strauss (1999), ^{87}Sr/^{86}Sr, δ^{13}C and δ^{18}O evolution of Phanerozoic seawater, Chemical Geology, *161*, 59–88, doi:10.1016/S0009-2541(99)00081-9.

Wacey, D., D. T. Wright, and A. J. Boyce (2007), A stable isotope study of microbial dolomite formation in the Coorong Region, South Australia, Chemical Geology, *244*(1–2), 155–174, doi:10.1016/j.chemgeo.2007.06.032.

Wang, Y., P. M. Sadler, S. Z. Shen, D. H. Erwin, Y. C. Zhang, X. D. Wang, W. Wang, J. L. Crowley, and C. M. Henderson (2014), Quantifying the process and abruptness of the end-Permian mass extinction, Paleobiology, *40*(1), 113–129, doi:10.1666/13022.

Wignall, P. B. (2007), The End-Permian mass extinction–how bad did it get? Geobiology, *5*(4), 303–309, doi:10.1111/j.1472-4669.2007.00130.x.

Wignall, P. B., and R. J. Twitchett (1996), Oceanic anoxia and the End Permian mass extinction, Science, *272*(5265), 1155–1158, doi:10.1126/science.272.5265.1155.

Wignall, P. B., and R. Newton (2003), Contrasting deep-water records from the Upper Permian and Lower Triassic of south Tibet and British Columbia: Evidence for a diachronous mass extinction, Palaios, *18*(2), 153–167, doi:10.1669/0883-13 51(2004)019⟨0101:CDRFTU⟩2.0.CO;2.

Woods, A. D. (2009), Anatomy of an anachronistic carbonate platform: Lower Triassic carbonates of the southwestern United States, Australian Journal of Earth Sciences, *56*(6), 825–839, doi:10.1080/08120090903002649.

Woods, A. D., D. J. Bottjer, M. Mutti, and J. Morrison (1999), Lower Triassic large sea-floor carbonate cements: Their origin and a mechanism for the prolonged biotic recovery from the end-Permian mass extinction, Geology, *27*(7), 645, doi:10.1130/0091-7613(1999)027⟨0645:LTLSFC⟩2.3.CO;2.

Yin, H., K. Zhang, J. Tong, Z. Yang, and S. Wu (2001), The global stratotype section and point (GSSP) of the Permian-Triassic boundary, Episodes, *24*(2), 102–114.

Zachos, J., M. Pagani, L. Sloan, E. Thomas, and K. Billups (2001), Trends, rhythms, and aberrations in global climate 65 Ma to present., Science, *292*(5517), 686–693, doi:10.1126/science.1059412.

Zheng, Y.-F. (1999), Oxygen isotope fractionation in carbonate and sulfate minerals, Geochemical Journal, *33*(1978), 109–126, doi:10.2343/geochemj.33.109.

Part IV
Mesozoic

10

Chemostratigraphy Across the Triassic–Jurassic Boundary

Christoph Korte[1], Micha Ruhl[2,3], József Pálfy[4,5], Clemens Vinzenz Ullmann[6], and Stephen Peter Hesselbo[6]

ABSTRACT

The Triassic-Jurassic transition (~201.5 Ma) is marked by one of the largest mass extinctions in Earth's history. This was accompanied by significant perturbations in ocean and atmosphere geochemistry, including the global carbon cycle, as expressed by major fluctuations in carbon isotope ratios. Central Atlantic Magmatic Province (CAMP) volcanism triggered environmental changes and played a key role in this biotic crisis. Biostratigraphic and chronostratigraphic studies link the end-Triassic mass extinction with the early phases of CAMP volcanism, and notable mercury enrichments in geographically distributed marine and continental strata are shown to be coeval with the onset of the extrusive emplacement of CAMP. Sulfuric acid induced atmospheric aerosol clouds from subaerial CAMP volcanism can explain a brief, relatively cool seawater temperature pulse in the mid-paleolatitude Pan-European seaway across the T–J transition. The occurrence of CAMP-induced carbon degassing may explain the overall long-term shift toward much warmer conditions. The effect of CAMP volcanism on seawater $^{87}Sr/^{86}Sr$ values might have been indirect by driving enhanced continental weathering intensity. Changes in ocean-atmosphere geochemistry and associated (causative) effects on paleoclimatic, paleoenvironmental, and paleoceanographic conditions on local, regional, and global scales are however not yet fully constrained.

10.1. INTRODUCTION

The base of the Jurassic system, and therewith the Triassic-Jurassic (T–J) boundary (TJB) (201.36 Ma, *Wotzlaw et al.*, 2014], is defined in the Global Boundary

[1] *Department of Geosciences and Natural Resource Management, University of Copenhagen, Copenhagen, Denmark*

[2] *Department of Geology, Trinity College Dublin, The University of Dublin, Dublin, Ireland*

[3] *Department of Earth Sciences, University of Oxford, Oxford, United Kingdom*

[4] *Deapartment of Geology, Eötvös University, Budapest, Hungary*

[5] *Research Group for Paleontology, Hungarian Academy of Sciences-Hungarian Natural History Museum-Eötvös University, Budapest, Hungary*

[6] *Camborne School of Mines and Environment and Sustainability Institute, University of Exeter, Cornwall, United Kingdom*

Stratotype Section and Point (GSSP) at Kuhjoch (Karwendel Mountains, Northern Calcareous Alps, Tyrol, Austria) [*Hillebrandt et al.*, 2013]. The TJB transition at Kuhjoch and at correlated successions in the same area [see *Hillebrandt et al.*, 2013, for overview] as well as many other coeval successions around the globe [see *Hesselbo et al.*, 2007, for overview] are sedimentologically, paleontologically, and chemostratigraphically well studied (Fig. 10.1). It has been long recognized that one of the most severe mass extinctions in Earth's history, affecting both the marine and continental biota, occurred at this time [e.g., *Newell*, 1967; *Raup and Sepkoski*, 1982; *Hallam and Wignall*, 1997; *McElwain et al.*, 1999, 2007]. The biotic crisis occurred in the latest Triassic (end-Triassic mass extinction) and is radioisotopically dated at 201.564 ± 0.150 Ma [*Blackburn et al.*, 2013; *Davies et al.*, 2017]. Several hypotheses have been proposed for the trigger of this biotic crisis, including extraterrestrial impact [*Olsen et al.*, 1987, 2002; *Morante and Hallam*, 1996],

Chemostratigraphy Across Major Chronological Boundaries, Geophysical Monograph 240, First Edition.
Edited by Alcides N. Sial, Claudio Gaucher, Muthuvairavasamy Ramkumar, and Valderez Pinto Ferreira.

Figure 10.1 Paleogeographic map of discussed and cited T–J boundary sections. Rocks of CAMP are marked in dark red, and the reconstructed CAMP area is colored pale and dark red and taken from *McHone* [2003]. 1: Fundy Basin [*Schoene et al.*, 2006; *Blackburn et al.*, 2013]; 2: Hartford Basin [*Whiteside et al.*, 2010]; 3: Newark Basin [*Whiteside et al.*, 2010; *Marzoli et al.*, 2011; *Schaller et al.*, 2011; *Blackburn et al.*, 2013]; 4: Culpeper Basin [*Marzoli et al.*, 2011]; 5: Argana Basin [*Deenen et al.*, 2010; *Blackburn et al.*, 2013]; 6: High Atlas Basin [*Marzoli et al.*, 2004]; 7: Northern Calcareous Alps [*Kuerschner et al.*, 2007; *Ruhl et al.*, 2009, 2011; *Ruhl and Kürschner*, 2011]; 8: Pelso Unit, Hungary [*Pálfy et al.*, 2001, 2007]; 9: Western Carpathians [*Michalík et al.*, 2007, 2010]; 10: Southern Alps [*Galli et al.*, 2007; *van de Schootbrugge et al.*, 2008; *Bachan et al.*, 2012]; 11: Apennines [*van de Schootbrugge et al.*, 2008; *Bachan et al.*, 2012]; 12: Southern Germany [*van de Schootbrugge et al.*, 2008; *Ruhl and Kürschner*, 2011]; 13: Polish Trough [*Pieńkowski et al.*, 2012]; 14: Northern Germany [*van de Schootbrugge et al.*, 2013]; 15: Danish Basin [*Lindström et al.*, 2012]; 16: Southwest Britain [*Hesselbo et al.*, 2002; *Korte et al.*, 2009; *Clémence et al.*, 2010; *Ruhl et al.*, 2010]; 17: East Greenland [*McElwain et al.*, 1999; *Hesselbo et al.*, 2002]; 18: Queen Charlotte Islands [*Pálfy et al.*, 2000; *Ward et al.*, 2001; *Williford et al.*, 2007; *Friedman et al.*, 2008]; 19: New York Canyon, Nevada [*Guex et al.*, 2004, 2012; *Ward et al.*, 2004; *Schoene et al.*, 2010; *Bartolini et al.*, 2012]; 20: Utcubamba Valley, Peru [*Schaltegger et al.*, 2008; *Schoene et al.*, 2010]; 21: Arroyo Malo, Argentina [*Damborenea and Manceñido*, 2012; *Percival et al.*, 2017]. Map modified after *Ruiz-Martínez et al.* [2012] and *Pálfy and Kocsis* [2014]. (*See electronic version for color representation of the figure.*)

extensive volcanism (triggering a whole range of paleoenvironmental changes including climate warming, euxinia, ocean acidification, etc.) in the Central Atlantic Magmatic Province (CAMP) [e.g., *Marzoli et al.*, 1999; *Schoene et al.*, 2010; *Whiteside et al.*, 2010; *Blackburn et al.*, 2013; *Davies et al.*, 2017; *Percival et al.*, 2017] and associated climate change [*McElwain et al.*, 1999; *van de Schootbrugge et al.*, 2009; *Ruhl et al.*, 2011], increased photic zone anoxia/euxinia and enhanced ocean stratification [*Richoz et al.*, 2012; *Jaraula et al.*, 2013; *Kasprak et al.*, 2015], or

ocean acidification [*Hautmann et al.*, 2008; *Hönisch et al.*, 2012; *Greene et al.*, 2012].

The T–J transition was accompanied by major changes in ocean and atmosphere geochemistry [e.g., *Hallam and Wignall*, 1997; *Pálfy et al.*, 2001; *Cohen and Coe*, 2002, 2007; *Hesselbo et al.*, 2002; *Pálfy*, 2003; *Tanner et al.*, 2004; *McElwain and Punyasena*, 2007; *Hautmann et al.*, 2008; *Kiessling et al.*, 2009; *Kiessling*, 2009; *Schaller et al.*, 2012; *Bottini et al.*, 2016], and strontium isotope data suggest a temporary reversal of the long-term decrease in marine $^{87}Sr/^{86}Sr$ values around the T–J transition [*Veizer et al.*, 1999]. A first indication for a negative carbon isotope excursion (CIE) at the TJB was reported from the Kendelbach section in the Northern Calcareous Alps (NCA) (Austria; Fig. 10.1), but the authors regarded this as likely diagenetically induced [*Hallam and Goodfellow*, 1990]. *McRoberts et al.* [1997] showed a negative $\delta^{13}C$ peak on bulk rock carbonates originating from the Lorüns section of the NCA; however, this was based only on a single sample that yielded a light value. *McElwain et al.* [1999] generated two very light $\delta^{13}C$ values in organic material from low stratigraphic resolution bulk rock samples from the terrestrial Astartekløft succession in Greenland (Fig. 10.1). Combined, these results were suggestive of a potential global carbon cycle perturbation at the T–J transition, and they provided the basis for researchers, including the working group from IGCP project 458 [see *Hesselbo et al.*, 2007], to evaluate the evolution of the global carbon cycle at this time.

Here we review and evaluate temporal fluctuations of isotope ratios and elements in geomaterials across the TJB, contributing to the development of a comprehensive chemostratigraphy that can be applied for transcontinental stratigraphic correlation and understanding the processes causing environmental and climatic perturbations that ultimately led to biotic change and mass extinction.

10.2. THE END-TRIASSIC MASS EXTINCTION AND POTENTIAL CAUSES

The end-Triassic biotic crisis is one of the big five mass extinction events in Earth's history [*Raup and Sepkoski*, 1982]. The taxa affected included ammonoids, scleractinian corals, and conodonts in the oceans [e.g., *Guex et al.*, 2004; *Alroy*, 2010] and megaflora, sporomorphs, and vertebrates on land [e.g., *Benton*, 1995; *McElwain et al.*, 1999, 2009; *van de Schootbrugge et al.*, 2009]. The temporal pattern of extinctions and originations through this event and their causes are controversial, and different scenarios have been suggested.

An early hypothesis for what might have caused the end-Triassic mass extinction is a potential impact of a celestial body. Such an event was proposed for the end-Cretaceous mass extinction after finding an iridium concentration in the event beds at Stevns Klint (Denmark) and Gubbio (Italy), reaching highest values of 3 ng/g at the latter locality [*Alvarez et al.*, 1980]. Iridium is rare in crustal rocks (<0.1 ng/g), but relatively enriched in certain meteorite types (>1 μg/g) [*Ehmann et al.*, 1970; *Crocket and Teruta*, 1977; *Crocket*, 1979]. An enrichment of Ir was also identified in the Grenzmergel of the TJB section at Kendelbach (Austria) and St Audrie's Bay (United Kingdom) [*Orth*, 1989; *McLaren and Goodfellow*, 1990], but with much lower concentrations compared to that

found in Cretaceous-Paleogene (K–Pg) boundary sections, and such small enrichment can also be explained by only invoking volcanic activity [*McCartney et al.*, 1990]. A clear spike of Ir with maximum values of 0.285 ng/g identified by *Olsen et al.* [2002] was found in a white clay layer between typical Triassic and typical Hettangian pollen and spore assemblages in the Jacksonwald syncline section of the Newark Basin. Even this Ir peak is much smaller compared to those of the K–Pg boundary successions [*Alvarez et al.*, 1980], but it is larger than expected from typical crustal concentration. No clear evidence of other indicators for an extraterrestrial impact, such as impact glass (microtektites, tektites), Ni-rich spinels, micro-spherules, or micro-diamonds, has been found so far at the TJB [e.g., *Tanner et al.*, 2004]. Reports of quartz with planar deformation features from the Grenzmergel at the Kendelbach T–J section in Austria [*Badjukov et al.*, 1987] are probably not impact indicators [*Hallam and Wignall*, 1997], but rather metamorphic features [*Mossman et al.*, 1998]. Furthermore, the Manicouagan impact structure of Quebec, with ~100 km diameter that represents one of the largest impacts known in the Phanerozoic [*Grieve*, 1998] and originally favored by *Olsen et al.* [1987] as that responsible for the end-Triassic mass extinction, is more than 10 million years older than the TJB [*Hodych and Dunning*, 1992]. Other Triassic craters [for review see *Tanner et al.*, 2004], such as the 80 km Puchezh-Katunki structure in Russia [*Ivanov*, 1992], are also not of the same age as the end-Triassic mass extinction [*Pálfy*, 2004; *Tanner et al.*, 2004]. In addition, an $^{187}Os/^{188}Os$ decrease that might be used to identify impact-related strata [*Sato et al.*, 2013] is identified already in Late Triassic Norian strata, therefore predating the TJB.

A more accepted scenario for the cause of the end-Triassic biotic crisis is extensive volcanism. It is well documented that the supercontinent Pangea, dominating the late Paleozoic and Triassic paleogeography for at least 100 million years, began to break up around the T–J transition, substantiated by extensive volcanism in the CAMP, which coincided with the opening of the central Atlantic Ocean [*Marzoli et al.*, 1999; *Schlische et al.*, 2003]. This province had an extent of about $1.1 \times 10^7 km^2$ including parts of North and South America, northeast Africa, and southwest Europe (Fig. 10.1; e.g., *Wignall*, 2001a; *McHone*, 2003; *Blackburn et al.*, 2013; *Merle et al.*, 2014; *Pálfy and Kocsis*, 2014]. CAMP comprises about three million cubic kilometers of basaltic lava, including continental flood basalts that flowed into the large rift basins, and mafic dikes and sills intruded into sedimentary deposits [*Marzoli et al.*, 1999, 2018; *McHone*, 2003; *Saunders*, 2005; *Davies et al.*, 2017]. Some CAMP basalt flows, volcanic ashes [*Schoene et al.*, 2010; *Blackburn et al.*, 2013; see also *Tanner et al.*, 2004], or sills

[*Davies et al.*, 2017] have been directly related to the end-Triassic mass extinction. It has been suggested that CAMP activity documented by seismites occurring worldwide and by mafic rocks in Morocco started already before the end-Triassic biotic event [*Dal Corso et al.*, 2014; *Lindström et al.*, 2015].

Certainly, vast amounts of CO_2 were exhaled by CAMP volcanism, and, in addition, CH_4 was likely generated by thermal metamorphism of organic-rich sediments [*McElwain et al.*, 2005; *Svensen et al.*, 2007; *Korte et al.*, 2009]. This CO_2 injection has been suggested to be responsible for an abrupt global warming [e.g., *Blackburn et al.*, 2013]. At TJB sections of the Western Carpathians, a sudden cessation of carbonate deposition of the Fatra Formation is evident at the "boundary shale" facies of the Kopieniec Formation, with an inferred increase of riverine influence. This sudden shift has been taken to indicate a sudden climatic change at the erosional TJB [*Michalík et al.*, 2007], potentially induced by an enhanced hydrological cycle upon rising temperatures. However, strong short-term fluctuations superimposed on an overall eustatic sea-level fall have been suggested based on North American and European successions, possibly linked to glacial eustacy and climatic cooling brought about by sulfur degassing in an early stage of the CAMP [*Guex et al.*, 2004, 2012; *Schoene et al.*, 2010]. However, the finding of photic zone anoxia/euxinia resulting from enhanced ocean stratification [*Kasprak et al.*, 2015], as well as the marine biodiversity drop [*van de Schootbrugge et al.*, 2009], has rather been attributed to warming global climates.

10.3. INTENSELY STUDIED TRIASSIC-JURASSIC BOUNDARY SUCCESSIONS

Geochemical and chemostratigraphic studies of marine and continental TJB successions have been conducted in recent years, including on some previously proposed GSSPs for the base of the Jurassic. Here we present a synoptic stratigraphic framework and discuss some of the outstanding problems (see Fig. 10.1).

10.3.1. Alpine Sections Including the GSSP Kuhjoch Succession (Austria)

The T–J transition has been extensively studied in paleo-marine shelf settings along the passive margin of the northwestern Tethys Ocean. TJB successions in the NCA in particular are stratigraphically expanded and highly fossiliferous and mainly studied in the Eiberg Basin, Austria, a Rhaetian intraplatform depression, which extends for over 200 km from the Salzkammergut in the east to the Lahnewiesgraben valley (Bavaria) in the west [*Hillebrandt et al.*, 2013]. The intraplatform Eiberg

Basin was bordered to the southeast by a broad Rhaetian carbonate platform (Dachstein platform), locally with fringing reefs, and to the southeast by an outer shelf (Hallstatt basin), transitional to the Tethys Ocean [*Hillebrandt et al.*, 2013]. Another partly terrestrially influenced carbonate platform, represented by the Oberrhaet Limestone, is located to the north. Intraplatform depressions within the Oberrhaet Limestone also accumulated sedimentary successions across the T–J transition, similar to those in the Eiberg Basin.

The Rhaetian Kössen Formation was deposited over the Hauptdolomit and is composed of limestone and argillaceous bioclastic rocks formed in subtidal systems. The sedimentary facies of the Rhaetian Kössen Formation change around the middle to upper Rhaetian boundary (base of *Choristoceras marshi* Zone), with the appearance of a basinal facies (Eiberg Member), following on from the shallow-water sequences of the Hochalm Member [*Golebiowski*, 1989]. The subsiding Eiberg Basin reached 150–200 m water depth in the late Rhaetian [*Golebiowski*, 1989; *Krystyn et al.*, 2005; *Mette et al.*, 2012; *Hillebrandt et al.*, 2013; and references therein]. Marine conditions therefore prevailed across the T–J transition. However, a distinct and abrupt lithological change from basinal carbonates of the Eiberg Member to marls and clays of the Tiefengraben Member (lower Kendlbach Formation) occurred, possibly in response to a sea-level drop. This siliciclastic stratigraphic unit has also been suggested to coincide with the onset of extrusive CAMP emplacement, partly based on elemental geochemistry and mineralogy for the bituminous topmost layer of the Kössen Formation [*Pálfy and Zajzon*, 2012]. Observed lithological change at this time may alternatively therefore also reflect changing climatic and environmental conditions, leading to increased precipitation and weathering, and the associated enhanced supply of siliciclastic sediments into the Eiberg Basin.

The lithological transition at the top of the Kössen Formation is marked by the development of a thin (1–5 cm) bituminous layer, with total organic carbon (TOC) values up to 10% [*Ruhl et al.*, 2009]. The overlying reddish gray and faintly laminated mudstones of the Schattwald beds are marked by low TOC and low carbonate concentrations and gradually transition upward into grayish brown marls for the remainder of the Tiefengraben Member [*Ruhl et al.*, 2009]. The Tiefengraben Member (and Kendlbach Formation) were succeeded by Lower Jurassic (upper Hettangian to Sinemurian) carbonate strata of the Adnet Formation, which were deposited under increasing water depths and greater pelagic influence [*Böhm et al.*, 1999].

All sections within the Eiberg Basin show similar sedimentary records across the TJB, with only minor variations in carbonate and clay content depending on more proximal versus more distal positions. The Karwendel syncline exposures provide some of the thickest and most expanded and complete marine TJB successions worldwide, which is one of the reasons the Kuhjoch locality was selected as GSSP for the base of the Jurassic system.

The abrupt lithological change from the predominantly carbonate Kössen Formation to the marly sediments of the lower Tiefengraben Mb (locally known as the "Grenzmergel," which includes the reddish gray colored Schattwald beds) was for long considered to represent the TJB [Golebiowski, 1990; Hallam and Goodfellow, 1990], as it is marked by the disappearance of typical Triassic ammonoid and conodont fossils. Recent studies, however, showed that the lowest meters of the Tiefengraben Member still yield Triassic microflora and nannoflora [Kuerschner et al., 2007].

Newly described psiloceratids in the TJB successions of the Eiberg Basin, including Psiloceras spelae tirolicum, stratigraphically precede the well-known earliest Psiloceras of England (Psiloceras erugatum, Psiloceras planorbis) and the Alps (Psiloceras calliphyllum) [Hillebrandt et al., 2013]. The first occurrence of P. spelae tirolicum was selected as marker for the base of the Jurassic system [Hillebrandt et al., 2013]. Within the GSSP succession, P. calliphyllum is correlated with the earliest Psiloceras in England [Page, 2003; Bloos, 2004; Hillebrandt and Krystyn, 2009; Page et al., 2010].

The GSSP for the base of the Jurassic and several other TJB sections of the Eiberg Basin were studied in detail for microfossil and macrofossil occurrences [McRoberts et al., 1997; Kuerschner et al., 2007; Bonis et al., 2009a, 2009b; Hillebrandt and Krystyn, 2009; Bonis and Kürschner, 2012; Hillebrandt et al., 2013; and references therein]. Notably, the first occurrence of terrestrial biological markers (pollen/spores) stratigraphically close to the base of the Jurassic allows also for correlation to continental sedimentary sequences of this age [Kuerschner et al., 2007; Bonis et al., 2009a; Bonis and Kürschner, 2012].

Carbon isotope analyses of bulk sedimentary organic matter as well as individual higher plant-derived leaf-wax n-alkanes show a pronounced $\delta^{13}C$ negative excursion of ~6 and ~8‰, respectively, in the studied sections of the Eiberg Basin [Ruhl et al., 2009, 2011]. This negative CIE occurs at the very base of the Tiefengraben Member, directly coinciding with the end-Triassic mass extinction, and it is marked by a TOC-rich bituminous black shale at its onset [Bonis et al., 2009b]. Further to the west, in the Lechtal nappe of the Bajuvaric nappe group, in the western NCA (western Austria), the T–J transition is recorded within broad carbonate platform sedimentary sequences [Felber et al., 2015]. Also here, the upper Triassic Kössen Formation was deposited on top of the Hauptdolomit (of which the upper part is informally known as "Plattenkalk") and followed by progradational siliciclastic

sediments [Berra et al., 2010]. The transitional Schattwald beds including the TJB are here followed by the Hettangian "Lorüns oolite" [Felber et al., 2015].

Specific clay minerals (low- to medium-charged smectite and Mg-vermiculite), altered but euhedral mafic minerals, and elevated heavy REE concentrations in the lower Tiefengraben Member of the Eiberg Basin were suggested to have been associated with the onset of CAMP volcanism and the associated atmospheric dispersal of volcanic ash and weathering processes [Pálfy and Zajzon, 2012]. The combined biostratigraphic and chemostratigraphic framework of the TJB sections in the Austrian Eiberg Basin provides an excellent framework for the stratigraphical correlation to marine and continental T–J sedimentary sequences elsewhere; for the study of oceanographic, climatic, and environmental changes at that time; and for understanding their temporal and potentially causative link to CAMP volcanism.

10.3.2. Bristol Channel Basin at St Audrie's Bay (United Kingdom)

The T–J sedimentary succession in the Bristol Channel Basin has long been studied for microfossil and macrofossil content, and the outcrops at St Audrie's Bay were proposed as GSSP for the base of the Hettangian stage (base of the Jurassic system) [Warrington et al., 1994, 2008, and references therein]. The recognition of a negative excursion in $\delta^{13}C_{TOC}$ coinciding with the end-Triassic mass extinction event in this sedimentary record sparked the study of the potential (temporal) link between CAMP volcanism, global carbon cycle change, and biotic response at this time interval [Hesselbo et al., 2002]. The ~5‰ negative CIE, informally called the "initial negative CIE," at the end-Triassic mass extinction level is followed by a return to more positive values and a subsequent long-term shift to yet more negative values (the "main CIE") [Hesselbo et al., 2002]. This evolution in $\delta^{13}C_{TOC}$ from the mass extinction level upward is also mirrored in the $\delta^{13}C_{fossil-calcite}$ record from the same and other nearby sections [Korte et al., 2009]. Facies analysis in the Bristol Channel Basin sedimentary succession suggests a sea-level lowstand stratigraphically just preceding the end-Triassic mass extinction and "initial" negative CIE [Hesselbo et al., 2004].

Paleomagnetic, chemostratigraphic, biostratigraphic, and cyclostratigraphic analyses of the sedimentary record at St Audrie's Bay have provided a detailed stratigraphic framework for correlation to marine and continental sedimentary records in other regions worldwide and to the CAMP magmatic records of Morocco and North America [Hounslow et al., 2004; Warrington et al., 2008; Bonis et al., 2010; Deenen et al., 2010; Ruhl et al., 2010; Bonis and Kürschner, 2012; Mander et al., 2013; Hüsing et al., 2014; Xu et al., 2017].

10.3.3. Csővár Section in the Transdanubian Range Unit (Hungary)

The sections near the village of Csővár are located in north-central Hungary (Fig. 10.1), near the northeastern end of the Transdanubian Range Unit segment of the Tethyan shelf [*Haas et al.*, 2010]. The T–J transition is recorded in carbonate rocks deposited in open marine, basinal to toe-of-slope settings of the Norian to Sinemurian Csővár Limestone Formation [*Pálfy and Haas*, 2012] deposited in a periplatform basin. Dark gray, bituminous calcarenite (containing shallow-water fossils and clasts supplied from the adjacent platform) and marl beds are predominant in the lower part, and well-bedded, pale yellow to pale brown micritic limestone and cherty limestone composes the upper part [*Haas et al.*, 1997; *Haas and Tardy-Filácz*, 2004] (note that *Kozur* [1993] suggested that the upper unit should be distinguished as a separate formation, the Várhegy Cherty Limestone Formation). Two different sections of partly overlapping age are exposed on both sides of the Pokol-völgy (Hell Valley) northwest of the Csővár village: the Pokol-völgy quarry and the Vár-hegy (Castle Hill) section (see map, e.g., in *Pálfy et al.*, 2007]. An uppermost Rhaetian marker bed containing abundant lithoclasts and platform-derived bioclasts helps to establish lithologic correlation between the two sections. Syn-sedimentary slump structures occur at several levels in the lowermost Jurassic and potentially complicate interpretation of the isotope curve.

Ammonoid biostratigraphy provides broad constraints for drawing the TJB between occurrences of Rhaetian *Choristoceras* sp. and an early Hettangian psiloceratid, with the intervening 17 m only yielding an ex situ specimen of *Nevadaphyllites* [*Pálfy and Dosztály*, 2000; *Pálfy et al.*, 2007]. Radiolarians indicating the *Globolaxtorum tozeri* and *Canoptum merum* zones (latest Rhaetian and earliest Hettangian, respectively) also bracket a ~25 m barren interval [*Pálfy et al.*, 2007]. The staggered disappearance of conodonts falls into this interval, subdivided into the *Misikella ultima* and *Neohindeodella* zones [*Pálfy et al.*, 2001, 2007; *Korte and Kozur*, 2011]. However, as *Kozur* [1993] already noted, the Csővár section may be unique in preserving the last and youngest conodont taxa, with rare survivors reaching into the earliest Jurassic. On palynological grounds, a synchronous marine and terrestrial TJB event is suggested by correlated spikes in the abundance of prasinophyte algae and fern spores [*Götz et al.*, 2009].

The Vár-hegy section provided one of the first detailed C isotope curves across the TJB and the first one where a T–J transition negative anomaly was simultaneously documented in both $\delta^{13}C_{carb}$ and $\delta^{13}C_{org}$, with a magnitude of −3.5 and −2‰, respectively, over a stratigraphic interval of ~2 m within the available biostratigraphic brackets of the system boundary [*Pálfy et al.*, 2001]. Despite some noise in the $\delta^{13}C_{carb}$ data attributed to minor diagenetic overprint, this anomaly stands out and is confirmed by the $\delta^{13}C_{org}$ data, lending support to its primary character. Carbon isotope ratios are not significantly correlated with either TOC or hydrogen index values. Another comparably minor negative anomaly is observed somewhat lower in the upper Rhaetian. A subsequent high-resolution study extended the $\delta^{13}C_{carb}$ curve for an additional 20 m into the middle Hettangian and focused on the TJB negative anomaly [*Pálfy et al.*, 2007]. Here the TJB anomaly appears to contain a series of short-term oscillations, whereas no clear trend was found after the curve leveled off returning to the pre-excursion values. A $\delta^{13}C_{carb}$ curve was also obtained from a 16 m thick upper Rhaetian part of the Pokol Valley quarry section by *Korte and Kozur* [2011]. Although a <1‰ negative excursion followed by a stepped ~2‰ positive shift is recognized, its correlation to the nearby Vár-hegy section is not straightforward. Additional geochemical investigations of Rhaetian conodonts from the Pokol Valley quarry are required to help refine the $^{87}Sr/^{86}Sr$ reference curve focused on the TJB [*Korte*, 1999; *Korte et al.*, 2003].

Although the Csővár sections yield important data for our understanding of the carbon cycle perturbation at the TJB, they pose a challenge regarding the curves obtained to date and the correlation of them with other sections. The main negative anomaly in the Vár-hegy section is best equated with the initial CIE, and a precursor anomaly also appears to be recorded. However, the main negative anomaly reported from other sections remains unproven here.

10.3.4. Kennecott Point in Haida Gwaii (Queen Charlotte Islands, Canada)

Studies of an expanded TJB section at Kennecott Point yielded $\delta^{13}C$ curves with one of the earliest recognized anomalies across the TJB [*Ward et al.*, 2001]. Haida Gwaii (also known as the Queen Charlotte Islands), a Pacific archipelago in British Columbia, forms part of the accreted terrane of Wrangellia and preserves a relatively continuous and unmetamorphosed Middle Triassic to Middle Jurassic sedimentary succession [*Lewis et al.*, 1991]. Kennecott Point is located on the northwestern shore of Graham Island, with excellent exposures on a wave-cut intertidal platform. Siliciclastic sediments (thin-bedded, organic-rich shale and siltstone with fine- to medium-grained sandstone interbeds) of the Rhaetian to Sinemurian Sandilands Formation were deposited in deep marine outer shelf to basinal systems [*Desrochers and Orchard*, 1991].

The TJB has been defined by integrated ammonoid, radiolarian, and conodont biostratigraphy [*Tipper et al.*, 1994]. Radiolarians occur abundantly throughout the section, enabling a highly resolved biostratigraphy. A marked faunal turnover leads to recognition of the boundary between the *Globolaxtorum tozeri* and *Canoptum merum* zones, which is also equated to the system boundary [*Carter and Hori,* 2005; *Longridge et al.,* 2007]. The Norian-Rhaetian boundary is approximated by the base of the *Parvicingula moniliformis* radiolarian zone [*Carter,* 1993] or the last appearance of the distinctive bivalve *Monotis* [*Ward et al.,* 2004]. Sparse latest Triassic ammonoids include *Choristoceras rhaeticum* and *Choristoceras nobile,* assigned to the North American *Choristoceras crickmayi* Zone [*Tozer,* 1994; *Ward et al.,* 2001], whereas the *Choristoceras minutus* Zone, the second lowest Jurassic ammonoid zone, is documented by the index species *Choristoceras minutus,* appearing ~8 m higher than the earliest Jurassic radiolarians, and followed by *Psiloceras* ex gr. *tilmanni,* indicative of the next higher *Psiloceras pacificum* Zone of the lower Hettangian [*Longridge et al.,* 2007]. The Rhaetian zonal index conodont *Misikella posthernsteini* was found in a single sample in the uppermost Triassic part of the section, and the highest conodont occurrence is recorded close to the radiolarian-defined TJB [*Tipper et al.,* 1994]. Significantly, there is no change in lithology at the TJB in the Kennecott Point section.

Ward et al. [2001] obtained a $\delta^{13}C_{org}$ curve from a ~120 m thick part of the section, from the uppermost Norian through the lowermost Hettangian. This study documented a significant, −2‰ excursion over <5 m of strata directly at the TJB, following the Rhaetian with no obvious trend in the C isotopic evolution. No significant correlation was found between TOC and $\delta^{13}C_{org}$ values, lending support to the interpretation as a primary signal. Subsequent work at higher sampling resolution confirmed the presence of the TJB negative anomaly and described it as a series of short-term oscillations with up to six local minima [*Ward et al.,* 2004]. On the other hand, isotopic values measured on bulk carbonate from limestone concretions and interbeds appeared diagenetically overprinted and were not considered for further interpretation. *Williford et al.* [2007] extended the isotopic analyses to the higher, Lower Jurassic part of the section, doubling the thickness of the sampled stratigraphic section to 250 m, through the entire Hettangian and well into the Sinemurian. However, as faulting is known to cause tectonic repetitions at Kennecott Point [*Pálfy et al.,* 1994; *Longridge et al.,* 2008], caution is needed as the measured and sampled section may include some overlooked tectonically duplicated parts. The key feature of the extended $\delta^{13}C_{org}$ curve is the presence of a pronounced 5‰ positive excursion in the Hettangian, closely following a transient return from the TJB negative spike to preexcursion values. Although correlation of this Lower Jurassic isotope curve together with ammonoid and

radiolarian biostratigraphic data is not tightly constrained, the positive anomaly appears early to mid-Hettangian in age. Post-excursion $\delta^{13}C_{org}$ values in the upper part of the Hettangian and Sinemurian section are ~1‰ more negative than the long-term Rhaetian average values and the values obtained for the transient return after the TJB negative anomaly.

Sulfur isotopic analyses of a suite of samples revealed a major positive $\delta^{34}S$ anomaly coincident with the Hettangian positive $\delta^{13}C_{org}$ excursion [*Williford et al.,* 2009]. On the other hand, the TJB negative CIE seems to correspond to a significant protracted negative $\delta^{34}S$ anomaly, as well as the occurrence of lipid biomarkers suggestive of photic zone euxinia [*Kasprak et al.,* 2015].

In summary, the Kennecott Point section with continuous deep marine sedimentation in the East Pacific realm provides a useful reference to define and understand perturbations of the global carbon, sulfur, and nitrogen cycles, whereas the fossil record is utilized for both biostratigraphic constraints and assessment of biotic changes during mass extinction.

10.3.5. New York Canyon Section (Nevada, United States)

One of the best known and most intensively studied TJB sections in North America (and the world) is at Ferguson Hill and Muller Canyon in the New York Canyon area of the Gabbs Valley Range (Mineral County, Nevada), ~170 km southeast of Reno. The significance of this continuous marine TJB section was first recognized by *Muller and Ferguson* [1936] who subdivided the predominantly dark shale, siltstone, and limestone strata across the system boundary into the Gabbs and Sunrise formations, with a gradational contact between them. Further lithostratigraphic subdivision led to the introduction of members [*Taylor et al.,* 1983], of which, in ascending order, the limestone-dominated Mount Hyatt and siltstone-dominated Muller Canyon members of the Gabbs Formation and the limestone-dominated Ferguson Hill Member of the Sunrise Formation were the subjects of several biostratigraphic and chemostratigraphic studies. *Hallam and Wignall* [2000] drew attention to the local expression of the extinction and discussed its relation to facies changes. The continuous shallow marine sedimentary succession was deposited in a foreland basin east of the Sonoma allochthon, following Permian-Triassic thrusting related to the Sonoma orogeny at the Cordilleran margin of North America [*Dickinson,* 2004]. Magmatism related to a later Cretaceous phase of tectonic evolution led to a low-grade metamorphic overprint, hindering magnetostratigraphic studies and destroying palynomorphs [*Lucas et al.,* 2007].

The macrofossil record across the TJB is particularly rich in ammonoids [*Guex,* 1995] and bivalves [*Laws,*

1982], which served as the basis for the GSSP candidacy of the section [*Guex et al.*, 1997; *Lucas et al.*, 2007; *McRoberts et al.*, 2007]. Latest Triassic *Choristoceras* spp. occur up to the lowermost Mount Hyatt Member, below a 7 m thick barren interval, followed by the first occurrence of *Psiloceras spelae* and *P. tilmanni*, regarded as the oldest Jurassic ammonoid species and zonal indices of the lowermost Hettangian ammonoid biozone. The first occurrence of pectinid bivalve *Agerchlamys boellingi* is also of stratigraphic significance near the system boundary, immediately below the first psiloceratid ammonoids [*McRoberts et al.*, 2007]. The age of latest Triassic strata is also well supported by conodont and radiolarian biostratigraphy [*Orchard et al.*, 2007]. A high-precision U-Pb zircon age of 201.33 ± 0.13 Ma provides a numeric tie point to calibrate other stratigraphic schemes and the geological time scale and to correlate with other radioisotopically dated sections [*Schoene et al.*, 2010].

The first $\delta^{13}C_{org}$ curve from the New York Canyon area was produced by *Guex et al.* [2003, 2004], documenting two negative excursions of similar ~2‰ amplitude, the first one near the last occurrence of *C. crickmayi* and the upper one between the first occurrence of *P. spelae*, *P. tilmanni*, and *P. pacificum*. Despite some scatter in the data, a return to pre-excursion values is observed between the anomalies. *Ward et al.* [2007] reported a new set of $\delta^{13}C_{org}$ data from an independently collected suite of samples. Although it confirmed the presence of two negative anomalies in a curve with less scatter, it also led to controversies regarding the position of the negative anomalies with regard to lithostratigraphy and biostratigraphic markers, recording a positive peak in the earliest Jurassic and an apparent offset in values of isotopic ratios compared to those reported in *Guex et al.* [2004]. *Guex et al.* [2009] argued that part of these discrepancies is explained by differing views on the definition of lithostratigraphic units and the overlooking of a fault by the other authors. Comparison of the stratigraphic positions of both negative anomalies in $\delta^{13}C_{org}$ therefore remains ambiguous. Repeated measurements on five samples yielded values >0.3‰ different from the originally reported ones. A third independent sampling was carried out, and new measurements were reported by *Thibodeau et al.* [2016], with the resultant curve in good agreement with a corrected version of that of *Ward et al.* [2007]. These data were obtained for a geochemical study which also documented elevated Hg concentrations, with a peak at the termination of the first negative C isotope anomaly, supporting the inference of a volcanic trigger for the environmental changes [*Thibodeau et al.*, 2016]. An additional 20 m of section was sampled up to the Hettangian-Sinemurian boundary, and the extended curve features a prominent 5‰ positive anomaly in the upper Hettangian [*Bartolini et al.*, 2012]. Thus, the New York Canyon section also contributes to our understanding of the Hettangian carbon isotope record, where correlation of positive anomalies remains controversial.

In summary, geochemical studies from the New York Canyon area (i) span the T–J transition and the entire Hettangian; (ii) benefited from good ammonoid biostratigraphical control; (iii) are important in establishing the succession of two separate negative carbon isotope anomalies, followed by a prominent positive anomaly; and (iv) are unparalleled in being based on three independent sets of samples and measurements by different research teams.

10.3.6. Astartekløft (East Greenland)

Astartekløft is the best studied of several localities in the Hurry Inlet (Scoresby Sund) area of east Greenland, part of the Jameson Land Basin, which have been of particular interest for their paleobotanical contents. Within the Jameson Land Basin, the TJB occurs in the fluviolacustrine strata of the Kap Stewart Group [*Surlyk*, 2003], which range from marginal fluvial sandstone to basin-center shale. The Hurry Inlet localities show a marginal fluvial succession comprising channels filled with coarse sand, commonly multistory, with thinner overbank deposits that include plant-bearing crevasse splays [*Dam and Surlyk*, 1992; *McElwain et al.*, 2007]. Initial thorough stratigraphic work by *Harris* [1937] documented in the Hurry Inlet localities a significant turnover of floras from the supposed Triassic *Lepidopteris* flora to the Jurassic *Thaumatopteris* flora. Harris' work was subsequently expanded upon by *McElwain et al.* [2007, 2009] who amassed large collections of plant macrofossils from Harris's plant beds at Astartekløft. Several studies have built upon this paleobotanical framework to suggest significant changes in plant ecosystems across the TJB in response to large igneous province (LIP) forcing [e.g., *Belcher et al.*, 2010; *Bacon et al.*, 2013; *Mander et al.*, 2013; *Steinthorsdottir et al.*, 2018], and the Astartekløft fossil plant (stomatal density/index) record has been the principal basis for reconstruction of atmospheric CO_2 changes across TJB [*McElwain et al.*, 1999; *Steinthorsdottir et al.*, 2011, 2012]. Carbon isotope chemostratigraphic correlation was used by *Hesselbo et al.* [2002] for correlation to St Audrie's Bay in southern England. The carbon isotope stratigraphy of *Hesselbo et al.* [2002], based on macrofossil wood, was corroborated by analysis of specifically identified leaf cuticles [*Bacon et al.*, 2011]. On the basis of detailed palynological study, *Mander et al.* [2013] suggested a revised correlation to St Audrie's Bay in which strata equivalent to the "initial" CIE are missing or undetected at Astartekløft.

10.4. CHEMOSTRATIGRAPHY

10.4.1. Carbon Isotope Stratigraphy and Total Organic Carbon (TOC) Variation

Following on the pioneering work by *McRoberts et al.* [1997] and *McElwain et al.* [1999], many studies investigated the carbon isotope trend across the TJB. Negative CIEs were confirmed from other marine [e.g., *Pálfy et al.*, 2001; *Ward et al.*, 2001; *Hesselbo et al.*, 2002; *Guex et al.*, 2004; *Galli et al.*, 2005; *Kuerschner et al.*, 2007; *McRoberts et al.*, 2007; *Williford et al.*, 2007; *Ruhl et al.*, 2009] and terrestrial [*Hesselbo et al.*, 2002; *Steinthorsdottir et al.*, 2011; *Pieńkowski et al.*, 2012] TJB sections and have been related to the end-Triassic mass extinction. In the western NCA, the isotope excursion is marked by a ~3‰ negative shift in bulk carbonate $\delta^{13}C$ [*Felber et al.*, 2015], which is similar to that recorded at a northern Italian section (3–5‰ negative CIE), the Budva Basin in Montenegro (1–2‰ negative CIE), and the United Arab Emirates (~5‰ negative CIE) [*Galli et al.*, 2005; *Crne et al.*, 2011; *Al-Suwaidi et al.*, 2016].

The negative CIE in the Eiberg Basin is followed by a return to pre-excursion values throughout the Schattwald beds (lower Tiefengraben Member) and the subsequent gray marls of the upper Tiefengraben Member (Fig. 10.2), with 1–2‰ more negative $\delta^{13}C$ values broadly coinciding with the first occurrence of *Psiloceras spelae tirolicum* at the base of the Jurassic [*Ruhl et al.*, 2009]. A shift to the continuously lighter $\delta^{13}C$ values of the Hettangian stage, as, for example, observed in the marine Bristol Channel Basin [*Hesselbo et al.*, 2002; *Korte et al.*, 2009; *Ruhl et al.*, 2010], the Danish Basin [*Lindström et al.*, 2012], in the New York Canyon section of Nevada [*Bartolini et al.*, 2012], and the continental Newark and Hartford basins [*Whiteside et al.*, 2010], occurs in the Eiberg Basin broadly at the level of the first occurrence of the *Psiloceras* cf. *pacificum* [*Hillebrandt et al.*, 2013].

Several studies have suggested a minor carbon cycle perturbation and associated negative CIE to precede the end-Triassic mass extinction (sometimes referred to as the precursor CIE) [*Deenen et al.*, 2011; *Ruhl & Kürschner*, 2011; *Lindström et al.*, 2012; *Dal Corso et al.*, 2014; *Davies et al.*, 2017]. The Upper Triassic sequences in the Eiberg Basin have only been studied extensively in the Eiberg quarry [*Korte et al.*, 2017]. Based on this succession, an even earlier carbon cycle perturbation was suggested to have occurred in the late Rhaetian (the late Rhaetian CIE; *Mette et al.*, 2012).

Observed trends in $\delta^{13}C$ records led to the suggestion of a characteristic $\delta^{13}C$ geometry for the TJB interval including a short-term "initial" negative excursion, followed by a longer-lasting positive excursion and the long-lasting "main" negative excursion in marine successions [*Hesselbo et al.*, 2002; *Kuerschner et al.*, 2007; *McRoberts et al.*, 2007; *Ward et al.*, 2007; *Williford et al.*, 2007; *Korte et al.*, 2009; *Ruhl et al.*, 2009, 2010, 2011; *Korte & Kozur*, 2011] and that the end-Triassic mass extinction is coeval with the "initial" negative peak [e.g., *Guex et al.*, 2004; *Hesselbo et al.*, 2007; *McRoberts et al.*, 2007; *Ruhl et al.*, 2009; *Korte & Kozur*, 2011]. This trend in $\delta^{13}C$ has also been identified in some terrestrial successions [*Deenen et al.*, 2010; *Whiteside et al.*, 2010; *Dal Corso et al.*, 2014], reinforcing that it is likely global in nature and that it reflects changes in $\delta^{13}C$ values of the global ocean-atmosphere system. Additional potential excursions predating that across the TJB have been also discovered [*Cleveland et al.*, 2008; *Ruhl & Kürschner*, 2011; *Schaller et al.*, 2011; *Mette et al.*, 2012; *Korte et al.*, 2017]. The clear major ("main") negative TJB excursion, however, has not be identified in all marine sections [e.g., see *Galli et al.*, 2007; *Pálfy et al.*, 2007; *van de Schootbrugge et al.*, 2008; *Götz et al.*, 2009]. This has led some others to suggest that the amplitude, shape, duration, and even the stratigraphic position of the "initial" $\delta^{13}C$ minimum are different between sections and therefore do not represent a clear chemostratigraphic marker [*Lindström et al.*, 2017]. This inference, however, largely stems from limited data on the abundance of scarce pollen and spore species and sometimes neglects all other available stratigraphic markers, such as ammonites and the first and last occurrences of specific palynomorphs. Detailed stratigraphic correlation between individual TJB successions can be achieved through an integrated biostratigraphic and chemostratigraphic framework (Figure 10.2b).

The release of isotopically light carbon, such as CO_2 from volcanic degassing, thermal metamorphism of organic-rich sediments, and/or biogenic methane likely resulted in a $\delta^{13}C$ negative shift in global exogenic carbon reservoirs. Because of this and the relatively short (10^4–10^5 yr) residence time of carbon in the ocean-atmosphere-biosphere, peaks and troughs in carbon isotope records should represent useful markers for transcontinental (and marine-terrestrial) correlations.

Only a limited amount of compound-specific $\delta^{13}C$ data has been published to date, only on leaf-wax-derived long-chain *n*-alkanes [*Whiteside et al.*, 2010; *Ruhl et al.*, 2011].

The observed changes in these organic geochemical compounds are suggested to reflect large changes in T–J atmospheric $\delta^{13}C$, and they do directly correspond to changes in bulk organic $\delta^{13}C$ ($\delta^{13}C_{TOC}$ and $\delta^{13}C_{wood}$) in the same sedimentary successions [*Whiteside et al.*, 2010; *Ruhl et al.*, 2011]. Further similar work spanning the Upper Triassic and Lower Jurassic is necessary, but it has been limited by, for example, thermal maturity, low sedimentary TOC values, and poor organic sedimentary preservation in available successions.

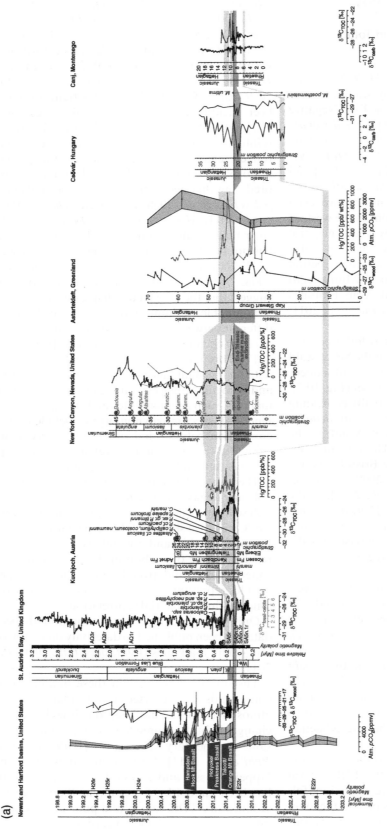

Figure 10.2 (a) Comparison of Triassic–Jurassic $\delta^{13}C_{carb}$, $\delta^{13}C_{TOC}$ and $\delta^{13}C_{wood}$ data from geographically distributed marine and terrestrial sections, with Hg/TOC ratios and inferred atmospheric pCO_2. Atmospheric pCO_2 estimates are based on stomatal density analyses (Astartekløft, Greenland) and $\delta^{13}C$ values of pedogenic carbonate sequences (Newark and Hartford basins). Note that this compilation is not exhaustive and that many more Triassic–Jurassic boundary sections have been studied. Data in (a) are from: Newark and Hartford basins, USA: *Kent et al.* [2017] and references therein, *Schaller et al.* [2011, 2012, 2015]; St Audrie's Bay, UK: *Hesselbo et al.* [2002], *Korte et al.* [2009], *Ruhl et al.* [2010], *Hüsing et al.* [2014], *Xu et al.* [2017]; *Kuhjoch, Austria: Ruhl et al.* [2009], *Hillebrandt et al.* [2013]; New York Canyon, Nevada, USA: *Bartolini et al.* [2012], *Thibodeau et al.* [2016]; Astartekløft, Greenland: *Hesselbo et al.* [2002], *Steinthorsdottir et al.* [2011], *Percival et al.* [2017]; Csővár, Hungary: *Pálfy et al.* [2001]; Canj, Montenegro: *Črne et al.* [2011]. (b) Highly resolved Triassic–Jurassic boundary integrated stratigraphic framework based on key European successions studied for $\delta^{13}C_{TOC}$ palynostratigraphy, and ammonite biostratigraphy. Data in (b) are from: Stenlille 1/2: *Lindström et al.* [2017] and references therein; St Audrie's Bay, UK: *Hesselbo et al.* [2002] and references therein, *Hounslow et al.* [2004], *Bonis and Kürschner* [2012]; Tiefengraben, Austria: *Kuerschner et al.* [2007], *Hillebrandt et al.* [2013]; Kuhjoch (integrated Kuhjoch East & West sections), Austria: *Bonis et al.* [2009a], *Ruhl et al.* [2009a], *Hillebrandt et al.* [2013]; Hochalplgraben, Austria: *Bonis et al.* [2009b], *Ruhl et al.* [2009], *Hillebrandt et al.* [2013].

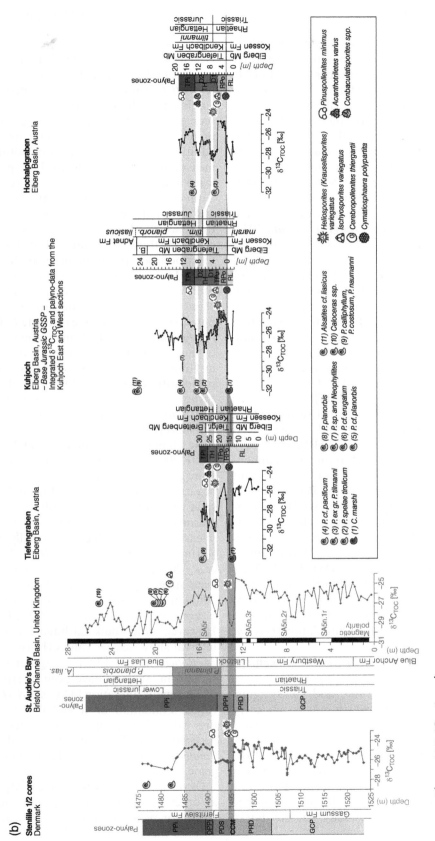

Figure 10.2 (*Continued*)

10.4.2. Mercury Chemostratigraphy

The concentration of Hg in the ocean-atmosphere system is largely controlled by emission as a trace volcanic gas from continental volcanoes or mid-ocean spreading ridge systems [*Pyle & Mather*, 2003; *Bowman et al.*, 2015]. Gaseous elemental Hg has a typical residence time of 0.5–2 years, allowing for global atmospheric dispersal before eventual drawdown into marine and continental sediments [*Blum et al.*, 2014]. Importantly, Hg is typically drawn down into sediments bound with organic matter [*Benoit et al.*, 2001; *Outridge et al.*, 2007], although chemical binding to sulfides and clays may also be of some importance [*Benoit et al.*, 1999; *Niessen et al.*, 2003; *Kongchum et al.*, 2011; *Bergquist*, 2017]. Sedimentary Hg concentrations are therefore typically normalized against TOC content of the sediment [*Percival et al.*, 2017]. The timing of activity and impact of several Phanerozoic LIPs and their potential relationships to global change (e.g., such as oceanic anoxic events (OAEs)) and mass extinction events, including the end-Permian, early Toarcian, mid-Cretaceous (Cenomanian-Turonian) OAE2, and end-Cretaceous events, were previously already studied [*Sanei et al.*, 2012; *Percival et al.*, 2015, 2016; *Font et al.*, 2016; *Sial et al.*, 2016; *Scaife et al.*, 2017].

The potential temporal correlation between events at the T–J transition, including the end-Triassic mass extinction, and the emplacement of CAMP was recently extensively studied in marine and continental sedimentary records from both hemispheres and from different paleolatitudes [*Thibodeau et al.*, 2016; *Percival et al.*, 2017]. An initial major pulse in sedimentary Hg concentrations, possibly reflecting the onset of Hg emissions, directly coincides with the end-Triassic mass extinction event [*Percival et al.*, 2017] (Fig. 10.2a). More importantly, however, individual major CAMP basalt flows preserved in the continental basins of North America (the United States and Canada) and North Africa (Morocco) can potentially be temporally (and perhaps causatively) linked to peaks in sedimentary Hg accumulation [*Percival et al.*, 2017].

A recent study suggests that extrusive emplacement of CAMP may have been preceded, by ~100 kyrs, by the intrusive emplacement of dike and sill systems, partly in sedimentary basins, which caused an early perturbation stage in the Earth's climate [*Davies et al.*, 2017]. The analyses of Hg concentration have so far been largely stratigraphically focused at the TJB. Future studies may provide further insights on Hg release, which may also have been generated from thermogenic processes around the sill complexes that intruded sedimentary basins already in the latest Triassic.

10.4.3. Oxygen Isotope Stratigraphy

The oxygen isotopic composition of marine carbonates is generally controlled by seawater temperature [*Urey et al.*, 1951] as well as by seawater $\delta^{18}O$ and pH [*Zeebe & Wolf-Gladrow*, 2001]. The temperature dependence, leading to lighter oxygen isotope ratios in carbonates secreted in warmer water and vice versa, enables the use of $\delta^{18}O_{carb}$ as paleothermometer for ancient oceans. In addition, the carbonate carbon isotope (Section 10.4.1) measurements are performed in parallel and on the same aliquots as those for the oxygen isotopes, allowing a direct comparison, for example, reconstructing temperature (climate) change together with atmosphere-ocean fluctuations of carbon dioxide. Oxygen isotope ratios, especially those from bulk carbonates, are, in comparison to carbon isotopes, more prone to diagenesis [e.g., *Veizer*, 1983; *Marshall*, 1992]. A careful evaluation of the data is therefore necessary to interpret seawater temperature changes of the past. On the basis of a dataset from which potentially altered bulk carbonate $\delta^{18}O$ data were culled [*Pálfy et al.*, 2001, 2007], an extreme temperature increase of more than 10 °C across the TJB at the Csővár section in Hungary is suggested, and this is quite dramatic for paleolatitudes of about 30°N (Fig. 10.1). A similar bulk carbonate $\delta^{18}O$ negative shift is also evidenced in other successions, such as the Doniford section in southwest England [*Clémence et al.*, 2010], suggesting marine climate change with a superregional extent. *Pálfy et al.* [2001, 2007] pointed out that this severe apparent temperature increase occurred during the period when the initial negative CIE happened (see Fig. 10.2) and during a time span when an injection of isotopically light carbon from volcanic exhalation of CO_2 and/or a methane hydrate release occurred, which potentially triggered a global warming [*Pálfy et al.*, 2001; *Hesselbo et al.*, 2002, *Ruhl et al.*, 2011]. This hypothesis of climate warming has also been suggested by model data [e.g., *Huynh & Poulsen*, 2005] and the findings of shelf-sea photic zone anoxia or even euxinia [*Jaraula et al.*, 2013; *Kasprak et al.*, 2015]. However, other bulk carbonate $\delta^{18}O$ datasets of this interval show heavier values (Lorüns section in Austria; *McRoberts et al.*, 1997], or lighter values (Kendelbach section in Austria) have been interpreted as the result of diagenetic alteration [*Morante & Hallam*, 1996]. For a robust evaluation of a global temperature rise, more proxy data from pristine samples across the TJB are still necessary, and this is in concert with the conclusion by *Tanner et al.* [2001], evaluating carbon isotopes on pedogenic carbonates.

Oxygen isotope values from low-Mg calcite fossils, such as brachiopods and oysters, represent a robust dataset useful for reconstruction of past seawater temperatures because this material is relatively resistant to diagenesis

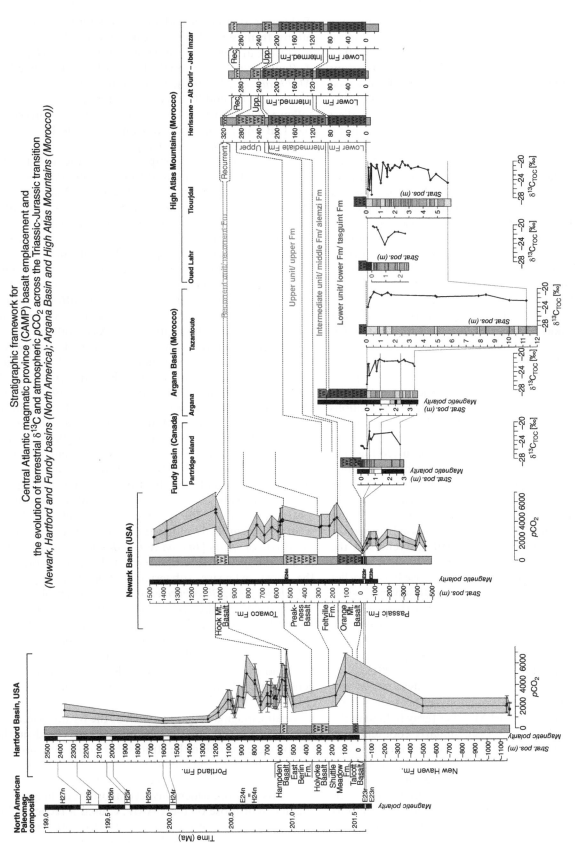

Figure 10.3 Compilation of continental Triassic–Jurassic sequences, which have been analyzed for $\delta^{13}C_{TOC}$ and $\delta^{13}C_{wood}$ and which record Central Atlantic Magmatic Province (CAMP) basalt emplacement. Data from: Hartford Basin, USA: *Schaller et al.* [2012], *Kent et al.* [2017] and references therein; Newark Basin, USA: *Schaller et al.* [2011], *Kent et al.* [2017] and references therein; Fundy Basin, Canada: *Deenen et al.* [2011]; Argana Basin, Morocco: *Deenen et al.* [2010]; High Atlas Mountains, Morocco: *Dal Corso et al.* [2014], *El Ghilani et al.* [2017]. (See insert for color representation of the figure.)

[e.g., *Popp et al.*, 1986; *Veizer et al.*, 1986] and because the alteration degree can be assessed by a multitude of physical and chemical techniques [see *Veizer*, 1983; *Marshall*, 1992; *Ullmann & Korte*, 2015 for reviews]. From different regions in Europe, $\delta^{18}O$ (and $\delta^{13}C$) data exist from well-preserved brachiopods of the Late Triassic [*Korte et al.*, 2005, 2017; *Mette et al.*, 2012] and pristine oysters of the earliest Jurassic [*Jones*, 1992; *Korte et al.*, 2009]. Unfortunately, no continuous dataset over the TJB is available from a single locality. A combination of datasets from different regions (with a lack of data in the latest Triassic and earliest Jurassic [*Korte et al.*, 2009, 2017]) is a relatively poor basis for reconstructing the temperature evolution across the whole time period of interest. However, the dataset from *Korte et al.* [2009] provides insights about the bottom water temperature changes during the earliest Jurassic in the Bristol Channel Basin, United Kingdom (Fig. 10.4). Relatively cool seawater (assuming a seawater $\delta^{18}O$ of $-1.2‰$ SMOW) temperatures between <7 and 14°C existed for (gray field in the biozone row of Fig 10.4) the period of deposition of the upper Langport Member around the TJB. The temperature increased then distinctly by more than 8°C, reaching values ~12 and 22°C in the first 3 ammonite subzones of the Jurassic. Extreme changes in seawater $\delta^{18}O$ (evaporation, melting of continental ice, meteoric water dilution) as a trigger for these severe oxygen isotope fluctuations have been excluded by *Korte et al.* [2009]. This oxygen isotope signal therefore potentially reflects a true climate signal in which the warming (Fig. 10.4) occurred together with the trend toward to lighter carbon isotope values (main negative CIE of *Hesselbo et al.* [2002]). The data are compatible with the suggestion that heavy carbon isotope values correspond to periods of lower pCO_2 contents and vice versa. However, the detailed comparison between the $\delta^{18}O$ and the $\delta^{13}C$ curves shows small differences in the positions of the positive peaks (Fig. 10.4). The $\delta^{18}O$ decrease (i.e., rise in temperature) in this case began earlier than the $\delta^{13}C$ decline and with it the inferred increase in CO_2 levels. This suggests that factors other than CO_2 may have contributed to the warming in this area, such as opening up of the seaway to marine currents derived from warmer, more equatorial waters as also suggested for the Laurasian seaway in the Aalenian [*Korte et al.*, 2015].

We note, however, that the relative heavy oxygen isotope values of nearly $+2‰$ (temperatures between <7 and 14°C) in the upper Langport Member (gray field in in the biozone row in Fig. 10.4) reflect similarly cool temperatures evidenced for the Aalenian stage in the Hebrides Basin of Scotland [*Korte et al.*, 2015; see also *Korte & Hesselbo*, 2011; *Korte et al.*, 2017]. These data rather indicate that a short-term "cold mode" interval existed around the TJB, at least equivalent to the upper Langport Member at mid-European paleolatitudes, which were south of 40°N (Fig. 10.1). Cool temperatures close to the T–J transition were also identified by sporomorph associations from several successions in central and northwestern Europe and in northeast Greenland [*Hubbard & Boulter*, 2000]. Although this was later challenged by *McElwain et al.* [1999], pollen data do suggest initial warming at the end-Triassic mass extinction interval to be followed by a short cooling phase at or directly preceding the TJB and a long-term warming into the Hettangian stage [*Bonis & Kürschner*, 2012].

These relatively cool temperatures at the TJB (upper Langport Member) occur during the interval of subaerial basaltic eruptions from the CAMP coeval with Hg peaks (Figs. 10.2a and 10.4; Section 10.4.2; see also Fig. 10.3) and suggest volcanism-induced cooling as a plausible explanation [*Guex et al.*, 2004, 2012; *Schoene et al.*, 2010]. It has been proposed that volcanic eruptions have triggered short-term climate cooling events [*Lamb*, 1970; *Kennett & Thunell*, 1975; *Rampino et al.*, 1988] when erupted ashes, sulfur dioxide, and hydrogen sulfide (hydrogen sulfide rapidly oxidizes to SO_2 in the atmosphere) reach the stratosphere. Ashes fall down rapidly, whereas SO_2 reacts with hydroxyl (OH^-) and water to form sulfuric acid and generates aerosol clouds which circulate for several years around the planet [*Cadle et al.*, 1976; *Pollack et al.*, 1976; *Rampino et al.*, 1988]. These clouds reduce the transparency for light and, in addition, reflect solar radiation, and these factors cause the troposphere and Earth's surface to cool down [*Devine et al.*, 1984; *Sigurdsson*, 1990]. Volcanic cooling events, induced by short-term felsic (high-SiO_2) or intermediate volcanic eruptions, are usually of short duration and lasting not longer than two years (e.g., Mt. Pinatubo eruption in 1991; *Genin et al.*, 1995], and it was therefore debated whether volcanically induced cool periods can last over hundreds of years [*Bryson & Goodman*, 1980; *Rampino et al.*, 1988]. CAMP emplacement, however, represents the most extensive subaerial LIP in Earth's history [*McHone*, 2003; *Blackburn et al.*, 2013]. Its emplacement spanned 600–800 kyr [*Kent et al.*, 2017 and references therein] and occurred likely recurrently and at multitude eruption centers along the Atlantic rift basin [*Davies et al.*, 2017]. Potentially, volcanic aerosol enrichments in the atmosphere maintained over a longer period and triggered the cool interval in the upper Langport Member (Figs. 10.2, 10.3, 10.4).

Basaltic magma eruptions, such as the fissure eruptions at Laki in Iceland, usually generate aerosol clouds in the troposphere, reaching the stratosphere only occasionally [*Devine et al.*, 1984; *Walker et al.*, 1984; *Palais & Sigurdsson*, 1989]. It could be shown, however, that even the Laki eruption, which was several orders of magnitude smaller than those of the CAMP, caused a recognizable cooler period over even several years [*Sigurdsson*, 1982; *Rampino et al.*, 1988; *Thordarson & Self*, 2003]. Moreover,

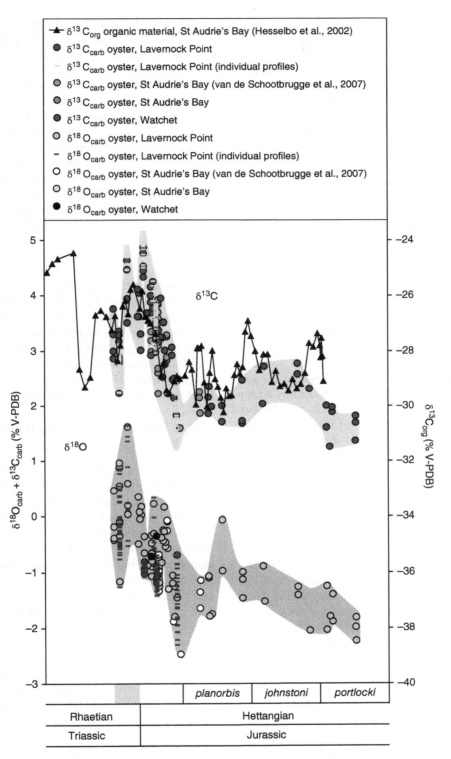

Figure 10.4 Oxygen and carbon isotope values from pristine oysters originating from the earliest Jurassic successions at Lavernock Point, St Audrie's Bay, and Watchet, United Kingdom, from *Korte et al.* [2009; including results from *van de Schootbrugge et al.,* 2007] and bulk organic δ^{13}C data from *Hesselbo et al.* [2002], plotted against stratigraphy. Modified after *Korte et al.* [2009]. Reproduced with permission of the Geological Society of London.

aerosols originating from basaltic eruptions contain 10–50 times more H_2SO_4 than those from felsic eruptions and influence climate much more effectively [*Rampino et al.*, 1988; *Palais & Sigurdsson*, 1989]. Taking this latter observations and evaluations into account, the ~2.5‰ decrease in $\delta^{18}O$ (~10 °C warming in the Early Jurassic) would then reflect a shift back from volcanically induced relatively cool temperatures to more normal climatic conditions.

10.4.4. Strontium and Osmium Isotope Stratigraphy

The TJB falls into an interval of a ~30 Myr long gradual decline of the marine $^{87}Sr/^{86}Sr$ curve which commenced in the Norian (Late Triassic) at a ratio near 0.7080 [*Korte et al.*, 2003]. This decreasing trend was terminated by a rebound to more radiogenic values in the earliest Toarcian (Early Jurassic) after reaching a ratio close to 0.7070 [*Jones et al.*, 1994a]. The finer structure of the marine $^{87}Sr/^{86}Sr$ curve across the TJB where values are near 0.7077 [*Jones et al.*, 1994a; *Korte et al.*, 2003] is not defined with great confidence at present.

It has been suggested that a rapid decrease of marine $^{87}Sr/^{86}Sr$ in the Rhaetian was briefly reversed in the latest Rhaetian and in the earliest Jurassic followed by a phase of zero change [*Cohen & Coe*, 2007] or even to brief increase [*Callegaro et al.*, 2012] that lasted throughout the entire Hettangian stage [*Cohen & Coe*, 2007]. This interpretation hinges on the accuracy of the geological time scale, the direct comparability of $^{87}Sr/^{86}Sr$ ratios from Late Triassic Austrian shelf seas to latest Triassic and earliest Jurassic UK shelf seas, and the pristine preservation state of two oyster specimens from the Rhaetian of the United Kingdom. Opposing this scenario stands the current marine Sr isotope curve [*McArthur et al.*, 2012] which adds four Rhaetian conodont values from the Hungarian Csővár section to Rhaetian brachiopod data from the Austrian Weißloferbach section [*Korte et al.*, 2003] but discards earliest Jurassic oyster data from *Jones et al.* [1994a]. A statistical fit through these data suggests a minor slowdown of the marine $^{87}Sr/^{86}Sr$ decline during the latest Rhaetian and Hettangian. This slowdown, however, is not constrained by any tie point in the critical time interval of the TJB due to the choice of which fossil materials can be regarded as trustworthy.

Regardless of the adopted model, at some point shortly preceding the TJB, the seawater Sr reservoir appears to have been affected by a more or less substantial shift in the balance of unradiogenic (mantle) and radiogenic (continental) Sr sources to the oceans. This shift has been tied to fluctuations in Os isotope ratios as measured in organic-rich mudrocks in the United Kingdom [*Cohen et al.*, 1999] and related to the emplacement of the CAMP [*Cohen & Coe*, 2007]. Such a shift in Sr isotope values could either be related to a stronger flux of continental Sr, for example, by way of globally increased weathering rate or by weaker riverine drainage from catchments with unradiogenic $^{87}Sr/^{86}Sr$ ratios. Alternatively, this effect could also have been brought about by a reduction of mid-ocean ridge activity [cf., *Jones et al.* 1994b]. CAMP rocks themselves, emplaced at low latitudes and likely highly susceptible to weathering [*Cohen & Coe*, 2007], would rather have counteracted the observed change in slope of the marine $^{87}Sr/^{86}Sr$ curve. CAMP weathering would have led to the injection of somewhat more unradiogenic Sr into the coeval seawater, as most reconstructed initial $^{87}Sr/^{86}Sr$ ratios for CAMP igneous rocks fall within a range from 0.705 to 0.708 [*Whalen et al.*, 2015], slightly lower than the ratio of coeval seawater. CAMP's direct influence on the marine $^{87}Sr/^{86}Sr$ must thus have been overwhelmed by more potent environmental changes brought about by CAMP emplacement and changes to the Earth surface system that happened simultaneously. Potentially enhanced hydrological cycling as well as elevated sulfuric acid rain may have increased weathering of more radiogenic continental/crustal rocks, delivering relatively Sr to the global ocean with higher $^{87}Sr/^{86}Sr$ ratios, resulting in an Early Jurassic (Hettangian) plateau in seawater $^{87}Sr/^{86}Sr$ [*Jenkyns et al.*, 2002 and references therein].

10.4.5. Redox Changes Across the Triassic-Jurassic Transition

Major global change events in Earth's history are often associated with OAEs, in which significant parts of globally distributed marine basins developed anoxic and/or euxinic conditions [*Jenkyns*, 2010; *Percival et al.*, 2016]. Much discussion has focused on the causes, consequences, and timing of Mesozoic OAEs and associated changes in global (bio)geochemical cycles, biotic and ecosystem response, and carbon drawdown on local, regional, and global scales [*Jenkyns*, 2010, and references therein]. The end-Triassic mass extinction and T–J transition interval have generally not been considered to be an OAE, but recent studies do suggest anoxic-euxinic conditions developing at this time. Several studies suggest the increased flux to and/or preservation of organic carbon in marine sediments to explain the deposition of laminated organic-rich black shales described from TJB strata [*Wignall*, 2001b; *Bonis et al.*, 2010; *Clemence et al.*, 2010; *Ruhl et al.*, 2010; *Richoz et al.*, 2012; *Thibodeau et al.*, 2016]. In addition, organic geochemical analyses of T–J deposits in multiple marginal marine basins suggest the increased abundance of isorenieratane biomarkers, derived from bacteria living in a euxinic photic zone, and gammacerane from the boundary between water masses in a stratified water column [*Richoz et al.*, 2012; *Jaraula et al.*, 2013; *Kasprak et al.*, 2015].

10.5. CONCLUSIONS

The T–J transition is marked by significant changes in δ^{13}C values in organic and inorganic substrates from marine and continental (terrestrial and lacustrine) sedimentary records. A (major) negative CIE predates the TJB by ~100–200 kyr and directly coincides with the end-Triassic mass extinction recorded in marine realms. The observed δ^{13}C signature at that time suggests the release of isotopically light carbon into the ocean-atmosphere system. The temporal relationship between T–J carbon cycle change and the emplacement of the CAMP suggests a potentially causative effect. Different mechanisms of carbon release have been proposed, including (i) carbon degassing directly from CAMP basalts, (ii) thermogenic carbon (methane) release from subsurface organic-rich sediments by intruding sills related to CAMP emplacement, and (iii) methane clathrate release from seafloor sediments in response to initial CAMP carbon release which might be associated with global warming. At this point, contrasts in timing and amplitude observed in T–J δ^{13}C curves between different localities, depositional environments, and proxy records do not allow a full consensus on carbon cycle evolution for the TJB.

Changes in the global exogenic carbon cycle across the T–J transition and related changes in atmospheric and oceanic pCO_2 did likely impact local, regional, and global climates, environments, and depositional conditions. Changes in marine δ^{18}O values are suggestive of global warming, and seawater [87]Sr/[86]Sr and [188]Os/[187]Os changes suggest increase in continental weathering rates, while other geochemical and sedimentological markers suggest changes in the redox state of, at least, marginal marine basins. The global warming in the Early Jurassic, however, could also reflect a shift back to normal conditions after a cool interval, the latter triggered by aerosol clouds originating from sulfuric acid exhaled from CAMP volcanism. This sulfuric acid would acidify the rain and could also explain enhanced weathering at that time.

Biostratigraphically and chemostratigraphically well constrained T–J sedimentary records, especially from the open marine realm, are relatively scarce compared to other Early Jurassic, Cretaceous, or Cenozoic global change events (such as OAEs). The T–J, however, stands out as it arguably has one of the best constrained stratigraphic frameworks linking continental LIP emplacement with the marine sedimentary environments.

Recent and increasing interest in the processes controlling events at the T–J transition and the increasing number of marine and terrestrial, and globally distributed, sedimentary records being studied strongly enhance our understanding of the drivers of marine and continental paleoclimatic, paleoenvironmental, and paleobiotic change across this highly enigmatic interval in Earth's history.

ACKNOWLEDGMENTS

We thank two anonymous reviewers for providing critical comments that helped to improve the quality of this review. This publication is MTA-MTM-ELTE Paleo contribution no. 259.

REFERENCES

Alroy, J. (2010). The shifting balance of diversity among major marine animal groups. Science, 329(5996), 1191–1194. doi:10.1126/science.1189910.

Al-Suwaidi, A. H., Steuber, T., & Suarez, M. B. (2016). The Triassic–Jurassic boundary event from an equatorial carbonate platform (Ghalilah Formation, United Arab Emirates). Journal of the Geological Society, 173, 949–953. doi:10.1144/jgs2015-102

Alvarez, L., Alvarez, W., Asaro, F., & Michel, H. (1980). Extraterrestrial cause for the Cretaceous-Tertiary extinction. Science, 208(4448), 1095–1108. doi:10.1126/science.208.4448.1095.

Bachan, A., van de Schootbrugge, B., Fiebig, J., McRoberts, C. A., Ciarapica, G., & Payne, J. L. (2012). Carbon cycle dynamics following the end-Triassic mass extinction: Constraints from paired δ^{13}C$_{carb}$ and δ^{13}C$_{org}$ records. Geochemistry, Geophysics, Geosystems, 13(9), Q09008. doi:10.1029/2012GC004150.

Bacon, K. L., Belcher, C. M., Hesselbo, S. P., & McElwain, J. C. (2011). The Triassic-Jurassic boundary carbon-isotope excursions expressed in taxonomically identified leaf cuticles. Palaios, 26(7–8), 461–469. doi:10.2110/palo.2010.p10-120r.

Bacon, K. L., Belcher, C. M., Haworth, M., & McElwain, J. C. (2013). Increased atmospheric SO_2 detected from changes in leaf physiognomy across the Triassic–Jurassic boundary interval of East Greenland. PLOS One, 8(4), e60614. doi:10.1371/journal.pone.0060614.

Badjukov, D. D., Lobitzer, H., & Nazarov, M. A. (1987). Quartz grains with planar features in the Triassic-Jurassic boundary sediments from Northern Calcareous Alps, Austria. Lunar and Planetary Science Letters, 18, 38–39.

Bartolini, A., Guex, J., Spangenberg, J., Schoene, B., Taylor, D., Schaltegger, U., & Atudorei, V. (2012). Disentangling the Hettangian carbon isotope record: Implications for the aftermath of the end-Triassic mass extinction. Geochemistry, Geophysics, Geosystems, 13(1), Q01007. doi:10.1029/2011GC003807.

Belcher, C. M., Mander, L., Rein, G., Jervis, F. X., Haworth, M., Hesselbo, S. P., … McElwain, J. C. (2010). Increased fire activity at the Triassic/Jurassic boundary in Greenland due to climate-driven floral change. Nature Geoscience, 3(6), 426–429. doi:10.1038/ngeo871.

Benoit, J. M., Gilmour, C. C., Mason, R. P., & Heyes, A. (1999). Sulfide controls on mercury speciation and bioavailability to methylating bacteria in sediment pore waters. Environmental Science & Technology, 33(6), 951–957. doi:10.1021/es9808200.

Benoit, J. M., Mason, R. P., Gilmour, C. C., & Aiken, G. R. (2001). Constants for mercury binding by dissolved organic matter isolates from the Florida Everglades. Geochimica et Cosmochimica Acta, 65(24), 4445–4451. doi:10.1016/S0016-7037(01)00742-6.

Benton, M. J. (1995). Diversification and extinction in the history of life. Science, 268(5207), 52–58. doi:10.1126/science.7701342.

Bergquist, B. A. (2017). Mercury, volcanism, and mass extinctions. Proceedings of the National Academy of Sciences of the United States of America, 114(33), 8675–8677. doi:10.1073/pnas.1709070114.

Berra, F., Jadoul, F., & Anelli, A. (2010). Environmental control on the end of the Dolomia Principale/Hauptdolomit depositional system in the central Alps: Coupling sea-level and climate changes. Palaeogeography, Palaeoclimatology, Palaeoecology, 290(1), 138–150. doi:10.1016/j.palaeo.2009.06.037.

Blackburn, T. J., Olsen, P. E., Bowring, S. A., McLean, N. M., Kent, D. V., Puffer, J., ... Et-Touhami, M. (2013). Zircon U-Pb geochronology links the end-Triassic extinction with the Central Atlantic Magmatic Province. Science, 340(6135), 941–945. doi:10.1126/science.1234204.

Bloos, G. (2004). Psiloceratids of the earliest Jurassic in the North-West European and Mediterranean Provinces: Remarks and new observations. Stuttgarter Beiträge zur Naturkunde, Serie B, 347, 1–15.

Blum, J. D., Sherman, L. S., & Johnson, M. W. (2014). Mercury isotopes in Earth and environmental sciences. Annual Review of Earth and Planetary Sciences, 42, 249–269. doi:10.1146/annurev-earth-050212-124107.

Böhm, F., Ebli, O., Krystyn, L., Lobitzer, H., Rakús, M., & Siblík, M. (1999), Fauna, stratigraphy and depositional environment of the Hettangian-Sinemurian (Early Jurassic) of Adnet (Salzburg, Österreich). Abhandlungen der geologischen Bundesanstalt, 56(2), 143–271.

Bonis, N. R., & Kürschner, W. M. (2012). Vegetation history, diversity patterns, and climate change across the Triassic/Jurassic boundary. Paleobiology, 38(2), 240–264. doi:10.1666/09071.1.

Bonis, N. R., Kürschner, W. M., & Krystyn, L. (2009a). A detailed palynological study of the Triassic-Jurassic transition in key sections of the Eiberg Basin (Northern Calcareous Alps, Austria). Review of Palaeobotany and Palynology, 156(3–4), 376–400. doi:10.1016/j.revpalbo.2009.04.003.

Bonis, N. R., Ruhl, M., & Kürschner, W. M. (2009b). Climate change driven black shale deposition during the end-Triassic in the western Tethys. Palaeogeography, Palaeoclimatology, Palaeoecology, 290(1–4), 151–159. doi:10.1016/j.palaeo.2009.06.016.

Bonis, N. R., Ruhl, M., & Kürschner, W. M. (2010). Milankovitch-scale palynological turnover across the Triassic-Jurassic transition at St. Audrie's Bay, SW UK. Journal of the Geological Society, 167(5), 877–888. doi:10.1144/0016-76492009-141.

Bottini, C., Jadoul, F., Rigo, M., Zaffani, M., Artoni, C., & Erba, E. (2016). Calcareous nannofossils at the Triassic/Jurassic boundary: Stratigraphic and paleoceanographic characterization. Rivista Italiana di Paleontologia e Stratigrafia, 122(3), 141–164. doi:10.13130/2039-4942/7726.

Bowman, K. L., Hammerschmidt, C. R., Lamborg, C. H., & Swarr, G. (2015). Mercury in the North Atlantic Ocean: The U.S. GEOTRACES zonal and meridional sections. Deep-Sea Research II, 116, 251–261. doi:10.1016/j.dsr2.2014.07.004.

Bryson, R. A., & Goodman, B. M. (1980). Volcanic activity and climatic changes. Science, 207(4435), 1041–1044. doi:10.1126/science.207.4435.1041.

Cadle, R. D., Kiang, C. S., & Louis, J.-F. (1976). The global scale dispersion of the eruption clouds from major volcanic eruptions. Journal of Geophysical Research, 81, 3125–3132. doi:10.1029/JC081i018p03125.

Callegaro, S., Rigo, M., Chiaradia, M., & Marzoli, A. (2012). Latest Triassic marine Sr isotopic variations, possible causes and implications. Terra Nova, 24(2), 130–135. doi:10.1111/j.1365-3121.2011.01046.x.

Carter, E. S. (1993). Biochronology and paleontology of uppermost Triassic (Rhaetian) radiolarians, Queen Charlotte Islands, British Columbia, Canada. Mémoires de Géologie (Lausanne), 11, 1–175.

Carter, E. S., & Hori, R. S. (2005). Global correlation of the radiolarian faunal change across the Triassic-Jurassic boundary. Canadian Journal of Earth Sciences, 42(5), 777–790. doi:10.1139/e05-020.

Clémence, M.-E., Bartolini, A., Gardin, S., Paris, G., Beaumont, V., & Page, K. N. (2010). Early Hettangian benthic-planktonic coupling at Doniford (SW England): Palaeoenvironmental implications for the aftermath of the end-Triassic crisis. Palaeogeography, Palaeoclimatology, Palaeoecology, 295(1–2), 102–115. doi:10.1016/j.palaeo.2010.05.021.

Cleveland, D. M., Nordt, L. C., Dworkin, S. I., & Atchley, S. C. (2008). Pedogenic carbonate isotopes as evidence for extreme climatic events preceding the Triassic-Jurassic boundary: Implications for the biotic crisis? Geological Society of America Bulletin, 120(11–12), 1408–1415. doi:10.1130/b26332.1.

Cohen, A. S. & Coe, A. L. (2002). New geochemical evidence for the onset of volcanism in the Central Atlantic magmatic province and environmental change at the Triassic–Jurassic boundary. Geology, 30(3), 267–270. doi:10.1130/0091-7613(2002)030<0267:NGEFTO>2.0.CO;2.

Cohen, A. S. & Coe, A. L. (2007). The impact of the Central Atlantic Magmatic Province on climate and on the Sr- and Os-isotope evolution of seawater. Palaeogeography, Palaeoclimatology, Palaeoecology, 244(1–4), 374–390. doi:10.1016/j.palaeo.2006.06.036.

Cohen, A. S., Coe, A. L., Bartlett, J. M., & Hawkesworth, C. J. (1999). Precise Re-Os ages of organic-rich mudrocks and the Os isotope composition of Jurassic seawater. Earth and Planetary Science Letters, 167, 159–173. doi:10.1016/S0012-821X(99)00026-6.

Crne, A. E., Weissert, H., Gorican, S., & Bernasconi, S. M. (2011). A biocalcification crisis at the Triassic-Jurassic boundary recorded in the Budva Basin (Dinarides, Montenegro). Geological Society of America Bulletin, 123(1–2), 40–50. doi:10.1130/b30157.1.

Crocket, J. H. (1979). Platinum-group elements in mafic and ultramafic rocks: a survey. The Canadian Mineralogist, 17(2), 391–402.

Crocket, J. H. & Teruta, Y. (1977). Palladium, iridium, and gold contents of mafic and ultramafic rocks drilled from the Mid-Atlantic Ridge, Leg 37, Deep Sea Drilling Project. Canadian Journal of Earth Sciences, 14(4), 777–784. doi:10.1139/e77-076.

Dal Corso, J., Marzoli, A., Tateo, F., Jenkyns, H. C., Bertrand, H., Youbi, N., … Cirilli, S. (2014). The dawn of CAMP volcanism and its bearing on the end-Triassic carbon cycle disruption. Journal of the Geological Society, 171(2), 153–164. doi:10.1144/jgs2013-063.

Dam, G. & Surlyk, F. (1992). Forced regressions in a large wave- and storm-dominated anoxic lake, Rhaetian-Sinemurian Kap Stewart Formation, East Greenland. Geology, 20(8), 749–752. doi:10.1130/0091-7613(1992)020<0749:FRIALW>2.3.CO;2.

Damborenea, S. E., & Manceñido, M. O. (2012). Late Triassic bivalves and brachiopods from southern Mendoza, Argentina. Revue de Paléobiologie, Genève, 11, 317–344.

Davies, J. H. F. L., Marzoli, A., Bertrand, H., Youbi, N., Ernesto, M., & Schaltegger, U. (2017). End-Triassic mass extinction started by intrusive CAMP activity. Nature Communications, 8, 15596. doi:10.1038/ncomms15596.

Deenen, M. H. L., Ruhl, M., Bonis, N. R., Krijgsman, W., Kuerschner, W. M., Reitsma, M., & van Bergen, M. J. (2010). A new chronology for the end-Triassic mass extinction. Earth and Planetary Science Letters, 291(1–4), 113–125. doi:10.1016/j.epsl.2010.01.003.

Deenen, M. H. L., Krijgsman, W., & Ruhl, M. (2011). The quest for chron E23r at Partridge Island, Bay of Fundy, Canada: CAMP emplacement postdates the end-Triassic extinction event at the North American craton. Canadian Journal of Earth Sciences, 48(8), 1282–1291. doi:10.1139/e11-015.

Desrochers, A., & Orchard, M. J. (1991). Stratigraphic revisions and carbonate sedimentology of the Kunga Group (Upper Triassic - Lower Jurassic), Queen Charlotte Islands, British Columbia. In G. J. Woodsworth (Ed.), Evolution and Hydrocarbon Potential of the Queen Charlotte Basin, British Columbia, Geological Survey of Canada (Paper no. 90–10, pp. 163–172). Geological Survey of Canada, Ottawa. doi:10.4095/131970.

Devine, J. D., Sigurdsson, H., Davis, A. N., & Self, S. (1984). Estimates of sulfur and chlorine yield to the atmosphere from volcanic eruptions and potential climatic effects. Journal of Geophysical Research, 89(B7), 6309–6325. doi:10.1029/JB089iB07p06309.

Dickinson, W. R. (2004). Evolution of the North American Cordillera. Annual Review of Earth and Planetary Sciences, 32, 13–45. doi:10.1146/annurev.earth.32.101802.120257.

Ehmann, W. D., Baedecker, P. A., & McKown, D. M. (1970). Gold and iridium in meteorites and some selected rocks. Geochimica et Cosmochimica Acta, 34(4), 493–507. doi:10.1016/0016-7037(70)90140-7.

El Ghilani, S., Youbi, N., Madeira, J., Chellai, E. H., López-Galindo, A., Martins, L., & Mata, J. (2017). Environmental implication of subaqueous lava flows from a continental Large Igneous Province: Examples from the Moroccan Central Atlantic Magmatic Province (CAMP). Journal of African Earth Sciences, 127, 211–221. doi:10.1016/j.jafrearsci.2016.07.021.

Felber, R., Weissert, H. J., Furrer, H., & Bontognali, T. R. R. (2015). The Triassic–Jurassic boundary in the shallow-water marine carbonates from the western Northern Calcareous Alps (Austria). Swiss Journal of Geosciences, 108(2–3), 213–224. doi:10.1007/s00015-015-0192-1.

Font, E., Adatte, T., Keller, G., Abrajevitch, A., Sial, A. N., de Lacerda, L. D., & Punekar, J. (2016). Mercury anomaly, Deccan volcanism and the end-Cretaceous mass extinction: REPLY. Geology, 44(3), e382. doi:10.1130/g37717y.1.

Friedman, R., Mundil, R., & Pálfy, J. (2008). Revised zircon U-Pb ages for the Triassic-Jurassic boundary and the earliest Jurassic employing the chemical abrasion pretreatment (CA-TIMS) technique. Geochimica et Cosmochimica Acta, 72(12), A284. doi:10.1016/j.gca.2008.05.009.

Galli, M. T., Jadoul, F., Bernasconi, S. M., & Weissert, H. (2005). Anomalies in global carbon cycling and extinction at the Triassic/Jurassic boundary: Evidence from a marine C-isotope record. Palaeogeography, Palaeoclimatology, Palaeoecology, 216(3–4), 203–214. doi:10.1016/j.palaeo.2004.11.009.

Galli, M. T., Jadoul, F., Bernasconi, S. M., Cirilli, S., & Weissert, H. (2007). Stratigraphy and palaeoenvironmental analysis of the Triassic-Jurassic transition in the western Southern Alps (Northern Italy). Palaeogeography, Palaeoclimatology, Palaeoecology, 244(1–4), 52–70. doi:10.1016/j.palaeo.2006.06.023.

Genin, A., Lazar, B., & Brenner, S. (1995). Vertical mixing and coral death in the Red Sea following the eruption of Mount Pinatubo. Nature, 377(6549), 507–510. doi:10.1038/377507a0.

Golebiowski, R. (1989). Stratigraphie und Biofazies der Kössener Formation (Obertrias, Nördliche Kalkalpen) (Doctoral dissertation). University of Vienna.

Golebiowski, R. (1990). Facial and faunistic changes from Triassic to Jurassic in the Northern Calcareous Alps (Austria). Les Cahiers de l'Université Catholique de Lyon, Série Sciences, 3, 175–184.

Götz, A. E., Ruckwied, K., Pálfy, J., & Haas, J. (2009). Palynological evidence of synchronous changes within the terrestrial and marine realm at the Triassic/Jurassic boundary (Csővár section, Hungary). Review of Palaeobotany and Palynology, 156(3–4), 401–409. doi:10.1016/j.revpalbo.2009.04.002.

Greene, S. E., Martindale, R. C., Ritterbush, K. A., Bottjer, D. J., Corsetti, F. A., & Berelson, W. M. (2012). Recognising ocean acidification in deep time: An evaluation of the evidence for acidification across the Triassic-Jurassic boundary. Earth-Science Reviews, 113(1–2), 72–93. doi:10.1016/j.earscirev.2012.03.009.

Grieve, R. A. F. (1998). Extraterrestrial impacts on earth: the evidence and the consequences. Geological Society, London, Special Publications, 140(1), 105–131. doi:10.1144/gsl.sp.1998.140.01.10.

Guex, J. (1995). Ammonites hettangiennes de la Gabbs Valley Range (Nevada, USA). Mémoires de Géologie (Lausanne), 27, 1–131.

Guex, J., Rakus, M., Taylor, D., & Bucher, H. (1997). Proposal for the New York Canyon area, Gabbs Valley Range (Nevada) USA. ISJS Newsletter, 24, 26–30.

Guex, J., Bartolini, A., Atudorei, V., & Taylor, D. (2003). Two negative $\delta^{13}C_{org}$ excursions near the Triassic-Jurassic boundary in the New York Canyon area (Gabbs Valley Range, Nevada). Bulletin de Géologie Lausanne, *360*, 4 unnumbered pages.

Guex, J., Bartolini, A., Atudorei, V., & Taylor, D. (2004). High-resolution ammonite and carbon isotope stratigraphy across the Triassic-Jurassic boundary at New York Canyon (Nevada). Earth and Planetary Science Letters, *225*(1–2), 29–41. doi:10.1016/j.epsl.2004.06.006.

Guex, J., Bartolini, A., Taylor, D., Atudorei, V., Thelin, P., Bruchez, S., … Lucas, S. (2009). Comment on: "The organic carbon isotopic and paleontological record across the Triassic-Jurassic boundary at the candidate GSSP section at Ferguson Hill, Muller Canyon, Nevada, USA" by Ward et al. (2007). Palaeogeography, Palaeoclimatology, Palaeoecology, *273*(1–2), 200–204. doi:10.1016/j.palaeo.2008.01.010.

Guex, J., Schoene, B., Bartolini, A., Spangenberg, J., Schaltegger, U., O'Dogherty, L., … Atudorei, V. (2012). Geochronological constraints on post-extinction recovery of the ammonoids and carbon cycle perturbations during the Early Jurassic. Palaeogeography, Palaeoclimatology, Palaeoecology, *346*, 1–11. doi:10.1016/j.palaeo.2012.04.030.

Haas, J., & Tardy-Filácz, E. (2004). Facies changes in the Triassic-Jurassic boundary interval in an intraplatform basin succession at Csővár (Transdanubian Range, Hungary). Sedimentary Geology, *168*(1–2), 19–48. doi:10.1016/j.sedgeo.2004.03.002.

Haas, J., Tardi-Filácz, E., Oravecz-Scheffer, A., Góczán, F., & Dosztály, L. (1997). Stratigraphy and sedimentology of an Upper Triassic toe-of-slope and basin succession at Csővár, Hungary. Acta Geologica Hungarica, *40*(2), 111–177.

Haas, J., Götz, A. E., & Pálfy, J. (2010). Late Triassic to Early Jurassic palaeogeography and eustatic history in the NW Tethyan realm: New insights from sedimentary and organic facies of the Csővár Basin (Hungary). Palaeogeography, Palaeoclimatology, Palaeoecology, *291*(3–4), 456–468. doi:10.1016/j.palaeo.2010.03.014.

Hallam, A., & Goodfellow, W. D. (1990). Facies and geochemical evidence bearing on the end-Triassic disappearance of the Alpine reef ecosystem. Historical Biology, *4*(2), 131–138. doi:10.1080/08912969009386538.

Hallam, A., & Wignall, P. B. (1997). Mass Extinctions and Their Aftermath. Oxford University Press, Oxford.

Hallam, A., & Wignall, P. B. (2000), Facies changes across the Triassic-Jurassic boundary in Nevada, USA. Journal of the Geological Society, London, *157*, 49–54. doi:10.1144/jgs.157.1.49.

Harris, T. M. (1937). The fossil flora of Scoresby Sound East Greenland. Part 5: Stratigraphic relations of the plant beds. Meddelelser om Groenland, *112*(2), 1–112.

Hautmann, M., Benton, M. J., & Tomasovych, A. (2008). Catastrophic ocean acidification at the Triassic-Jurassic boundary. Neues Jahrbuch für Geologie und Paläontologie - Abhandlungen, *249*(1), 119–127. doi:10.1127/0077-7749/2008/0249-0119.

Hesselbo, S. P., Robinson, S. A., Surlyk, F., & Piasecki, S. (2002). Terrestrial and marine mass extinction at the Triassic–Jurassic boundary synchronized with major carbon-cycle perturbation: A link to initiation of massive volcanism? Geology, *30*(3), 251–254. doi:10.1130/0091-7613(2002)030<0251:TAMEAT>2.0.CO;2.

Hesselbo, S. P., Robinson, S. A., & Surlyk, F. (2004). Sea-level change and facies development across potential Triassic–Jurassic boundary horizons, SW Britain. Journal of the Geological Society, London, *161*(3), 365–379. doi:10.1144/0016-764903-033.

Hesselbo, S. P., McRoberts, C. A., & Pálfy, J. (2007). Triassic–Jurassic boundary events: Problems, progress, possibilities. Palaeogeography, Palaeoclimatology, Palaeoecology, *244*(1–4), 1–10. doi:10.1016/j.palaeo.2006.06.020.

Hillebrandt, A. v., & Krystyn, L. (2009). On the oldest Jurassic ammonites of Europe (Northern Calcareous Alps, Austria) and their global significance. Neues Jahrbuch für Geologie und Paläontologie - Abhandlungen, *253*, 163–195. doi:10.1127/0077-7749/2009/0253-0163.

Hillebrandt, A. v., Krystyn, L., Kürschner, W. M., Bonis, N. R., Ruhl, M., Richoz, S., … Tomãsových, A. (2013). The Global Stratotype Sections and Point (GSSP) for the base of the Jurassic System at Kuhjoch (Karwendel Mountains, Northern Calcareous Alps, Tyrol, Austria). Episodes, *36*(3), 162–198.

Hodych, J. P., & Dunning, G. R. (1992). Did the Manicouagan impact trigger end-of-Triassic mass extinction? Geology, *20*(1), 51–54. doi:10.1130/0091-7613(1992)020<0051:DTMITE>2.3.CO;2.

Hönisch, B., Ridgwell, A., Schmidt, D. N., Thomas, E., Gibbs, S. J., … Williams, B. (2012). The geological record of ocean acidification. Science, *335*(6072), 1058–1063. doi:10.1126/science.1208277.

Hounslow, M. W., Posen, P. E., & Warrington, G. (2004). Magnetostratigraphy and biostratigraphy of the Upper Triassic and lowermost Jurassic succession, St. Audrie's Bay, UK. Palaeogeography, Palaeoclimatology, Palaeoecology, *213*(3–4), 331–358. doi:10.1016/j.palaeo.2004.07.018.

Hubbard, R. N. L. B., & Boulter, M. C. (2000). Phytogeography and paleoecology in western Europe and eastern Greenland near the Triassic-Jurassic boundary. Palaios, *15*, 120–131. doi:10.1669/0883-1351(2000)015<0120:PAPIWE>2.0.CO;2.

Hüsing, S. K., Beniest, A., van der Boon, A., Abels, H. A., Deenen, M. H. L., Ruhl, M., & Krijgsman, W. (2014). Astronomically-calibrated magnetostratigraphy of the Lower Jurassic marine successions at St. Audrie's Bay and East Quantoxhead (Hettangian-Sinemurian; Somerset, UK). Palaeogeography, Palaeoclimatology, Palaeoecology, *403*, 43–56. doi:10.1016/j.palaeo.2014.03.022.

Huynh, T. T., & Poulsen, C. J. (2005). Rising atmospheric CO_2 as a possible trigger for the end-Triassic mass extinction. Palaeogeography, Palaeoclimatology, Palaeoecology, *217*(3–4), 223–242. doi:10.1016/j.palaeo.2004.12.004.

Ivanov B. A. (1992). Geomechanical models of impact cratering: Puchezh-Katunki structure. In B. O. Dressier, R. A. F. Grieve, & V. L. Sharpton (Eds.), Large Meteorite Impacts and Planetary Evolution, *Geological Society of America Special Paper* (Vol. 293, pp. 81–91). GeoScienceWorld, McLean, VA. doi:10.1130/SPE293-p81.

Jaraula, C. M. B., Grice, K., Twitchett, R. J., Böttcher, M. E., LeMetayer, P., Dastidar, A. G., & Opazo, L. F. (2013). Elevated pCO_2 leading to Late Triassic extinction, persistent photic zone euxinia, and rising sea levels. Geology, *41*(9), 955–958. doi:10.1130/g34183.1.

Jenkyns, H. C. (2010). Geochemistry of oceanic anoxic events. Geochemistry, Geophysics, Geosystems, *11*, Q03004. doi:10.1029/2009GC002788.

Jenkyns, H. C., Jones, C. E., Gröcke, D. R., Hesselbo, S. P., & Parkinson, D. N. (2002). Chemostratigraphy of the Jurassic System: Applications, limitations and implications for palaeoceanography. Journal of the Geological Society, London, *159*, 351–378. doi:10.1144/0016-764901-130.

Jones, C. E., 1992. Strontium isotopes in Jurassic and Early Cretaceous seawater (PhD dissertation). Oxford University.

Jones, C. E., Jenkyns, H. C., & Hesselbo, S. P. (1994a). Strontium isotopes in Early Jurassic seawater. Geochimica et Cosmochimica Acta, *58*, 1285–1301. doi:10.1016/0016-7037(94)90382-4.

Jones, C. E., Jenkyns, H. C., Coe, A. L., & Hesselbo, S. P. (1994b). Strontium isotopic variations in Jurassic and Cretaceous seawater. Geochimica, Cosmochimica Acta, *58*, 3061–3074. doi:10.1016/0016-7037(94)90179-1.

Kasprak, A. H., Sepúlveda, J., Price-Waldman, R., Williford, K. H., Schoepfer, S. D., Haggart, J. W., … Whiteside, J. H. (2015). Episodic photic zone euxinia in the northeastern Panthalassic Ocean during the end-Triassic extinction. Geology *43*(4), 307–310. doi:10.1130/g36371.1.

Kennett, J. P., & Thunell, R. C. (1975). Global increase in quaternary explosive volcanism. Science, *187*(4176), 497–502. doi:10.1126/science.187.4176.497.

Kent, D. V., Olsen, P. E., & Muttoni, G. (2017). Astrochronostratigraphic polarity time scale (APTS) for the Late Triassic and Early Jurassic from continental sediments and correlation with standard marine stages. Earth-Science Reviews, *166*, 153–180. doi:10.1016/j.earscirev.2016.12.014.

Kiessling, W. (2009). Geologic and biologic controls on the evolution of reefs. Annual Review of Ecology, Evolution, and Systematics, *40*(1), 173–192. doi:10.1146/annurev. ecolsys.110308.120251.

Kiessling, W., Roniewicz, E., Villier, L., Leonide, P., & Struck, U. (2009). An Early Hettangian coral reef in southern France: Implications for the End-Triassic reef crisis. Palaios, *24*(9–10), 657–671. doi:10.2110/palo.2009.p09-030r.

Kongchum, M., Hudnall, W. H., & Delaune, R. D. (2011). Relationship between sediment clay minerals and total mercury. Journal of Environmental Science and Health, Part A, *46*(5), 534–539. doi:10.1080/10934529.2011.551745.

Korte, C. (1999). $^{87}Sr/^{86}Sr$-, $\delta^{18}O$- und $\delta^{13}C$-Evolution des triassischen Meerwassers: Geochemische und stratigraphische Untersuchungen an Conodonten und Brachiopoden. Bochumer geologische und geotechnische Arbeiten, *52*, 1–171.

Korte, C., & Hesselbo, S. P. (2011). Shallow marine carbon and oxygen isotope and elemental records indicate icehouse–greenhouse cycles during the Early Jurassic. Paleoceanography, *26*, PA4219. doi:10.1029/2011PA002160.

Korte, C., & Kozur, H. W. (2011). Bio- and chemostratigraphic assessment of carbon isotope records across the Triassic–Jurassic boundary at Csővár quarry (Hungary) and Kendlbachgraben (Austria) and implications for global correlations. Bulletin of the Geological Society of Denmark, *59*, 101–115.

Korte, C., Kozur, H.W., Bruckschen, P., & Veizer, J. (2003). Strontium isotope evolution of Late Permian and Triassic seawater. Geochimica et Cosmochimica Acta, *67*(1), 47–62. doi:10.1016/S0016-7037(02)01035-9.

Korte, C., Kozur, H. W., & Veizer, J. (2005). $\delta^{13}C$ and $\delta^{18}O$ values of Triassic brachiopods and carbonate rocks as proxies for coeval seawater and palaeotemperature. Palaeogeography, Palaeoclimatology, Palaeoecology, *226*(3–4), 287–306. doi:10.1016/j.palaeo.2005.05.018.

Korte, C., Hesselbo, S. P., Jenkyns, H. C., Rickaby, R. E. M., & Spötl, C. (2009). Palaeoenvironmental significance of carbon- and oxygen-isotope stratigraphy of marine Triassic–Jurassic boundary sections in SW Britain. Journal of the Geological Society, London, *166*, 431–445. doi:10.1144/0016-76492007-177.

Korte, C., Hesselbo, S. P., Ullmann, C. V., Dietl, G., Ruhl, M., Schweigert, G., & Thibault, N. (2015). Jurassic climate mode governed by ocean gateway. Nature Communications, *6*, 10015. doi:10.1038/ncomms10015.

Korte, C., Thibault, N., Ullmann, C. V., Clémence, M.-E., Mette, W., Olsen, T. K. … Ruhl, M. (2017). Brachiopod biogeochemistry and isotope stratigraphy from the Rhaetian Eiberg section in Austria: Potentials and limitations. Neues Jahrbuch für Geologie und Paläontologie Abhandlungen, *284*(2), 117–138. doi:10.1127/njgpa/2017/0651.

Kozur, H. (1993). First evidence of Liassic in the vicinity of Csővár (Hungary), and its paleogeographic and paleotectonic significance. Jahrbuch der Geologischen Bundesanstalt, *136*(1), 89–98.

Krystyn, L., Böhm, F., Kürschner, W., & Delecat, S. (2005). The Triassic–Jurassic boundary in the Northern Calcareous Alps. In J. Pálfy, & P. Ozsvárt (Eds.), Program, Abstracts and Field Guide. 5th Field Workshop of IGCP 458 Project (Tata and Hallein) (pp. A1–A14).

Kuerschner, W. M., Bonis, N. R., & Krystyn, L. (2007). Carbon-isotope stratigraphy and palynostratigraphy of the Triassic–Jurassic transition in the Tiefengraben section: Northern Calcareous Alps (Austria). Palaeogeography, Palaeoclimatology, Palaeoecology, *244*(1–4), 257–280. doi:10.1016/j.palaeo.2006.06.031.

Lamb, H. H. (1970). Volcanic dust in the atmosphere; with a chronology and assessment of its meteorological significance. Philosophical transactions of the Royal Society of London A, *266*(1178), 425–533. doi:10.1098/rsta.1970.0010.

Laws, R. A. (1982). Late Triassic depositional environments and molluscan associations from west-central Nevada. Palaeogeography, Palaeoclimatology, Palaeoecology, *37*, 131–148. doi:10.1016/0031-0182(82)90036-0.

Lewis, P. D., Haggart, J. W., Anderson, R. G., Hickson, C. J., Thompson, R. I., Dietrich, J. R., & Rohr, K. M. M. (1991). Triassic to Neogene geologic evolution of the Queen Charlotte region. Canadian Journal of Earth Sciences, *28*, 854–869. doi:10.1139/e91-07.

Lindström, S., van de Schootbrugge, B., Dybkjaer, K., Pedersen, G. K., Fiebig, J., Nielsen, L. H., & Richoz, S. (2012). No causal link between terrestrial ecosystem change and methane release during the end-Triassic mass extinction. Geology, *40*(6), 531–534. doi:10.1130/g32928.1.

Lindström, S., Pedersen, G. K., van de Schootbrugge, B., Hansen, K. H., Kuhlmann, N., … Tegner, C. (2015). Intense and widespread seismicity during the end-Triassic mass extinction due to emplacement of a large igneous province. Geology, *43*(5), 387–390. doi:10.1130/G36444.1.

Lindström, S., van de Schootbrugge, B., Hansen, K. H., Pedersen, G. K., Alsen, P., Thibault, N., ... Nielsen, L. H. (2017). A new correlation of Triassic–Jurassic boundary successions in NW Europe, Nevada and Peru, and the Central Atlantic Magmatic Province: A time-line for the end-Triassic mass extinction. Palaeogeography, Palaeoclimatology, Palaeoecology, *478*, 80–102. doi: 10.1016/j.palaeo.2016.12.025.

Longridge, L. M., Carter, E. S., Smith, P. L., & Tipper, H. W. (2007). Early Hettangian ammonites and radiolarians from the Queen Charlotte Islands, British Columbia and their bearing on the definition of the Triassic-Jurassic boundary. Palaeogeography, Palaeoclimatology, Palaeoecology, *244* (1–4), 142–169. doi:10.1016/j.palaeo.2006.06.027.

Longridge, L. M., Pálfy, J., Smith, P. L., & Tipper, H. W. (2008). Middle and late Hettangian (Early Jurassic) ammonites from the Queen Charlotte Islands, British Columbia, Canada. Revue de Paléobiologie, *27*(1), 191–248.

Lucas, S. G., Taylor, D. G., Guex, J., Tanner, L. H., & Krainer, K. (2007). Updated proposal for Global Stratotype Section and Point for the base of the Jurassic System in the New York Canyon area, Nevada, USA. ISJS Newsletter, *34*(1), 34–42.

Mander, L., Kürschner, W. M., & McElwain, J. C. (2013). Palynostratigraphy and vegetation history of the Triassic–Jurassic transition in East Greenland. Journal of the Geological Society, *170*(1), 37–46. doi:10.1144/jgs2012-018.

Marshall, J. D. (1992). Climatic and oceanographic isotopic signals from the carbonate rock record and their preservation. Geological Magazine, *129*(2), 143–160. doi:10.1017/S0016756800008244.

Marzoli, A., Renne, P. R., Piccirillo, E. M., Ernesto, M., Bellieni, G., & De Min, A. (1999). Extensive 200-million-year-old continental flood basalts of the Central Atlantic Magmatic Province. Science, *284*(5414), 616–618. doi:10.1126/science.284.5414.616.

Marzoli, A., Bertrand, H., Knight, K. B., Cirilli, S., Buratti, N., Vérati, C., ... Bellieni, G. (2004). Synchrony of the Central Atlantic magmatic province and the Triassic-Jurassic boundary climatic and biotic crisis. Geology, *32*(11), 973–976. doi:10.1130/G20652.1.

Marzoli, A., Jourdan, F., Puffer, J. H., Cuppone, T., Tanner, L. H., Weems R. E., ... De Min, A. (2011). Timing and duration of the Central Atlantic magmatic province in the Newark and Culpeper basins, Eastern U.S.A. Lithos, *122*(3–4), 175–188. doi:10.1016/j.lithos.2010.12.013.

Marzoli, A., Callegaro, S., Dal Corso, J., Davies, J.H.F.L., Chiaradia, M., Youbi N., ... Jourdan, F. (2018). The Central Atlantic Magmatic Province (CAMP): A Review. In L.H. Tanner (Ed.), The Late Triassic World: Earth in a Time of Transition, *91–125*. Springer International Publishing, Cham. doi:10.1007/978-3-319-68009-5_4.

McArthur, J. M., Howarth, R. J., & Shields, G. A. (2012). Strontium Isotope Stratigraphy. In M. Gradstein, J. G. Ogg, M. D. Schmitz, & G. M. Ogg (Eds.), The Geologic Time Scale (pp. 127–144). Elsevier, Boston. doi:10.1016/B978-0-444-59425-9.00007-X.

McCartney, K., Huffman, A. R., & Tredoux, M. (1990). A paradigm for endogenous causation of mass extinctions. In V. L. Sharpton, & P. D. Ward (Eds.), Global Catastrophes in Earth History: An Interdisciplinary Conference on Impacts, Volcanism, and Mass Mortality (pp. 125–138).

Geological Society of America, Boulder, CO. doi:10.1130/SPE247-p125.

McElwain, J. C., & Punyasena, S. W. (2007). Mass extinction events and the plant fossil record, Trends in Ecology and Evolution, *22*(10), 548–557. doi:10.1016/j.tree.2007.09.003.

McElwain, J. C., Beerling, D. J., & Woodward, F. I. (1999). Fossil plants and global warming at the Triassic-Jurassic Boundary. Science, *285*(5432), 1386–1390. doi:10.1126/science.285.5432.1386.

McElwain, J. C., Wade-Murphy, J., & Hesselbo, S. P. (2005). Changes in carbon dioxide during an oceanic anoxic event linked to intrusion into Gondwana coals. Nature, *435*(7041), 479–482. doi:10.1038/nature03618.

McElwain, J. C., Popa, M. E., Hesselbo, S. P., Haworth, M., & Surlyk, F. (2007). Macroecological responses of terrestrial vegetation to climatic and atmospheric change across the Triassic/Jurassic boundary in East Greenland. Paleobiology, *33*(4), 547–573. doi:10.1666/06026.1.

McElwain, J. C., Wagner, P. J., & Hesselbo, S. P. (2009). Fossil plant relative abundances indicate sudden loss of Late Triassic biodiversity in east Greenland. Science, *324*(5934), 1554–1556. doi:10.1126/science.1171706.

McHone, J. G. (2003). Volatile emissions from Central Atlantic Magmatic Province basalts: Mass assumptions and environmental consequences. In W. E. Hames, J. G. McHone, P. R. Renne, & C. Ruppel (Eds.), The Central Atlantic Magmatic Province: Insights from Fragments of Pangea (pp. 241–254). American Geophysical Union, Washington, DC. doi:10.1029/136GM013.

McLaren, D. J., & Goodfellow, W. D. (1990). Geological and biological consequences of giant impacts. Annual Review of Earth and Planetary Sciences, *18*, 123–171. doi:10.1146/annurev.ea.18.050190.001011.

McRoberts, C. A., Furrer, H., & Jones, D. S. (1997). Palaeo-environmental interpretation of a Triassic-Jurassic boundary section from Western Austria based on palaeoecological and geochemical data. Palaeogeography, Palaeoclimatology, Palaeoecology, *136*, 79–95. doi:10.1016/S0031-0182(97)00074-6.

McRoberts, C. A., Ward, P. D., & Hesselbo, S. (2007). A proposal for the base Hettangian Stage (= base Jurassic System) GSSP at New York Canyon (Nevada, USA) using carbon isotopes. International Subcommission on Jurassic Stratigraphy Newsletter, *34*(1), 43–49.

Merle, R., Marzoli, A., Reisberg, L., Bertrand, H., Nemchin, A., Chiaradia, M., ... McHone, J. G. (2014). Sr, Nd, Pb and Os isotope systematics of CAMP tholeiites from eastern North America (ENA): Evidence of a subduction-enriched mantle source. Journal of Petrology, *55*(1), 133–180. doi:10.1093/petrology/egt063.

Mette, W., Elsler, A., & Korte, C. (2012). Palaeoenvironmental changes in the Late Triassic (Rhaetian) of the Northern Calcareous Alps: Clues from stable isotopes and microfossils. Palaeogeography, Palaeoclimatology, Palaeoecology, *350–352*, 62–72. doi:10.1016/j.palaeo.2012.06.013.

Michalík, J., Lintnerová, O., Gazdzicki, A., & Soták, J. (2007). Record of environmental changes in the Triassic-Jurassic boundary interval in the Zliechov Basin, Western Carpathians. Palaeogeography, Palaeoclimatology, Palaeoecology, *244*(1–4), 71–88. doi:10.1016/j.palaeo.2006.06.024.

Michalík, J., Biroň, A., Lintnerová, O., Götz, A. E., & Ruckwied, K. (2010). Climate change at the Triassic/Jurassic boundary in the northwestern Tethyan realm, inferred from sections in the Tatra Mountains (Slovakia). Acta Geologica Polonica, 60(4), 535–548.

Morante, R., & Hallam, A. (1996). Organic carbon isotopic record across the Triassic-Jurassic boundary in Austria and its bearing on the cause of the mass extinction. Geology, 24(5), 391–394. doi:10.1130/0091-7613(1996)024<0391: OCIRAT> 2.3.CO;2.

Mossman, D. J., Grantham, R. G., & Langenhorst, F. (1998). A search for shocked quartz at the Triassic-Jurassic boundary in the Fundy and Newark basins of the Newark Supergroup. Canadian Journal of Earth Sciences, 35(2), 101–109. doi:10.1139/e97-101.

Muller, S. W., & Ferguson, H. G. (1936). Triassic and Lower Jurassic formations of west central Nevada. Geological Society of America Bulletin, 47, 241–252. doi:10.1130/gsab-47-241.

Newell, N. D. (1967). Revolutions in the history of life. Geological Society of America Special Paper, 89, 63–91.

Niessen, S., Foucher, D., Clarisse, O., Fischer, J. C., Mikac, N., Kwokal, Z., … Horvat, M. (2003). Influence of sulphur cycle on mercury methylation in estuarine sediment (Seine estuary, France). Journal de Physique IV France, 107, 953–956. doi:10.1051/jp4:20030456.

Olsen, P. E., Shubin, N. H., & Anders, M. H. (1987). New Early Jurassic tetrapod assemblages constrain Triassic-Jurassic tetrapod extinction event. Science, 237, 1025–1029. doi:10.1126/science.3616622.

Olsen, P. E., Kent, D. V., Sues, H.-D., Koeberl, C., Huber, H., Montanari, A., … Hartline, B. W. (2002). Ascent of dinosaurs linked to an iridium anomaly at the Triassic-Jurassic boundary. Science, 296(5571), 1305–1307. doi:10.1126/science.1065522.

Orchard, M. J., Carter, E. S., Lucas, S. G., & Taylor, D. G. (2007). Rhaetian (Upper Triassic) conodonts and radiolarians from New York Canyon, Nevada, USA. Albertiana, 35, 59–65.

Orth, C. J. (1989). Geochemistry of the bio-event horizons. In S. K. Donovan (Ed.), Mass Extinctions: Processes and Evidence (pp. 37–72). Belhaven Press, London.

Outridge, P. M., Sanei, H., Stern, G. A., Hamilton, P. B., & Goodarzi, F. (2007). Evidence for control of mercury accumulation rates in Canadian high Arctic Lake sediments by variations of aquatic primary productivity. Environmental Science and Technology, 41(15), 5259–5265. doi:10.1021/es070408x.

Page, K. (2003). The Lower Jurassic of Europe: Its subdivision and correlation. Geological Survey of Denmark and Greenland Bulletin, 1, 23–59.

Page, K., Clémence, M., & Bloos, G. (2010). The Tilmanni Chronozone in NW Europe: Re-correlating the Base of the Jurassic System. Earth Science Frontiers, 17, 8–9.

Palais, J. M., & Sigurdsson, H. (1989). Petrologic evidence of volatile emissions from major historic and pre-historic volcanic eruptions. In A. Berger, R. E. Dickinson, & J. W. Kidson (Eds.), Understanding Climate Change, Geophysical Monograph Series (Vol. 52, pp. 31–53). American Geophysical Union, Washington, DC. doi:10.1029/GM052p0031.

Pálfy, J. (2003). Volcanism of the Central Atlantic Magmatic Province as a potential driving force in the end-Triassic mass extinction. In W. E. Hames, J. G. McHone, P. R. Renne, & C. Ruppel (Eds.), The Central Atlantic Magmatic Province: Insights from Fragments of Pangea, Geophysical Monograph Series (Vol. 136, pp. 255–267). American Geophysical Union, Washington, DC. doi:10.1029/136GM14.

Pálfy, J. (2004). Did the Puchezh-Katunki impact trigger an extinction? In H. Dypvik, M. Burchell, & P. Claeys (Eds.), Cratering in Marine Environments and on Ice (pp. 135–148). Springer, Berlin. doi:10.1007/978-3-662-06423-8_8.

Pálfy, J., & Dosztály, L. (2000). A new marine Triassic-Jurassic boundary section in Hungary: Preliminary results. In R. L. Hall, & P. L. Smith (Eds.), Advances in Jurassic Research 2000 (pp. 173–179). TransTech Publications, Zurich.

Pálfy, J., & Haas, J. (2012). Csővári Mészkő Formáció [Csővár Limestone Formation], in Magyarország litosztratigráfiai alapegységei. In I. Főzy (Ed.), Jura [Lithostratigraphic Units of Hungary. Jurassic] (pp. 34–36). Magyarhoni Földtani Társulat [Hungarian Geological Society], Budapest.

Pálfy, J., & Zajzon, N. (2012). Environmental changes across the Triassic-Jurassic boundary and coeval volcanism inferred from elemental geochemistry and mineralogy in the Kendlbach-graben section (Northern Calcareous Alps, Austria). Earth and Planetary Science Letters, 335, 121–134. doi:10.1016/j.epsl.2012.01.039.

Pálfy, J., & Kocsis, T. Á. (2014). Volcanism of the Central Atlantic Magmatic Province as the trigger of environmental and biotic changes around the Triassic-Jurassic boundary. In G. Keller, A. C. Kerr (Eds.), Volcanism, Impacts and Mass Extinctions: Causes and Effects (pp. 245–261). Geological Society of America, Boulder, CO. doi:10.1130/2014.2505(12).

Pálfy, J., Smith, P. L., & Tipper, H. W. (1994). Sinemurian (Lower Jurassic) ammonoid biostratigraphy of the Queen Charlotte Islands, western Canada. Géobios Mémoire Spécial, 17, 385–393. doi:10.1016/S0016-6995(94)80158-4.

Pálfy, J., Mortensen, J. K., Carter, E. S., Smith, P. L., Friedman, R. M., & Tipper, H. W. (2000). Timing the end-Triassic mass extinction: First on land, then in the sea? Geology, 28(1), 39–42. doi:10.1130/0091-7613(2000)28<39: TTEMEF>2.0.CO;2.

Pálfy, J., Demény, A., Haas, J., Hetényi, M., Orchard, M., & Vető, I. (2001). Carbon isotope anomaly and other geochemical changes at the Triassic-Jurassic boundary from a marine section in Hungary. Geology, 29(11), 1047–1050. doi:10.1130/0091-7613(2001)029<1047:CIAAOG>2.0.CO;2.

Pálfy, J., Demény, A., Haas, J., Carter, E. S., Görög, Á., Halász, D., … Zajzon, N. (2007). Triassic-Jurassic boundary events inferred from integrated stratigraphy of the Csővár section, Hungary. Palaeogeography, Palaeoclimatology, Palaeoecology, 244(1-4), 11–33. doi:10.1016/j.palaeo.2006.06.021.

Percival, L. M. E., Witt, M. L. I., Mather, T. A., Hermoso, M., Jenkyns, H. C., Hesselbo, S. P., … Ruhl, M. (2015). Globally enhanced mercury deposition during the end-Pliensbachian extinction and Toarcian OAE: A link to the Karoo–Ferrar Large Igneous Province. Earth and Planetary Science Letters, 428, 267–280. doi:10.1016/j.epsl.2015.06.064.

Percival, L. M. E., Cohen, A. S., Davies, M. K., Dickson, A. J., Hesselbo, S. P., Jenkyns, H. C., … Xu, W. (2016). Osmium isotope evidence for two pulses of increased continental weathering linked to Early Jurassic volcanism and climate change. Geology, *44*(9), 759–762. doi:10.1130/g37997.1.

Percival, L. M. E., Ruhl, M., Hesselbo, S. P., Jenkyns, H. C., Mather, T. A., & Whiteside, J. H. (2017). Mercury evidence for pulsed volcanism during the end-Triassic mass extinction. Proceedings of the National Academy of Sciences of the United States of America, *114*(30), 7929–7934. doi:10.1073/pnas.1705378114.

Pieńkowski, G., Niedzwiedzki, G., & Waksmundzka, M. (2012). Sedimentological, palynological and geochemical studies of the terrestrial Triassic-Jurassic boundary in northwestern Poland. Geological Magazine, *149*(2), 308–332. doi:10.1017/S0016756811000914.

Pollack J. B., Toon O. B., Sagan C., Summers A., Baldwin B., Van Camp W. (1976). Volcanic explosions and climatic change: A theoretical assessment. Journal of Geophysical Research, *81*, 1071–1083. doi:10.1029/JC081i006p01071.

Popp, B. N., Anderson, T. F., & Sandberg, P. A., (1986). Brachiopods as indicators of original isotopic compositions in some Paleozoic limestones. Geological Society of America Bulletin, *97*, 1262–1269.

Pyle, D. M., & Mather, T. A. (2003). The importance of volcanic emissions for the global atmospheric mercury cycle. Atmospheric Environment, *37*(36), 5115–5124. doi:10.1016/j.atmosenv.2003.07.011.

Rampino, M. R., Self, S., & Stothers, R. B. (1988). Volcanic winters. Annual Review of Earth and Planetary Sciences, *16*, 73–99. doi:10.1146/annurev.ea.16.050188.000445.

Raup, D. M., & Sepkoski, J. J., Jr. (1982). Mass extinctions in the marine fossil record. Science *215*(4539), 1501–1503. doi:10.1126/science.215.4539.1501.

Richoz, S., van de Schootbrugge, B., Pross, J., Püttmann, W., Quan, T. M., Lindström, S., … Wignall, P. B. (2012). Hydrogen sulphide poisoning of shallow seas following the end-Triassic extinction. Nature Geoscience, *5*, 662–667. doi:10.1038/ngeo1539.

Ruhl, M., & Kürschner, W. M. (2011). Multiple phases of carbon cycle disturbance from large igneous province formation at the Triassic-Jurassic transition. Geology, *39*(5), 431–434. doi:10.1130/g31680.1.

Ruhl, M., Kuerschner, W. M., & Krystyn, L. (2009). Triassic-Jurassic organic carbon isotope stratigraphy of key sections in the western Tethys realm (Austria). Earth and Planetary Science Letters, *281*(3–4), 169–187. doi:10.1016/j.epsl.2009.02.020.

Ruhl, M., Deenen, M. H. L., Abels, H. A., Bonis, N. R., Krijgsman, W., & Kürschner, W. M. (2010). Astronomical constraints on the duration of the early Jurassic Hettangian stage and recovery rates following the end-Triassic mass extinction (St Audries Bay/East Quantoxhead, UK). Earth and Planetary Science Letters, *295*(1–2), 262–276. doi:10.1016/j.epsl.2010.04.008.

Ruhl, M., Bonis, N. R., Reichart, G. J., Damste, J. S. S., & Kurschner, W. M. (2011). Atmospheric carbon injection linked to end-Triassic mass extinction. Science, *333*(6041), 430–434. doi:10.1126/science.1204255.

Ruiz-Martínez, V. C., Torsvik, T. H., van Hinsbergen, D. J. J., & Gaina, C. (2012). Earth at 200 Ma: Global palaeogeography refined from CAMP palaeomagnetic data. Earth and Planetary Science Letters, *331–332*, 67–79. doi:10.1016/j.epsl.2012.03.008.

Sanei, H., Grasby, S. E., & Beauchamp, B. (2012). Latest Permian mercury anomalies. Geology, *40*(1), 63–66. doi:10.1130/G32596.1.

Sato, H., Onoue, T., Nozaki, T., & Suzuki, K. (2013). Osmium isotope evidence for a large Late Triassic impact event. Nature Communications, *4*, 2455. doi:10.1038/ncomms3455.

Saunders, A. D. (2005). Large igneous provinces: Origin and environmental consequences. Elements, *1*(5), 259–263. doi:10.2113/gselements.1.5.259.

Scaife, J. D., Ruhl, M., Dickson, A. J., Mather, T. A., Jenkyns, H. C., Percival, L. M. E., … Minisini, D. (2017). Sedimentary mercury enrichments as a marker for submarine large igneous province volcanism? Evidence from the Mid-Cenomanian event and Oceanic Anoxic Event 2 (Late Cretaceous). Geochemistry, Geophysics, Geosystems, *18*, 4253–4275. doi:10.1002/2017GC007153.

Schaller, M. F., Wright, J. D., & Kent, D. V. (2011). Atmospheric pCO_2 perturbations associated with the Central Atlantic Magmatic Province. Science, *331*(6023), 1404–1409. doi:10.1126/science.1199011.

Schaller, M. F., Wright, J. D., Kent, D. V., & Olsen, P. E. (2012). Rapid emplacement of the Central Atlantic Magmatic Province as a net sink for CO_2. Earth and Planetary Science Letters, *323–324*, 27–39. doi:10.1016/j.epsl.2011.12.028.

Schaller, M. F., Wright J. D., & Kent, D. V. (2015). A 30 Myr record of Late Triassic atmospheric pCO_2 variation reflects a fundamental control of the carbon cycle by changes in continental weathering. Geological Society of America Bulletin, *127*(5–6), 661–671. doi:10.1130/B31107.1.

Schaltegger, U., Guex, J., Bartolini, A., Schoene, B., & Ovtcharova, M. (2008). Precise U-Pb age constraints for end-Triassic mass extinction, its correlation to volcanism and Hettangian post-extinction recover. Earth and Planetary Science Letters, *267*(1–2), 266–275. doi:10.1016/j.epsl.2007.11.031.

Schlische, R. W., Withjack, M. O., & Olsen, P. E. (2003). Relative timing of CAMP, rifting, continental breakup, and basin inversion: Tectonic Significance. In W. Hames, J. G. Mchone, P. Renne, & C. Ruppel (Eds.), The Central Atlantic Magmatic Province: Insights from Fragments of Pangea, *Geophysical Monograph Series* (Vol. *136*, pp. 33–59). American Geophysical Union, Washington, DC. doi:10.1029/136GM03.

Schoene, B., Crowley, J. L., Condon, D. J., Schmitz, M. D., & Bowring, S. A. (2006). Reassessing the uranium decay constants for geochronology using ID-TIMS U-Pb data. Geochimica et Cosmochimica Acta *70*(2), 426–445. doi:10.1016/j.gca.2005.09.007.

Schoene, B., Guex, J., Bartolini, A., Schaltegger, U., & Blackburn, T. J. (2010). Correlating the end-Triassic mass extinction and flood basalt volcanism at the 100 ka level. Geology, *38*(5), 387–390. doi:10.1130/G30683.1.

Sial, A. N., Chen, J., Lacerda, L. D., Frei, R., Tewari, V. C., Pandit, M. K., … Pereira, N. S. (2016). Mercury enrichment and Hg isotopes in Cretaceous–Paleogene boundary successions: Links to volcanism and palaeoenvironmental impacts. Cretaceous Research, *66*, 60–81. doi:10.1016/j.cretres.2016.05.006.

Sigurdsson, H. (1982). Volcanic pollution and climate: The 1783 Laki eruption. Eos, Transactions American Geophysical Union, *63*, 601–602.

Sigurdsson, H. (1990). Evidence of volcanic loading of the atmosphere and climate response. Palaeogeography, Palaeoclimatology, Palaeoecology, 89(3), 277–289. doi:10.1016/0031-0182(90)90069-J.

Steinthorsdottir, M., Jeram, A. J., & McElwain, J. C. (2011). Extremely elevated CO_2 concentrations at the Triassic/Jurassic boundary. Palaeogeography, Palaeoclimatology, Palaeoecology, 308(3–4), 418–432. doi:10.1016/j.palaeo.2011.05.050.

Steinthorsdottir, M., Woodward, F. I., Surlyk, F., & McElwain, J. C. (2012). Deep-time evidence of a link between elevated CO_2 concentrations and perturbations in the hydrological cycle via drop in plant transpiration. Geology, 40(9), 815–818. doi:10.1130/G33334.1.

Steinthorsdottir, M., Elliott-Kingston, C., & Bacon, K. L. (2018). Cuticle surfaces of fossil plants as a potential proxy for volcanic SO_2 emissions: Observations from the Triassic–Jurassic transition of East Greenland. Palaeobiodiversity and Palaeoenvironments, 98(1), 49–69. doi:10.1007/s12549-017-0297-9.

Surlyk, F. (2003). The Jurassic of East Greenland: A sedimentary record of thermal subsidence, onset and culmination of rifting. In J. R. Ineson, F. Surlyk (Eds.), The Jurassic of Denmark and Greenland, Geological Survey of Denmark and Greenland, Bulletin (Vol. 1, pp. 659–722). Danmarks og Grønlands Geologiske Undersøgelse (GEUS), Copenhagen.

Svensen, H., Planke, S., Chevallier, L., Malthe-Sorenssen, A., Corfu, F., & Jamtveit, B. (2007). Hydrothermal venting of greenhouse gases triggering Early Jurassic global warming. Earth and Planetary Science Letters, 256(3–4), 554–566. doi:10.1016/j.epsl.2007.02.013.

Tanner, L. H., Hubert, J. F., Coffey, B. P., & McInerneyk, D. P. (2001). Stability of atmospheric CO_2 levels across the Triassic/Jurassic boundary. Nature, 411(6838), 675–677. doi:10.1038/35079548.

Tanner, L. H., Lucas, S. G., & Chapman, M. G. (2004). Assessing the record and causes of Late Triassic extinctions. Earth-Science Reviews, 65(1–2), 103–139. doi:10.1016/s0012-8252(03)00082-5.

Taylor, D. G., Smith, P. L., Laws, R. A., & Guex, J. (1983). The stratigraphy and biofacies trends of the Lower Mesozoic Gabbs and Sunrise formations, west-central Nevada. Canadian Journal of Earth Sciences, 20(10), 1598–1608.

Thibodeau, A. M., Ritterbush, K., Yager, J. A., West, A. J., Ibarra, Y., Bottjer, D. J., ... Corsetti, F. A. (2016). Mercury anomalies and the timing of biotic recovery following the end-Triassic mass extinction. Nature Communications, 7, 11147. doi:10.1038/ncomms11147.

Thordarson, T., & Self, S. (2003). Atmospheric and environmental effects of the 1783–1784 Laki eruption: A review and reassessment. Journal of Geophysical Research, 108(D1), 4011. doi:10.1029/2001JD002042.

Tipper, H. W., Carter, E. S., Orchard, M. J., & Tozer, E. T. (1994). The Triassic-Jurassic (T-J) boundary in Queen Charlotte Islands, British Columbia defined by ammonites, conodonts and radiolarians. Geobios Mémoire Spécial, 17, 485–492. doi:10.1016/S0016-6995(94)80170-3.

Tozer, E. T. (1994). Canadian Triassic ammonoid faunas. Geological Survey of Canada Bulletin, 467, 1–663. doi:10.4095/194325.

Ullmann, C. V., & Korte, C. (2015). Diagenetic alteration in low-Mg calcite from macrofossils: A review. Geological Quarterly 59(1), 3–20. doi:10.7306/gq.1217.

Urey, H. C., Lowenstam, H. A., Epstein, S., & McKinney, C. R. (1951). Measurement of paleotemperatures and temperatures of the Upper Cretaceous of England, Denmark, and the southeastern United States. Geological Society of America Bulletin, 62(4), 399–416. doi:10.1130/0016-7606(1951)62[399:MOPATO]2.0.CO;2.

van de Schootbrugge, B., Tremolada, F., Rosenthal, Y., Bailey, T. R., Feist-Burkhardt, S., Brinkhuis, H., ... Falkowski, P. G. (2007). End-Triassic calcification crisis and blooms of organic-walled 'disaster species'. Palaeogeography, Palaeoclimatology, Palaeoecology, 244, 126–141. doi:10.1016/j.palaeo.2006.06.026.

van de Schootbrugge, B., Payne, J. L., Tomasovych, A., Pross, J., Fiebig, J., Benbrahim, M., ... Quan, T. M. (2008). Carbon cycle perturbation and stabilization in the wake of the Triassic-Jurassic boundary mass-extinction event. Geochemistry, Geophysics, Geosystems, 9(4), 1–16. doi:10.1029/2007GC001914.

van de Schootbrugge, B., Quan, T. M., Lindström, S., Püttmann, W., Heunisch, C., Pross, J., ... Falkowski, P. G. (2009). Floral changes across the Triassic/Jurassic boundary linked to flood basalt volcanism. Nature Geoscience, 2, 589–594. doi:10.1038/ngeo577.

van de Schootbrugge, B., Bachan, A., Suan, G., Richoz, S., & Payne, J. L. (2013). Microbes, mud and methane: Cause and consequence of recurrent Early Jurassic anoxia following the end-Triassic mass extinction. Palaeontology, 56(4), 685–709. doi:10.1111/pala.12034.

Veizer, J. (1983) Chemical diagenesis of carbonates: Theory and application of trace element technique. In M. A. Arthur et al. (Eds.), Stable Isotopes in Sedimentary Geology, Society of Economic Palaeontologists and Mineralogists Short Course (Vol. 10, pp. 3/1–3/100). SEPM, Tulsa.

Veizer, J., Fritz, P., & Jones, B. (1986). Geochemistry of brachiopods: Oxygen and carbon isotopic records of Paleozoic oceans. Geochimica et Cosmochimica Acta, 50(8), 1679–1696. doi:10.1016/0016-7037(86)90130-4.

Veizer, J., Ala, D., Azmy, K., Bruckschen, P., Buhl, D., Bruhn, F., ... Strauss, H. (1999). $^{87}Sr/^{86}Sr$, $\delta^{13}C$ and $\delta^{18}O$ evolution of Phanerozoic seawater. Chemical Geology, 161, 59–88. doi:10.1016/S0009-2541(99)00081-9.

Walker, G. P. L., Self, S., & Wilson, L. (1984). Tarawera 1886, New Zealand: A basaltic plinian fissure eruption. Journal of Volcanology and Geothermal Research. 21(1–2), 61–78. doi:10.1016/0377-0273(84)90016-7.

Ward, P. D., Haggart, J. W., Carter, E. S., Wilbur, D., Tipper, H. W., & Evans, T. (2001). Sudden productivity collapse associated with the Triassic-Jurassic boundary mass extinction. Science, 292(5519), 1148–1151. doi:10.1126/science.1058574.

Ward, P. D., Garrison, G. H., Haggart, J. W., Kring, D. A., & Beattie, M. J. (2004). Isotopic evidence bearing on Late Triassic extinction events, Queen Charlotte Islands, British Columbia, and implications for the duration and cause of the Triassic/Jurassic mass extinction. Earth and Planetary Science Letters, 224(3–4), 589–600. doi:10.1016/j.epsl.2004.04.034.

Ward, P. D., Garrison, G. H., Williford, K. H., Kring, D. A., Goodwin, D., Beattie, M. J., & McRoberts, C. A. (2007). The organic carbon isotopic and paleontological record across the Triassic-Jurassic boundary at the candidate GSSP section at Ferguson Hill, Muller Canyon, Nevada, USA. Palaeogeography, Palaeoclimatology, Palaeoecology, *244*(1–4), 281–289. doi:10.1016/j.palaeo.2006.06.042.

Warrington, G., Cope, J. C. W., & Ivimey-Cook, H. C. (1994). St Audrie's Bay, Somerset, England: A candidate Global Stratotype Section and Point for the base of the Jurassic System. Geological Magazine, *131*(2), 191–200. doi:10.1017/S0016756800010724.

Warrington, G., Cope, J. C. W., & Ivimey-Cook, H. C. (2008). The St Audries Bay – Doniford Bay section, Somerset, England: Updated proposal for a candidate Global Stratotype Section and Point for the base of the Hettangian Stage, and of the Jurassic System. International Subcommission on Jurassic Stratigraphy Newsletter, *35*(1), 2–66.

Whalen, L., Gazel, E., Vidito, C., Puffer, J., Bizimis, M., Henika, W., & Caddick, M. J. (2015). Supercontinental inheritance and its influence on supercontinental breakup: The Central Atlantic Magmatic Province and the breakup of Pangea. Geochemistry, Geophysics, Geosystems, *16*(10), 3532–3554. doi:10.1002/2015GC005885.

Whiteside, J. H., Olsen, P. E., Eglinton, T., Brookfield, M. E., & Sambrotto, R. N. (2010). Compound-specific carbon isotopes from Earth's largest flood basalt eruptions directly linked to the end-Triassic mass extinction. Proceedings of the National Academy of Sciences of the United States of America, *107*(15), 6721–6725. doi:10.1073/pnas.1001706107.

Wignall, P. B. (2001a). Large igneous provinces and mass extinctions. Earth-Science Reviews, *53*(1–2), 1–33. doi:10.1016/S0012-8252(00)00037-4.

Wignall, P. B. (2001b). Sedimentology of the Triassic-Jurassic boundary beds in Pinhay Bay (Devon, SW England). Proceedings of the Geologists' Association, *112*(4), 349–360. doi:10.1016/S0016-7878(01)80014-6.

Williford, K. H., Ward, P. D., Garrison, G. H., & Buick, R. (2007). An extended organic carbon-isotope record across the Triassic-Jurassic boundary in the Queen Charlotte Islands, British Columbia, Canada. Palaeogeography, Palaeoclimatology, Palaeoecology, *244*(1–4), 290–296. doi:10.1016/j.palaeo.2006.06.032.

Williford, K. H., Foriel, J., Ward, P. D., & Steig, E. J. (2009). Major perturbation in sulfur cycling at the Triassic-Jurassic boundary. Geology, *37*(9), 835–838. doi:10.1130/g30054a.1.

Wotzlaw, J.-F., Guex, J., Bartolini, A., Gallet, Y., Krystyn, L., McRoberts, C. A., … Schaltegger, U. (2014). Towards accurate numerical calibration of the Late Triassic: High-precision U-Pb geochronology constraints on the duration of the Rhaetian. Geology, *42*, 571–574. doi:10.1130/G35612.1.

Xu, W., Ruhl, M., Hesselbo, S. P., Riding, J. B., & Jenkyns, H. C. (2017). Orbital pacing of the Early Jurassic carbon cycle, black-shale formation and seabed methane seepage. Sedimentology, *64*(1), 127–149. doi:10.1111/sed.12329.

Zeebe, R., & Wolf-Gladrow, D. (2001). CO_2 in seawater: Equilibrium, kinetics, isotopes, 347.

11

Jurassic-Cretaceous Carbon Isotope Geochemistry–Proxy for Paleoceanography and Tool for Stratigraphy

Helmut Weissert

ABSTRACT

The discovery of isotope geochemistry as a powerful tool in oceanography and paleoclimatology is dating back into middle of the twentieth century. Carbon isotope geochemistry was identified as a potentially powerful tool for paleoceanography only in the late 1960s, when Helen Tappan published her fundamental study on "primary production, isotopes, extinction and the atmosphere." It took another 10 years until carbonate carbon isotope data were accepted as a proxy of the global carbon cycle. Changes in the marine carbon pool are regarded as synchronous on a global scale due to short mixing time of the oceans in the order of 1000 years. Based on this, fluctuations in the C isotope record of marine carbonates can be used as accurate stratigraphic markers. Pelagic carbonates continue to serve as reference archives for C isotope stratigraphy. Projection of pelagic C isotope curves into hemipelagic or neritic environments asks for a thorough understanding of environmental conditions during formation of these sediments and of diagenesis affecting the isotopic composition of marine sediments. Times of high-amplitude changes in the Mesozoic C isotope record contrast with times of only low-amplitude changes in the C isotope curve as, for example, the Jurassic-Cretaceous transition.

11.1. INTRODUCTION

"The understanding of the inorganic and biological carbon cycles and their interrelationships through geological time is of fundamental importance for geochemistry" [*Craig and Boato*, 1955]. Craig and Boato realized as early as 1955 that the investigation of the C isotope geochemistry of inorganic and organic carbon compounds will be of importance also in investigations of the "path of carbon" in the biosphere. In 1953, Harmon Craig had presented a thorough inventory of distribution of stable carbon isotopes in earth materials [*Craig*, 1953]. Interest in carbon isotope geochemistry remained limited in the years following Craig's fundamental study. Fractionation factors between dissolved carbon and calcium carbonate were identified, and a few early studies

provided information on the distribution of carbon isotopes in limestones and in organic matter through geological time. Fractionation processes during photosynthesis favoring C-12 fixation into plants were studied by a number of authors [e.g., *Park and Epstein*, 1960]. A link between the carbon cycle and carbon isotopes was recognized by *Wickman* [1956]. *Keith and Weber* [1964] presented the first comprehensive dataset of carbon and oxygen isotope analyses of 500 limestone samples from the Phanerozoic. Carbon isotope compositions in sediments were, at that time, seen as indicators of marine or lacustrine depositional environment. Interest in carbon isotope geochemistry remained, however, limited in the 1960s; stable isotope research focused on the use of oxygen isotope geochemistry in Earth sciences. Oxygen isotope composition in marine calcite shells was recognized as proxy of past climate and as thermometer for past temperatures. As part of his PhD study, Cesare Emiliani could, for the first time, document a cyclic

Department of Earth Sciences, ETH Zurich, Zurich, Switzerland

Chemostratigraphy Across Major Chronological Boundaries, Geophysical Monograph 240, First Edition.
Edited by Alcides N. Sial, Claudio Gaucher, Muthuvairavasamy Ramkumar, and Valderez Pinto Ferreira.

pattern in Pleistocene climate based on oxygen isotope analyses of planktic foraminifera [*Emiliani*, 1955]. The work of Emiliani represented a breakthrough in paleoclimatology. Paleoclimatologists soon realized that the oxygen isotope-paleotemperature link recognized by Cesare Emiliani turned out to be more complicated than seen in the very first paleoclimate studies. Changes in the isotopic composition of seawater due to ice volume variation, changing fractionation due to biological or "vital" effects, and variations in evaporation rates all had an impact on the oxygen isotope composition of marine calcite [*Berger*, 1979]. Improved oxygen isotope records were combined with biostratigraphy and magnetostratigraphy [*Shackleton and Opdyke*, 1973]. The construction of more and more accurate stratigraphic age models opened the door for correlating oxygen isotope records with the "orbital pulse" as it was presented first by Milutin Milankovitch in his early studies on orbital climate forcing [*Hays et al.*, 1976]. Carbon isotope data were not considered as of major relevance for paleoclimate studies, and hence, these data rarely were even published in early paleoclimate studies.

Helen Tappan, an eminent micropaleontologist of the twentieth century, introduced the concept of isotope geochemistry into geology. In her study on "production, isotopes, extinctions and the atmosphere" [*Tappan*, 1968], she devoted one chapter to a discussion of the potential of carbon isotope geochemistry as a tool in paleoproductivity studies. She hypothesized that "...marked changes in phytoplankton production would affect carbon isotope ratios of carbonates and petroleum...." And she assumed that "an exceptionally heavy period of productivity ... would result in 13-C enrichment of the oceanic reservoir. Such a fractionation may explain the changing isotopic composition of limestones...." Helen Tappan formulated these hypotheses at a time when still very few data on fossil carbonate were published in geological literature. It took another 10 years until carbon isotope studies confirmed some of Tappan's hypotheses. *Scholle and Arthur* [1980] published a first comprehensive study on the variability of carbon isotope compositions in Cretaceous limestones of the Peregrina Canyon (Mexico). In a combined sedimentological and geochemical study, *Weissert et al.* [1979] found evidence for a link between ocean productivity, anoxia, and C isotope values in marine carbonates as suggested by *Tappan* [1968]. In the 1970s and early 1980s, both carbon isotope and oxygen isotope geochemistry developed into most important tools in paleoceanography and paleoclimatology. The orbital pulse preserved in oxygen isotope composition of planktic and benthic foraminifera was identified as an excellent stratigraphic tool in Neogene marine sediments. Carbon isotope geochemistry evolved from a paleoproductivity into a carbon cycle proxy, and at this stage, the potential of carbon isotope geochemistry as a new stratigraphic tool in Earth's history was discovered.

11.2. CARBON ISOTOPE GEOCHEMISTRY: FROM A PALEOCLIMATIC TO A STRATIGRAPHIC TOOL

At the time, when Helen Tappan published her study on productivity fluctuations through geological time, she had no information on the carbon isotopic composition of modern seawater and on the impact of marine production on the isotopic signature of seawater. Peter Kroopnick published his first paper on the carbon isotopic composition of seawater in 1974 [*Kroopnick*, 1974]. His data mark a turning point in paleoceanography and in the application of C isotope geochemistry in paleoceanography. Kroopnick succeeded in documenting how productivity in surface water shifts the carbon isotope composition of dissolved inorganic carbon (DIC) to more positive values and how the oxidation of decaying organic matter in deeper water masses results in a shift of deepwater carbon isotope values to low numbers near +1‰. Of comparable importance for the further development of C isotope geochemistry were experimental studies on fractionation of carbon isotopes [e.g., *Emrich et al.*, 1970]. These studies confirmed that calcite precipitating from seawater shows only a small fractionation of around 1‰ at temperature around 20 °C and that fractionation of C isotopes between calcite and DIC is around 10 times less sensitive to changes in temperature than fractionation of oxygen isotopes between calcite and water.

Berger et al. [1978] were among the first authors applying C isotope compositions of foraminifera as a tool in paleoceanography. They identified changes in upwelling pattern along the northwestern margin of Africa across the Pleistocene-Holocene transition. They interpreted the C isotope composition of planktic foraminifera as a water mass proxy controlled by varying upwelling intensity across the Pleistocene-Holocene transition. *Boersma and Shackleton* [1977] constructed an early Cenozoic C isotope curve derived from isotope analyses of benthic foraminifera from Rio Grande Rise Deep Sea Drilling Project (DSDP) Site 357. The benthic C isotope curve was constructed with very few data points. Yet, Boersma and Shackleton already proposed that some of the fluctuations they observe in their data may record global changes in the isotopic composition of seawater. They wrote: "... At this stage we would only point out, that at least some of the events depicted in Figure 3 do probably reflect worldwide events" [*Boersma and Shackleton*, 1977]. *Cita et al.* [1977], *Vergnaud-Grazzini* [1978], *Létolle et al.* [1979], and *Weissert et al.* [1979] traced changes in paleoceanography, in paleoproductivity, and in the development of anoxia using C isotope geochemistry. *Weissert* [1979] recognized a trend in the Early Cretaceous C isotope record toward more positive values; he explained this trend as a mirror of increasing and widespread anoxia in mid-Cretaceous oceans. He compared isotopic patterns in black shale-limestone successions with isotopic fluctuations in sapropel successions in the Mediterranean, and, in agreement with interpretations of sapropel deposits by *Cita et al.* [1977], he proposed that productivity variations resulted in mid-Cretaceous black shale deposition. The comprehensive C isotope study of the Cretaceous Peregrina Canyon section in Mexico by *Scholle and Arthur* [1980] provided, for the first time, a long and detailed C isotope curve through Lower and middle Cretaceous pelagic limestones. They proposed, again in agreement with earlier authors, that changes in C isotope patterns measured in bulk carbonate store the isotopic composition of past seawater and that changes in past seawater isotope composition are recording changes in oceanography.

Two developments in the investigation of the history of the global carbon cycle were of importance for the application of C isotope geochemistry in paleoceanography and stratigraphy: (i) The early 1980s were a time of increasing awareness that changes in the chemistry of the atmosphere have an impact on climate. *Berner et al.* [1978] discovered that past atmospheric carbon dioxide levels are preserved in air bubbles of ice. Early results from ice cores indicated that pCO_2 in the atmosphere changed across glacial-interglacial cycles. This discovery also had an impact on geology. Long-neglected hypotheses on changes in atmospheric CO_2 concentrations in the history of Earth's climate [*Chamberlin*, 1898] were rediscovered. The search for a history of the global carbon cycle developed into a prominent research theme. (ii) The DSDP provided new archives on pre-Cenozoic paleoceanography. The discovery of widespread or even global "oceanic anoxic events" (OAEs) [*Schlanger and Jenkyns*, 1976] stimulated new research in "Mesozoic paleoceanography," a term coined by *Hsü* [1976]. C isotopes were identified as a potential proxy for tracing changes in paleoceanography during OAEs [e.g., *Scholle and Arthur*, 1980]. The recognition of black shale episodes (OAEs) triggered interest not only in tracing changes in paleoceanography through time but also in investigating links between the global carbon cycle, atmospheric CO_2 concentrations, and the biosphere through time. Changes in the isotopic composition of seawater beyond the residence time of carbon in the oceans (about 50 kyrs) were interpreted as indicators of changes in carbon flux out of or into the marine carbon reservoir. Positive excursions in the C isotope record were seen as evidence for altered partitioning between organic carbon and carbonate carbon in the export of carbon from the ocean pool into the sediments [e.g., *Weissert*, 1989, for a review]. Elevated organic carbon burial rates triggered by perturbations of the carbon cycle left their signature in the marine carbon pool and in biogenic calcite, the proxy of DIC in seawater. These changes in DIC, triggered by altered Corg/Ccarb burial rates, were, at ocean mixing times of a few thousand years, of global extent. Therefore, changes in the carbon isotope composition of pelagic carbonate at time scales of 50 kyrs or more were considered as changes of global dimension, and, therefore, these changes can be translated into stratigraphic markers. Carbon isotope stratigraphy was born at the moment when C isotope records were positively identified as proxies of past changes in the global carbon cycle [e.g., *Boersma and Shackleton*, 1977; *Scholle and Arthur*, 1980; *Arthur et al.*, 1985; *Weissert et al.*, 1985].

11.3. CARBON ISOTOPE STRATIGRAPHY: WIGGLES, EXCURSIONS, AND SPIKES

In the last decades, numerous C isotope records throughout the Phanerozoic and into the Proterozoic were established, and they successfully were used as stratigraphic tools. Positive "excursions" and negative "spikes" recording changes in the global marine carbon pool are widely accepted as global stratigraphic markers. Informal reference stratigraphies were established based on C isotope studies of pelagic limestones in the Mesozoic and/or of analyses of shell calcite in Paleozoic successions (see, e.g., data by Michael Joachimski in *Weissert et al.*, 2008). Only in a second step were C isotope variations, identified in pelagic records, traced into shallow water and into hemipelagic environments and later also into continental settings [e.g., *Gröcke et al.*, 2005]. Around 20 years after the successful introduction of oxygen isotope geochemistry into stratigraphy [*Emiliani*, 1955], C isotope geochemistry also developed into a highly valuable tool in stratigraphy and in Earth systems history. Mass spectrometry and sample preparation techniques were improved over the decades, and new laboratories were established at earth science departments all over the world. An exponential increase in measuring capacity and in data production is reflected in an exponentially growing number of publications containing C isotope data. Google Scholar counts around 19,000 results for "carbon isotopes" between 1980 and 1990, while between 2000 and 2010, already 280,000 counts are given for "carbon isotopes." High-resolution carbonate carbon and organic carbon isotope records are published at an accelerated rate. "High resolution" in the Phanerozoic should correspond to one data point per 10^3–10^4 years. Stratigraphic resolution is pushed toward limits, and wiggle matching became an important task for chemostratigraphers. However, the question is asked if small-scale changes of <0.5‰ measured in marine sedimentary successions are of global dimension and if fluctuations at time scales <50 kyrs, that is, shorter than residence time of carbon in seawater, can be used for global correlation [*Jarvis et al.*, 2015]. Graphic displays of carbon isotope curves are widely used for wiggle matching. However, scaling of the graphs is of relevance. Graphic representation of data can result in "hiding" a C isotope excursion if the C isotope scale is strongly compressed [e.g., *van de Schootbrugge et al.*, 2005]. Or it can be used to make excursions better visible by stretching the scale, even if these "excursions" are of small amplitude of <0.4‰ [e.g., *Godet et al.*, 2006]. As a rule of thumb, I propose that the diameter of the dots in a graph roughly corresponds to the size of the error bar of duplicate sample measurement. Based on this rule, reasonable scales in C isotope plots can be applied.

11.4. CARBONATE CARBON AND ORGANIC CARBON ISOTOPE STRATIGRAPHY

The interpretation of the carbon isotopic composition of bulk organic carbon is less straightforward than the interpretation of carbonate carbon isotope data. Bulk organic carbon often has both a marine and a terrestrial source. Variations in the isotopic composition between samples in a stratigraphic succession may simply result from changes in the ratio of marine to terrestrial organic carbon [e.g., *Sanson-Barrera et al.*, 2015; *Suan et al.*, 2015]. Changes in fractionation due to changing nutrient levels in the ambient water mass, due to changes in algal growth rate, or due to changes in atmospheric CO_2 concentrations may affect the isotopic composition of marine organic carbon [e.g., *Hayes et al.*, 1989, 1999; *Freeman and Hayes*, 1992]. However, numerous studies on C-cycle changes in the geological past show that paired analyses of inorganic and organic carbon provide insight into functioning of the carbon cycle. Changes in atmospheric CO_2 will result in changing isotopic fractionation between marine organic carbon and sedimentary carbonate. Increasing pCO_{2atm} is reflected in an increased fractionation as demonstrated in simulations by *Kump and Arthur* [1999] and in interpretation of Phanerozoic paired carbon isotope records [*Hayes et al.*, 1999]. If bulk organic carbon data are used for reconstruction of changes in carbon cycling, additional geochemical and palynological information concerning the composition of organic matter is essential. Another approach used in paired organic and carbon isotope studies across critical intervals of major C isotope anomalies is based on the analysis of the isotopic composition of marine and terrestrial biomarkers [e.g., *Hayes et al.*, 1990; *Hayes*, 1993; *Damsté and Köster*, 1998; *Pancost and Damsté*, 2003, *Méhay et al.*, 2009]. *Méhay et al.* [2009] succeeded in demonstrating that the negative carbonate carbon isotope spike

defining the base of Oceanic Anoxic Event 1a (OAE1a) is reflected in a large negative spike of up to several ‰ in a variety of molecular fossils. A corresponding increase in the fractionation between individual biomarker values and the isotopic composition of carbonate carbon records a substantial atmospheric pCO_2 pulse at the beginning of OAE1a [*Méhay et al.*, 2009]. Paired organic carbon and carbonate carbon curves do not always show parallel trends [*Kump and Arthur*, 1999]. *Louis-Schmid et al.* [2007] documented how a positive carbonate carbon isotope excursion of Late Jurassic age (Oxfordian) is not accompanied by an organic carbon isotope excursion. An increase of more than 1‰ observed in the carbonate C isotope curve within the *transversarium* ammonite zone coincides with decreasing Corg isotope values. The authors demonstrated that the divergence of the carbonate carbon and the organic carbon curve is not the result of changing composition of the bulk organic carbon or of peculiar diagenetic conditions. Simulations of the Oxfordian carbon isotope anomaly indicate that a major carbon dioxide pulse triggered an important change in fractionation between marine organic carbon and biotic calcite. The authors presented simulations where an increase in Oxfordian atmospheric CO_2 concentrations was triggered by an extraordinary increase in shallow-water carbonate production as documented in reef growth records of the Late Jurassic [*Leinfelder*, 2002]. The change in atmospheric carbon dioxide induced by excessive reef growth in the Oxfordian may have triggered the observed change in fractionation between organic carbon and marine sedimentary carbonate. Paired organic carbon-carbonate carbon curves can further help to identify possible diagenetic overprint under peculiar conditions causing a shift of the carbonate carbon isotopic composition to more negative values. *Wohlwend et al.* [2016] studied the C isotopic pattern in shallow-water carbonates of Cenomanian age formed along the northeastern margin of the Arabian platform (Adam Foothills, Oman). They studied shallow-water carbonates belonging to the Natih Formation, which contains carbonaceous intervals (e.g., Natih B member). These intervals are known as major petroleum source rocks in the United Arab Emirates and in Oman. They were accumulated in an intraplatform basin under restricted circulation conditions. Elevated organic carbon contents (up to more than 10%) in the sediments of the intraplatform basin had an impact on early diagenesis. Under sulfate-reducing conditions, isotopically light carbon was added to the pore fluids. During diagenesis, these C-13-depleted pore fluids left their imprint on newly formed calcite cement. However, low C isotope values of Natih B cannot be exclusively explained as result of diagenetic overprint due to elevated organic carbon content. Rather, the negative $\delta^{13}C$ shift is linked to the distinct lithology of Natih B, characterized by alternating limestones and organic carbon-enriched carbonate mud-stones. Diagenesis of this succession is controlled by both argillaceous sediments and interbedded neritic carbon-ates. Argillaceous sediments enriched in organic carbon experienced sulfate reduction and anaerobic oxidation of methane during early diagenesis. Both processes resulted in alkaline pore fluids, and authigenic carbonate depleted in C-13 was precipitated from these supersaturated pore fluids. Bulk sediment consisting of marine and isotopi-cally depleted diagenetic calcite is depleted in C-13, and a negative C isotope excursion is mimicked in the strati-graphic plot [see also *Fisher et al., 2005*].

11.5. CARBON ISOTOPE STRATIGRAPHY: TERMINOLOGY

The importance of C isotope stratigraphy as a correla-tion instrument in chronostratigraphy has been docu-mented in a study by *Hennig et al.* [1999]. These authors first succeeded in calibrating C isotope stratigraphy with standard ammonite stratigraphy of the Valanginian and Hauterivian in southern France. In a next step, they used their C isotope curve as a correlation tool between ammo-nite standard zones defined in southern France and nan-nofossil and magnetostratigraphy established in southern Tethyan sections, today outcropping in northern Italy. Because of diagenetic overprint, magnetostratigraphy failed to be established in the ammonite standard section. The C isotope data published by *Hennig et al.* [1999] provided the framework for a new combined magneto-stratigraphy and biostratigraphy, which differed signifi-cantly by about one magnetozone from earlier published correlations. This revised chronostratigraphy of the Valanginian-Hauterivian was integrated into the geologic time scale in the first years of the 21st century. *Gradstein et al.* [2004] propose to use C isotope stratigraphy as a tool for the first time in their 2004 version of the geologic time scale.

C isotope excursions and C isotope spikes serve as markers in stratigraphy; however, no terminology at all has so far been established. In contrast to oxygen isotope-based climatostratigraphy, carbon isotope stratigraphers still have an anarchic freedom when they establish their own terminologies. Most remarkable seems the history of the so-called C terminology used in C isotope stratig-raphy across OAE1a in the Aptian (Lower Cretaceous). *Menegatti et al.* [1998] measured their detailed C isotope curve at the locality Cismon in the southern Alps of northern Italy. In order to simplify the discussion of their bulk isotope curve established in pelagic limestones and black shales, they combined the letter "C" from the locality Cismon with a set of numbers given to several segments of their C isotope curve. Start and end of individual C isotope curve segments are defined by major turning points in the curve established. The terminology used by *Menegatti et al.* [1998] soon was applied in other studies on Aptian paleoceanography and chemostratigra-phy. *Wissler et al.* [2002] made an attempt to change from a "C terminology" toward a more general terminology combining the first letter of a studied stage with num-bered wiggle patterns. These authors investigated the Barremian reference section near Angles (S. France), and they used C isotope stratigraphy for correlating the Angles section with the magnetostratigraphically and biostratigraphically dated section at Cismon (N. Italy). In their study, they proposed a new terminology for the Barremian-Aptian C isotope stratigraphy. Their codes for the Barremian (B1–B5) and for the Aptian (A1–A2), however, were not applied in any further studies. Researchers preferred to stick to the C terminology intro-duced by *Menegatti et al.* [1998] into the literature. Another approach for a terminology in Cretaceous C iso-tope stratigraphy was chosen by Ian Jarvis and colleagues. These authors decided to give names to individual C isotope anomalies [*Jarvis et al., 2006*]. In most of the C isotope studies, however, no specific terminology is applied at all. The International Commission on Stratigraphy (ICS) will have to decide if, in analogy to isotope stages in oxygen isotope stratigraphy, a termi-nology will be introduced for C isotope stratigraphy and if "reference" sections for C isotope stratigraphies should be identified. In their presentation of new C isotope stra-tigraphies, researchers tend to choose those sections pub-lished in the literature which seem to best fit their own dataset. *Naafs et al.* [2016] presented a C isotope curve through OAE1a in a section from Spain. In their study, they applied the "C terminology" derived from the locality Cismon. However, the authors decided to com-pare their curve with a section from the Vocontian Trough and not with the Cismon curve. The Vocontian pattern seems to better agree with their curve, while the Cismon data contain a "negative spike" which so far has not been recognized in this new detailed C isotope stratigraphy. If pelagic reference sections have to be established for C isotope stratigraphy, these will have to be decided by ICS.

One of the problems we are confronted with, when we apply C isotope geochemistry in stratigraphy, is the vari-ability of amplitude of C isotope excursions in different paleoenvironments [see, e.g., *Immenhauser et al., 2002; Herrle et al., 2015*]. Shallow-water carbonate successions often record C isotope excursions with larger amplitude than seen in pelagic "reference sections." The Aptian C isotope excursion in the section Cismon has an amplitude near 3‰ and reaches peak values near 4.5‰. A corresponding shallow-water carbonate curve from the Basque-Cantabrian Basin is marked by positive C isotope values of more than +5‰ [*Millan et al., 2009*]. Another problem concerns the absolute values measured in C isotope records. Due to diagenesis and due to source of sediments, C isotope curves can shift in the negative or positive direction still preserving characteristic C isotope curve patterns. Shedding of aragonite from carbonate platforms into basinal settings will affect the isotopic composition of bulk carbonate as shown by *Swart and Eberli* [2005].

11.6. CARBON ISOTOPE STRATIGRAPHY: A PALEOCEANOGRAPHIC TOOL

Carbon isotope geochemistry was successfully developed as a proxy in paleoceanography before its potential as a stratigraphic tool was discovered. Changes in the isotopic composition of the marine carbon pool, recorded in biogenic carbonate precipitated in seawater, were recognized as stratigraphic markers of global scale. In the last decades, C isotope stratigraphy if combined with biostratigraphy, magnetostratigraphy, and cyclostratigraphy has developed into a powerful correlation tool for solving paleoceanographic problems in Earth systems history. One of the recurring questions in paleoceanography concerns the continuity of sedimentary archives. Pelagic sediments were long regarded as the product of continuous pelagic rain, resulting in equally continuous archives of ocean history not affected by erosive processes. New studies on deepwater currents, on benthic storms, and on contour currents [*Hüneke et al.*, 2011; *Gambacorta et al.*, 2016] teach us that pelagic sediments also may be punctuated by discontinuities in sedimentation. Times of erosion and winnowing of sediments are often but not always accompanied by mineralization processes at the sediment-water interface resulting in firm- or hardground deposits [*Heim*, 1924; *Föllmi*, 2016]. The identification of sedimentary gaps in pelagic successions without any sedimentological evidence remains challenging (Fig. 11.1a, b). Traces of current reworking can be identified with a detailed sedimentological and/or geochemical investigation of pelagic successions [*Giorgioni et al.*, 2012; *Gambacorta et al.*, 2014]. C isotope stratigraphy developed into a useful tool for identification of hidden sedimentary gaps in pelagic sediments [e.g., *Hadji*, 1991; *Herrle et al.*, 2015]. The Cretaceous period may provide an example of successful application of C isotope stratigraphy for paleoceanography. Simulations of physical oceanography in the Cretaceous provide information on current intensity, on circulation patterns, and on the distribution of condensed sediments in past oceans [e.g., *Barron and Peterson*, 1990, *Föllmi and Delamette*, 1991]. Plate configuration in the Cretaceous seems to have favored the establishment of strong shelf and even deepwater currents defining a circum-equatorial circulation system [*Hotinski and Toggweiler*, 2003]. Therefore, it seems not surprising

that Cretaceous sedimentary successions formed along an east-west trending low-latitude ocean are marked by numerous sedimentary gaps [*Föllmi*, 1990]. During early Aptian OAE1a, northern Tethyan shelf regions were affected by a crisis in biocalcification, which was followed by widespread drowning of carbonate ramps and platforms [*Weissert et al.*, 1997, *Wissler et al.*, 2002]. Drowned carbonate platforms experienced reduced sediment accumulation during and after OAE1a. Strong currents left their imprint in condensed shelf successions often marked by phosphorite mineralization [*Föllmi*, 1990]. Phosphorite mineralization of Aptian age was not restricted to northern Tethyan shelf regions, but it is also described from seamounts of central and southern Tethys [*Weissert and Wohlwend*, 2014]. While the coincidence of OAE1a with condensation along the northern Tethys shelf is well documented, no major phosphorite hardgrounds were formed during OAE2. Based on C isotope stratigraphy, deep shelf sedimentation across OAE2 was interpreted as continuous and not affected by shelf current activity [*Westermann et al.*, 2010]. *Wohlwend et al.* [2015] revised the C isotope stratigraphy of these northern Tethyan pelagic limestone successions, and their data unhide a major sedimentary gap which coincides with OAE2. Accurate chemostratigraphy facilitated recognition of sedimentological evidence for erosive current activity in these pelagic limestones. *Wohlwend et al.* [2015] described glauconite-enriched limestones interbedded with pelagic limestones which provide evidence for intensified shelf current activity, resulting in the condensed sedimentation during and after OAE2 [see also *Gambacorta et al.*, 2016]. The study by *Wohlwend et al.* [2015] serves as an example for a successful combined sedimentological and geochemical approach, resulting in an improved understanding of physical oceanography in Mesozoic oceans.

11.7. C ISOTOPE STRATIGRAPHY AND THE JURASSIC-CRETACEOUS BOUNDARY

Cretaceous oceans responded with increased productivity and burial rates of organic carbon to multiple perturbations of the carbon pool, and these perturbations are preserved in the C isotope record. Fluctuations in the C isotope record, caused by global perturbations of the carbon cycle, serve as excellent chemostratigraphic markers [*Weissert and Erba*, 2004]. Times of multiple perturbations of the carbon cycle in the geological past contrast with periods of stable carbon cycle often coinciding with stable oceanic conditions. The Jurassic-Cretaceous transition serves as a prominent example of stable low-latitude oceanography at a time of a stable carbon cycle reflected in an equally stable carbon isotope record [*Weissert*, 2011]. Latest Jurassic low-latitude

Figure 11.1 (a) Example of current-reworked radiolarian sand, today preserved as siliceous limestone, with evidence for cross-bedding. Pelagic Maiolica Formation, Lower Cretaceous. S. Alps. Italy. (b) Bioturbation features are used as indicator for winnowing in a pelagic setting. Bioturbated pelagic limestone reflects continuous sedimentation, and non-bioturbated upper part was redeposited by currents. (c) Lower Cretaceous Maiolica Formation, Central Tethys Ocean (S. Alps, Italy). The picture illustrates the change from a white almost pure and up to several dm-bedded nannofossil limestone (>90% $CaCO_3$) to a darker thin-bedded limestone. Transition is dated as lower Valanginian [*Weissert*, 1979]. (d) Limestone-chert succession in pelagic limestones of earliest Cretaceous age (Maiolica Formation, S. Alps, N. Italy). Chert layers are interpreted as indicators of better surface water mixing and increased productivity and radiolarian flux into the pelagic sediment. Photo: H. Weissert.

oceans record a remarkable transition to pelagic carbonate production with calcareous nannoplankton as major constituent in pelagic and deep-sea carbonates [e.g., *Celestino et al.,* 2017]. The expansion and diversi-

fication of calcareous nannoplankton in the latest Jurassic was favored by stable and oligotrophic conditions in low-latitude oceans [*Erba*, 2004]. *Weissert and Channell* [1989] documented a gradually decreasing C isotope trend across the Jurassic-Cretaceous transition starting with $\delta^{13}C$ values of around 2.5‰ in the Kimmeridgian and decreasing to values near +1.0‰ in the late Tithonian to early Berriasian. The gradual transition to lower $\delta^{13}C$ values was identified to occur within magnetozones M18-M17 and within the B/C calpionellid zone [*Weissert and Channell*, 1989]. Numerous succeeding C isotope studies of Jurassic-Cretaceous transitions confirmed the pattern with gradually decreasing C isotope values marking the Jurassic-Cretaceous transition. *Price et al.* [2016] provide an excellent summary of all available C isotope curves covering the Jurassic-Cretaceous transition [for references, see *Price et al.*, 2016]. These authors propose that the gradually declining C isotope curve reflects increased burial of pelagic carbonate across this interval. They also suggest that the opening of the Hispanic corridor explains the gradual change in the observed C isotope pattern and in oceanography. However, pelagic sedimentation along the low-latitude central Atlantic and Tethys oceans suggests that oceanographic conditions in equatorial settings remained stable from Tithonian up to the earliest Valanginian [*Weissert*, 1979] (Fig. 11.1c). White nannofossil carbonate facies documents oligotrophic surface water conditions in the equatorial Atlantic-Tethys ocean [*Weissert*, 1979; *Celestino et al.*, 2017]. Only chert layers intercalated irregularly into the nannofossil limestone succession are indicators of episodically changing surface water conditions (Fig. 11.1d). Times of intensified surface water mixing resulted in an increase of siliceous organisms in surface water and in an increased flux of radiolarians to the seafloor, preserved today as chert layers [*Weissert*, 2011]. If the sedimentary record of the North American Basin is chosen as a proxy for changes in ocean circulation, a remarkable shift in pelagic sedimentary facies from white nannofossil limestones to alternating white limestones and dark claystones (DSDP Leg 11; *Bernoulli*, 1972) occurs in the Valanginian, suggesting a major change in oceanography [*Stein et al.*, 1986; *Weissert*, 2011]. The change may be related to further deepening of an already open Hispanic corridor. This deepening may also be reflected in radiolarian fauna, as reported by *Baumgartner et al.* [2013]. Deepening of the Hispanic corridor coincided with the establishment of a transequatorial current and with strong equatorial upwelling [*Hotinski and Toggweiler*, 2003]. These oceanic conditions favored episodic increase in burial of organic carbon coinciding with times of Early Cretaceous C-cycle perturbation. The first of these perturbations dated as late Valanginian in age corresponds to the first major C isotope anomaly of the Cretaceous [e.g., *Celestino et al.*, 2017]. This anomaly also marks the end of the gradually changing C isotope pattern of the uppermost Jurassic and the lowermost Cretaceous. Oceanography explains

why C isotope stratigraphy may not be very useful as a tool when defining GSSP of the Jurassic-Cretaceous boundary.

ACKNOWLEDGMENTS

The author thanks ETH Zurich and the Swiss Science Foundation for continuous and generous research support. He also acknowledges the many stimulating discussions with numerous PhD and master students over the last decades.

REFERENCES

Arthur, M.A., Dean, W.E., and Schlanger, S.E., 1985, Variations in the global carbon cycle during the Cretaceous related to climate, volcanism and changes in atmospheric CO2, in: Sundquist, E.T. and Broecker, W.S. (Eds.), The Carbon Cycle and Atmospheric CO2: Natural Variations Archean to the Present, American Geophysical Union, Washington, DC, 504–530.

Barron, E.J. and Peterson, W.H., 1990, Mid-Cretaceous ocean circulation: Results from model sensitivity studies, *Paleoceanography 5*, 319–337

Baumgartner, P.O., Rojas-Agramonte, Y., Sandoval-Gutierrez, M., Urbani, F., García-Delgado, D., Garban, G., and Pérez Rodríguez, M., 2013, Late Jurassic breakup of the Proto-Caribbean and circum-global circulation across Pangea, Proceedings EGU General Assembly Conference Abstracts 2013, Volume 15, 13408.

Berger, W., 1979, Stable isotopes in foraminifera, *SEPM Short Course*, *6*, 105–155.

Berger, W., DiesterHaass, L., and Killingley, J., 1978, Upwelling off northwest Africa-Holocene decrease as seen in carbon isotopes and sedimentological indicators, *Oceanologica Acta*, *1*, 3–7.

Berner, W., Stauffer, B., and Oeschger, H., 1978, Past atmospheric composition and climate, gas parameters measured on ice cores, *Nature*, *276*, 53–55

Bernoulli, D., 1972, North Atlantic and Mediterranean Mesozoic facies: A comparison, in: Hollister, C.D., Ewing, J.I., et al. (Eds.), Initial Reports of the Deep Sea Drilling Project, Vol. *11*. U.S. Government Printing Office, Washington, DC, 801–879.

Boersma, A. and Shackleton, N., 1977, Tertiary oxygen and carbon isotope stratigraphy, Site 357 (mid latitude South Atlantic), in: Supko, P.R., Perch-Nielsen, K. et al. (Eds.), Initial Reports of the Deep Sea Drilling Project, Vol. *39*, U.S. Government Printing Office, Washington, DC, 911–924.

Celestino, R., Wohlwend, S., Reháková, D., and Weissert, H., 2017, Carbon isotope stratigraphy, biostratigraphy and sedimentology of the Upper Jurassic–Lower Cretaceous Rayda Formation, Central Oman Mountains, *Newsletters on Stratigraphy*, *50*, 1, 91–109.

Chamberlin, T.C., 1898, The influence of great epochs of limestone formation upon the constitution of the atmosphere, *The Journal of Geology*, *6*, 609–621.

Cita, M.B., Vergnaud-Grazzini, C., Robert, C., Chamley, H., Ciaranfi, N., and d'Onofrio, S., 1977, Paleoclimatic record of a long deep-sea core from the eastern Mediterranean, *Quaternary Research*, *8*, 205–235.

Craig, H., 1953, The geochemistry of stable carbon isotopes, *Geochimica et Cosmochimica Acta*, *3*, 53–92.

Craig, H. and Boato, G., 1955, Isotopes, *Annual Review of Physical Chemistry*, *6*, 403–432.

Damsté, J.S.S. and Köster, J., 1998, A euxinic southern North Atlantic Ocean during the Cenomanian/Turonian oceanic anoxic event, *Earth and Planetary Science Letters*, *158*, 165–173.

Emiliani, C., 1955, Pleistocene temperatures, *The Journal of Geology*, *63*, 538–578.

Emrich, K., Ehhalt, D., and Vogel, J., 1970, Carbon isotope fractionation during the precipitation of calcium carbonate, *Earth and Planetary Science Letters*, *8*, 363–371.

Erba, E., 2004, Calcareous nannofossils and Mesozoic oceanic anoxic events, *Marine Micropaleontology*, *52*, 85–106.

Fisher, J.K., Price, G.D., Hart, M.B., and Leng, M.J., 2005, Stable isotope analysis of the Cenomanian–Turonian (Late Cretaceous) oceanic anoxic event in the Crimea, *Cretaceous Research*, *26*(6), 853–863.

Föllmi, K.B., 1990, Condensation and phosphogenesis: Example of the Helvetic mid-Cretaceous (northern Tethyan margin), *Geological Society, London, Special Publications*, *52*(1), 237–252.

Föllmi, K.B., 2016, Sedimentary condensation, *Earth-Science Reviews*, *152*, 143–180.

Föllmi, K.B. and Delamette, M., 1991, Model simulation of mid-Cretaceous ocean circulation, *Science (New York, NY)*, *251*, 94.

Freeman, K.H. and Hayes, J.M., 1992, Fractionation of carbon isotopes by phytoplankton and estimates of ancient CO_2 levels, *Global Biogeochemical Cycles*, *6*, 185–198.

Gambacorta, G., Bersezio, R., and Erba, E., 2014, Sedimentation in the Tethyan pelagic realm during the Cenomanian: Monotonous settling or active redistribution?, *Palaeogeography, Palaeoclimatology, Palaeoecology*, *409*, 301–319.

Gambacorta, G., Bersezio, R., Weissert, H., and Erba, E., 2016, Onset and demise of Cretaceous oceanic anoxic events: The coupling of surface and bottom oceanic processes in two pelagic basins of the western Tethys, *Paleoceanography*, *31*(6), 732–757.

Giorgioni, M., Weissert, H., Bernasconi, S.M., Hochuli, P.A., Coccioni, R., and Keller, C.E., 2012, Orbital control on carbon cycle and oceanography in the mid-Cretaceous greenhouse, *Paleoceanography*, *27*, 12.

Godet, A., Bodin, S., Follmi, K., Vermeulen, J., Gardin, S., Fiet, N., Adatte, T., Berner, Z., Stuben, D., and Vandeschootbrugge, B., 2006, Evolution of the marine stable carbon-isotope record during the early Cretaceous: A focus on the late Hauterivian and Barremian in the Tethyan realm, *Earth and Planetary Science Letters 242*, 254–271.

Gradstein, F.M., Ogg, J.G., and Smith, A.G., 2004, A Geologic Time Scale 2004, Cambridge University Press, Cambridge.

Gröcke, D.R., Price, G.D., Robinson, S.A., Baraboshkin, E.Y., Mutterlose, J., and Ruffell, A.H., 2005, The Upper Valanginian (Early Cretaceous) positive carbon–isotope event recorded in terrestrial plants, *Earth and Planetary Science Letters*, *240*, 495–509.

Hadji, S., 1991, Stratigraphie isotopique des carbonates pelagiques (Jurassique supérieur-Crétacé inférieur) du Bassin d'Ombrie- Marches [Ph.D. thesis], Paris, University of Pierre et Marie Curie, 118 p.

Hayes, J.M., 1993, Factors controlling 13C contents of sedimentary organic compounds: Principles and evidence, *Marine Geology*, *113*(1–2), 111–125.

Hayes, J.M., Popp, B.N., Takigiku, R., and Johnson, M.W., 1989, An isotopic study of biogeochemical relationships between carbonates and organic carbon in the Greenhorn Formation, *Geochimica Cosmochimica Acta*, *53*, 2961–2972.

Hayes, J., Freeman, K.H., Popp, B.N., and Hoham, C.H., 1990, Compound-specific isotopic analyses: A novel tool for reconstruction of ancient biogeochemical processes, *Organic Geochemistry*, *16*, 1115–1128.

Hayes, J.M., Strauss, H., and Kaufman, A.J., 1999, The abundance of 13C in marine organic matter and isotopic fractionation in the global biogeochemical cycle of carbon during the past 800 Ma, *Chemical Geology*, *161*, 103–125.

Hays, J.D., Imbrie, J., and Shackleton, N.J., 1976, Variations in the Earth's orbit: Pacemaker of the ice ages, *Science*, *194*(4270), 1121–1132.

Heim, A., 1924, Über submarine Denudation und chemische Sedimente, *Geologische Rundschau*, *15*, 1–47.

Hennig, S., Weissert, H., and Bulot, L., 1999, C-isotope stratigraphy, a calibration tool between ammonite-and magnetostratigraphy: The Valanginian-Hauterivian transition, *Geologica Carpathica*, *50*, 91–96.

Herrle, J.O., Schröder-Adams, C.J., Davis, W., Pugh, A.T., Galloway, J.M., and Fath, J., 2015, Mid-Cretaceous High Arctic stratigraphy, climate, and oceanic anoxic events, *Geology*, *43*(5), 403–406.

Hotinski, R.M. and Toggweiler, J.R., 2003, Impact of a Tethyan circumglobal passage on ocean heat transport and climates, *Paleoceanography*, *18*, 1007.

Hsu, K.J., 1976, Paleoceanography of the Mesozoic Alpine Tethys, *Special Papers - Geological Society of America*, *110*, 44 pp.

Hüneke, H., Henrich, R., Hüneke, H., and Mulder, T., 2011, Pelagic Sedimentation in Modern and Ancient Oceans, Developments in Sedimentology, *63*, Elsevier, Amsterdam, 750p.

Immenhauser, A., Kenter, J.A., Ganssen, G., Bahamonde, J.R., Van Vliet, A., and Saher, M.H., 2002, Origin and significance of isotope shifts in Pennsylvanian carbonates (Asturias, NW Spain), *Journal of Sedimentary Research*, *72*, 82–94.

Jarvis, I., Gale, A.S., Jenkyns, H.C., and Pearce, M.A., 2006, Secular variation in Late Cretaceous carbon isotopes: A new $\delta^{13}C$ carbonate reference curve for the Cenomanian-Campanian (99.6-70.6 Ma), *Geological Magazine, London*, *143*, 561.

Jarvis, I., Trabucho-Alexandre, J., Gröcke, D.R., Uličný, D., and Laurin, J., 2015, Intercontinental correlation of organic carbon and carbonate stable isotope records: Evidence of climate and sea-level change during the Turonian (Cretaceous), *The Depositional Record*, *1*(2), 53–90.

Keith, M. and Weber, J., 1964, Carbon and oxygen isotopic composition of selected limestones and fossils, *Geochimica et Cosmochimica Acta*, *28*, 1787–1816.

Kroopnick, P.M., 1974, The dissolved O_2-CO_2-^{13}C system in the eastern equatorial Pacific, *Deep-Sea Research*, *21*, 211–227.

Kump, L.R. and Arthur, M.A., 1999, Interpreting carbon-isotope excursions: Carbonates and organic matter, *Chemical Geology*, *161*, 181–198.

Leinfelder, R.R., 2002, Jurassic reef patterns: The expression of a changing globe, *SEPM Special Publication*, *72*, 465–520.

Létolle, R., Vergnaud-Grazzini, C., and Pierre, C., 1979, Oxygen and carbon isotopes from bulk carbonates and foraminiferal shells at DSDP Sites 400, 401, 402, 403, and 406, *Initial Reports DSDP*, *48*, 741–755.

Louis-Schmid, B., Rais, P., Schaeffer, P., Bernasconi, S.M., and Weissert, H., 2007, Plate tectonic trigger of changes in pCO2 and climate in the Oxfordian (Late Jurassic): Carbon isotope and modeling evidence, *Earth and Planetary Science Letters*, *258*, 44–60.

Méhay, S., Keller, C.E., Bernasconi, S.M., Weissert, H., Erba, E., Bottini, C., and Hochuli, P.A., 2009, A volcanic CO_2 pulse triggered the Cretaceous Oceanic Anoxic Event 1a and a biocalcification crisis, *Geology*, *37*, 819–822.

Menegatti, A.P., Weissert, H., Brown, R.S., Tyson, R.V., Farrimond, P., Strasser, A., and Caron, M., 1998, High-resolution $\delta^{13}C$ stratigraphy through the early Aptian (Livello Selli) of the Alpine Tethys, *Paleoceanography*, *13*, 530–545.

Millan, M.I., Weissert, H.J., Fernandez-Mendiola, P.A., and Garcia-Mondojar, J., 2009, Impact of Early Aptian carbon cycle perturbations on evolution of a marine shelf system in the Basque-Cantabrian Basin (Aralar, N Spain), *Earth and Planetary Science Letters*, *287*, 392–401.

Naafs, B., Castro, J., De Gea, G., Quijano, M., Schmidt, D., and Pancost, R., 2016, Gradual and sustained carbon dioxide release during Aptian Oceanic Anoxic Event 1a, *Nature Geoscience*, *9*(2), 135.

Pancost, R.D. and Damsté, J.S.S., 2003, Carbon isotopic compositions of prokaryotic lipids as tracers of carbon cycling in diverse settings, *Chemical Geology*, *195*, 29–58.

Park, R. and Epstein, S., 1960, Carbon isotope fractionation during photosynthesis, *Geochimica et Cosmochimica Acta*, *21*, 110–126.

Price, G.D., Főzy, I., and Pálfy, J., 2016, Carbon cycle history through the Jurassic–Cretaceous boundary: A new global δ 13 C stack, *Palaeogeography, Palaeoclimatology, Palaeoecology*, *451*, 46–61.

Sanson-Barrera, A., Hochuli, P.A., Bucher, H., Schneebeli-Hermann, E., Weissert, H., Adatte, T., and Bernasconi, S.M., 2015, Late Permian–earliest Triassic high-resolution organic carbon isotope and palynofacies records from Kap Stosch (East Greenland), *Global and Planetary Change*, *133*, 149–166.

Schlanger, S.O. and Jenkyns, H.C., 1976, Cretaceous anoxic events: Causes and consequences, *Geologie en Mijnbouw*, *55*, 179–184.

Scholle, P.A. and Arthur, M.A., 1980, Carbon isotope fluctuations in Cretaceous pelagic limestones: Potential stratigraphic and petroleum exploration tool, *American Association of Petroleum Geologists Bulletin*, *64*, 67–87.

Shackleton, N.J. and Opdyke, N.D., 1973, Oxygen isotope and palaeomagnetic stratigraphy of Equatorial Pacific core V28-238: Oxygen isotope temperatures and ice volumes on a 10^5 year and 10^6 year scale, *Quaternary Research*, *3*, 39–55.

Stein, R., Rullkötter, J., and Welte, D.H., 1986, Accumulation of organic-carbon-rich sediments in the Late Jurassic and Cretaceous Atlantic Ocean: A synthesis, *Chemical Geology*, *56*(1), 1–32.

Suan, G., van de Schootbrugge, B., Adatte, T., Fiebig, J., and Oschmann, W., 2015, Calibrating the magnitude of the Toarcian carbon cycle perturbation, *Paleoceanography*, *30*, 495–509.

Swart, P.K. and Eberli, G., 2005, The nature of the δ^{13} C of periplatform sediments: Implications for stratigraphy and the global carbon cycle, *Sedimentary Geology*, *175*, 115–129.

Tappan, H., 1968, Primary production, isotopes, extinctions and the atmosphere, *Palaeogeography, Palaeoclimatology, Palaeoecology*, *4*, 187–210.

van de Schootbrugge, B., McArthur, J.M., Bailey, T., Rosenthal, Y., Wright, J., and Miller, K., 2005, Toarcian oceanic anoxic event: An assessment of global causes using belemnite C isotope records, *Paleoceanography*, *20*, PA3008.

Vergnaud-Grazzini, C., 1978, Miocene and Pliocene oxygen and carbon isotopic changes at DSDP sites 372, 374 and 375: Implications for pre-Messinian history of the Mediterranean, in: Kidd, R.B. and Worstell, P.J. (Eds.), Initial Reports of the Deep Sea Drilling Project, *42*, U.S. Government Printing Office, Washington, DC, 829–836.

Weissert, H.J., 1979, Die Paläoozeanographie der südwestlichen Tethys in der Unterkreide, Diss. Naturwiss. ETH Zürich, Nr. 6349, 189p.

Weissert, H., 1989, C-isotope stratigraphy, a monitor of paleoenvironmental change: A case study from the Early Cretaceous, *Surveys in Geophysics*, *10*, 1–61.

Weissert, H., 2011, Mesozoic pelagic sediments: Archives for ocean and climate history during green-house conditions, in: Hüneke, H. and Mulder, T. (Eds.), Deep-Sea Sediments, Developments in Sedimentology, Elsevier, Amsterdam, 765–792.

Weissert, H. and Channell, J.E.T., 1989, Tethyan carbonate carbon isotope stratigraphy across the Jurassic-Cretaceous boundary: An indicator of decelerated carbon cycling, *Paleoceanography*, *4*, 483–494.

Weissert, H. and Erba, E., 2004, Volcanism, CO2 and palaeoclimate: A Late Jurassic-Early Cretaceous carbon and oxygen isotope record, *Journal of the Geological Society*, *161*, 695–702.

Weissert, H. and Wohlwend, S., 2014, Impact of Cretaceous Climate on Upwelling and Erosive Capacity of Transequatorial Current in the Tethys Seaway-Tales from Deep-Sea Sediments, paper presented at AGU Fall Meeting Abstracts, San Francisco, CA.

Weissert, H., McKenzie, J., and Hochuli, P., 1979, Cyclic anoxic events in the Early Cretaceous Tethys ocean, *Geology*, *7*, 147–151.

Weissert, H., McKenzie, J.A., and Channell, J.E.T., 1985, Natural variations in the carbon cycle during the Early Cretaceous, in: Sundquist, E.T. and Broecker, W. (Eds.), The Carbon Cycle and Atmospheric CO2: Natural Variations

Archean to the Present, American Geophysical Union, Washington, DC, 531–545.

Weissert, H., Lini, A., Föllmi, K., and Kuhn, O., 1997, Correlation of Early Cretaceous carbon isotope stratigraphy and platform drowning events: A possible link? *Palaeogeography, Palaeoecology, Palaeoclimatology, 137,* 189–203.

Weissert, H., Joachimski, M., and Sarnthein, M., 2008, Chemostratigraphy, *Newsletters on Stratigraphy, 42,* 145–179.

Westermann, S., Caron, M., Fiet, N., Fleitmann, D., Matera, V., Adatte, T., and Föllmi, K., 2010, Evidence for oxic conditions during oceanic anoxic event 2 in the northern Tethyan pelagic realm, *Cretaceous Research, 31,* 500–514.

Wickman, F.E., 1956, The cycle of carbon and the stable carbon isotopes, *Geochimica et Cosmochimica Acta, 9,* 136–153.

Wissler, L., Weissert, H., Masse, J.P., and Bulot, L., 2002, Chemostratigraphic correlation of Barremian and lower Aptian ammonite zones and magnetic reversals, *International Journal of Earth Sciences, 91,* 272–279.

Wohlwend, S., Hart, M., and Weissert, H., 2015, Ocean current intensification during the Cretaceous oceanic anoxic event 2–evidence from the northern, *Tethys Terra Nova, 27,* 147–155.

Wohlwend, S., Hart, M., and Weissert, H., 2016, Chemostratigraphy of the Upper Albian to mid-Turonian Natih Formation (Oman)–how authigenic carbonate changes a global pattern, *The Depositional Record, 2*(1), 97–117.

12

Chemostratigraphy Across the Cretaceous-Paleogene (K-Pg) Boundary: Testing the Impact and Volcanism Hypotheses

Alcides Nobrega Sial[1], Jiubin Chen[2], Luis Drude Lacerda[3], Robert Frei[4,5], John A. Higgins[6], Vinod Chandra Tewari[7], Claudio Gaucher[8], Valderez Pinto Ferreira[1], Simonetta Cirilli[9], Christoph Korte[4], José Antonio Barbosa[10], Natan Silva Pereira[1,11], and Danielle Santiago Ramos[6]

ABSTRACT

The mass extinction which marks the K-Pg has been linked to a catastrophic event. Cr and Os isotopes and ^{3}He/^{4}He ratios of He encapsulated in fullerenes within Ir-rich K-Pg layer point to an extraterrestrial cause, while Hg/TOC spikes across the K-Pg boundary suggest Hg loading from the Deccan volcanism. Three Hg/TOC spikes are present in some classical K-Pg sections: (i) spike I within the CF2 planktic foraminiferal biozone, (ii) spike II at the K-Pg boundary layer, and (iii) spike III within the P1a planktic foraminiferal subzone. The spike II has, perhaps, resulted from Hg loading from the asteroid impact and volcanism. We suggest that higher ΣREE+Y values in the K-Pg layers are, perhaps, related to Deccan volcanism or to sea-level fluctuations, coeval to the K-Pg transition that enhanced continental influx. True negative Ce anomaly suggests predominance of oxidized surface waters during the K-Pg transition.

In a δ^{202}Hg versus Δ^{201}Hg plot, samples from the spike II and from Bidart-France lie within the Hg volcanic emission box. Samples from spikes I and III from Bidart lie within the volcanic emission/chondrite box. Small positive Δ^{201}Hg favors long-term atmospheric transport and supports Hg loading to the environment by Deccan phase 2 in three distinct episodes.

12.1. INTRODUCTION

The Cretaceous-Paleogene (K-Pg) boundary, formerly known as the Cretaceous-Tertiary (KT) boundary, is a geological feature represented by a thin clay layer. *K*, the first letter of the German word *Kreide* (chalk), is the traditional abbreviation for the Cretaceous period, and *Pg* is the abbreviation for the Paleogene period. This boundary has its Global Boundary Stratotype Section and Point (GSSP) defined as the base of the boundary

[1] NEG–LABISE, Department of Geology, Federal University of Pernambuco, Recife, PE, Brazil

[2] State Key Laboratory of Environmental Geochemistry, Institute of Geochemistry, Chinese Academy of Sciences, Guiyang, China

[3] LABOMAR, Institute of Marine Sciences, Federal University of Ceará, Fortaleza, Brazil

[4] Department of Geosciences and Natural Resource Management, University of Copenhagen, Copenhagen, Denmark

[5] Nordic Center for Earth Evolution (NordCEE), University of Southern Denmark, Odense, Denmark

[6] Department of Geosciences, Princeton University, Princeton, NJ, USA

[7] Department of Geology, Sikkim University, Gangtok, SK, India

[8] Instituto de Ciencias Geológicas, Facultad de Ciencias, Universidad de la República, Montevideo, Uruguay

[9] Department of Physics and Geology, University of Perugia, Perugia, Italy

[10] LAGESE, Department of Geology, Federal University of Pernambuco, Recife, PE, Brazil

[11] Department of Biology, State University of Bahia, Paulo Afonso, Brazil

Chemostratigraphy Across Major Chronological Boundaries, Geophysical Monograph 240, First Edition.
Edited by Alcides N. Sial, Claudio Gaucher, Muthuvairavasamy Ramkumar, and Valderez Pinto Ferreira.

clay at the section near El Kef, Tunisia, since approval by the International Commission on Stratigraphy (ICS) in 1990 and ratification by the IUGS in 1991 [*Molina et al.*, 2006]. It has been accepted that this boundary is marked by the moment one giant asteroid impacted the Earth's surface at Chicxulub, Yucatán, Mexico. In distal areas to this impact site, the K-Pg boundary coincides with a millimeter-thick rusty layer, and in proximal, it correlates with a meter-thick clastic unit, a thick calcareous breccia [*Molina et al.*, 2009]. Radiogenic dating yielded an age of 66.043 ± 0.011 Ma for the K-Pg boundary [*Renne et al.*, 2013].

It is of general agreement that major catastrophic events led to massive biologic mass extinctions in the Phanerozoic whose nature and causes have been debated for decades. Five major mass extinctions are known in this eon, among which the one which marked the K-Pg boundary destroyed the majority of the world's Mesozoic species, including all dinosaurs except for birds, and it has been the most intensely studied. It is estimated that 75% or more of all species were made extinct by the K-Pg extinction event [*MacLeod*, 2012].

12.2. CAUSE FOR MASSIVE EXTINCTION AT THE CRETACEOUS-PALEOGENE BOUNDARY: IMPACT OR VOLCANISM OR BOTH?

A single or multi-bolide impact and flood basalt eruptions from large igneous provinces (LIPs) or a combination of the two events are widely discussed causes for the K-Pg massive extinction at this boundary, yet their contributions remain debated. These hypotheses are discussed in detail below.

12.2.1. The Asteroid Impact Event

The large mass extinction which marked the K-Pg has been ascribed to a huge asteroid impact that caused a global dust cloud and that persisted for several years [*Alvarez et al.*, 1980; *Smit and Hertogen*, 1980], blocking out the sunlight, along with global fires and catastrophic acid rains, causing disruption of food chains and driving mass extinction. The first problem that supporters of this hypothesis faced was spotting where the giant non-avian dinosaur-killer asteroid had struck the Earth's surface. For some time, it was generally agreed that this bolide had impacted the ocean crust [e.g., *DePaolo et al.*, 1983], but the presence of shocked quartz in K-Pg sediments suggested formation of a contemporaneous crater on continent. The twin Kara and Ust-Kara impact craters on the shore of the Kara Sea in the Arctic Ocean, northern Russia, were regarded as the possible K-Pg continental impact craters [*Koeberl et al.*, 1990]. The Kara crater is entirely located on land close to the estuary of the Kara River, and the Ust-Kara crater is partially

exposed on the shoreline of the Cape Polkovnik and partially underwater. A K-Ar age determination of impactites from the Kara crater pointed to 66.1 ± 0.8 Ma [*Kolesnikov et al.*, 1988; *Nazarov et al.*, 1989], but analyses of several $^{40}Ar/^{39}Ar$ age spectra have established that these two craters are older than 70 Ma [*Koeberl et al.*, 1990].

The Chicxulub multi-ring crater in the Yucatán Peninsula, Mexico, became the most probable locality for the asteroid impact since the work by *Hildebrand et al.* [1991] and *Swisher et al.* [1992] who determined an $^{40}Ar/^{39}Ar$ age of 64.98 ± 0.05 Ma for crater glassy melt rock of andesite composition and K-Pg tektite. It is the only known impact structure on Earth with an equivocal peak ring, buried and only accessible by drilling [*Morgan et al.*, 2016]. Massive deposits in deep waters of the Gulf of Mexico reported by *Denne et al.* [2013] substantiate widespread slope failure induced by the Chicxulub impact, providing further evidence for a single impact coincident with the K-Pg mass extinction.

Some studies have raised the possibility that the colossal Chicxulub impact may have predated the K-Pg [e.g., *Keller et al.*, 2004a, 2004b]. This contention was challenged by $^{40}Ar/^{39}Ar$ age dating of tektites and bentonite beds that revealed that the K-Pg mass extinction and the Chicxulub impact were synchronous to within 32,000 years [*Renne et al.*, 2013]. The discovery of a smaller crater with a similar age at Boltysh in Ukraine has raised the possibility that a series of asteroids or comets impacted the Earth close to the K-Pg, but palynological and C isotope analyses of the post-impact flora led to the conclusion that the Boltysh crater predated Chicxulub by ~2–5 ky [*Jolley et al.*, 2010].

The presence of high-temperature/pressure (HTP) phase of fullerenes (complex C molecules) in deposits associated with events involving the impact of a large bolide provides further evidence for the role of an asteroid impact in the K-Pg mass extinction [*Becker et al.*, 2000a]. *Parthasarathy et al.* [2008] reported the presence of the HTP fullerene C_{60} in carbonaceous matter of iridium-rich intertrappean sediments of the Anjar K-Pg site, Kutch, India, and assumed that an energetic impact event had provided the necessary HTP regime for the formation of HTP fullerene phase. The cage structure of fullerene is able to encapsulate and retain noble gases (He, Ar, Xe), the isotopic composition of which can only be described as typical of extraterrestrial origin [e.g., Stevns Klint; *Becker et al.*, 2000a, 2000b; Anjar, *Parthasarathy et al.*, 2008].

It has been proposed that the thermal radiation released by the K-Pg asteroid impact could have ignited wildfires locally or even globally [*Melosh et al.*, 1990; *Kring and Durda*, 2002]. In this case, forests could have been ignited as a consequence of thermal radiation by the reentering ejecta from the impact

[*Melosh et al.*, 1990]. However, multi-method and multi-proxy tests have demonstrated that extensive K/Pg wildfires were unlikely [*Belcher*, 2009].

The main evidence quoted in support of the asteroid impact model was the elevated concentration of iridium at K-Pg boundary clays in Gubbio, Italy, Stevns Klint, Denmark, and Woodside Creek, New Zealand [e.g., *Alvarez et al.*, 1980; *Ganapathy*, 1980; *Smit and Ten Kate*, 1982; *Claeys et al.*, 2002] because this element is enriched in meteorites (>10 ppm) and rare in common crustal rocks (<1 ppb) [*Crocket et al.*, 1978]. From the total amount of iridium in the K-Pg layer, and assuming that the asteroid contained about the percentage of iridium found in chondrites, the size of the asteroid was calculated to be around 10 km in diameter [*Courtillot*, 1990]. Such a large impact would have had approximately the energy of 100 trillion tons of TNT, 2 million times greater than the most powerful thermonuclear bomb ever tested [*Hanski*, 2016]. A paleogeographic map of the distribution of K-Pg ejecta, including microkrystites, (micro)tektites, and iridium anomaly at 66 Ma, is given in *Smit* [1999; fig. 1], and a selection of global K-Pg boundary sites is given in *Nichols and Johnson* [2008; fig. 2.3], *Schulte et al.* [2010; fig. 2a], *Vajda and Bercovici* [2014; fig. 3], and *Punekar et al.* [2014; fig. 7].

Chromium isotopes suggest a cosmic origin of the platinum-group elements (PGE) in clays in the K-Pg boundary at Stevns Klint [*Shukolyukov and Lugmair*, 1998; *Frei and Frei*, 2002]. Osmium isotopes deposited on the Earth's surface 66 Ma ago support the contention that an asteroid impact contributed to mass extinctions at the end of the Cretaceous [*Kerr*, 1983]. For example, *Frei and Frei* [2002] have explained the sudden drop of $^{187}Os/^{188}Os$ ratios from 0.210 to 0.160 as being derived from global fallout of extraterrestrial matter; the present $^{186}Os/^{188}Os$ ratio of 0.119836 ± 0.000004 measured in the basal clay layer of the K-Pg boundary is a chondritic signature. Osmium isotopic ratios most likely resulted from some asteroid impact, but cannot exclude that a huge volcanic eruption spewed iridium-laden volcanic debris derived from the mantle [*Kerr*, 1983].

Nimura et al. [2016] have reported iridium in an ~5 m thick section of pelagic sediment cored in the deep-sea floor in the Pacific Ocean, along with a distinct iridium spike at the K-Pg boundary related to the Chicxulub, assumed to be from the asteroid impact. They have proposed a way to distinguish the contribution of extraterrestrial matter in sediments from geogenic matter through a Co-Ir diagram called "extraterrestrial index" f_{EX}. This new index has revealed a broad iridium anomaly around the Chicxulub spike and led to the assertion that no mixture of materials on the Earth's surface could explain such a broad iridium component.

An attempt to model environmental changes that followed the Chicxulub impact, exploring longer-lasting cooling due to sulfate aerosols, was made by *Brugger et al.* [2017]. From their model, they concluded that depending on the aerosol stratospheric residence time, global annual mean temperature lowered by at least 26 °C, with up to 16-year subfreezing temperatures and recovery time >30 years, implying ocean mixing and plankton bloom caused by nutrient upwelling. These results have convinced them on the pivotal role of an asteroid impact for the end-Cretaceous mass extinction, although one cannot discard the possible role of sulfate aerosols from concomitant Deccan eruptions.

12.2.2. Volcanism: Deccan Province and the K-Pg Boundary Event

Radiometric age dating has gradually revealed the ties between LIP volcanic activities and major mass extinctions in the Phanerozoic eon [*Courtillot et al.*, 1986; *Jourdan et al.*, 2014; *Keller and Kerr*, 2014; *Percival et al.*, 2015, 2017; *Keller et al.*, 2016]. Several kill mechanisms leading to a complex web of catastrophic environmental effects are associated with LIPs, including oceanic anoxia, ocean acidification, sea-level changes, toxic metal input, and essential nutrient decrease [*Ernst and Youbi*, 2017]. The Deccan Traps eruptions, an extended period of end-Cretaceous volcanism, have been considered responsible for the mass extinction documented in intertrappean beds in India [e.g., *McLean*, 1985; *Keller et al.*, 2011, 2012; and references therein] and may offer an explanation for K-Pg geochemical anomalies.

Investigation on C, O, and Sr isotopes and PGE patterns in sections across two K-Pg classical localities (Stevns Klint and Woodside Creek) and at Richards Bay, South Africa, has revealed marked differences between these localities, suggesting that PGE anomalies are not unambiguous identifiers of large impact events concurrent with Phanerozoic mass extinctions [*Tredoux et al.*, 1989]. The presence of bentonite layers and Pt- and Pd-dominated PGE anomalies below and above eight K-Pg boundary sites in Mexico have been interpreted as indicating volcanic activity [*Stüben et al.*, 2005]. Moreover, *Zoller et al.* [1983] observed that hot spot volcanism can produce siderophile element-enriched aerosols (including iridium), and this has led *Hallam* [1987] to argue that possible terrestrial components of siderophile elements in the K-Pg clays had not been adequately identified. *Shukla et al.* [2001] observed that Deccan flows at Anjar, Kutch, western India, have iridium concentrations ranging from 2 to 178 pg.g^{-1} which are higher by an order of magnitude when compared to iridium concentrations in other basalts in the Deccan province (95% of the Deccan lavas are of tholeiitic basalts, followed by minor alkali basalts, nephelinites, lamprophyres, and carbonatites). Three phases of continental flood basalts are known in the Deccan lavas with a total of 6% volume erupted in phase 1 (C30n–C31n), while 80% was erupted in phase 2 (C29r), and 14% in phase 3 (C29n) [*Chenet et al., 2009*]. The Deccan phase 1 of volcanism is separated from Deccan phase 2 by about 850,000-year quiescence. Phase 2 is represented by the longest lava flows in the Phanerozoic [*Self et al., 2008*] when about 1.1 million km³ of lavas was erupted [*Schoene et al., 2015*]. This phase started at 66.288 ± 0.027 Ma (U-Pb zircon dating; *Schoene et al., 2015*) about 250,000 years before the K-Pg boundary (65.968 ± 0.085 Ma; *Renne et al., 2013*), encompassing the age interval of the CF2 and CF1 planktic foraminiferal biozones, and ended at 500,000 years after the K-Pg boundary (65.552 ± 0.026 Ma).

The Deccan phase 2 volcanism seems to have severely affected the flora in India. Shortly after the onset of this volcanic phase, gymnosperms and angiosperms were decimated as attested by a sharp decrease in pollen and spores coupled with the appearance of fungi, which mark increasing stress conditions as a direct response to this volcanic activity [*Adatte et al., 2015; Fantasia et al., 2016*]. Gases such as CO_2 and SO_2 accompanying an eruptive event may cause devastating impacts on Earth's environment and life [*Svensen et al., 2009; Bond and Wignall, 2014*]. It is believed that the Deccan Traps contained magmatic sulfur concentrations as high as 1900 ppm [*Callegaro et al., 2014*], much higher than in basalts from some other similarly sized igneous provinces (e.g., Paraná-Etendeka). Therefore, Deccan eruptions may have caused globally extensive acidic rains before and during the K-Pg boundary, resulting in weathering and dissolution effects on land. Dissolution of foraminifers [e.g., *Gertsch et al., 2011*] and of iron oxide minerals (biomagnetite and detrital magnetite) is, perhaps, evidence of acidification of oceans during this period [*Font et al., 2014*]. High chemical index of alteration (CIA) values in intertrappean sediments deposited coevally with the Deccan phase 2 volcanism record the role of acid rains formed as a consequence of SO_2 emissions. Lava flows separated by red weathered horizons ("the red boles") are observed near to the Deccan eruption center and suggest quiescent periods between basalt flows [*Adatte et al., 2015*].

Mercury (Hg) is regarded as a proxy of massive volcanism in some recent studies, allowing for a new insight into the relationship between LIP activity, abrupt environmental changes, and mass extinctions [e.g., *Hildebrand and Boynton, 1989; Nascimento-Silva et al., 2011, 2013; Sanei et al., 2012; Grasby et al., 2013, 2015, 2017; Sial et al., 2013, 2014, 2016, 2017; Adatte et al., 2015; Percival et al., 2015, 2017; Font et al., 2016, 2018; Thibodeau et al., 2016; Bond and Grasby, 2017; Charbonnier et al., 2017; Jones et al., 2017; Khozyem et al., 2017; Thibodeau and*

Bergquist, 2017]. It is possible that a huge amount of Hg was released to the atmosphere by the Deccan eruptions in an analogous way to that proposed for the Siberian Traps which emitted 3.8×10^9 ton Hg [*Sanei et al., 2012*] at the Permian-Triassic boundary. Moreover, it has been speculated that some of the observed Hg enrichments across some chronostratigraphical boundaries may represent true Hg loading to the environment, recorded by an increase in the Hg/TOC ratio [e.g., *Grasby et al., 2016, 2017; Percival et al., 2015, 2017; Font et al., 2016; Sial et al., 2016; Charbonnier et al., 2017; Jones et al., 2017; Thibodeau and Bergquist, 2017*].

12.2.3. Triggering of the Deccan Phase 2 Eruption by the Chicxulub Impact

The Deccan Traps underwent a massive rise in lava flows around the time of impact of Yucatán's Chicxulub crater (66 Ma ago), and this reinforces the possibility that the K-Pg extinction may have been caused by both volcanism and impact events [e.g., *Bhandari et al., 1996; Rampino and Brothers, 1998a, 1998b; Keller et al., 2009; Rampino, 2010; Schulte et al., 2010; Keller, 2014; Renne et al., 2015; Richards et al., 2015, 2017; Schoene et al., 2015*]. The connection among these three events, namely, the impact, the mass extinction, and the major pulse of the Deccan volcanism, seems to become clearer as high-precision dates continue to pin them [*Richards et al., 2015*] and reveal that they may have occurred within less than about one hundred thousand years of each other. Therefore, bolide impact and the Deccan flood volcanism, although more than 13,000 km away from each other, may have competed as cause of the end-Cretaceous mass extinction [*Renne et al., 2015*], a hypothesis that needs further support from high-precision dating of Deccan phase 2 flood basalt formations. The Chicxulub impact might have triggered the enormous Poladpur, Ambenali, and Mahabaleshwar (Wai Subgroup) lava flows, which together account for more than 70% of the Deccan phase 2 eruptions [*Richards et al., 2015*].

Richards et al. [2015] suggested that seismic waves resulting from the Chicxulub impact could have affected the magmatic system responsible for the Deccan volcanism. This geophysical model is based on two assumptions: (a) the contact between the Lonavala Subgroup (uppermost Bushe Formation) and Wai Subgroup (lowermost Poladpur Formation) marks a change in volcanic style and eruptive rates in the Deccan volcanic province, and (b) this change was coeval with the Chicxulub impact. The contact between the Bushe and Poladpur formations (Lonavala-Wai transition) was assumed to represent a hiatal disconformity based on (i) planation surfaces, (ii) non-penetration of structural features from lower (Bushe) flows into upper flows of the Wai Subgroup, and (iii) an inferred "flow contact" marked by a laminated depositional layer of unknown origin resting over a weathered lava flow of the Bushe Formation.

This model has been challenged by *Dole et al.* [2017] who argued that the evidences for the postulated disconformity pointed by *Richards et al.* [2015] were based on superficial observations that led them to think that this transition and the Chicxulub impact may have occurred within a time interval of ≲100,000 yr. However, as pointed out by *Richards et al.* [2017], subsequent geochronological work [*Renne et al.*, 2015] supports an increase in mean magma eruption rate at this transition, there is no evidence for a hiatus, and the K-Pg boundary, the Chicxulub impact, and the Lonavala-Wai transition likely occurred within a time interval of only ~50,000 yr.

In this chapter, Hg variation patterns as volcanogenic tracers in sections straddling the K-Pg in well-documented sites in the Northern Hemisphere and Southern Hemisphere are discussed. By examining the Hg stratigraphic patterns across these K-Pg sites, we will investigate worldwide fingerprints of Deccan eruptions and bolide impact in the climate reorganization in this most famous paleontological murder riddle. Hg isotopes from the K-Pg layer and rocks spanning the K-Pg are examined to evaluate how much Hg loading to the environment was geogenic (from the Deccan volcanism) and how much was extraterrestrial (bolide impact).

12.3. SELECTED CRETACEOUS-PALEOGENE BOUNDARY SECTIONS

Three distal K-Pg boundary sections relative to the Deccan Traps volcanic center (Denmark, Italy, and Brazil) and one proximal section (India) will be the focus of this chapter (Fig. 12.1). They represent well-documented sites in the Northern Hemisphere (Højerup at Stevns Klint, Denmark, and Bottaccione at Gubbio, Italy) and in the Southern Hemisphere (Um Sohryngkew, Meghalaya, India, and a Poty section from a continuous drill core near Recife, Brazil). The Um Sohryngkew section is located at proximal distance to the Deccan volcanic province and is of paramount importance for the assessment of fingerprints of Deccan eruptions in the period of the end-Cretaceous mass extinction.

In selecting the sections to be discussed in this chapter, attention was paid to stratigraphic completeness of the sedimentary successions and the geographic distribution of these sites. The Højerup, Bottaccione, and Um Sohryngkew sections are regarded as complete, based on the study of planktic foraminiferal biozones. The Højerup section is, perhaps, the best preserved and most laterally continuous K-Pg succession (UNESCO World Heritage Site, since June 2014), and all foraminiferal biozones are present in the uppermost Maastrichtian to lowermost

Danian succession [*Rasmussen et al.*, 2005; *Surlyk et al.*, 2006; *Thibault et al.*, 2015]. Refined magnetostratigraphy and planktic foraminiferal and calcareous nannofossil biostratigraphies for the stratigraphic succession at Bottaccione have been reported [e.g., *Coccioni et al.*, 2010; *Coccioni and Premoli Silva*, 2015], and CF1, CF2, and CF3 foraminiferal biozones were identified in the uppermost Maastrichtian, encompassing the Scaglia Bianca and Scaglia Rossa formations up to the K-Pg boundary. In the Um Sohryngkew section, Meghalaya, the presence of the planktic foraminiferal biozones CF4, CF3, CF2, and CF1 in the upper Maastrichtian part and the P0 and Pα zones and P1a subzone in the lower Danian part assures biostratigraphically continuous records across the K-Pg boundary [*Mukhopadhyay*, 2008, 2017; *Gertsch et al.*, 2011; *Pal et al.*, 2015; *Mukhopadhyay et al.*, 2017].

The Poty section at about 7800 km from the Yucatán, northeastern Brazil, is the most distant K-Pg boundary site with published accounts of the Chicxulub bolide impact [*Gertsch et al.*, 2013]. A hiatus in the early Danian that caused the reworking of Maastrichtian and Danian sediments is evidenced by microfossils in the conglomeratic bed that marks the K-Pg boundary in the Paraíba Basin [*Gertsch et al.*, 2013]. This short hiatus of ~200 kyr affected the late Maastrichtian planktic foraminiferal biozone CF1 and the early Danian subzone P1a(1), with the K-Pg boundary clay (zone P0), the Ir anomaly, and the characteristic negative $\delta^{13}C$ excursion missing [*Gertsch et al.*, 2013]. Even so, this Maastrichtian-Danian succession is, perhaps, the best preserved one in South America, and for this reason it is presented in this chapter.

12.3.1. The Danish Basin, Stevns Klint, Denmark

The Danish Basin offers one of the best records of the K-Pg boundary in the world. In a section straddling the K-Pg boundary in this basin, it is possible to see a layer beneath a topographic overhang separating the lowermost Danian Cerithium Limestone Member from the overlying lower Danian bryozoan limestone of the Stevns Klint Formation [*Surlyk*, 1997; *Surlyk et al.*, 2006, 2013]. The K-Pg boundary clay is usually 5 cm thick (Fiskeler Member of the Rødvig Formation), but reaches about 40 cm at Kulstirenden, in the northernmost part of the cliff [*Hart et al.*, 2004]. There was a shallowing in the sedimentation in the latest Maastrichtian before the K-Pg boundary, and there is clear evidence for seawater temperature fluctuations [*Surlyk*, 1997; *Hart et al.*, 2004; *Thibault et al.*, 2015].

It is possible to see the stratigraphic evolution of the Danish Basin, from the latest Cretaceous across the K-Pg boundary into the early Paleogene, in a 45 m thick succession exposed at Stevns Klint, located about 45 km south of Copenhagen, Denmark, on the island of Sjælland. Comprehensive stratigraphic investigations across the K-Pg

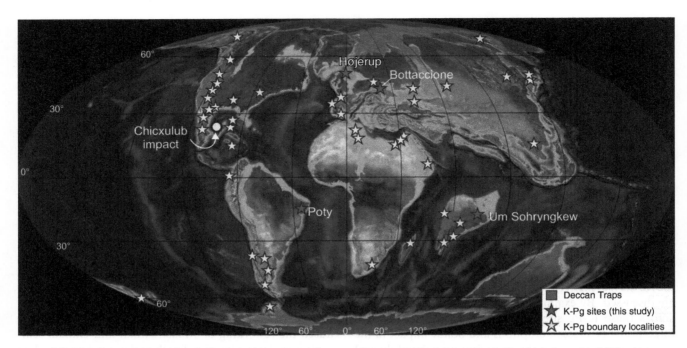

Figure 12.1 Paleomap at 66 Ma (Based on *Scotese* [2013]) showing paleogeography and a selection of published global terrestrial and nearshore marine K-Pg boundary sites (yellow star) and the four K-Pg boundary sections focused in this chapter (red star). Location of K-Pg boundary sites has been compiled from *Smit* [1999], *Nichols and Johnson* [2008], *Schulte et al*. [2010], *Vajda and Bercovici* [2014], and *Punekar et al*. [2014]. The location of the Chicxulub impact site is shown as a yellow dot. (*See electronic version for color representation of the figure*.)

boundary in this basin are found in studies conducted by *Surlyk* [1997], *Rasmussen et al.* [2005], and *Surlyk et al.* [2006, 2013]. One of the best sections is that of the Højerup coastal cliff (55°16′46.9″N and 12°26′46.32″E; Fig. 12.1), in which the Maastrichtian Møns Klint Formation (Sigerslev and Højerup members; *Surlyk et al.*, 2013; *Hansen and Surlyk*, 2014) is overlain by the lower Danian Rødvig Formation (Fiskeler and Cerithium Limestone members) which in turn is covered by the lower to middle Danian Korsnæb Member of the Stevns Klint Formation [*Surlyk et al.*, 2006]. The diachronous Cerithium Limestone becomes gradually younger from the southern part of Stevns Klint toward the northern part, and a hiatus including all the *Parvorugoglobigerina eugubina* Zone, present at the Fiskeler-Cerithium Limestone transition in the northern part of the cliff, is absent in the southern part [*Rasmussen et al.*, 2005]. Outlined by black flint bands, Danian bryozoan limestone mounds were formed shortly after the K-Pg boundary mass extinction [*Surlyk*, 1997; *Surlyk et al.*, 2006; *Bjerager and Surlyk*, 2007].

12.3.2. The Umbria-Marche Succession, Central Apennines, Italy

The Gubbio mountains are part of the Umbria-Marche Basin with two well-known K-Pg boundary sections, the Bottaccione Gorge and the Contessa Highway, in two

parallel valleys and a third less known section at Monte Acuto (Perugia). The Bottaccione section starts just outside Gubbio and cuts through pelagic Middle Jurassic to upper Eocene sedimentary rocks, including the Calcari Diasprigni, Maiolica, and Marne a Fucoidi formations and the Scaglia Group (Scaglia Bianca, Scaglia Rossa, Scaglia Variegata, and Scaglia Cinerea formations). This section represents the magnetostratigraphic standard for the Upper Cretaceous-Eocene interval [e.g., *Galeotti et al.*, 2015]. Similar to Stevns Klint, this section contains a 1–2 cm thick clay layer at the K-Pg boundary, within the Scaglia Rossa Formation, enriched in Ir [*Alvarez et al.*, 1980], but with a very short lateral extent as the succession is tilted.

The contact between the Scaglia Bianca and the overlying Scaglia Rossa formations is marked by the alternation of gray and pink colored beds. About 8 m below this contact, a 1.5 m thick marker horizon (corresponding to the Oceanic Anoxic Event 2, Livello Bonarelli, Cenomanian-Turonian boundary) is present in all these K-Pg sections.

The K-Pg boundary was recorded within the Scaglia Rossa Formation which exhibits bedding thicknesses between 10 and 20 cm and is formed by predominantly pink to red pelagic limestones with cherty nodules and calcareous marls. Apparently, this succession represents a continuous stratigraphic record across the K-Pg boundary, and based on planktic and benthic foraminifers and calcareous nannofossils, it has been dated as Late Cretaceous to early/middle Eocene [*Premoli Silva and Sliter*, 1994; *Alvarez*, 2009; *Coccioni and Premoli Silva*, 2015; and references therein].

The K-Pg boundary crops out on the eastern side of the main road (43°13′39.78″N and 12°20′14.28″E), at about 240 m above the base of the Scaglia Rossa Formation. In this section, a 30–40 cm thick whitish limestone bed represents the topmost Cretaceous which is overlain by a 2 cm thick dark clay K-Pg boundary layer [*Premoli Silva and Sliter*, 1994]. The lowermost Paleocene is correlated with the occurrence of *Globigerina eugubina* foraminiferal zone [=*P. eugubina*; *Luterbacher and Premoli Silva*, 1964].

12.3.3. Meghalaya, South Shillong Plateau, NE India

The Um Sohryngkew K-Pg section is situated in India, north of Bangladesh, at about 800–1000 km from the Deccan volcanic province (25°11′18″N and 91°45′54″W; Fig. 12.1). It is exposed along the Um Sohryngkew River in Meghalaya, next to Terria village, East Khasi Hills District, south Shillong Plateau, in the eastern Himalayas. In this section, it is possible to examine one of the best exposures in India of a complete marine succession bracketing the K-Pg boundary [e.g., *Mukhopadhyay*, 2008; *Gertsch et al.*, 2011; and references therein]. This section exhibits strong evidence of mass extinction patterns (e.g., planktic foraminifers and larger ammonoids), a preserved K-Pg boundary layer, the first appearance of Danian foraminifers, and indications for sea-level change [*Keller et al.*, 2008, 2009; *Tewari et al.*, 2010a, 2010b; *Gertsch et al.*, 2011; and references therein].

The K-Pg boundary at the Um Sohryngkew River is marked by a thin red clay layer enriched in Ir and other PGE [*Bhandari et al.*, 1993, 1994; *Garg et al.*, 2006; *Pandey*, 1990] with abundant subangular quartz grains in a brown matrix [*Gertsch et al.*, 2011]. This section consists of a continuous Campanian-Eocene succession characteristic of coastal, estuarine, and nearshore environments [*Krishnan*, 1968; *Banerji*, 1981; *Tewari et al.*, 2010a, 2010b] with marine shelf sediments including thick sandstone layers, marl, shale, and carbonates. Based on the distribution of zonal indices, *Mukhopadhyay* [2008] and *Pal et al.* [2015] have recognized seven successive planktic foraminiferal zones across the K-Pg boundary at the Um Sohryngkew River section. These zones are, in stratigraphic order, CF4, CF3, CF2, and CF1 in the upper Maastrichtian part and P0 and Pα and P1a in the lower Danian part, thus representing a biostratigraphically continuous succession across the K-Pg boundary.

Recent biostratigraphical study of continuous sections across the K-Pg boundary of upper Maastrichtian shelf successions of the Therriaghat and Mahadeo blocks, Meghalaya, has provided further details on their foraminiferal zones and subzones [*Mukhopadhyay*, 2017]. The tectonic development of these unstable blocks led to local tectono-eustatic differences in lithostratigraphy and biostratigraphy between them and changes in sea-level and depositional environments. The succession of the Therriaghat block is more complete (4 foraminiferal zones, 11 subzones) than the succession of the Mahadeo block (3 zones, 6 subzones), according to *Mukhopadhyay* [2017]. Five transgressive-regressive cycles have been recognized, and documentation of local sea-level changes has been achieved based on planktic and benthic foraminiferal ratios.

Neither *Mukhopadhyay* [2008] nor *Gerstch et al.* [2011] have provided geographic coordinates of Um Sohryngkew K-Pg boundary sections they have studied. Therefore, *Sial et al.* [2016], aiming at investigating true Hg loading to the environment by the coeval Deccan phase 2 volcanism, have used a different section, located at 25°11′18″N and 91°45′54″W, which is shown in this chapter. They have focused only on 2.5 m of the Maastrichtian-Danian interval of this section, bracketing the K-Pg boundary.

12.3.4. The Paraíba Basin, Brazil

The Maastrichtian-Danian transition has been recorded in a carbonate succession represented by the Maastrichtian Gramame and Danian Maria Farinha formations in the Paraíba Basin in northeastern Brazil (Fig. 12.1). This basin occupies an area of about 7600 km² on land and 31,400 km² offshore, extending into the continental shelf and down to 3000 m depth. A preserved succession of almost continuous sediment deposition across the K-Pg boundary is observed at around 30 km to the north of Recife at the Poty quarry (7°52′58″S, 34°51′11″W) and also at the Ponta do Funil beach further north. The sedimentary succession at the Poty quarry encompasses marl, marly limestone, and limestone deposited on a shallow platform, with a gentle gradient, continuously affected by storms and bottom winnowing. This succession includes deposits formed in a highstand event (Gramame Formation (Maastrichtian)) and a regressive event (Maria Farinha Formation (Danian)) [*Neumann et al.*, 2009 and references therein]. The fossil assemblage suggests that the depositional environment for the Gramame and Maria Farinha formations varied from middle to deep neritic conditions [*Koutsoukos*, 1998; *Rodrigues et al.*, 2014; and references therein].

A reduction of the fauna in the late Maastrichtian in this basin has been observed by *Stinnesbeck and Keller* [1996]. A series of sedimentological and micropaleontological studies [e.g., *Albertão and Martins*, 1996 and references therein] have brought the fossil record data into discussion demonstrating that a microplankton reduction followed sedimentological changes across the K-Pg boundary. However, the question of the K-Pg boundary location in this basin remained open as an iridium anomaly was measured above a conglomeratic limestone that marks the Maastrichtian-Danian transition and contains late Maastrichtian and early Danian microfossils [*Albertão et al.*, 1994]. This iridium anomaly led them to suspect that a second impact may have occurred in the Danian. Based on microfossil studies in this basin, *Koutsoukos* [1998] suggested multiple comet impacts for hundreds of thousands of years across the K-Pg boundary. At the same time, studies concerning the sedimentological and paleontological record of the K-Pg boundary at the Poty section claimed that a hiatus which had affected the K-Pg record was probably caused by a sea-level fall during the early Danian [*Keller and Stinnesbeck*, 1996a]. Other studies focusing on the micropaleontological record around the K-Pg boundary in this basin have encompassed dinoflagellates [*Sarkis et al.*, 2002], ostracods [*Fauth et al.*, 2005], calcareous nannofossils [*Lima and Koutsoukos*, 2006], and foraminifera [*Gertsch et al.*, 2013]. These studies have shown a faunal turnover across the K-Pg boundary, with recovery marked by the continuity of few resistant species and establishment of new species from the early Danian.

Morgan et al. [2006] performed a global study of K-Pg boundary sites to investigate the dispersion of the Chicxulub ejecta and have included some sections of the Paraíba Basin. They found no evidence of spherules and shocked quartz in the K-Pg sections of the Poty quarry and Ponta do Funil in this basin. *Gertsch et al.* [2013] found evidence of a hiatus in the early Danian that caused the reworking of Maastrichtian and Danian sediments, indicated by microfossils in the conglomeratic bed that marks the K-Pg boundary in this basin. This hiatus affected the upper Maastrichtian biozone CF1 and the lower Danian subzone P1a(1), ~200 kyr. The loss of part of zone P1a probably has caused the loss of the most important layer related to the critical period of the Maastrichtian-Danian transition. *Gertsch et al.* [2013] found no evidence of impact-related features such as iridium anomalies and spherules in this basin.

12.4. ELEMENTAL AND ISOTOPE CHEMOSTRATIGRAPHY

12.4.1. Carbon Isotope Chemostratigraphy and Total Organic Carbon (TOC)

The covariation between $\delta^{13}C_{carb}$ and $\delta^{13}C_{org}$ records can help establish whether observed variations reflect changes in the isotopic composition of the ancient dissolved inorganic carbon pool [e.g., *Oehlert and Swart*, 2014, and references therein]. Covariant $\delta^{13}C_{carb}$ and $\delta^{13}C_{org}$ records demonstrate that both carbonate and organic matter were originally produced in ocean surface waters and have retained their original $\delta^{13}C$ composition [e.g., *Korte and Kozur*, 2010; *Meyer et al.*, 2013], as probably no secondary process is able to shift $\delta^{13}C_{carb}$ and $\delta^{13}C_{org}$ in the same direction at the same rate [*Knoll et al.*, 1986]. Diagenetic alteration can lead to decoupled $\delta^{13}C_{carb}$ and $\delta^{13}C_{org}$ records [e.g., *Grotzinger et al.*, 2011; *Meyer et al.*, 2013]. A noise in the $\delta^{13}C_{org}$ record may result also from local syn-sedimentary processes [*Maloof et al.*, 2010].

Global $\delta^{13}C_{carb}$ and $\delta^{13}C_{org}$ records of bulk sediment comparing C isotope pathways across K-Pg boundary sections at Højerup, Bottaccione, and Um Sohryngkew localities were published by *Sial et al.* [2016]. $\delta^{13}C_{carb}$ pathways obtained from drill core samples from three drill holes in the Paraíba Basin are found in *Nascimento-Silva et al.* [2011, 2013] and *Sial et al.* [2013]. $\delta^{13}C_{carb}$ and $\delta^{13}C_{org}$ data for the Højerup, Bottaccione, Um Sohryngkew, and Poty sections are found here in Tables 12.1 through 12.4 and are plotted in Figs. 12.2a, b and 12.3a, b.

Three among the four sections discussed here display negative isotope excursions immediately above the K-Pg boundary, represented by a shift to lighter $\delta^{13}C$ values. The important factor driving this negative excursion is the collapse of bioproductivity at the boundary, with a gradual recovery in the early Danian. An abrupt decrease of $\delta^{13}C_{carb}$ at the K-Pg boundary is noticeable at the Højerup, Bottaccione, and Um Sohryngkew sections as discussed by *Sial et al.* [2016].

The $\delta^{13}C_{carb}$ stratigraphic curve for the Poty section (Fig. 12.3b) is based on new data (Table 12.4), with a higher resolution than the $\delta^{13}C_{carb}$ pathway published by *Nascimento-Silva et al.* [2011, 2013]. The absence of the sedimentary record corresponding to part of the CF1 and the P0, Pα, and part of P1a biozones precludes one to fully picture the $\delta^{13}C_{carb}$ pathway across the K-Pg boundary. The $\delta^{13}C_{carb}$ values display a roughly steady increase from 1.54 to 2.27‰ in the uppermost 2.5 m of the Gramame Formation followed by a drop of $\delta^{13}C_{carb}$ values from +2.3 to +1.9‰ in the reworked layer on top of the K-Pg erosional contact (Fig. 12.3b).

The total organic carbon in these four sections, expressed as percent TOC, is <1.0% (Tables 12.1 through 12.4), and the majority of samples yield values <0.2%. The highest TOC values were recorded in the Højerup section at the K-Pg boundary layer (~0.9%) and in the uppermost Maastrichtian and Danian portions of the Um Sohryngkew section (0.3–0.6%), while much lower TOC values are observed in the whole Bottaccione section (0.01–0.09%). At Højerup, the TOC stratigraphic variation curve shows a very monotonous pattern with

Table 12.1 C, O isotopes and Ca isotopes, total organic carbon, Hg, Mo/Al, and Y/Ho ratios across the K/Pg boundary in a section at Højerup, Stevns Klint, Denmark.

Formation	Sample	Height (cm)	δ13C‰VPDB	δ13Corg‰VPDB	δ18O‰VPDB	δ44/40Ca‰	Hg (ng·g−1)	TOC (%)	Hg/TOC	Mo (ppm)	Al (%)	Mo/Al	Y/Ho
Danian Stevns Klint Formation	SKc	1050	1.4	–	−1.9	–	1.8	–	–	–	–	–	–
	N7	990	1.19	−20.85	−2.33	−1.19	1.0	0.010	100.0	0.06	0.02	3.00	45.71
	N6	960	1.38	−25.76	−1.61	−1.20	48.2	0.029	1662.06	0.11	0.07	1.57	43.98
	N5	930	1.28	−26.82	−1.8	−1.26	6.8	0.020	340.00	0.08	0.02	4.00	47.70
	N4	900	1.23	−21.68	−1.87	−1.23	11.7	0.013	900.00	0.11	0.03	3.66	44.86
	N3	870	1.16	−21.57	−1.81	−1.28	3.0	0.044	68.18	0.23	0.03	7.66	41.90
	N2	840	1.28	–	−1.93	–	6.6	0.030	220.00	0.15	0.04	3.75	47.22
Danian Rødvig Formation	D3	830	1.4	–	−1.6	–	4.55	–	–	–	–	–	–
	D2	815	1.6	–	−1.5	–	4.51	–	–	–	–	–	–
	N1	810	1.15	−24.94	−1.87	−1.23	4.5	0.006	750.00	0.15	0.04	3.75	42.25
	No. 3	805	1.7	–	−1.5	–	2	0.030	66.66	0.14	0.06	2.33	–
	FC-1	804	–	–	–	–	127.72	0.859	148.68	1.83	1.14	1.60	43.40
	FCD	803	1.9	−25.36	−0.5	−1.14	67.9	0.944	71.92	3.67	2.28	1.60	0.00
	FCC	802	1.7	−26.52	−0.9	–	257.94	0.859	300.27	3.67	2.28	1.60	
	FCB	801	1.3	–	−1.7	–	194.53	0.944	206.06	3.67	2.28	1.60	
	FCA	800	1.3	–	−1.7	–	108.76	0.944	115.21	3.67	2.28	1.60	
Maastrichtian Møns Klint Formation	M2	785	1.8	–	−1.41	–	9.09	–	–	–	–	–	–
	N(-1)	770	0.71	−24.94	−1.72	−1.23	–	0.014	–	0.02	0.01	2.00	45.81
	M1A	770	1.7	–	−1.4	–	0.88	0.014	62.85	0.02	0.01	2.00	
	No. 2	760	1.8	–	−3.61	–	2.3	–	–	–	–	–	
	N(-2)	740	1.71	−24.20	−2	–	<1.26	0.041	30.73	0.08	0.01	8.00	47.60
	N(-3)	685	1.94	−23.03	−1.7	−1.22	4.6	–	–	–	–	–	47.73
	N(-4)	660	1.93	−22.80	−1.59	−1.19	6.5	0.015	433.3	0.01	0.01	1.00	47.93
	No. 1	0	1.7	–	−1.61	–	11.3	–	–	–	–	–	–

Table 12.2 C, O and Ca isotopes, total organic carbon, Hg, Mo/Al, and Y/Ho ratios across the K/Pg boundary at Bottaccione, Gubbio, Italy.

Lithology	Sample	Height (cm)	$\delta^{13}C_{\text{‰VPDB}}$	$\Delta^{13}C_{\text{org‰VPDB}}$	$\delta^{18}O_{\text{‰VPDB}}$	$\delta^{44/40}Ca_{\text{‰}}$	Hg (ng·g⁻¹)	TOC (%)	Hg/TOC	Mo (ppm)	Al (%)	Mo/Al	Y/Ho
Marl	GP-28	1685.4	1.26	−25.32	−3.29		3.44	0.064	53.75	0.19	0.32	0.59	49.10
	GP-27	1565.4	1.27	−24.82	−3.18		1.62	0.066	24.54	0.15	0.31	0.48	48.29
	GP-26	1495.4	1.38	−20.97	−2.60	−1.35	3.79	0.094	40.31	0.16	0.41	0.39	51.97
	GP-25	1395.4	1.18	−27.34	2.36		1.51	0.064	23.59	0.18	0.31	0.58	49.96
	GP-24	1295.4	1.43	−27.97	−2.53		1.55	0.039	39.74	0.19	0.22	0.86	47.74
	GP-23	1195.4	1.34	−26.96	−2.55	−1.30	1.83	0.030	61.00	0.17	0.25	0.68	49.66
	GP-22	1095.4	1.28	−28.21	−2.82	−1.13	0.87	0.042	20.71	0.24	0.39	0.61	48.89
	GP-21	995.4	1.48	−25.49	−2.36		0.42	0.022	19.09	0.15	0.20	0.75	44.92
Marl (rhytmites)	GP-20	895.4	1.58	−25.02	−2.04	−1.17	0.68	0.045	15.11	0.21	0.30	0.70	46.29
	GP-19	875.4	1.77	−27.25	−1.69	−1.30	0.57	0.033	17.27	0.34	0.57	0.91	44.24
	GP-18	845.4	2.02	–	−1.57		0.72	0.035	20.57	0.39	0.32	1.21	46.36
	GP-17	815.4	2.08	–	−1.99	−1.12	0.44	0.023	19.13	0.12	0.15	0.80	44.79
	GP-16	795.4	2.03	−24.39	−1.79	−1.14	0.75	0.044	17.04	0.21	0.17	1.23	45.85
Clay layer	GP-15	KPg-B	–	−26.78		−1.35	5.23	0.027	193.70	0.49	0.64	0.29	44.04
Mudstone with planktic foraminifera	GP-14	794	2.28	−25.36	−2.83		0.22	0.018	12.22	0.09	0.37	2.67	42.03
	GP-13	790	2.32	−24.24	−2.57	−1.23	0.16	–	–	0.07	0.26	0.26	42.11
	GP-12	780	2.29	−25.55	−2.68	−1.25	0.35	0.026	13.46	0.08	0.25	0.32	45.82
	GP-11	770	2.27	−25.35	−2.67	−1.36	0.22	0.059	3.72	0.05	0.30	0.16	40.83
	GP-10	760	2.24	−25.55	−2.85	−1.31	0.37	0.041	9.02	0.06	0.21	0.28	37.50
	GP-9	730	2.17	–	−2.75	−1.32	0.38	0.045	8.44	0.20	0.34	0.58	36.82
	GP-8	700	2.18	–	−2.80	−1.30	0.47	0.049	9.59	0.17	0.26	0.65	39.61
	GP-7	600	2.07	−25.19	−3.22		0.42	0.027	15.55	0.23	0.39	0.58	39.29
	GP-6	500	2.15	–	−2.82	−1.29	0.88	0.068	12.94	0.17	0.23	0.73	39.24
	GP-5	400	2.32	−24.83	−3.46	−1.33	0.48	0.017	28.23	0.30	0.59	0.50	36.59
	GP-4	300	2.13	−23.32	−2.57		0.97	0.015	64.66	0.23	0.43	0.53	35.81
	GP-3	200	2.15	−27.13	−2.32	−1.27	1.38	0.036	38.33	0.18	0.37	0.48	35.59
	GP-2	100	2.23	−27.18	−2.39		1.75	0.033	53.03	0.58	0.57	1.01	36.00
	GP-1	0	2.29	−25.11	−2.30	−1.39	0.37	0.091	4.06	0.26	0.78	0.33	35.36

Table 12.3 C, O and Ca isotopes, total organic carbon, Hg, Mo/Al, and Y/Ho ratios across the K/Pg boundary in the Um Sohryngkew section, Meghalaya, India.

Lithology	Sample	Height (cm)	$\delta^{13}C_{‰VPDB}$	$\delta^{13}C_{org‰VPDB}$	$\delta^{18}O_{‰VPDB}$	$\delta^{44/40}Ca_{‰}$	Hg (ng.g^{-1})	TOC (%)	Hg/TOC	Mo (ppm)	Al (%)	Mo/Al	Y/Ho
Light gray shale	KT25	250	0.1	−23.78	−11.9	–	10.3	0.042	245.23	–	–	–	–
Dark gray shale	KT24	240	1.1	−25.36	−8.5		9.8	0.039	251.28	–	–	–	–
	KT23	230	0.4	−24.96	−10.7		7.4	0.040	185.00	–	–	–	–
	KT22	220	0.6	−24.35	−9.5		11.5	0.039	294.87	–	–	–	–
	KT21	210	−0.1	−24.93	−11.8		2.4	0.042	57.14	–	–	–	–
	KT 20	200	0.4	−24.91	−11.0		7.8	0.036	216.66	–	–	–	–
	KT19	190	−0.1	−24.42	−12.6		3.9	0.044	88.63	–	–	–	
	KT18	180	−0.1	−24.43	−11.9		12.1	0.041	295.12	–	–	–	
	KT17	170	0.6	−24.38	−10.8		6.2	0.065	95.38	–	–	–	
	KT16	160	1.3	−24.59	−7.8		6.7	0.039	171.79	–	–	–	
	KT15	150	1.4	−24.79	−7.8		10.0	0.035	285.71	–	–	–	
	KT14	140	1.1	−24.77	−7.8		11.8	0.033	357.57	–	–	–	
	KT13	130	−0.0	−24.49	−9.9		4.3	0.039	110.25				
Light gray shale	KT12	120	−0.1	−24.56	−12.7	−1.28	1.6	0.627	2.55	2.05	1.13	1.81	32.29
	KT11	110	0.1	−21.69	−11.8	−1.33	2.8	0.571	4.90	2.93	1.05	2.79	33.02
	KT10	100	0.1	−21.99	−13.0	−1.44	0.0	0.447	0.00	0.71	1.39	0.51	29.32
	KT9	90	0.5	−20.22	−11.0	−1.31	0.0	0.468	0.00	0.53	1.16	0.45	29.11
K/Pg layer	K/Pg	80	−8.13	−23.12	−4.89		6.3	0.059	16.94	3.11	1.81	1.71	44.86
Greenish sandstone and shale	KT8	70	−0.4	−23.93	−13.0		6.3	–	–	–	–	–	–
	KT7	60	0.4	−23.91	−9.9		3.2	–	–	–	–	–	–
	KT6	50	−0.0	−23.91	−12.7	−1.46	6.2	0.070	88.57	0.66	1.50	0.44	30.59
	KT5	40	0.7	−24.14	−10.7	−1.45	2.9	0.052	55.76	0.59	1.45	0.40	29.32
	KT4	30	0.2	−24.82	−10.8	−1.36	13.1	0.335	39.10	1.67	1.22	1.36	29.75
	KT3	20	0.5	−23.81	−11.6	−1.44	98.6	0.424	232.54	1.21	1.53	0.79	30.26
	KT2	10	0.3	−24.49	−11.8	−1.28	31.4	0.337	93.17	1.88	1.18	1.59	29.86
	KT1	0	1.0	−24.54	−10.3	−1.33	6.9	0.349	19.77	0.91	1.12	0.81	29.86

Table 12.4 C-, O-, and Ca-isotope analyses, Hg and TOC concentrations and Mo/Al and Y/Ho ratios for the Poty Quarry section, Paraíba Basin, Brazil.

Formation	Litholoy	Sample	Height (m)	δ¹³C‰VPDB	δ¹³C$_{org}$‰VPDB	δ¹⁸O‰VPDB	δ⁴⁴/⁴⁰Ca (‰)	Hg (ng.g⁻¹)	TOC (%)	Hg/TOC	Mo (ppm)	Al (%)	Mo/Al	Y/Ho
Maria Farinha Formation	Limestone	PO 96	11.4	2.60	−23.38	−2.74	−1.32	2.71	0.23	11.78	1.24	0.06	17.43	32.15
		PO 97	11.5	2.57	−24.30	−2.76	−1.35	1.80	0.18	10.00	0.99	0.07	14.92	33.07
		PO 98	11.6	2.52	−23.46	−3.01	−1.20	2.33	0.17	13.71	1.25	0.04	19.67	32.69
	Marl	PO 99	11.8	2.53	−23.73	−3.21	−1.20	2.35	0.21	11.19	1.25	0.03	17.09	34.13
	Limestone	PO 100	12.0	2.51	−23.76	−3.64	−1.11	2.39	0.15	15.93	0.92	0.03	22.28	36.09
		PO 101	12.3	2.44	−24.02	−4.12	−1.25	1.42	0.14	10.14	0.84	0.02	27.77	37.84
		PO 102	12.4	2.36	−24.28	−3.96	−1.26	0.86	0.09	9.56	0.40	0.06	15.61	40.00
		PO 103	12.6	2.55	−24.50	−4.60	−1.38	1.31	0.10	13.10	0.27	0.01	11.78	40.38
		PO 104	12.7	2.51	−23.88	−3.96	−1.23	2.40	0.16	15.00	3.74	0.01	57.67	36.92
		PO 105	12.9	2.92	−24.17	−2.56	−1.28	0.87	0.06	14.50	0.33	0.01	38.51	37.38
		PO 106	13.1	2.67	−24.23	−3.99	−1.15	0.51	0.05	10.20	0.17	0.01	21.94	43.57
	Intercalation of marl-limestone	PO 107	13.3	2.68	−24.43	−3.73	−1.08	1.81	0.14	12.93	0.22	0.02	20.69	46.91
		PO 108	13.4	2.74	−24.42	−3.33	−1.00	0.64	0.08	8.00	0.15	0.03	16.37	47.36
	Reworked limestone	PO 109	13.6	2.42	−23.80	−4.56	−0.95	0.84	0.10	8.40	0.67	0.05	32.14	49.05
		PO 110	13.9	2.06	−23.78	−4.99	−1.32	1.07	0.13	8.23	1.20	0.05	40.13	47.79
		PO 111	14.0	1.89	−23.71	−4.83	−1.35	1.47	0.18	8.17	2.07	0.06	41.50	48.42
Conglomerate (erosional contact)														
Gramame Formation	Marl	PO 112	14.4	2.27	−23.98	−3.17	−0.97	1.12	0.13	8.62	1.43	0.07	12.91	31.23
		PO 113	14.6	2.15	−24.13	−3.56	−0.71	1.17	0.33	3.55	1.75	0.11	19.99	36.72
		PO 114	14.6	2.23	−24.14	−3.40	−1.05	1.02	0.19	5.37	3.36	0.09	20.52	33.05
	Marly limestone	PO 115	14.8	2.14	−24.09	−3.25	−0.90	0.78	0.30	2.60	2.01	0.16	9.84	31.15
		PO 116	14.9	2.13	−23.92	−2.65	−0.86	1.24	0.38	3.26	1.66	0.20	8.63	29.92
		PO 117	15.0	2.14	−23.97	−2.72	−0.89	1.30	0.34	3.82	1.49	0.19	8.93	32.82
	Marl	PO 118	15.2	1.69	−23.92	−3.59	−1.04	2.47	0.28	8.82	0.81	0.17	11.15	35.24
		PO 119	15.3	1.99	−24.13	−3.62	−0.88	2.20	0.16	13.75	1.68	0.07	11.80	33.67
		PO 120	15.4	1.76	−24.08	−3.76	−1.11	2.28	0.24	9.50	1.24	0.14	10.09	33.54
		PO 121	15.7	1.80	−24.23	−3.91	−1.04	2.14	0.25	8.56	1.56	0.12	13.92	32.82
		PO 122	15.9	1.78	−24.26	−5.04	−1.31	2.09	0.25	8.36	1.86	0.11	18.10	32.42
		PO 123	16.0	1.77	−24.06	−4.06	−0.91	2.13	0.22	9.68	1.82	0.10	15.44	34.71
		PO 124	16.2	1.74	−24.38	−3.87	−1.07	2.38	0.24	9.92	1.39	0.12	78.04	35.90
		PO 125	16.4	1.76	−24.43	−3.88	−1.02	2.21	0.36	6.14	1.64	0.02	67.45	35.77
		PO 126	16.5	1.68	−24.58	−3.87	−0.89	2.77	0.23	12.04	0.61	0.02	2.99	31.84
	Marly limestone	PO 127	18.2	1.51	−24.80	−3.04	−0.83	3.60	0.42	8.57	0.93	0.20	5.02	32.14
		PO 128	18.3	1.58	−25.00	−2.92	−1.08	3.54	0.39	9.08	0.72	0.18	4.02	30.64
		PO 129	18.5	1.54	−24.66	−3.33	−1.03	3.17	0.30	10.57	1.43	0.18	12.91	31.23

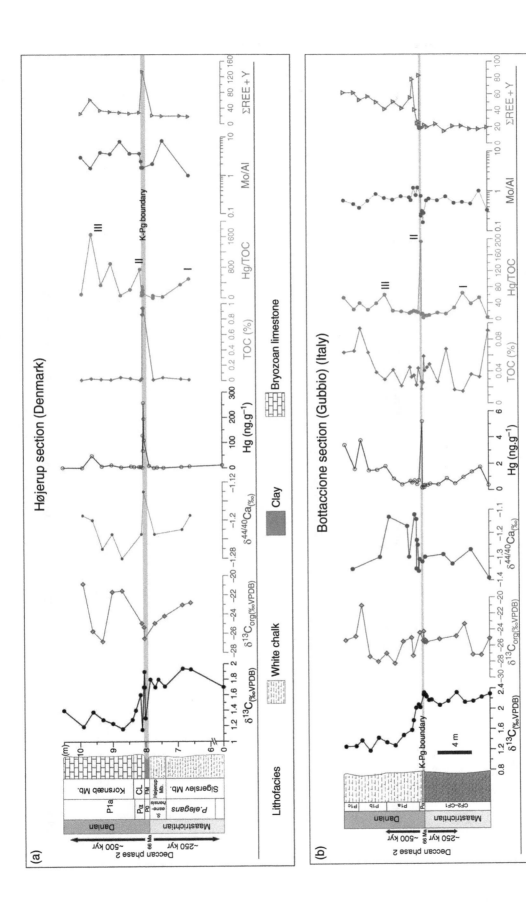

Figure 12.2 $\delta^{13}C_{carb}$, $\delta^{13}C_{org}$, $\delta^{44/40}Ca$, Hg/TOC, and Mo/Al variation patterns for (a) the Højerup section, Stevns Klint (Maastrichtian planktic foraminiferal biostratigraphy from *Surlyk et al.* [2006], and Danian from *Rasmussen et al.* [2005]). (b) The Bottaccione (Gubbio) section (Planktic foraminiferal biostratigraphy from *Coccioni et al.* [2010] and *Coccioni and Premoli Silva* [2015]).

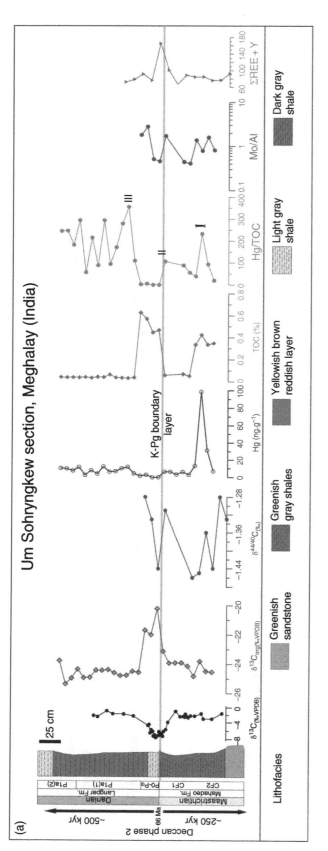

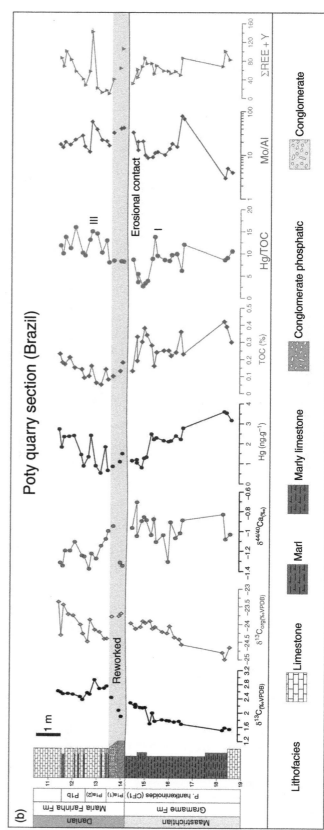

Figure 12.3 $\delta^{13}C_{carb}$, $\delta^{13}C_{org}$, $\delta^{44/40}Ca$, Hg/TOC, and Mo/Al variation patterns for (a) an Um Sohryngkew River section ($\delta^{13}C_{carb}$ and planktic foraminiferal biostratigraphy from *Gertsch et al.* [2011]). (b) A Poty quarry drill hole (Planktic foraminiferal biostratigraphy from *Gertsch et al.* [2013]).

almost no oscillation, except for a prominent positive shift at the K-Pg boundary layer (Fig. 12.2a). At Bottaccione, however, the TOC stratigraphic pathway shows vigorous oscillations with a negative shift about 4 m below the K-Pg boundary (within the CF2 biozone), a small negative excursion at the K-Pg boundary, and a prominent positive shift about 8 m above the K-Pg boundary (within the P1a biozone).

At the Um Sohryngkew section, the TOC stratigraphic variation curve shows two markedly positive excursions, about one meter below the K-Pg boundary (CF2 biozone) and another at the K-Pg boundary and in lowermost Danian (P0 biozone) [*Sial et al.*, 2016]. At the Poty section, the absence of the P0, Pα, and part of the CF1 biozones as well as part of the P1a subzone precludes a full picture of the TOC variation across the K-Pg boundary. An enrichment of TOC is observed within 1.0 m below the K-Pg transition, followed upsection by a decrease of the TOC values within 40 cm above the erosional contact and a progressively recovery upward.

The $\delta^{13}C_{org}$ curve for organic matter in the Højerup section shows a gradual decrease from values of −23‰ toward the K-Pg boundary layer, with a minimum of −27‰ at this boundary (Fig. 12.2a), followed by a strong increase to −22‰ upsection and again by a negative shift to −27‰ and a positive shift to −22‰. The $\delta^{13}C_{org}$ curve shows a less pronounced decrease toward the K-Pg boundary (CF2-CF1 biozones) in the Bottaccione section (Fig. 12.2b), a discrete decrease at the K-Pg boundary, and two positive shifts within the P1a biozone. The Um Sohryngkew section displays a slightly different pattern (Fig. 12.3a), and no decrease of $\delta^{13}C_{org}$ toward the K-Pg boundary is evident. A maximum (−21‰) in the $\delta^{13}C_{org}$ pathway is shown within the P0 biozone, followed upsection by a monotonous isotope curve within the P1a(1) subzone and a small increase to −24‰ within the P1a(2) subzone. A covariance between $\delta^{13}C_{carb}$ and $\delta^{13}C_{org}$ in the Højerup and Bottaccione sections is depicted in Fig. 12.2a, b, providing evidence that the carbonate and organic matter were produced in the surface waters of the ocean and have probably retained their original $\delta^{13}C$ compositions. The $\delta^{13}C_{carb}$ and $\delta^{13}C_{org}$ curves at the Um Sohryngkew section are decoupled around the K-Pg boundary, and $\delta^{13}C_{org}$ displays a large positive excursion.

The $\delta^{13}C_{org}$ curve for the Poty section (Fig. 12.3b) shows a certain covariation with the $\delta^{13}C_{carb}$ curve up to the erosional contact between the uppermost Maastrichtian and lowermost Danian sedimentary layers. For the rest of the Danian samples, these two curves depart from each other with a marked increase of the $\delta^{13}C_{org}$ values.

12.4.2. Hg and Hg/TOC Chemostratigraphy

Sial et al. [2016] have investigated Hg/TOC spikes across the K-Pg boundary in the three continuous K-Pg sections reported here, two of which are distal in relation to the Deccan volcanic center and one is proximal. We identify three Hg/TOC spikes across the K-Pg boundary in these three sections, which are possibly associated with the Deccan phase 2 volcanism, an assertion that seems to find further support in Hg isotopes in these Hg/TOC spikes, as discussed later in this chapter. In addition, we investigate here the behavior of the Hg/TOC from the Poty section, Brazil, a distal site to the Deccan volcanic center. Two Hg/TOC spikes are revealed, one about 90 cm below the K-Pg transition and the other one, a double spike, at about 1 m above (Fig. 12.3b). If a Hg/TOC spike did exist in the K-Pg boundary at the Poty section, as observed in the three other sections under consideration, the sedimentary layer which recorded it has been eroded away.

At Bidart, France, high Hg contents of likely volcanic origin characterize the sedimentary interval immediately below the iridium-rich layer where low magnetic susceptibility was thought to have resulted from paleoenvironmental turnover associated with the Deccan phase 2 [*Font et al.*, 2016]. These elevated Hg concentrations (Hg/TOC spikes) correlate with high shell fragmentation and dissolution effects in planktic and benthic foraminifera, suggesting that ocean acidification drastically affected marine biodiversity [*Adatte et al.*, 2015; *Font et al.*, 2016].

12.4.3. Ca Isotope Chemostratigraphy and REE and Ce, Eu, and Y Anomalies

12.4.3.1. Ca Isotope Chemostratigraphy

Marine calcifying planktic communities were highly impacted during the K-Pg transition, with substantial net losses of planktic foraminifer and calcareous nannoplankton genera and species [*Thierstein*, 1982]. The loss of primary producers, especially the loss of marine calcifying nannoplankton, seems to have ultimately affected the marine C cycle, as suggested by a pronounced $\delta^{13}C$ negative anomaly displayed by carbonates spanning the K-Pg transition [*Zachos and Arthur*, 1986]. Although large, the C-cycle perturbation at the K-Pg transition is unlikely to have resulted in a resolvable (>0.15‰) perturbation to the global Ca cycle in seawater [*Komar and Zeebe*, 2015]. Rather, $\delta^{44/40}Ca$ (0‰ = modern seawater) values in bulk carbonate sediments across the K-Pg transition are more likely to reflect changes in carbonate sedimentation and diagenesis [*Fantle and Higgins*, 2014; *Griffith et al.*, 2015] associated with changes in the magnitude and/or composition of the ocean's biological (CaCO$_3$ and C$_{org}$) pump.

Ca and C isotope compositions of marine carbonates spanning the K-Pg transition such as in the Højerup, Bottaccione, and Um Sohryngkew sections yield $\delta^{44/40}Ca$ values that range from -1.4 to -1.1% well within the range of bulk pelagic carbonates from the Mesozoic and Cenozoic [*Gothmann et al.*, 2016]. Although the magnitude of the variability is small, many sections are characterized by a shift toward higher $\delta^{44/40}Ca$ values across the K-Pg boundary or in its aftermath (Figs. 12.2a, b and 12.3a). Similar behavior has been documented for marine carbonates spanning the K-Pg transition at the ODP Site 149A [Blake Nose; *Silva-Tamayo*, 2015]. This positive shift in Ca isotopes partially parallels a negative excursion on the carbonate $\delta^{13}C$ values in these figures. One possible explanation is changes in biological fractionation of Ca isotopes associated with the dramatic biotic turnover at the K-Pg.

Another possibility is that the shift in Ca isotopes observed in these three sections is due to global changes in carbonate sedimentation and diagenesis across the K-Pg and in the early Danian. In this case, the increase in sedimentary $\delta^{44/40}Ca$ values is consistent with increasing diagenetic alteration under seawater-buffered conditions in the aftermath of the K-Pg boundary. One possible mechanism for an increase in seawater-buffered diagenetic alteration is a local decline in carbonate sedimentation associated with a global ocean acidification event [e.g., *Griffith et al.*, 2015]. A similar Ca isotope shift (in sign) was observed by *Griffith et al.* [2015] during the Paleocene-Eocene Thermal Maximum at sites where local carbonate sedimentation declines in response to a global shoaling of the carbonate compensation depth in response to ocean acidification by CO_2. Thus, although further high-resolution studies are needed to tease out small but systematic changes in bulk sediment $\delta^{44/40}Ca$ values, the evolution and duration of these Ca and C isotope anomalies are consistent with major perturbations to the marine carbonate budget linked to ocean acidification during a period of rapid and massive volcanic CO_2 outgassing across the K-Pg boundary.

The Poty section exhibits a distinct behavior with a clear shift across the K-Pg transition toward lower $\delta^{44}Ca$ values by $\sim 0.3\%$ (between 13 and 14 m; Fig. 12.3b). This decrease across the K-Pg transition is probably not caused by global perturbation of the Ca cycle but to more diagenetic alteration of Maastrichtian sediments than their Danian counterparts.

12.4.3.2. Rare Earth Elements and Ce, Eu, and Y Anomalies

Rare earth elements (REEs) and yttrium (inserted between Dy and Ho) are indicators of environmental conditions such as marine anoxia [e.g., *Liu et al.*, 1988; *German and Elderfield*, 1990; *Murray et al.*, 1991] and

paleoredox conditions in general [e.g., *Elderfield and Pagett*, 1986; *Wang et al.*, 1986; *Liu et al.*, 1988; *Kato et al.*, 2002; *Madhavaraju and Lee*, 2009; *Madhavaraju and González-León*, 2012], and $\Sigma REE/Al$ can be used as a detrital input proxy [e.g., *Sosa-Montes et al.*, 2017]. Among them, only Ce and Eu show potential variations as a function of redox conditions observed in the ocean. The Ce anomaly was first proposed by Ederfield and Greaves [1982] and has been used as an empirical technique to estimate fine-scale sea-level variation [e.g., *Wilde et al.*, 1996]. It results from oxidation of trivalent Ce to Ce^{4+} and subsequent decoupling of Ce from the other REEs due to formation of less soluble Ce^{4+} species and/or preferential adsorption of Ce^{4+} species on particle surfaces [*Bau and Dulski*, 1996]. Development of Ce anomalies appears to be restricted to oxidizing environments of high complex-forming capacity, and therefore, occurrence of negative Ce anomalies argues in favor of the presence of oxidized surface waters during the time of deposition [*Towe*, 1991; *Bau and Dulski*, 1996]. Under anoxic/reducing depositional conditions, Ce occurs in the dissolved Ce^{+3} state; therefore, marine carbonates precipitated in reducing/anoxic conditions, with no negative Ce anomaly, are important indicators of paleoredox condition of the depositional environment [*Wang et al.*, 1986]. Cerium is depleted in open ocean waters but not in shallow sea waters [*Shimizu and Masuda*, 1977].

Sometimes Ce anomalies arise as artifacts resulting from unusual La enrichments when calculated using the neighbor abundances of La and Pr. To avoid this problem, *Bau and Dulski* [1996] have suggested to compare $(Pr/Pr^*)n$ ratios $[(Pr/Pr^*)n = Prn/(0.5Cen + 0.5Ndn)]$ with the $(Ce/Ce^*)n$ signatures. If neither Prn nor Ndn anomalies are evident, a negative Cen anomaly inevitably results in $(Pr/Pr^*)n > 1$, whereas a positive Cen anomaly generates $(Pr/Pr^*)n < 1$. A combination of $(Ce/Ce^*)n < 1$ and $(Pr/Pr^*)n \sim 1$ indicates a positive Lan anomaly. The Eu anomaly is often calculated by using the equation $Eu/Eu^* = Eun/[(Smn)(Gdn)]^{1/2}$ or $Eu/Eu^* = 2Eun/[(Smn + Gdn)]$ as reported by *Kato et al.* [2006]. The Y anomalies are a typical signature of seawater REE+Y patterns (Y inserted between Dy and Ho), and Y fractionates from the HREEs in estuaries [*Bau et al.*, 1995; *Lawrence and Kamber*, 2006]. Y/Ho anomalies, therefore, are not a result of either addition or removal of Ho relative to other HREE, but the Y behavior relative to the HREE is what drives the marine Y/Ho ratio. When riverine particles are destabilized by salt in the estuary, Y is preferentially released from these particles, increasing the Y/Ho ratio.

New REE+Y data are reported in this chapter, measured in samples spanning the K-Pg boundary from the four sections under consideration (Tables 12.5 through 12.8), and presented as post–Archean Australian Shale

Table 12.5 Rare earth elements (ppm) and PAAS-normalized Y/Ho, Th/U, Ce/Ce*, Pr/Pr*, and Eu/Eu* for the Højerup section, Stevns Klint, Denmark.

	La	Ce	Pr	Nd	Sm	Eu	Gd	Tb	Dy	Y	Ho	Er	Tm	Yb	Lu	ΣREE+Y	Y/Ho	Th/U	Ce/Ce*	Pr/Pr*	Eu/Eu*
SK-KTB-7	5.32	3.96	0.97	3.93	0.77	0.18	0.87	0.13	0.80	7.77	0.17	0.48	0.07	0.42	0.06	25.91	45.71	2.31	0.40	1.33	1.14
SK-KTB-6	12.03	8.64	2.57	10.71	2.16	0.50	2.43	0.35	1.98	17.59	0.40	1.11	0.14	0.85	0.12	61.56	43.98	4.72	0.36	1.37	1.12
SK-KTB-5	7.29	4.26	1.29	5.31	1.04	0.24	1.24	0.18	1.09	10.97	0.23	0.67	0.09	0.57	0.08	34.56	47.70	1.43	0.32	1.39	1.09
SK-KTB-4	6.49	3.87	1.14	4.61	0.91	0.21	1.08	0.16	0.98	9.87	0.22	0.63	0.09	0.53	0.08	30.86	44.86	1.42	0.33	1.40	1.10
SK-KTB-3	5.81	3.99	1.16	4.79	0.93	0.23	1.08	0.16	0.96	8.38	0.20	0.57	0.08	0.49	0.07	28.89	41.90	1.36	0.35	1.38	1.16
SK-KTB-2	5.32	3.39	0.99	4.07	0.78	0.19	0.92	0.14	0.86	8.50	0.18	0.53	0.08	0.46	0.07	26.47	47.22	1.10	0.34	1.37	1.15
SK-KTB-1	5.91	3.82	1.15	4.74	0.95	0.22	1.02	0.16	0.93	8.45	0.20	0.58	0.08	0.52	0.08	28.81	42.25	1.43	0.34	1.39	1.12
SK-KTB-FK-1	25.55	20.34	5.72	23.78	4.66	1.09	5.21	0.74	4.27	36.89	0.85	2.31	0.29	1.62	0.23	133.56	43.40	0.53	0.39	1.35	1.14
SK-KTB-(-1)	4.19	2.48	0.76	3.20	0.65	0.15	0.76	0.11	0.71	7.33	0.16	0.45	0.06	0.40	0.06	21.49	45.81	0.31	0.32	1.38	1.11
SK-KTB-(-2)	3.93	2.33	0.64	2.75	0.53	0.13	0.65	0.10	0.65	7.14	0.15	0.44	0.06	0.39	0.06	19.97	47.60	0.34	0.33	1.32	1.13
SK-KTB-(-3)	3.99	2.40	0.68	2.89	0.54	0.14	0.71	0.11	0.68	7.16	0.15	0.44	0.07	0.40	0.06	20.4	47.73	0.34	0.33	1.33	1.14
SK-KTB-(-4)	3.60	2.09	0.60	2.55	0.48	0.12	0.62	0.10	0.61	6.71	0.14	0.41	0.06	0.35	0.05	18.5	47.93	0.29	0.32	1.35	1.16

Table 12.6 Rare earth elements (ppm) and PAAS-normalized Y/Ho, Th/U, Ce/Ce*, Pr/Pr*, and Eu/Eu* for the Bottaccione section, Gubbio, Italy.

	La	Ce	Pr	Nd	Sm	Eu	Gd	Tb	Dy	Y	Ho	Er	Tm	Yb	Lu	ΣREE+Y	Y/Ho	Th/U	Ce/Ce*	Pr/Pr*	Eu/Eu*
GP-28	17.38	16.50	2.90	12.13	2.58	0.62	2.98	0.44	2.64	19.39	0.55	1.52	0.22	1.28	0.20	61.95	35.10	13.65	0.53	1.16	1.15
GP-27	17.34	17.68	2.82	11.67	2.40	0.59	2.86	0.42	2.48	19.08	0.53	1.48	0.21	1.24	0.19	61.91	36.17	13.97	0.57	1.13	1.16
GP-26	15.32	14.19	2.48	10.24	2.13	0.51	2.43	0.37	2.23	16.73	0.47	1.32	0.19	1.15	0.18	53.21	35.92	13.53	0.52	1.17	1.14
GP-25	15.59	15.30	2.75	11.39	2.36	0.56	2.80	0.41	2.50	18.62	0.52	1.46	0.21	1.25	0.19	57.29	35.52	14.03	0.53	1.18	1.13
GP-24	13.57	12.54	2.49	10.35	2.18	0.53	2.51	0.39	2.30	17.93	0.49	1.38	0.20	1.24	0.19	50.35	36.44	14.17	0.49	1.22	1.15
GP-23	11.64	10.13	2.07	8.49	1.75	0.41	2.11	0.32	1.95	16.48	0.42	1.18	0.18	1.02	0.15	41.83	38.95	11.97	0.47	1.24	1.11
GP-22	13.39	11.82	2.67	11.10	2.27	0.55	2.70	0.41	2.40	20.04	0.51	1.41	0.20	1.14	0.16	50.72	39.54	12.13	0.45	1.27	1.14
GP-21	12.53	9.92	2.07	8.59	1.76	0.43	2.20	0.33	1.99	17.03	0.43	1.20	0.17	1.01	0.14	42.77	39.73	11.54	0.44	1.24	1.15
GP-20	15.54	13.71	2.73	11.31	2.34	0.57	2.81	0.41	2.50	20.25	0.55	1.53	0.21	1.25	0.19	55.66	37.09	13.57	0.48	1.22	1.15
GP-19	19.89	19.00	4.14	17.67	3.67	0.89	4.14	0.61	3.49	27.00	0.72	1.90	0.25	1.46	0.21	78.04	37.38	15.69	0.48	1.23	1.17
GP-18	11.30	9.85	1.96	8.24	1.69	0.42	2.07	0.32	2.00	17.96	0.44	1.27	0.18	1.09	0.16	41.00	40.75	9.46	0.48	1.21	1.14
GP-17	6.87	5.58	1.15	4.77	0.97	0.24	1.23	0.19	1.25	13.29	0.29	0.88	0.14	0.82	0.13	24.50	45.12	5.94	0.45	1.23	1.10
GP-16	6.40	5.96	1.11	4.52	0.92	0.23	1.12	0.18	1.09	10.53	0.25	0.74	0.11	0.71	0.10	23.45	42.90	7.68	0.51	1.20	1.16
GP-15	17.07	20.56	4.63	20.74	4.56	1.10	5.17	0.72	3.95	33.20	0.79	1.97	0.24	1.27	0.18	82.96	42.12	11.47	0.53	1.20	1.19
GP-14	6.20	7.24	1.31	5.46	1.14	0.27	1.32	0.20	1.15	10.57	0.24	0.69	0.10	0.60	0.09	25.99	43.30	10.75	0.59	1.17	1.14
GP-13	4.91	5.78	0.98	4.10	0.83	0.21	0.96	0.15	0.90	8.71	0.19	0.56	0.08	0.49	0.08	20.23	44.73	8.60	0.61	1.14	1.16
GP-12	4.57	5.27	0.95	3.94	0.79	0.20	0.96	0.15	0.85	8.51	0.19	0.53	0.07	0.46	0.07	19.00	45.35	8.78	0.58	1.18	1.19
GP-11	4.56	5.33	0.99	4.13	0.86	0.20	0.97	0.15	0.90	8.81	0.19	0.53	0.08	0.45	0.07	19.40	46.28	9.00	0.58	1.19	1.11
GP-10	4.62	5.39	0.93	3.84	0.80	0.19	0.92	0.14	0.85	7.96	0.18	0.51	0.07	0.45	0.07	18.97	43.49	9.31	0.60	1.16	1.13
GP-09	5.25	6.48	1.09	4.60	0.94	0.22	1.08	0.16	0.96	9.34	0.20	0.58	0.08	0.49	0.08	22.21	46.07	9.14	0.62	1.14	1.13
GP-08	4.60	5.55	1.01	4.25	0.88	0.21	1.01	0.15	0.88	8.54	0.19	0.52	0.08	0.46	0.07	19.86	45.44	9.01	0.59	1.18	1.13
GP-07	5.40	6.57	1.17	5.00	1.05	0.25	1.24	0.18	1.06	11.64	0.24	0.66	0.10	0.57	0.09	23.57	49.40	11.95	0.60	1.15	1.14
GP-06	3.53	4.25	0.73	3.09	0.65	0.15	0.72	0.11	0.66	6.95	0.14	0.40	0.06	0.34	0.05	14.89	48.68	6.94	0.61	1.15	1.11
GP-05	4.63	6.10	1.06	4.56	0.93	0.22	1.04	0.16	0.89	9.07	0.19	0.52	0.08	0.45	0.07	20.88	48.84	10.71	0.64	1.14	1.12
GP-04	4.61	6.17	1.09	4.55	0.94	0.23	1.04	0.16	0.91	9.49	0.19	0.53	0.07	0.46	0.07	21.02	50.10	9.73	0.64	1.16	1.17
GP-03	4.05	4.93	0.88	3.71	0.79	0.18	0.87	0.13	0.81	8.84	0.17	0.49	0.07	0.45	0.07	17.60	50.59	8.53	0.60	1.17	1.09
GP-02	3.71	5.14	0.89	3.73	0.76	0.18	0.85	0.13	0.76	7.73	0.16	0.44	0.06	0.40	0.06	17.27	48.69	8.38	0.65	1.15	1.13
GP-01	4.17	5.85	1.02	4.21	0.85	0.20	0.93	0.14	0.82	8.35	0.17	0.47	0.07	0.46	0.07	19.43	49.07	7.48	0.65	1.17	1.14

Table 12.7 Rare earth elements and PAAS-normalized Y/Ho, Th/U, Ce/Ce*, Pr/Pr*, and Eu/Eu* for the Um Sohryngkew section, Meghalaya, India.

	La	Ce	Pr	Nd	Sm	Eu	Gd	Tb	Dy	Y	Ho	Er	Tm	Yb	Lu	ΣREE+Y	Y/Ho	Th/U	Ce/Ce*	Pr/Pr*	Eu/Eu*
KT-12	10.52	23.35	2.84	11.66	2.53	0.57	2.44	0.37	2.09	13.2	0.41	1.13	0.16	1	0.14	72.45	32.06	1.63	0.98	1.01	1.13
KT-11	11.48	25.68	3.06	12.66	2.68	0.6	2.7	0.4	2.23	14.5	0.44	1.21	0.17	1.02	0.15	79.01	32.72	1.69	1.00	1.00	1.13
KT-10	12.79	29.9	3.77	15.72	3.37	0.77	3.31	0.5	2.77	15.3	0.52	1.39	0.21	1.21	0.18	91.66	29.36	3.14	0.99	1.02	1.14
KT-9	10.63	24.97	3.14	13.03	2.86	0.64	2.73	0.41	2.25	12.8	0.44	1.19	0.17	1	0.15	76.42	29.09	3.10	0.99	1.02	1.13
KTB (bulk)	31.41	22.88	6.85	28.86	5.74	1.35	6.51	0.92	5.32	48.9	1.09	2.89	0.38	2.1	0.31	165.51	44.98	0.36	0.36	1.36	1.15
KT-6	12.55	29.79	4.09	17.94	3.96	0.93	4.04	0.6	3.41	19.6	0.64	1.68	0.22	1.28	0.18	100.89	30.38	1.75	0.95	1.02	1.17
KT-5	8.8	21.05	2.72	11.68	2.54	0.59	2.55	0.39	2.13	12	0.41	1.1	0.16	0.92	0.13	67.19	29.41	2.60	0.98	1.01	1.16
KT-4	12.39	27.77	3.57	15.21	3.3	0.72	3.17	0.48	2.67	15.8	0.53	1.44	0.21	1.3	0.2	88.73	29.74	2.58	0.96	1.01	1.11
KT-3	12.15	27.29	3.4	14.04	2.96	0.68	2.92	0.44	2.47	14.8	0.49	1.39	0.19	1.22	0.18	84.65	29.97	2.59	0.97	1.02	1.15
KT-2	12.38	27.7	3.4	13.91	2.94	0.66	2.94	0.44	2.51	14.6	0.49	1.31	0.19	1.15	0.18	84.83	29.76	2.30	0.98	1.01	1.12
KT-1	11.44	25.23	3.04	12.6	2.62	0.6	2.55	0.38	2.14	12.8	0.43	1.15	0.17	1.03	0.15	76.37	29.94	2.24	0.98	1.00	1.15
KT-0	11.25	25.86	3.16	12.97	2.74	0.61	2.6	0.38	2.12	12.5	0.42	1.15	0.16	0.98	0.14	77	29.83	2.15	1.00	1.01	1.15
KT-(-1)	13.59	31.44	3.94	16.11	3.38	0.74	3.21	0.47	2.55	14.1	0.49	1.29	0.18	1.08	0.16	92.77	28.84	2.11	0.99	1.02	1.13

Table 12.8 Rare earth elements (ppm), PAAS-normalize U/Th, Ce/Ce*, Pr/Pr* and Eu/Eu* for the Poty Quarry section, Brazil.

	La	Ce	Pr	Nd	Sm	Eu	Gd	Tb	Dy	Y	Ho	Er	Tm	Yb	Lu	ΣREE+Y	Y/Ho	Th/U	Ce/Ce*	Pr/Pr*	Eu/Eu*
PO96	18.47	32.03	4.03	14.80	2.36	0.48	2.07	0.28	1.39	8.42	0.26	0.66	0.09	0.53	0.08	85.95	32.1	1.48	0.86	1.09	1.10
PO97	14.11	25.17	3.19	11.66	1.91	0.39	1.70	0.24	1.15	7.10	0.21	0.56	0.08	0.46	0.06	67.99	33.1	1.29	0.87	1.09	1.09
PO98	20.90	39.06	4.73	17.31	2.72	0.53	2.28	0.32	1.51	9.33	0.29	0.75	0.10	0.65	0.09	100.57	32.7	1.20	0.91	1.07	1.07
PO99	17.93	30.62	3.77	13.72	2.23	0.46	1.94	0.27	1.29	8.71	0.26	0.66	0.09	0.50	0.07	82.52	34.1	0.90	0.86	1.08	1.11
PO100	12.88	20.36	2.64	9.75	1.51	0.30	1.34	0.19	0.91	6.43	0.18	0.46	0.07	0.37	0.05	57.44	36.1	0.80	0.80	1.10	1.07
PO101	9.45	13.48	1.89	7.06	1.08	0.22	1.03	0.14	0.66	5.03	0.13	0.36	0.05	0.28	0.04	40.9	37.8	0.45	0.73	1.13	1.06
PO102	6.80	8.28	1.20	4.45	0.69	0.15	0.68	0.09	0.49	4.13	0.10	0.29	0.04	0.23	0.04	27.66	40.0	0.40	0.66	1.16	1.13
PO103	13.28	15.79	2.51	9.60	1.62	0.36	1.62	0.23	1.18	9.60	0.24	0.65	0.09	0.51	0.08	57.36	40.4	0.37	0.63	1.18	1.13
PO104	24.52	42.66	6.03	24.09	4.85	1.06	4.61	0.65	3.44	25.83	0.70	1.78	0.24	1.37	0.21	142.04	36.9	0.59	0.81	1.10	1.13
PO105	4.90	6.25	0.91	3.41	0.57	0.13	0.55	0.08	0.45	3.51	0.09	0.26	0.04	0.24	0.04	21.43	37.4	0.26	0.68	1.15	1.12
PO106	3.07	3.05	0.49	1.84	0.28	0.06	0.30	0.04	0.22	2.36	0.05	0.14	0.02	0.11	0.02	12.05	43.6	0.14	0.57	1.19	1.13
PO107	4.10	4.14	0.71	2.66	0.43	0.09	0.41	0.06	0.35	3.29	0.07	0.18	0.02	0.16	0.02	16.69	46.9	0.17	0.56	1.23	1.10
PO108	2.27	2.24	0.37	1.44	0.23	0.05	0.24	0.03	0.19	2.04	0.04	0.11	0.01	0.09	0.01	9.36	47.4	0.11	0.55	1.19	1.02
PO109	8.35	9.23	1.61	6.38	1.11	0.25	1.19	0.16	0.96	9.96	0.20	0.56	0.08	0.43	0.07	40.54	49.1	0.15	0.58	1.19	1.14
PO110	12.31	13.67	2.46	9.84	1.80	0.42	1.97	0.28	1.60	17.26	0.36	1.01	0.14	0.79	0.13	64.04	47.8	0.20	0.57	1.21	1.14
PO111	19.79	22.45	3.96	16.17	3.04	0.69	3.28	0.47	2.68	29.01	0.60	1.66	0.22	1.29	0.19	105.5	48.4	0.22	0.58	1.18	1.12
PO112	6.35	10.64	1.41	5.17	0.88	0.18	0.78	0.11	0.57	4.16	0.11	0.30	0.04	0.28	0.04	31.02	38.6	0.49	0.82	1.11	1.09
PO113	12.08	22.36	2.97	10.99	1.83	0.37	1.60	0.23	1.08	6.49	0.21	0.54	0.07	0.43	0.06	61.31	31.2	1.16	0.86	1.11	1.07
PO114	8.80	15.46	2.01	7.49	1.25	0.27	1.18	0.16	0.81	5.68	0.15	0.43	0.06	0.35	0.06	44.16	36.7	0.57	0.85	1.10	1.14
PO115	12.66	24.53	3.23	12.04	2.10	0.41	1.81	0.25	1.24	7.61	0.23	0.58	0.08	0.49	0.07	67.33	33.1	0.65	0.88	1.10	1.04
PO116	16.27	32.58	4.32	16.07	2.78	0.52	2.21	0.31	1.43	8.27	0.27	0.69	0.09	0.52	0.08	86.41	31.2	1.04	0.90	1.11	1.03
PO117	14.42	28.07	3.65	13.22	2.22	0.43	1.90	0.26	1.20	7.11	0.24	0.60	0.08	0.48	0.07	73.95	29.9	1.07	0.89	1.11	1.05
PO118	14.74	27.80	3.61	13.31	2.29	0.42	1.86	0.24	1.20	7.39	0.23	0.60	0.09	0.47	0.07	74.32	32.8	1.25	0.88	1.10	1.01
PO119	10.52	18.85	2.38	8.88	1.49	0.30	1.28	0.18	0.85	5.99	0.17	0.46	0.07	0.36	0.05	51.83	35.2	0.83	0.87	1.08	1.06
PO120	14.09	26.03	3.29	12.04	2.08	0.39	1.71	0.24	1.17	7.37	0.22	0.59	0.08	0.46	0.06	69.82	33.7	1.24	0.88	1.09	1.03
PO121	11.66	20.75	2.75	10.51	1.75	0.34	1.53	0.21	1.04	6.59	0.20	0.53	0.07	0.43	0.06	58.42	33.5	1.07	0.85	1.09	1.05
PO122	11.50	20.45	2.76	10.25	1.75	0.34	1.51	0.22	1.07	6.73	0.21	0.54	0.08	0.44	0.06	57.91	32.8	1.09	0.84	1.12	1.03
PO123	10.47	18.55	2.51	9.38	1.58	0.33	1.36	0.19	0.98	6.28	0.19	0.51	0.07	0.40	0.06	52.86	32.4	0.99	0.84	1.11	1.11
PO124	11.23	20.24	2.70	10.19	1.72	0.35	1.47	0.21	1.05	6.80	0.20	0.53	0.07	0.41	0.06	57.23	34.7	1.18	0.85	1.10	1.09
PO125	9.61	17.19	2.23	8.42	1.45	0.30	1.28	0.18	0.95	6.51	0.18	0.47	0.07	0.38	0.06	49.28	35.9	1.48	0.86	1.09	1.10
PO126	10.84	19.47	2.60	9.79	1.72	0.36	1.56	0.21	1.09	7.61	0.21	0.57	0.08	0.43	0.06	85.95	35.8	1.29	0.85	1.10	1.11
PO127	18.77	35.95	4.53	16.54	2.82	0.53	2.30	0.33	1.53	9.23	0.29	0.76	0.10	0.59	0.08	67.99	31.8	1.20	0.90	1.09	1.03
PO128	18.03	34.58	4.38	15.88	2.67	0.52	2.23	0.32	1.50	9.22	0.29	0.74	0.10	0.61	0.09	100.57	32.1	0.90	0.90	1.10	1.05
PO129	17.60	33.14	4.23	15.34	2.69	0.48	2.12	0.30	1.41	8.50	0.28	0.68	0.09	0.53	0.08	82.52	30.6	0.80	0.89	1.10	0.99

(PAAS)-normalized patterns [*Nance and Taylor*, 1976; *McLennan*, 1989] in Figures 12.4 and 12.5a, b. The Højerup section shows ΣREE+Y values from 18.5 to 61.6 ppm, with Maastrichtian samples displaying higher values than Danian ones and the K-Pg boundary layer exhibiting a much higher value (134 ppm; Table 12.5; Fig. 12.2a). PAAS-normalized REE patterns are slightly fractionated and display markedly negative Ce (Ce/Ce* from 0.32 to 0.40) and positive Y (Y/Ho from 42 to 47) anomalies and slightly positive to absent Eu anomalies (Eu/Eu* from 1.09 to 1.16); REE values in the K-Pg boundary layer, except for these anomalies, approach PAAS values (Fig. 12.6a).

Samples from the Bottaccione section exhibit ΣREE+Y values from 15 to 78 ppm, with the highest value at the K-Pg layer (83 ppm; Table 12.6; Fig. 12.2a), and differ from the Højerup section in the sense that Maastrichtian samples exhibit lower values than Danian ones. Consistent with PAAS-normalized REE patterns in the Højerup section, Bottaccione REE+Y patterns are little fractionated, with well-defined negative Ce (Ce/Ce* from 0.44 to 0.65) and positive Y (Y/Ho from 35 to 51) anomalies and slightly positive to absent Eu (Eu/Eu* from 1.09 to 1.19).

Otherwise, REE+Y values in the K-Pg boundary layer also approach PAAS values (Fig. 12.6b).

Twelve samples from the Um Sohryngkew section display ΣREE+Y values from 67 to 101 ppm and much higher values in the K-Pg layer (165 ppm; Table 12.7; Fig. 12.3a). PAAS-normalized patterns do not show much variability and differ from the patterns seen in the Højerup and Bottaccione sections. On the contrary, Ce anomaly is absent in the Um Sohryngkew section (Ce/Ce* ~1.0), except for the sample from the K-Pg layer, a shale with some carbonate that yields a PAAS-normalized REE+Y pattern analogous to the one seen in the Højerup K-Pg layer with pronounced negative Ce anomaly. The PAAS-normalized REE+Y patterns are slightly upwarped with maximum around Eu, and Eu/Eu* exhibits a very discrete positive anomaly varying from 1.11 to 1.17. Neither La nor (Pr/Pr*) anomalies are observed. As shales predominate in the section straddling the K-Pg boundary, their patterns reflect altogether the REE+Y patterns of their detrital source.

A 200 kyr hiatus in the sedimentary record in the K-Pg transition in the Paraíba Basin precludes a full picture of the REE+Y pattern across this transition. Analyses of 28

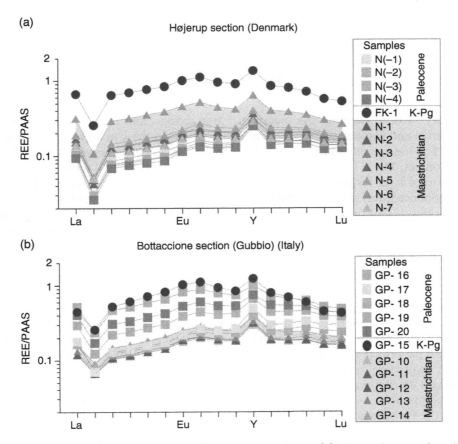

Figure 12.4 PAAS-normalized REE patterns for (a) the Højerup section and (b) Bottaccione section. *(See insert for color representation of the figure.)*

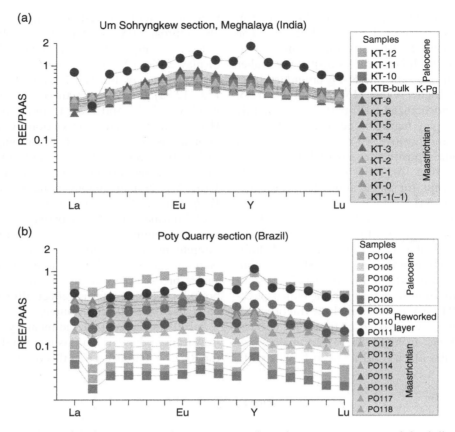

Figure 12.5 PAAS-normalized REE patterns for (a) an Um Sohryngkew River section and (b) drill core samples from a Poty drill hole. *(See insert for color representation of the figure.)*

samples from the Poty section spanning the preserved uppermost Maastrichtian and Danian samples are found in Table 12.8 (PO-111 is the sample immediately below the K-Pg erosional contact; PO-111, 110, and 109 samples are from the reworked layer resting on this contact). They show more pronounced, less systematic ΣREE+Y variation (9–142 ppm; Fig. 12.3b) than samples from the other three K-Pg sections examined here. PAAS-normalized REE+Y patterns are in general little fractionated, with a negative Ce anomaly (Ce/Ce* from 0.55 to 0.91), a markedly positive Y anomaly (Y/Ho from 32 to 49), and slightly positive to absent Eu/Eu* anomaly (Eu/Eu* varies from 0.99 to 1.14); together these suggest slightly reducing conditions around the K-Pg transition. Maastrichtian marls to marly carbonate rocks in this section display slightly depleted HREE patterns in relation to the LREE ones. These patterns may have been affected by the REE+Y of contaminant detrital material, likely with an important felspathic component.

To better visualize the Ce behavior by eliminating the influence of La, Ce/Ce* and Pr/Pr* were plotted against each other, following *Bau and Dulski* [1996] (Fig. 12.6a). All Maastrichtian and Danian samples from the Højerup,

Bottaccione, and Poty sections exhibit true negative Ce anomaly. In the Um Sohryngkew section, only the K-Pg layer displays conspicuous negative Ce anomaly. In a Ce anomaly [log(Ce/Ce*)] versus Nd (ppm) plot (Fig. 12.6b), following *Wang et al.* [2014; modified from *Elderfield and Pagett*, 1986; *Wright et al.*, 1987], all samples from the Højerup and Bottaccione sections plot within the oxic field (Fig. 12.6b). All samples from the uppermost Maastrichtian in the Poty section are barely anoxic (Ce anomaly between zero and −0.1), while samples from the lowermost Danian including the reworked layer resting on the K-Pg transition erosional contact (samples PO-111 through PO-100; Table 12.4) plot within the oxic field, except for sample PO-104 (80 cm above the erosional contact), barely anoxic and with much higher ΣREE+Y (Table 12.8) and Hg and Mo contents (Table 12.4). All samples from the Um Sohryngkew section are barely anoxic, except for the K-Pg layer which lies within the oxic field (Fig. 12.6b).

Sea-level fluctuations with lowstand at about 25–100 kyr below the K-Pg boundary, marked by increased detrital influx that culminated at the K-Pg transition, have been reported from K-Pg successions elsewhere [*Keller and*

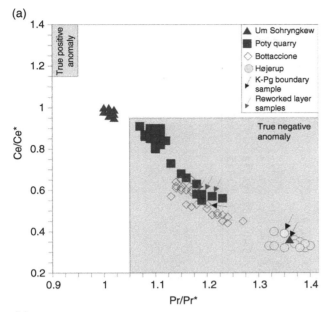

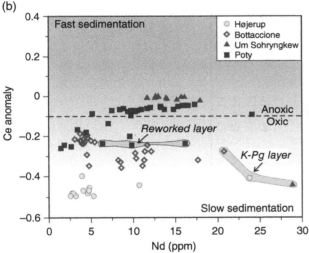

Figure 12.6 (a) Cross-plot of Ce/Ce* versus Pr/Pr* after *Bau and Dulski* [1996]. Samples in this study show true negative Ce anomaly, except those from the Um Sohryngkew section. (b) log (Ce/Ce*) versus Nd (ppm) plot (Modified from *Wang et al.* [2014]) in which all samples plot above the anoxic field. The K-Pg layers exhibit much higher Nd values.

Stinnesbeck, 1996a, 1996b; *Adatte et al.*, 2002]. Possibly sea-level oscillations across the K-Pg transition in the examined sections here have allowed for important continental detrital influx and subsequent ΣREE+Y enrichment (Figs. 12.2a, b and 12.3a, b).

The K-Pg layer in each of these four sections displays the highest ΣREE+Y and Nd (ppm) values and the lowest Ce/Ce* ratio. An Nd value of 24ppm and a Y/Ho ratio of 43 were determined for the K-Pg layer at the Højerup section, while for the Bottaccione section, K-Pg layer

values of 21ppm and Y/Ho ratio of 42 were measured. For the Um Sohryngkew section, 29ppm and 45 were determined for the K-Pg layer. The reworked layer on the K-Pg transition erosional contact in the Poty section yielded values of 16 and 48ppm, respectively. These Nd values are higher than the corresponding mean Nd composition for ocean ridge basalts of *Gale et al.* [2013] and within the range for Nd in Deccan basalts of *Alexander and Gibson* [1977].

12.4.3.3. Mo/Al, U/Mo, and Th/U Ratios

Molybdenum in sedimentary rocks seems to be an important elemental proxy for anoxia [e.g., *Bond et al.*, 2015; *Grasby et al.*, 2016]. *Lyons et al.* [2003] found a high correlation between TOC content and Mo/Al ratio in euxinic sediments from the Cariaco Basin, north-central coast of Venezuela. Given this empirical correlation, TOC compositions could potentially be calculated from Mo/Al trends. However, this method needs to be refined, and no universal elemental proxy for determining TOC has been established [*Wilde et al.*, 2004].

Molybdenum is a redox-sensitive element, readily scavenged from seawater into sediments in the form of $MoSxO_{4-}x^{2-}$ under anoxic conditions [*Wen et al.*, 2015]. The concentrations of Mo in the seawater column of anoxic basins are usually lower than in oxic seawater because of preferential uptake into anoxic and sulfidic sediments [*Emerson and Huested*, 2015]. Mo/Al ratios are used for discerning the original compositions of rocks that were once subjected to later diagenetic, low-grade metamorphism or weathering [*Grasby et al.*, 2016].

Patterns of U/Mo covariation in marine environment are regarded as a proxy for paleoredox depositional conditions [*Algeo and Rowe*, 2012; *Tribovillard et al.*, 2012], as well as the enrichment factors (EFs) of U and Mo [*Tribovillard et al.*, 2012; *Zhou et al.*, 2012]. In marine sediments, U enrichment is usually due to authigenic uptake from seawater and is observed at the Fe(II)-Fe(III) redox boundary (suboxic condition), while Mo enrichment requires the presence of H_2S (euxinic condition). Besides, the uptake of aqueous Mo into sediments can be increased by means of metal-oxyhydroxide particulate shuttles, while aqueous U is not affected by this process [*Tribovillard et al.*, 2012]. Therefore, a U/Mo ratio increase suggests oxic conditions [e.g., *Sosa-Montes et al.*, 2017].

Molybdenum concentrations measured in samples from the sections under consideration here have been normalized to corresponding Al values and plotted in Figs. 12.2a, b and 12.3a, b. At Højerup, no correlation is observed between the Mo/Al ratio and TOC, and low values of Mo/Al ratio characterize the uppermost Maastrichtian (0.50cm below the K-Pg boundary). Within the Fiskeler Member clays, a positive shift in TOC values is observed, and Mo/Al ratios are enhanced,

Table 12.9 Mo and U enrichment factors for some Højerup, Bottaccione and Um Sohryngkew samples in this study.

Højerup							Bottaccione						Um Sohryngkew						Poty				
Samples	Distance (cm)	Mo/Al	U/Al	Mo_{EF}	U_{EF}	Samples	Distance (cm)	Mo/Al	U/Al	Mo_{EF}	U_{EF}	Samples	Distance (cm)	Mo/Al	U/Al	Mo_{EF}	U_{EF}	Samples	Distance (cm)	Mo/Al	U/Al	Mo_{EF}	U_{EF}
N4	100	3.66	0.18	9.15	0.29	GP-19	80	0.91	0.09	2.28	0.15	KT 12	40	1.81	1.89	4.53	3.05	PO-107	70	20.69	106	51.75	170
N3	70	7.66	0.22	19.15	0.35	GP-18	50	1.21	0.19	3.03	0.31	KT 11	30	2.79	2.05	6.98	3.31	PO-108	60	16.37	35	40.92	66
N2	40	3.75	0.20	9.37	0.32	GP-17	20	0.80	0.38	2.00	0.61	KT 10	20	0.51	0.9	1.28	1.45	PO-109	40	32.14	101.6	80.35	103
N1	6	3.75	0.22	9.37	0.35	GP-16	2	1.23	0.32	3.08	0.52	KT 9	10	0.45	0.9	1.13	1.45	PO-110	10	40.13	129.8	100.32	209
K/Pg layer	0	1.60	3.75	4.00	6.05	K/Pg layer	0	0.29	0.12	0.73	0.19	K/Pg layer	0	1.71	3.05	4.28	4.92	PO-111	0	41.50	151.5	103.75	244
N(-1)	-30	2.00	0.28	5.00	0.45	GP-14	-5	2.67	0.89	6.68	1.44	KT6	-30	0.44	0.68	1.10	1.10	PO-112	-40	12.91	58.28	32.27	94
N(-2)	-60	8.00	0.23	20.00	0.37	GP-13	-15	0.26	0.31	0.65	0.50	KT5	-40	0.40	0.54	1.00	0.87	PO-113	-60	19.99	28.18	49.97	45
N(-4)	-140	1.00	0.24	2.50	0.39	GP-12	-25	0.32	0.26	0.8	0.42	KT4	-50	1.36	1.17	3.40	1.89	PO-114	-70	20.32	50.77	50.80	81
						GP-11	-35	0.16	0.34	0.4	0.55	KT3	-60	0.79	0.97	1.98	1.56	PO-115	-80	9.54	39.25	23.85	63

while $\delta^{13}C_{org}$ increase. In the lower Danian, no correlation between TOC and Mo/Al is apparent in the Cerithium Limestone and Stevns Klint formations, and Mo/Al ratios are higher than in the Fiskeler Member [*Sial et al.*, 2016].

At Bottaccione, Mo/Al ratios are low, with a monotonous stratigraphic curve in the uppermost Maastrichtian, but a negative shift is seen about 0.50 cm below the K-Pg boundary, coinciding with vigorous positive-negative shifts in TOC values followed by a positive one of Mo/Al just above the K-Pg boundary [*Sial et al.*, 2016]. At the Um Sohryngkew section, Mo/Al ratios are low in the uppermost Maastrichtian, but similar to the Højerup and Bottaccione sections, it shifts to lower values within 20 cm below the K-Pg boundary, in contrast to increasing TOC values. At the Bottaccione and Um Sohryngkew sections, minimum Mo/Al ratios do not coincide with $\delta^{13}C_{org}$ peaks. The Højerup section is the only exception from this rule.

Mo/Al ratio curve for the Poty section contains three positive excursions. The first one at about 25 m below the K-Pg transition is followed upsection by a negative shift within the CF1 foraminiferal biozone. A second positive shift (P1a(1) foraminiferal biozone) is observed within the reworked layer that rests on top of the K-Pg erosional contact. Finally, a third positive excursion of Mo/Al values is observed within the P1a(2) foraminiferal biozone.

The behavior of Mo and U across the K-Pg transition is further examined here by considering the U/Mo ratio

and the variation of the Mo_{EF} and U_{EF} within narrow intervals bracketing the K-Pg boundary: (i) from 60 cm below to 70 cm above the K-Pg layer at Højerup, (ii) from 35 cm below to 80 cm above at Bottaccione, (iii) from 60 cm below to 40 cm above at Um Sohryngkew, and (iv) from 80 cm below to 70 cm above K-Pg erosional contact at Poty (Table 12.9). The U/Mo ratio in the K-Pg boundary layer increases substantially in relation to the uppermost Maastrichtian and lowermost Paleogene deposits in the Højerup and Um Sohryngkew sections, suggesting deposition under oxic condition. Opposite behavior is observed in the K-Pg boundary layer in the Bottaccione section, where the U/Mo ratios decrease in relation to the uppermost Maastrichtian and lowermost Paleogene deposits, suggesting anoxic conditions during its deposition. These sections display distinct Mo_{EF} versus U_{EF} covariation trends, which are shown in Figure 12.7. Values of Mo_{EF} and U_{EF} for the K-Pg boundary layer in all four cases markedly deviate from the corresponding covariation trend. The reason why the U_{EF} for the K-Pg clay layer in the Bottaccione section differs so much from those in the Højerup and Um Sohryngkew sections is not clear. Leaching in the narrow K-Pg clay layer at Bottaccione, possibly facilitated by its tilted position, has, perhaps, affected initial U values more significantly than initial Mo compositions, leading to decreases in both U/Mo ratio and U_{EF}.

In a similar investigation, *Sosa-Montes et al.* [2017] have reported that at a Bottaccione section (from 22 cm below to 41 cm above the K-Pg boundary), Mo/Al and

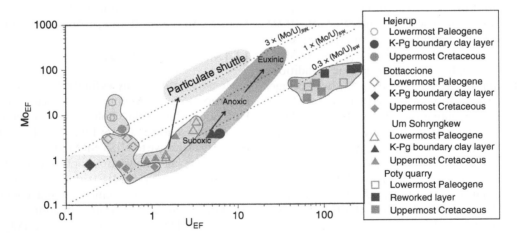

Figure 12.7 Mo_{EF} versus U_{EF} covariation for the Højerup, Bottaccione, Um Sohryngkew, and Poty K-Pg boundary sections (Based on *Tribovillard et al.* [2012] and *Sosa-Montes et al.* [2017]), in which $Mo_{EF} = [(Mo/Al)_{sample}/(Mo/Al)_{PAAS}]$ and $U_{EF} = [(U/Al)_{sample}/(U/Al)_{PAAS}]$. The PAAS composition used is from *Taylor and McLennan* [1985]. The diagonal lines represent multiples of the Mo/U ratio of the present-day seawater. The orange field represents a general pattern of Mo_{EF} versus U_{EF} covariation in unrestricted marine trend for modern eastern tropical Pacific (From *Tribovillard et al.* [2012], modified by *Sosa-Montes et al.* [2017]), and the yellow field (From *Tribovillard et al.* [2012]) represents the particulate shuttle trend in which intense cycling of metal oxyhydroxide occurs within water column. *(See insert for color representation of the figure.)*

Table 12.10 Hg isotopes (‰ relative to NIST SRM 3133) in samples with Hg enrichments: spike I (Um Sohryngkew, Poty); spike II (Højerup, Bottaccione, Um Sohryngkew), spike III (Poty).

Samples	$\delta^{199}Hg$	$\delta^{200}Hg$	$\delta^{201}Hg$	$\delta^{202}Hg$	$\Delta^{199}Hg$	$\Delta^{200}Hg$	$\Delta^{201}Hg$
Mass dependent fractionation (MDF)					Mass independent fractionation (MIF)		
(a) K/Pg boundary layer at Hojurup (Fiskeler Member), Denmark							
FCA	−0.51	−1.06	−1.76	−2.34	0.08	0.11	−0.01
FCB	−0.28	−0.74	−1.18	−1.62	0.13	0.07	0.04
FCC	−0.29	−0.55	−0.86	−1.06	−0.02	−0.02	−0.06
FCD	−0.43	−0.92	−1.44	−1.93	0.06	0.05	0.01
(b) K/Pg boundary layer, Bottaccione, Italy							
GP-15	−0.26	−0.59	−0.93	−1.28	0.06	0.05	0.03
(c) K/Pg boundary layer at Um Sohryngkew, Meghalaya, India							
KT-3	−0.35	−0.78	−1.16	−1.61	0.06	0.03	0.05
KT-bulk	−0.64	−0.94	−1.58	−1.89	−0.16	0.01	−0.16
(d) Poty Quarry drill hole, Brazil							
PO-104	−0.17	−0.97	−1.29	−1.87	0.30	−0.03	0.12
PO-105	−0.03	−0.62	−0.89	−1.36	0.31	0.07	0.13
PO-119	−0.31	−1.15	−1.59	−2.45	0.30	0.08	0.25
(e) Bidart, France							
K/Pg boundary layer	−0.04	−0.34	−0.42	−0.74	−1.12	−0.15	0.03
BI-9.28.5	−0.11	−0.57	−0.66	−1.44	−1.56	0.25	0.15
BI-9.34.2	0.05	−0.11	−0.13	−0.31	−0.73	0.12	0.04
BI-9.36.2	0.08	−0.10	−0.07	−0.25	−0.61	0.15	0.03
BI-10.35.1	−0.33	−1.28	−1.68	−2.66	−4.30	0.33	0.06
BI-11.7.3	0.13	−0.16	−0.04	−0.66	−0.73	0.30	0.17

Modifed from *Sial et al.* [2016].
Analyses from spikes I, II and II from Bidart were added for comparison.

U/Al ratios did not show much variation in deposits prior and after the K-Pg event. The slight enrichment of these ratios above the K-Pg boundary was interpreted as an indication of lower oxygenation. The U_{EF} versus Mo_{EF} covariation was thought documenting oxic conditions prior to and after the K-Pg boundary at Bottaccione, bracketing anoxic deposition of the K-Pg boundary layer.

In many cases, a negative correlation between Th/U ratios and TOC contents in sedimentary rocks has been recognized [e.g., *Hofer et al.*, 2013]. At Højerup, the Th/U ratios in the uppermost Maastrichtian rocks (0.29–0.34) undergo an increase from the K-Pg boundary (0.53) to the Danian rocks (1.1–4.7). At Bottaccione, a less systematic behavior is observed, Th/U ratios are much higher than at Højerup, and values in the Maastrichtian rocks, mostly in the 6.94–9.31 range, are replaced upward by more elevated ones in the Danian rocks (9.46–15.69). At Um Sohryngkew, Maastrichtian rocks display a gradual upward increase of Th/U ratio values (2.11–2.60) followed by a negative excursion at the K-Pg boundary (0.36) and by higher values in the Danian rocks (3.10). At the Poty section, Th/U ratios are much lower than at the Bottaccione and Um

Sohryngkew sections, and a shift to lower values was recorded in the K-Pg transition reworked layer.

12.4.4. Mercury Isotopes

Mercury isotope analyses for the Hg spikes detected in the Højerup (Fiskeler Member), Bottaccione (Scaglia Rossa Formation), and Um Sohryngkew sections have been reported by *Sial et al.* [2016; Table 12.10] in delta notation in per mil (‰) relative to NIST SRM 3133 Hg standard. $\delta^{202}Hg$ (MDF) for all analyzed samples were plotted against corresponding $\Delta^{201}Hg$ (MIF) values in Figure 12.8, in which the ranges for volcanic emission and chondrite/volcanic emission Hg are also indicated. In addition, three analyses for the two Hg peaks observed at the Poty section (this study) and six analyses from a section at Bidart, Basque Basin, France, a well-known complete K-Pg boundary succession [e.g., *Bonté et al.*, 1984; *Galbrun and Gardin*, 2004; *Font et al.*, 2014, 2016], were also plotted for comparison.

Two samples from the Um Sohryngkew section yielded $\delta^{202}Hg$ values of −1.61‰ (spike I) and −1.89‰ (spike II) and $\Delta^{201}Hg$ close to 0.0‰. Three among four of the analyzed samples (spike II) from the Fiskeler Member at

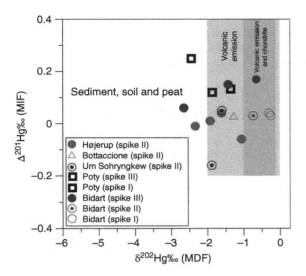

Figure 12.8 In a $\delta^{202}Hg$ (MDF)-$\Delta^{201}Hg$ (MIF) plot, modified from *Sial et al.* [2016], samples from the K-Pg layer (spike II) from the Højerup, Bottaccione, and Um Sohryngkew sections lie within the range for volcanogenic Hg. One sample from the K-Pg layer (spike II), two samples from the spike I, and one from the spike III, from the Bidart section, added for comparison, lie within the range for chondrite/volcanogenic emission. One sample from the spike III of this section lies within the volcanogenic emission field, and another within the sediment, soil, and peat range. Ranges for volcanogenic and chondritic Hg are from *Bergquist and Blum* [2009] and are shown as vertical bars. *(See insert for color representation of the figure.)*

Højerup yielded $\delta^{202}Hg$ between −1.00 and −2.00‰, and two of them display positive $\Delta^{201}Hg$ or negative but very close to 0.0‰. One sample from the K-Pg boundary layer at the Bottaccione section (spike II) yielded a $\delta^{202}Hg$ value of −1.28‰ and positive $\Delta^{201}Hg$. Hg isotopes analyzed in six samples from Bidart yielded $\delta^{202}Hg$ values from −0.25 to −2.66 and all positive $\Delta^{201}Hg$ values. A sample from the K-Pg boundary layer from which *Font et al.* [2016] determined a true Hg spike (corresponding to spike II in *Sial et al.* [2016]) yielded a $\delta^{202}Hg$ value of −0.74 compatible with volcanic or chondritic Hg source. In the present study, one sample from 90 cm below the K-Pg transition at the Poty quarry (spike I) yielded a value of 2.45‰, and two from 1 m above (spike III) yielded values of −1.87 and −1.36‰. So, in the light of these Hg isotope data, a volcanic source for Hg in the K-Pg boundary (spike II) is likely in these three classical K-Pg localities.

Most of the analyzed samples reported by *Sial et al.* [2016] show slightly (but significantly higher than the analytical precision of 0.04‰) positive $\Delta^{201}Hg$ signatures. Since most continental samples (soils, sediments, and land plants) usually display negative or close to zero $\Delta^{201}Hg$, postdepositional and diagenetic processes likely would not induce odd MIF, and the positive $\Delta^{201}Hg$ values observed in most samples reported by *Sial et al.* [2016] would likely indicate a long-term atmospheric transport prior to deposition. During transportation in the atmosphere, photoreduction of gaseous oxidized Hg (GOM) would enrich odd isotopes (^{199}Hg and ^{201}Hg) in water droplets and particles (e.g., by adsorption), thus triggering positive $\Delta^{201}Hg$ values during final deposition [*Chen et al.*, 2012; *Blum et al.*, 2014]. The results reported by *Sial et al.* [2016] point to Hg isotopes as a promising way for identification of the Hg source.

12.5. DISCUSSION AND CONCLUSIONS

Sea-level fluctuations with lowstand at about 25–100 kyr before the K-Pg boundary and increased detrital influx that culminates at the K-Pg transition are known to have occurred, leading up to the Paleogene [e.g., *Adatte et al.*, 2002]. The Poty, among the sections under consideration, seems to have been affected by a sea-level fall around the K-Pg transition, leading to erosion of part of the sedimentary record of the CF1, P0, and Pα biozones and P1a subzone. This sea-level fall has allowed for continental influx during the K-Pg transition, lowering Mo/Al ratios.

The $\Sigma REE+Y$ (ppm) values in the K-Pg layers examined here are much higher than in the Maastrichtian and Danian samples and lie within the range for Deccan basalt values. Continental detrital influx due to sea-level fluctuations during the K-Pg transition may have further contributed to raise the $\Sigma REE+Y$ values in the K-Pg boundary. Moreover, a slight increase in detrital input may have followed the Deccan volcanism or asteroid/comet impact, associated with a greater chemical alteration of continental areas. The present study demonstrates that the stratigraphical variation of ΣREE coupled with their normalized REE patterns and Mo/Al may help identify anoxia and sea-level fluctuations.

The possibility that anoxic conditions prevailed during the K-Pg transition is not supported by the stratigraphic record of Mo/Al ratio redox proxy. True negative Ce anomaly across the K-Pg boundary in the studied sections attests the predominance of oxidized surface waters in the Northern Hemisphere (e.g., Højerup and Bottaccione sites). Different conditions may have predominated in bottom waters, since the Fiskeler Member exhibits high TOC and is finely laminated, suggesting anoxic or at least dysoxic conditions at the seafloor. In the Southern Hemisphere, barely anoxic conditions were recorded in the uppermost Maastrichtian, and oxidized surface waters predominated in the lowermost Danian (e.g., at Poty) or were only recorded at the K-Pg layer (e.g., Um Sohryngkew).

The available chemostratigraphic Hg/TOC data for the three continuous K-Pg boundary sections discussed in this chapter support the following conclusions:

a. The three Hg/TOC spikes present in the Højerup, Bottaccione, and Um Sohryngkew sections are probably associated with the Deccan phase 2 eruptions, as well as the two spikes at the Poty section. One of these spikes (spike I) is situated within the CF2 foraminiferal biozone (e.g., Um Sohryngkew), within the 250 ky (beginning of Deccan phase 2) to 160 ky (CF2–CF1 biozones boundary) interval before the K-Pg boundary and after carbon dioxide, sulfuric aerosols, and other toxic agents had reached a critical threshold. The second spike (spike II), at the K-Pg boundary, is also concomitant with the Deccan phase 2 (e.g., Højerup and Bottaccione), and a third one (spike III), within the P1a foraminiferal biozone, is likely related to late Deccan phase 2 eruptions (within the 220–500 kyr interval after the K-Pg boundary). These three periods of anomalous Hg deposition identified here suggest events of enhanced Hg deposition over broad areas in a global scale.

b. The possibility that Hg enhancements below and at the K-Pg boundary layer (spikes I and II, respectively) are postdepositional, resulting from scavenging by anoxia on seafloor and transported downward into the uppermost 50 cm layer, is neither confirmed by the stratigraphic record of Mo/Al redox proxy, nor by the observed true negative Ce anomaly, both of which indicate sediment deposition in oxic to barely anoxic surface waters. Besides, the laminated boundary clay acted as an impermeable layer.

c. Difference in sedimentation rates, in distance relative to the landmass versus marine realm, and partial leaching of Hg during weathering and/or diagenesis are possible explanations for the differences in magnitude in measured Hg concentrations.

d. For confirmation of true volcanogenic Hg loading to the environment, normalization of Hg to corresponding TOC% values is commonly applied. Inaccuracies in measurements, however, may lead to unrealistic Hg/TOC spikes in cases where extremely low TOC values (<0.2%) are used.

e. The δ^{202}Hg data for clays of the K-Pg boundary layer (spike II) from the Højerup, Bottaccione, Um Sohryngkew, and Bidart sections as well as for the spike III of the Poty section lie within the range of volcanogenic Hg compositions. δ^{202}Hg values for spikes I and II from Bidart, however, lie within the volcanic emission/chondritic field. Long-range atmospheric transport of Hg is supported by small and positive Δ^{201}Hg.

Although these seem to be stimulating results, they should be taken with caution, before the limited amount of Hg isotope data available. Besides, eruption, transportation, and deposition may have led to isotope fractionation or to postdepositional changes of the original Hg isotope signals. Currently, there is a growing agreement that large asteroid/comet impact and volcanism concurred as main causes of the end-Cretaceous mass extinction.

ACKNOWLEDGMENTS

Grants to ANS (CNPq471036/2012-9, FACEPE APQ-1059-9.05/12, and APQ 1073-1.07/15), to JBC (NSFC 41625012, NSFC41273023, NSFC U1301231, and NSFC41561134017), to LDL (CNPq INCT-TMCOcean 573.601/ 2008-9, CNPq 576.601/2009-1), and to VPF (CNPq 471034/2012-6, FACEPE APQ1738-1.07/12) provided financial support to chemical or isotope analyses. Financial support through the Danish Agency for Science, Technology, and Innovation (grant n. 11-103378) to RF and through the Danish National Research Foundation's center of excellence NordCEE (DNRF grant n. DNRF53) is highly appreciated. Drilling at the Poty site was supported by the Paraíba Drilling Project/UFPE/CNPq/Princeton University. VCT is grateful to the Brazilian Council for Scientific and Technological Development (CNPq) for three-month visiting professor fellowship in the LABISE, Brazil, and thankful to Dr. A.K. Gupta, director of WIHG, for collaborative project between this institution and LABISE. We are grateful to Prof. Finn Surlyk (University of Copenhagen) and to an anonymous reviewer whose comments and suggestions on an earlier version of the manuscript greatly contributed to improve it. This is the NEG-LABISE contribution n. 284.

REFERENCES

Adatte, A., Keller, G., Stinnesbeck, W., 2002. Late Cretaceous to early Paleocene climate and sea-level fluctuations: the Tunisian record. Palaeogeography, Palaeoclimatology, Palaeoecology 2754, 1–32.

Adatte, T., Keller, G., Schoene, B., Samperton, K.M., Font, E., Sial, A.N., Lacerda, L.D., Punekar, J., Fantasia, A., Khadri, S., 2015. Paleoenvironmental influence of Deccan volcanism relative to the KT extinction. Baltimore, Geological Society of America Abstracts with Programs 47, No. 7, 210.

Albertão, G.A., Martins, P.P., Jr., 1996. A possible tsunami deposit at the Cretaceous–Tertiary boundary in Pernambuco, Northeastern Brazil. Sedimentary Geology 104, 189–201.

Albertão, G.A., Koutsoukos, E.A.M., Regali, M.P.S., Attrep, M., Jr., Martins, P.P., Jr., 1994. The Cretaceous–Tertiary boundary in southern low-latitude regions: preliminary study in Pernambuco, northeastern Brazil. Terra Nova 6, 366–375.

Alexander, P.O., Gibson, I.L., 1977. Rare earth abundances in Deccan trap basalts. Lithos 10, 143–147.

Algeo, T.J., Rowe, H., 2012. Paleoceanographic applications of trace-metal concentration data. Chemical Geology 324/325, 6–18.

Alvarez, W., 2009. The historical record in the Scaglia limestone at Gubbio: magnetic reversals and the Cretaceous–Tertiary mass extinctions. Sedimentology 56, 137–148.

Alvarez, L.W., Alvarez, W., Asaro, F., Michel, H.V., 1980. Extraterrestrial cause for the Cretaceous–Tertiary extinction. Science 208, 1095–1108.

Banerji, R.K., 1981. Cretaceous–Eocene sedimentation, tectonism and biofacies in the Bengal basin, India. Palaeogeography, Palaeoclimatology, Palaeoecology 34, 57–85.

Bau, M., Dulski, P., 1996. Distribution of yttrium and rare earth elements in the Penge and Kuruman Iron Formation, Transvaal Supergroup, South Africa. Precambrian Research 79, 37–55.

Bau, M., Dulski, P., Möller, P., 1995. Yttrium and holmium in South Pacific seawater: vertical distribution and possible fractionation mechanisms. Chemie der Erde, 55, 1–15.

Becker, L., Poreda, R.J., Bunch, T.E., 2000a. Fullerenes: an extraterrestrial carbon carrier phase for noble gases. Proceedings of the National Academy of Sciences, USA 97, 2979–2983.

Becker, L., Poreda, R.J., Bunch, T.E., 2000b. The origin of fullerenes in the 65 Myr old Cretaceous/Tertiary 'K/T' boundary. Lunar Planetary Science 31, 1832.

Belcher, C.M., 2009. Reigniting the Cretaceous–Palaeogene firestorm debate. Geology 37, 1147–1148.

Bergquist, B.A., Blum, J.D., 2009. The odds and evens of mercury isotopes: applications of mass-dependent and mass-independent isotope fractionation. Elements 5, 353–357.

Bhandari, N., Gupta, M., Shukla, P. N., 1993. Deccan volcanic contribution of Ir and other trace elements near the K/T boundary. Chemical Geology 103, 129–139.

Bhandari, N., Gupta, M., Pandey, J., Shukla, P.M., 1994. Chemical profiles in K/T boundary section of Meghalaya, India: cometary, asteroidal or volcanic. Chemical Geology 113, 45–60.

Bhandari, N., Shukla, P.N., Ghevariya, Z.G., Sundaram, S., 1996. K/T boundary layer in Deccan intertrappeans at Anjar, Kutch. In: Ryder, G., Fastovsky, D.E., and Gartner, S. (eds.), The Cretaceous–Tertiary Event and Other Catastrophes in Earth History: Geological Society of America Special Paper 307, 417–424. Boulder, CO: Geological Society of America.

Bjerager, M., Surlyk, F., 2007. Danian cool-water bryozoan mounds at Stevns Klint, Denmark: a new class of non-cemented skeletal mounds. Journal of Sedimentary Research 77, 634–660.

Blum, J.D., Sherman, L.S., Johnson, M.W., 2014. Mercury isotopes in earth and environmental sciences. Annual Reviews of Earth Planetary Science 42, 249–269.

Bond, D.P.G., Wignall, P.B., 2014. Large igneous provinces and mass extinctions: an update. Geological Society of America Special Paper 505, 29–55.

Bond, D.P.G., Grasby, S.E., 2017. On the causes of mass extinctions. Palaeogeography, Palaeoclimatology, Palaeoecology 478, 3–29.

Bond, D.P.G., Wignall, P.B., Joachimski, M., Sun, Y., Savov, I., Grasby, S.E., Beauchamp, B., Blomeier, D.P.G., 2015. An abrupt extinction in the Middle Permian (Capitanian) of the Boreal Realm (Spitsbergen). Bulletin of the Geological Society of America 127, 1411–1421, doi:10.1130/B31216.1.

Bonté, P., Delacotte, O., Renard, M., Laj, C., Boclet, D., Jehanno, C., Rocchia, R., 1984. An iridium rich layer at the Cretaceous–Tertiary boundary in the Bidart section (southern France). Geophysical Research Letters 11, 473–476.

Brugger, J., Feulner, G., Petri, S., 2017. Baby, it's cold outside: climate model simulations of the effects of the asteroid impact at the end of the Cretaceous. Geophysical Research Letters 44, 419–427.

Callegaro, S., Baker, D.R., De Min, A., Marzoli1, A., Geraki, K., Bertrand, H., Viti, C., Nestola, F., 2014. Microanalyses link sulfur from large igneous provinces and Mesozoic mass extinctions. Geology 42, 895–898.

Charbonnier, G., Morales, C., Duchamp-Alphonse, S., Westermann, S., Adatte, T., Föllmi, K.B., 2017. Mercury enrichment indicates volcanic triggering of Valanginian environmental change. Scientific Reports 7, 40808, doi:10.1038/srep40808.

Chen, J.-B., Hintelmann, H., Feng, X.-B., Dimock, B., 2012. Unusual fractionation of both odd and even mercury isotopes in precipitation from Peterborough, ON, Canada. Geochimica et Cosmochimica Acta 90, 33–46.

Chenet, A.L., Courtillot, V., Fluteau, F., Gerard, M., Quidelleur, X., Khadri, S.F.R., Subbarao, K.V., Thordarson, T., 2009. Determination of rapid Deccan eruptions across the Cretaceous–Tertiary boundary using paleomagnetic secular variation: 2. Constraints from analysis of eight new sections and synthesis for a 3500-m-thick composite section. Journal of Geophysical Research 114, 1–38.

Claeys, P., Kiessling, W., Alvarez, W., 2002. Distribution of Chicxulub ejecta at the Cretaceous–Tertiary boundary. In: Koeberl, C., and MacLeod, K.G. (eds.), Catastrophic Events and Mass Extinctions: Impacts and Beyond: Geological Society of America Special Paper 356, 55–68. Boulder, CO: Geological Society of America.

Coccioni, R., Premoli Silva, I., 2015. Revised Upper Albian–Maastrichtian planktonic foraminiferal biostratigraphy and magneto-stratigraphy of the classical Tethyan Gubbio section (Italy). Newsletters on Stratigraphy 48, 47–90.

Coccioni, R., Frontalini, F., Bancalà, G., Fornaciari, E., 2010. The Dan-C2 hyperthermal event at Gubbio (Italy): global implications, environmental effects, and cause(s). Earth and Planetary Science Letters 297, 298–305.

Courtillot, V., 1990. A volcanic eruption. Scientific American 263, 85–92.

Courtillot, V., Besse, J., Vadamme, D., Montigny, R., Jaeger, J.J., Capetta, H., 1986. Deccan flood basalts at the Cretaceous/Tertiary boundary? Earth and Planetary Science Letter 80, 361–374.

Crocket, J.H., Fleet, M.E., and Stone, W.E., 1978. Experimental partitioning of osmium, iridium and gold between basalt melt and sulphide liquid at 1300°C, between olivine and silicate melt. Earth and Planetary Science Letters 40, 203–219.

Denne, R.A., Scott, E.D., Eickhoff, D.P., Kaiser, J.S., Hill, R.J., Spaw, J.M., 2013. New evidence for widespread Chicxulub-induced slope failure Massive Cretaceous–Paleogene boundary deposit, deep-water Gulf of Mexico: new evidence for widespread Chicxulub-induced slope failure. Geology 41, 983–986.

DePaolo, D.J., Kyte, F.T., Marshall, B.D., O'Neil, J.R., Smit, J., 1983. Rb–Sr, Sm–Nd, K–Ca, O and H isotopic study of Cretaceous–Tertiary boundary sediments, Caravaca, Spain: evidence for an oceanic impact site. Earth Planetary Science Letters 64, 356–373.

Dole, G., Pillai, S.P., Upasani, D., Kale, V.S., 2017. Triggering of the largest Deccan eruptions by the Chicxulub impact: comment. Geological Society of America Bulletin 129, 253–255.

Elderfield, H., Greaves, M.J., 1982. The rare-earth elements in seawater. Nature 296, 214–219.

Elderfield, H., Pagett, R., 1986. Rare earth elements in icthyoliths: variations with redox conditions and depositional environments. Science of the Total Environment 49, 175–197.

Emerson, S.R., Huested, S.S., 2015. Ocean anoxia and the concentrations of molybdenum and vanadium in seawater. Marine Chemistry 34, 177–196.

Ernst, R.E., Youbi, N., 2017. How Large Igneous Provinces affect global climate, sometimes cause mass extinctions, and represent natural markers in the geological record. Palaeogeography, Palaeoclimatology, Palaeoecology 478, 30–52, doi:10.1016/j.palaeo.2017.03.014.

Fantasia, A., Adatte, T., Spangenberg, J.E., Font, E., 2016. Palaeoenvironmental changes associated with Deccan volcanism, examples from terrestrial deposits from Central India, Palaeogeography, Palaeoclimatology, Palaeoecology 441, 165–180.

Fantle, M.F., Higgins, J.A., 2014. The effects of diagenesis and dolomitization on Ca and Mg isotopes in marine platform carbonates: implications for the geochemical cycles of Ca and Mg. Geochimica et Cosmochimica Acta 142, 458–481.

Fauth, G., Colin, J., Koutsoukos, E.A.M., Bengtson, P., 2005. Cretaceous–Tertiary boundary ostracods from the Poty Quarry, Pernambuco, Northeastern Brazil. Journal of South American Earth Sciences 19, 285–305.

Font, E., Fabre, F., Nédélec, A., Adatte, T., Keller, G., Veiga-Pires, C., Ponte, J., Mirão, J., Khozyem, H., Spangenberg, J., 2014. Atmospheric halogen and acid rains during the main phase of Deccan eruptions: magnetic and mineral evidence. Geological Society of America Special paper 505, 1–16.

Font, E., Adatte, T., Sial, A.N., Lacerda, L.D., Keller, G., Punekar, J., 2016. Mercury anomaly, Deccan volcanism and the end-Cretaceous mass extinction. Geology 44, 171–174.

Font, E., Adatte, T., Andrade, M., Keller, G., Mbabi Bitchong, A., Carvallo, C., Ferreira, J., Diogo, Z., Mirão, J., 2018. Deccan volcanism induced high-stress environment during the Cretaceous–Paleogene transition at Zumaia, Spain: evidence from magnetic, mineralogical and biostratigraphic records. Earth and Planetary Science Letters 484, 53–66.

Frei, R., Frei, K.M., 2002. A multi-isotopic and trace element investigation of the Cretaceous–Tertiary boundary layer at Stevns Klint, Denmark: inferences for the origin and nature of siderophile and lithophile element geochemical anomalies. Earth and Planetary Science Letters 203, 691–708.

Galbrun, B., Gardin, S., 2004. New chronostratigraphy of the Cretaceous–Paleogene boundary interval at Bidart (France). Earth and Planetary Science Letters 224, 19–32.

Gale, A., Dalton, C.A., Langmuir, C.H., Su, Y., Schilling, J.G., 2013. The mean composition of ocean ridge basalts. Geochemistry, Geophysics, Geosystems 14 (3), 489–518. doi:10.1029/2012GC004334.

Galeotti, S., Moretti, M., Cappelli, C., Phillips, J., Lanci, L., Littler, K., Monechi, S., Petrizzo, M.R., Premoli Silva, I., Zachos, J.C., 2015. The Bottaccione section at Gubbio, central Italy: a classical Paleocene Tethyan setting revisited. Newsletters on Stratigraphy 48, 325–339.

Ganapathy, R., 1980. A major meteoritic impact on the Earth 65 million years ago: evidence from the Cretaceous–Tertiary boundary clay. Science 209, 921–923.

Garg, R., Ateequzzaman, K., Prasad, V., 2006. Significant dinoflagellate cyst biohorizons in the Upper Cretaceous–Paleocene succession of the Khasi Hills, Meghalaya. Journal Geological Society of India 67, 737–747.

German, C.R., Elderfield, H., 1990. Application of the Ce anomaly as a paleoredox indicator: the ground rules. Paleoceanography 5, 823–833.

Gertsch, B., Keller, G., Adatte, T., Garg, R., Prasad, V., Berner, Z., Fleitmann, D.S., 2011. Environmental effects of Deccan volcanism across the Cretaceous–Tertiary transition in Meghalaya, India. Earth and Planetary Science Letters 310, 272–285.

Gertsch, B., Keller, G., Adatte, T., Berner, Z., 2013. The Cretaceous–Tertiary boundary (KTB) transition in NE Brazil. Journal of the Geological Society, London 170, 249–262.

Gothmann, A.M., Bender, M.L., Blättler, C.L., Swart, P.K., Giri, S.J., Adkins, J.F., Stolarski, J., Higgins, J.A., 2016. Calcium isotopes in scleractinian fossil corals since the Mesozoic: implications for vital effects and biomineralization through time. Earth and Planetary Science Letters 444, 205–214.

Grasby, S.E., Sanei, H., Beauchamp, B., Chen, Z. 2013. Mercury deposition through the Permo–Triassic Biotic Crisis Chemical Geology 351, 209–216.

Grasby, S.E., Beauchamp, B., Bond, D.P.G., Wignal, P., Talavera, C., Galloway, J.M., Piepjohn, K., Reinhardt, L., Blomeier, D., 2015. Progressive environmental deterioration in northwestern Pangea leading to the latest Permian extinction. Geological Society of America Bulletin 127, 1311–1347.

Grasby, S.E., Beauchamp, B., Bond, D.P.G., Wignall, P.B., Sanei, H. 2016. Mercury anomalies associated with three extinction events (Capitanian Crisis, Latest Permian Extinction and the Smithian/Spathian Extinction) in NW Pangea. Geological Magazine 153, 285–297. doi:10.1017/S0016756815000436.

Grasby, S.E., Shen, W., Yin, R., Gleason, J.D., Blum, J.D., Lepak, R.F., Hurley, J.P., Beauchamp, B., 2017. Isotopic signatures of mercury contamination in latest Permian oceans. Geology 45, 55–58.

Griffith, E.M., Fantle, M.S., Eisenhauer, A., Paytan, A., Bullen, T.D., 2015. Effects of ocean acidification on the marine calcium isotope record at the Paleocene-Eocene Thermal Maximum. Earth and Planetary Science Letters 419, 81–92.

Grotzinger, J.P., Fike, D.A., Fischer, W.W., 2011. Enigmatic origin of the largest known carbon isotope excursion in Earth's history. Nature Geosciences 4, 285–292.

Hallam, A., 1987. End-Cretaceous mass extinction event: argument for terrestrial causation. Science 238, 1237–1242.

Hansen, T., Surlyk, F., 2014. Marine macrofossil communities in the uppermost Maastrichtian chalk of Stevns Klint, Denmark. Palaeogeography, Palaeoclimatology, Palaeoecology 399, 323–344.

Hanski, I., 2016. Messages from the Islands: A Global Biodiversity Tour. Chicago: Chicago Press, Chapter 1, 42–75.

Hart, M.B., Feist, S.E., Price, G.D., Leng, M.J., 2004. Reappraisal of the K–T boundary succession at Stevns Klint, Denmark. Journal of the Geological Society of London 161, 1–8.

Hildebrand, A.R., Boynton, W.V., 1989. Hg anomalies at the K/T boundary: evidence for acid rain? Meteoritics 24, 277–278.

Hildebrand, A.R., Penfield, G.T., Kring, D.A., Pilkington, M., Camargo, Z.A., Jacobsen, S.B., Boynton, W.V., 1991. Chicxulub Crater: a possible Cretaceous/Tertiary boundary impact crater on the Yucatán Peninsula, Mexico. Geology 19, 867–871.

Hofer, G., Wagreich, M., Neuhuber, S., 2013. Geochemistry of fine-grained sediments of the upper Cretaceous to Paleogene Gosau Group (Austria, Slovakia): implications for paleoenvironmental and provenance studies. Geoscience Frontiers 4, 449–468.

Jolley, D., Gilmour, I., Gurov, E., Kelley, S., Watson, J., 2010. Two large meteorite impacts at the Cretaceous–Paleogene boundary. Geology 38, 835–838.

Jones, D.S., Martini, A.M., Fike, A., Kaiho, K., 2017. A volcanic trigger for the Late Ordovician mass extinction? Mercury data from south China and Laurentia. Geology 45, 631–634, doi:10.1130/G38940.1.

Jourdan, F., Hodges, K., Sell, B., Schaltegger, U., Wingate, M.T.D., Evins, L.Z., Soderlund, U., Haines, P.W., Phillips, D., Blenkinsop, T., 2014. High-precision dating of the Kalkarindji large igneous province, Australia, and synchrony with the Early–Middle Cambrian (Stage 4–5) extinction. Geology 42, 543–546.

Kato, Y., Nakao, K., Isozaki, Y., 2002. Geochemistry of Late Permian Triassic pelagic cherts from southwest Japan: implications for an oceanic redox change. Chemical Geology 182, 15–34.

Kato, Y., Yamaguchi, K.E., Ohmoto, H., 2006. Rare earth elements in Precambrian banded iron formation: secular changes of Ce and Eu anomalies and evolution of atmosphere oxygen. Geological Society of America, Memoir 198, 269–280.

Keller, G., 2014. Deccan volcanism, the Chicxulub impact, and the end-Cretaceous mass extinction: coincidence? Cause and effect? In: Keller, G., and Kerr, A. (eds.), Volcanism, Impacts, and Mass Extinctions: Causes and Effects: GSA Special Paper 505, 29–55. Boulder, CO: Geological Society of America.

Keller, G., Stinnesbeck, W., 1996a. Sea level changes, clastic deposits and megatsunamis across the Cretaceous–Tertiary boundary. In: MacLeod, N., and Keller, G. (eds.), The Cretaceous–Tertiary Mass Extinction: Biotic and Environmental Events, 415–450. New York: Norton Press.

Keller, G., Stinnesbeck, W., 1996b. Near K/T age of clastic deposits from Texas to Brazil: impact, volcanism and/or sea-level lowstand? Terra Nova 8, 277–285.

Keller, G., Kerr, A.C., 2014. Foreword. In: Keller, G., and Kerr, A.C. (eds.), Volcanism, Impacts, and Mass Extinctions: Causes and Effects: Geological Society of America Special Paper 505, v–ix. Boulder, CO: Geological Society of America.

Keller, G., Adatte, T., Stinnesbeck, W., Rebolledo-Vieyra, M., Fucugauchi, J.U., Kramar, U., Stüben, D., 2004a. Chicxulub impact predates the K–T boundary mass extinction. Proceedings of the National Academy of Sciences 101, 3753–3758.

Keller, G., Adatte, T., Stinnesbeck, W., Stüben, D., Berner, Z., Kramar, U., Harting, M., 2004b. More evidence that the Chicxulub impact predates the K/T mass extinction. Meteoritics and Planetary Science 39, 1127–1144.

Keller, G., Adatte, T., Gardin, S., Bartolini, A., Bajpai, S., 2008. Main Deccan volcanism phase ends near the K–T boundary: evidence from the Krishna–Godavari Basin, SE India. Earth Planetary Science Letters 268, 293–311.

Keller, G., Sahni, A., Bajpai, S., 2009, Deccan volcanism, the KT mass extinction and dinosaurs. Journal of Bio-Science 34, 709–728.

Keller, G., Bhowmick, P.K., Upadhyay, H., Dave, A., Reddy, A.N., Jaiprakash, B.C., Adatte, T., 2011. Deccan volcanism linked to the Cretaceous–Tertiary boundary mass extinction: new evidence from ONGC wells in the Krishna-Godavari Basin. Journal of the Geological Society of India 78, 399–428.

Keller, G., Adatte, T., Bhowmick, P.K., Upadhyay, H., Dave, A., Reddy, A.N., Jaiprakash, B.C., 2012. Nature and timing of extinctions in Cretaceous–Tertiary planktic foraminifera preserved in Deccan intertrappean sediments of the Krishna-Godavari Basin, India. Earth Planetary Science Letters 341, 211–221.

Keller, G., Punekar, J., Mateo, P., 2016. Upheavals during the Late Maastrichtian: volcanism, climate and faunal events preceding the end-Cretaceous mass extinction. Palaeogeography, Palaeoclimatology, Palaeoecology 441, 137–151.

Kerr, R.A., 1983. Isotopes add support for asteroid impact: osmium isotope analysis supports an asteroid impact 65 million years ago but cannot exclude a huge volcanic eruption. Science 222, 603–604.

Khozyem, H., Adatte, T., Mbabi Bitchong, A., Mohamed, A., Keller, G., 2017. The role of volcanism (North Atlantic igneous province) in the PETM events revealed by mercury anomalies. GSA Annual Meeting in in Seattle, Washington, Paper No. 228-5. Geological Society of America Abstracts with Programs 49(6). doi:10.1130/abs/2017AM-302839.

Knoll, A.H., Hayes, J.M., Kaufman, A.J., Swett, K., Lambert, I.B., 1986. Secular variation in carbon isotope ratios from upper Proterozoic successions of Svalbard and East Greenland. Nature 321, 832–838.

Koeberl, C., Sharpton, V.L., Harrison, T.M., Sandwell, D., Murali, A.V., Burke, K., 1990. The Kara/Ust Kara twin impact structure; a large-scale impact event in the Late Cretaceous. In: Sharpton, V.L. and Ward, P.D. (eds.), Global Catastrophes in Earth History: An Interdisciplinary Conference on Impacts, Volcanism and Mass Mortality: Geological Society of America, Special Paper 247, 233–238. Boulder, CO: Geological Society of America.

Kolesnikov, E.M., Nazarov, M.A., Badjukov, D.D., Shukolyukov, Y.A., 1988. The Karskiy craters are probable records of catastrophe at the Cretaceous–Tertiary boundary. Abstract, Global catastrophes in Earth history; An interdisciplinary conference on impacts, volcanism, and mass mortality, Houston, Texas, Lunar and Planetary Institute Contribution 673, 99–100.

Komar, N., Zeebe, R.E., 2015. Calcium and calcium isotope changes during carbon cycle perturbations at the end-Permian. Paleoceanography 31, 115–130.

Korte, C., Kozur, H.W., 2010. Carbon-isotope stratigraphy across the Permian–Triassic boundary: a review. Journal of Asian Earth Sciences 39, 215–235.

Koutsoukos, E.A.M. 1998. An extraterrestrial impact in the Early Danian: a secondary K/T boundary event? Terra Nova 10, 68–73.

Kring, D.A., Durda, D.D., 2002. Trajectories and distribution of material ejected from the Chicxulub impact crater: implications for post impact wildfires. Journal of Geophysical Research 107, 6–22.

Krishnan, M.S. 1968. Geology of India and Burma. Madras: Higginbothams, 536 pp.

Lawrence, M.G., Kamber, B.S., 2006. The behaviour of the rare earth elements during estuarine mixing revisited. Marine Chemistry 100, 147–161.

Lima, F.H.O., Koutsoukos, E.A.M., 2006. Towards an integrated stratigraphy of the Gramame Formation (Maastrichtian), CIPASA Quarry, Pernambuco–Paraíba Basin, NE Brasil. Anuário do Instituto de Geociências, Universidade Federal do Rio de Janeiro 29, 81–94.

Liu, Y.G., Miah, M.R.U., Achmitt, R.A., 1988. Cerium: a chemical tracer for paleo-oceanic redox conditions. Geochimica et Cosmochimica Acta 52, 1361–1371.

Luterbacher, H.P., Premoli Silva, I., 1964. Biostratigrafia del limite Cretaceo–Terziario nell'Appennino centrale. Rivista Italiana di Paleontologia e Stratigrafia 70, 67–128.

Lyons, T.W., Werne, J.P., Hollander, D.J., Murray, R.W., 2003. Contrasting sulfur geochemistry and Fe/Al and Mo/Al ratios across the last oxic-to-anoxic transition in the Cariaco Basin, Venezuela. Chemical Geology 195, 131–157.

MacLeod, N., 2012. Extinction: K–Pg Mass Extinction. Chichester:eLS.Wiley.www.els.net,doi:10.1002/9780470015902. a0001656.pub3.

Madhavaraju, J., Lee, Y.I., 2009. Geochemistry of the Dalmiapuram Formation of the Uttatur Group (Early Cretaceous), Cauvery Basin, southeastern India: implications on provenance and paleoredox conditions. Revista Mexicana de Ciencias Geológicas 26, 380–394.

Madhavaraju, J., González-León, C.M., 2012. Depositional conditions and source of rare earth elements in carbonate strata of the Aptian–Albian Mural Formation, Pitaycachi section, northeastern Sonora, Mexico. Revista Mexicana de Ciencias Geológicas 29, 478–491.

Maloof, A.C., Porter, S.M., Moore, J.H., Dudás, F.O., Bowring, S.A., Higgins, J.A., Fike, D.A., Michael, E.P., 2010. The earliest Cambrian record of animals and ocean geochemical change. Geological Society of America Bulletin 122, 1731–1774.

McLean, D.M., 1985.Deccan Traps mantle degassing in the terminal Cretaceous marine extinctions. Cretaceous Research 6, 235–259.

McLennan, S.M., 1989. Rare earth elements in sedimentary rocks: influence of provenance and sedimentary processes. Reviews in Mineralogy 21, 169–200.

Melosh, H.J., Schneider, N.M., Zahnle, K.J., Latham, D., 1990. Ignition of global wildfires at the Cretaceous/Tertiary boundary. Nature 343, 251–254.

Meyer, K.M., Yu, M., Lehrmann, D., van de Schootbrugge, B., Payne, J.L., 2013. Constraints on early Triassic carbon cycle dynamics from paired organic and inorganic carbon isotope records. Earth Planetary Science Letters 361, 429–435.

Molina, E., Alegret, L., Arenillas, I., Arz, J.A., Gallala, N., Hardenbol, J., Von Salis, K., Steurbaut, E., Vandenberghe, N., Zaghbib-Turki, D., 2006. The global boundary Section and Point for the base of the Danian Stage (Paleocene, Paleogene, "Tertiary", Cenozoic) at El Kef, Tunisia– Original definition and revision. Episodes 29, 263–273.

Molina, E., Alegret, L., Arenillas, E., Arz, J.A., Gallala, N., Grajales-Nishimura, J.M., Murillo-Muñetón, G., Zaghbib-Turki, D., 2009. The Global Boundary Stratotype Section and Point for the base of the Danian Stage (Paleocene, Paleogene, "Tertiary", Cenozoic): auxiliary sections and correlation. Episodes 32, 84–95.

Morgan, J.V., Lana, C., Kearsley, A., Coles, B., Belcher, C., Montanari, S., Diaz-Martinez, E., Barbosa, A., Neumann, V., 2006. Analyses of shocked quartz at the global K–P boundary indicate an origin from a single, high-angle, oblique impact at Chicxulub. Earth and Planetary Science Letters 251, 264–279.

Morgan, J.V. et al., 2016. The formation of peak rings in large impact craters. Science 354, 878–882.

Mukhopadhyay, S.K., 2008. Planktonic foraminiferal succession in late Cretaceous to Early Paleocene strata in Meghalaya, India. Lethaia 41, 71–84.

Mukhopadhyay, S.K., 2017. Planktonic foraminiferal zonation and sea-level changes in the upper Maastrichtian-middle Danian successions of Meghalaya, India. Stratigraphy 13(4), 245–276.

Mukhopadhyay, S.K., Pal, S., Shrivastava, J.P., 2017. Comments on the paper by Sial et al. (2016) Mercury enrichments and Hg isotopes in Cretaceous–Paleogene boundary successions: links to volcanism and palaeoenvironmental impacts. Cretaceous Research 66, 45–81. Cretaceous Research 71, 81–83, doi:10.1016/j.cretres.2016.12.006.

Murray, R.W., Buchholtz Brink, M.R., Brumsack, H.J., Gerlach, D.C., Russ, G.P., III, 1991. Rare earth elements in Japan sea sediments and diagenetic behavior of Ce/Ce*: results from ODP Leg 127. Geochimica Cosmochimica Acta 55, 2453–2466.

Nance, W.B., Taylor, S.R., 1976. Rare earth element patterns and crustal evolution, I. Australian post-Archean sedimentary rocks. Geochimica et Cosmochimica Acta 40, 1539–1551.

Nascimento-Silva, V.M., Sial, A.N., Ferreira, V.P., Neumann, V.H., Barbosa, J.A., Pimentel, M.M., Lacerda, L.D., 2011. Cretaceous–Paleogene transition at the Paraíba Basin, northeastern, Brazil: carbon-Isotope and mercury subsurface stratigraphies. Journal of South American Earth Sciences *32*, 379–392.

Nascimento-Silva, M.V., Sial, A.N., Ferreira, V.P., Barbosa, J.A., Neumann, V.H., Pimentel, M.M., Lacerda, L.D., 2013. Carbon isotopes, rare-earth elements and mercury behavior of Maastrichtian–Danian carbonate succession of the Paraíba Basin, Northeastern Brazil. In: Bojar, A.V., Melinte-Dobrinescu, M.C., and Smit, J. (eds.), Isotopic Studies in Cretaceous Research: Geological Society, London, Special Publications *382*, 85–104. London: Geological Society of London.

Nazarov, M.A., Kolesmikov, E.M., Badjukov, D.D., Masaitis, V.L., 1989. Potassium-argon age of the Kara impact event. Lunar and Planetary Science *20*, 766–767.

Neumann, V.G., Barbosa, J.A., Nascimento Silva, M.V., Sial, A.N., Lima Filho, M. 2009. Sedimentary development and isotope analysis of deposits at the Cretaceous/Paleogene transition in the Paraíba Basin, NE Brazil. Geologos *15*, 103–113.

Nichols, D.J., Johnson, K.R., 2008. Plants and the K–T Boundary. Cambridge: Cambridge University Press, 280 pp. doi:10.1017/CBO9780511535536.

Nimura, T., Ebisuzaki, T., Maruyama, S., 2016. End-cretaceous cooling and mass extinction driven by a dark cloud encounter. Gondwana Research *37*, 301–307.

Oehlert, A.M., Swart, P.K., 2014. Interpreting carbonate and organic carbon isotope covariance in the sedimentary record. Nature Communications *5*, 4672, doi:10.1038/ncomms5672.

Pal, S., Srivastava, J.P., Mukhopadhyay, S.K., 2015. Polycyclic aromatic hydrocarbon compound excursions and K/Pg transition in the late Cretaceous–Palaeogene succession of the Um Sohryngkew River section, Meghalaya. Current Science *109*, 1140–1150.

Pandey, J., 1990. Cretaceous/Tertiary boundary, iridium anomaly and foraminifer break in the Um Sohryngkew River section. Current Science *59*, 570–575.

Parthasarathy, G., Bhandari, N., Vairamani, M., Kunwar, A.C., 2008. High-pressure phase of natural fullerene C_{60} in iridium-rich Cretaceous–Tertiary boundary layers of Deccan inter-trappean deposits, Anjar, Kutch, India. Geochimica et Cosmochimica Acta *72*, 978–987.

Percival, L.M.E., Witt, M.L.I., Mather, T.A., Hermoso, M., Jenkyns, H.C., Hesselbo, S.P., Al-Suwaidi, A.H., Storm, M.S., Xu, W., Ruhl, M., 2015. Globally enhanced mercury deposition during the end-Pliensbachian extinction and Toarcian OAE: a link to the Karoo–Ferrar Large Igneous Province. Earth and Planetary Science Letters *428*, 267–280.

Percival, L.M.E., Ruhl, M., Hesselbo, S.P., Jenkyns, H.C., Mather, T.A., Whiteside, J.H., 2017. Mercury evidence for pulsed volcanism during the end-Triassic mass extinction. PNAS Early edition; www.pnas.org/cgi/doi/10.1073/pnas.1705378114.

Premoli, S., Sliter, W.V., 1994. Cretaceous planktonic foraminiferal biostratigraphy and evolutionary trends from the Bottaccione section, Gubbio, Italy. Palaeontographica Italica *82*, 1–89.

Punekar, J., Mateo, P., Keller, G., 2014. Effects of Deccan volcanism on paleoenvironment and planktic foraminifera: a global survey. Geological Society of America Special Papers *505*, 91–116.

Rampino, M., 2010. Mass extinctions of life and catastrophic flood basalt volcanism. Proceedings of the National Academy of Science *107*, 6555–6556.

Rampino, M., Brothers, R., 1998a. Flood basalt volcanism during the past 250 million years. Science *241*, 663–668.

Rampino, M., Brothers, R., 1998b. Mass extinctions, comet impacts, and the galaxy. In: Andersen, J. (ed.), Highlights of Astronomy 11 A, 246–251. New York: Springer.

Rasmussen, J.A., Heinberg, C., Hankasson, E., 2005. Planktonic foraminifers, biostratigraphy and the diachronous nature of the lowermost Danian Cerithium Limestone at Stevns Klint, Denmark. Bulletin of the Geological Society Denmark *52*, 113–131.

Renne, P.R., Deino, A.L., Hilgen, F.J., Kuiper, D.F., Mark, D.F., Mitchell, W.S., 3rd, Morgan, E., Mundil, R., Smit, J., 2013. Time scales of critical events around the Cretaceous–Paleogene boundary. Science *339*, 684–687.

Renne, P., Sprain, C.J., Richards, M.A., Self, S., Vanderkluysen, L., Pande, K., 2015. State shift in Deccan volcanism at the Cretaceous–Paleogene boundary, possibly induced by impact. Science *350*, 76–78.

Richards, M.A., Alvarez, W., Self, S., Karlstrom, L., Renne, P.R., Manga, M., Sprain, C.J., Smit, J., Vanderkluysen, L., Gibson, S.A., 2015. Triggering of the largest Deccan eruptions by the Chicxulub impact. Geological Society of America Bulletin *127*, 1507–1520.

Richards, M.A., Alvarez, W., Self, S., Karlstrom, L., Renne, P.R., Manga, M., Sprain, C.J., Smit, J., Vanderkluysen, L., Gibson, S.A., 2017. Triggering of the largest Deccan eruptions by the Chicxulub impact: reply. Geological Society of America Bulletin *129*, 256.

Rodrigues, G.B., Fauth, G., Santos, R.V., Koutsoukos, E.A.M., Colin, J.P., 2014. Tracking paleoecological and isotopic changes through the K–Pg boundary from marine ostracods: the Poty quarry section, northeastern Brazil. Cretaceous Research *47*, 105–116.

Sanei, H., Grasby, S.E., Beauchamp, B., 2012. Latest Permian mercury anomalies. Geology *40*, 63–66.

Sarkis. M.F., Arai, M., Koutsoukos, E.A.M., 2002. Dinoflagellates of the Cretaceous–Tertiary (KT) Boundary in the Poty Quarry Pernambuco–Paraíba Basin, Northeast Brazil. Boletim do 6° Simpósio sobre o Cretáceo do Brasil e II Simpósio sobre el Cretácico de América del Sur. São Pedro, São Paulo, Boletim de Resumos, 271–277.

Schoene, B., Samperton, K.M., Eddy, M.P., Keller, G., Adatte, T., Bowring, S., Khadri, F.R., Gertsch, B., 2015. U–Pb geochronology of the Deccan Traps and relation to the end-Cretaceous mass extinction. Science *347*, 182–184.

Schulte, P. et al., 2010. The Chicxulub asteroid impact and mass extinction at the Cretaceous–Paleogene boundary. Science *327*, 1214–1218.

Scotese, C.R., 2013. Map Folio 16, KT Boundary (65.5 Ma, latest Maastrichtian), PALEOMAP PaleoAtlas for ArcGIS, volume 2, Cretaceous, PALEOMAP Project, Evanston, IL.

Self, S., Jay, A.E., Widdowson, M., Keszthelyi, L.P., 2008. Correlation of the Deccan and Rajahmundry Trap lavas: are these the longest and largest lava flows on Earth? Journal of Volcanology and Geothermal Research *172*, 3–19.

Shimizu, H., Masuda, A. 1977. Cerium in cherts as an indication of marine environment of its formation. Nature *266*, 346–348.

Shukla, A.D., Bhandari, N., Kusumgar, S., Shukla, P.N., Ghevariya, Z.G., Gopalan, K., Balaram, V., 2001. Geochemistry and magnetostratigraphy of Deccan flows at Anjar, Kutch. Proceedings of the Indian Academy of Science (Earth Planetary Sciences) *110*, 111–132.

Shukolyukov, A., Lugmair, G.W., 1998. Isotopic evidence for the Cretaceous–Tertiary impactor and its type. Science *282*, 927–929.

Sial, A.N., Lacerda, L.D., Ferreira, V.P., Frei, R., Marquillas, R.A., Barbosa, J.A., Gaucher, C., Windmöller, C.C., Pereira, N.S., 2013. Mercury as a proxy for volcanic activity during extreme environmental turnover: the Cretaceous–Paleogene transition. Palaeogeography, Palaeoclimatology, Palaeoecology *387*, 153–164.

Sial, A.N., Chen, J.-B., Lacerda, L.D., Peralta, S., Gaucher, C., Frei, R., Cirilli, S., Ferreira, V.P., Marquillas, R.A., Barbosa, J.A., Pereira, N.S., Belmino, I.K.C., 2014. High-resolution Hg chemostratigraphy: a contribution to the distinction of chemical fingerprints of the Deccan volcanism and Cretaceous–Paleogene boundary impact event. Palaeogeography, Palaeoclimatology, Palaeoecology *414*, 98–115.

Sial, A.N., Chen, J., Lacerda, L.D., Frei, R., Tewari, V.C., Pandit, M.K., Gaucher, C., Ferreira, V.P., Cirilli, S., Peralta, S., Korte, C., Barbosa, J.A., Pereira, N.S., 2016. Mercury enrichment and mercury isotopes in Cretaceous–Paleogene boundary successions: links to volcanism and palaeoenvironmental impacts. Cretaceous Research *66*, 60–81.

Sial, A.N., Chen, J., Lacerda, L.D., Frei, R., Tewari, V.C., Pandit, M.K., Gaucher, C., Ferreira, V.P., Cirilli, S., Peralta, S., Korte, C., Barbosa, J.A., Pereira, N.S., 2017. Reply to comments by Sanjay K. Mukhopadhyay, Sucharita Pal, J. P. Shrivastava on the paper by Sial et al. (2016) Mercury enrichments and Hg isotopes in Cretaceous–Paleogene boundary successions: links to volcanism and palaeoenvironmental impacts. Cretaceous Research 66, 60: 45–81. Cretaceous Research *78*, 84–88. doi:10.1016/j.cretres.2017.03.004.

Silva-Tamayo, J.C., 2015. Ca-isotope evidence of ocean acidification along the K–T transition. Prague, Goldschmidt Abstracts, 2899.

Smit, J., 1999. The global stratigraphy of the Cretaceous–Tertiary boundary impact ejecta. Annual Reviews of Earth and Planetary Sciences *17*, 75–113.

Smit, J., Hertogen, J., 1980. An extraterrestrial event at the Cretaceous–Tertiary boundary. Nature *285*, 198–200.

Smit, J., ten Kate, W.G.H.Z., 1982. Trace element patterns at the Cretaceous–Tertiary boundary: consequences of a large impact. Cretaceous Research *3*, 307–332.

Sosa-Montes, C., Rodríguez-Tovar, F.J., Martínez-Ruiz, F., Monaco, P., 2017. Paleoenvironmental conditions across the Cretaceous–Paleogene transition at the Apennines sections (Italy): an integrated geochemical and ichnological approach. Cretaceous Research *71*, 1–13, doi:10.1016/j.cretres.2016.11.005.

Stinnesbeck, W., Keller, G., 1996. Environmental changes across the Cretaceous–Tertiary boundary in northeastern Brazil. In:

MacLeod, N., and Keller, G. (eds.), The Cretaceous–Tertiary Mass Extinction: Biotic and Environmental Effects, 451–470. New York: Norton Press.

Stüben, D., Kramar, U., Harting, M., Stinnesbeck, W., Keller, G. 2005. High-resolution geochemical record of Cretaceous–Tertiary boundary sections in Mexico: new constraints on the K/T and Chicxulub events. Geochimica et Cosmochimica Acta *69*, 2559–2579.

Surlyk, F., 1997. A cool-water carbonate ramp with bryozoans mounds: late Cretaceous–Danian of the Danish Basin. In: James, N.P., and Clarke, J.D.A. (eds.), Cool Water Carbonates: SEPM Special Publication *56*, 293–307. Tulsa, OK: Society for Sedimentary Geology.

Surlyk, F., Damholt, T., Bjerager, M. 2006. Stevns Klint, Denmark: uppermost Maastrichtian chalk, Cretaceous–Tertiary boundary, and lower Danian bryozoans mound complex. Bulletin of the Geological Society of Denmark *54*, 1–48.

Surlyk, F., Rasmussen, S.L., Boussaha, M., Schiøler, P., Schovsbo, N.H., Sheldon, E., Stemmerik, L., Thibault, N., 2013. Upper Campanian-Maastrichtian holostratigraphy of the eastern Danish Basin. Cretaceous Research *46*, 232–256.

Svensen, H., Planke, S., Polozov, A.G., Schmidbauer, N., Corfu, F., Podladchikov, Y.Y., Jamtveit, B., 2009. Siberian gas venting and the end-Permian environmental crisis. Earth and Planetary Science Letters *277*, 490–500.

Swisher, C.C., III, Grajales-Nishimura, J.M., Montanari, A., Margolis, S.V., Claeys, P., Alvarez, W., Renne, P., Cedillo-Pardoa, E., Maurrasse, F.J.-M., Curtis, G.H., Smit, J., McWilliams, M.O., 1992. Coeval ^{40}Ar/^{39}Ar Ages of 65.0 million years ago from Chicxulub Crater melt rock and Cretaceous–Tertiary boundary tektites. Science *257*, 954–958.

Taylor, S.R., McLennan, S.M., 1985. The Continental Crust: Its Composition and Evolution. Carlton: Blackwell Scientific Publication, 312 p.

Tewari, V.C., Lokho, K., Kumar, K., Siddaiah, N.S., 2010a. Late Cretaceous–Paleocene Basin architecture and evolution of the Shillong shelf sedimentation, Meghalaya, Northeast India. Journal of Indian Geological Congress *22*, 61–73.

Tewari, V.C., Kumar, K., Lokho, K., Siddaiah, N.S., 2010b. Lakadong Limestone: Paleocene–Eocene boundary carbonate sedimentation in Meghalaya, northeastern India. Current Science *98*, 88–94.

Thibault, N., Harlou, R., Schovsbo, N.H., Stemmerik, L., Surlyk, F., 2015. Late Cretaceous (Late Campanian-Maastrichtian) sea surface temperature record of the Boreal Chalk Sea. Climate of the Past Discussion *11*, 5049–5071. doi:10.5194/cpd-11-5049-2015.

Thibodeau, A.M., Bergquist, B.A., 2017. Do mercury isotopes record the signature of massive volcanism in marine sedimentary records? Geology *45*, 95–96.

Thibodeau, A.M., Ritterbush, K., Yager, J.A., West, A.J., Ibarra, Y., Bottjer, D.J., Berelson, W.M., Bergquist, B.A., Corsetti, F.A., 2016. Mercury anomalies and the timing of biotic recovery following the end-Triassic mass extinction. Nature Communications *7*, 1–8.

Thierstein, H.R., 1982. Terminal Cretaceous plankton extinctions: a critical assessment. Special Paper Geological Society of America *190*, 385–399.

Towe, K.M., 1991. Aerobic carbon cycling and cerium oxidation: significance for Archean oxygen levels and banded iron-formation deposition. Palaeogeography, Palaeoclimatology, Palaeoecology 97, 113–123.

Tredoux, M., De Wit, M.J., Hart, R.J., Lindsay, M., Verhagen, B., Sellschop, J.P.F., 1989. Chemostratigraphy across the Cretaceous–Tertiary Boundary and a critical assessment of the iridium anomaly. Journal of Geology 97, 585–605.

Tribovillard, N., Algeo, T.J., Baudin, F., Riboulleau, A., 2012. Analysis of marine environmental conditions based on molybdenum-uranium covariation-applications to Mesozoic paleoceanography. Chemical Geology 324/325, 46–58.

Vajda, V., Bercovici, A., 2014. The global vegetation pattern across the Cretaceous–Paleogene mass extinction interval: a template for other extinction events. Global and Planetary Change 122 (2014), 29–49.

Wang, Y.L., Liu, Y.G., Schmitt, R.A., 1986. Rare earth geochemistry of south Atlantic deep sediments, Ce anomaly change at 54 My. Geochimica et Cosmochimica Acta 50, 337–1355.

Wang, S., Yan, W., Chen, Z., Zhang, N., Chen, H., 2014. Rare earth elements in cold seep carbonates from the southwestern Dongsha area, northern South China Sea. Marine and Petroleum Geology 57 482–493.

Wen, H., Fan, F., Zhang, Y., Cloquet, C., Carignan, J., 2015. Reconstruction of early Cambrian ocean chemistry from Mo isotopes. Geochimica et Cosmochimica Acta 164, 1–16.

Wilde, P., Quinby-Hunt, M.S., Erdtmann, B.D., 1996. The whole-rock cerium anomaly: a potential indicator of eustatic sea-level changes in shales of the anoxic events. Sedimentary Geology 101, 43–53.

Wilde, P., Lyons, T.W., Quinby-Hunt, M.S., 2004. Organic carbon proxies in Black shales: molybdenum. Chemical Geology 206, 167–176.

Wright, J., Schrader, H., Holser, W.T., 1987. Paleoredox variations in ancient oceans recorded by rare earth elements in fossil apatite. Geochimica et Cosmochimica Acta 51, 631–644.

Zachos, J.C., Arthur, M.A., 1986. Paleoceanography of the Cretaceous/Tertiary boundary event: inferences from stable isotopic and other data. Paleoceanography 1, 5–26.

Zhou, L., Paul, B., Wignall, P.B., Su, J., Feng, Q., Xie, S., Zhao, L., Huan, J., 2012. U/Mo ratios and $\delta^{98/95}$Mo as local and global redox proxies during mass extinction events. Chemical Geology 324/325, 99–107.

Zoller, W.H., Parrington, J.R., Phelan Korta, J.M., 1983. Iridium enrichment in airborne particles from Kilauea Volcano: January 1983. Science 222, 1118–1121.

Part V
Cenozoic

13

Cenozoic Chemostratigraphy: Understanding the Most Recent Era of the Earth's History

Priyadarsi Debajyoti Roy[1], Muthuvairavasamy Ramkumar[2], and Ramasamy Nagarajan[3]

ABSTRACT

Chemostratigraphy relies on the temporal variations of elemental and isotopic compositions of sedimentary and other rocks for geochemical characterization and correlation. The Cenozoic era is one of the intensively studied time slices of Earth's history. It witnessed dramatic changes in the interactions of hydrosphere-atmosphere-biosphere and resultant modifications in topography, climate, and stratigraphic records. These in turn provided background data for backward-forward models. In this chapter, we reviewed the geological, climatic, and paleobiotic events of the Cenozoic era from the literature, with special emphasis on chemostratigraphic indices and markers, to identify the gaps in our understanding and to suggest future research trends. While the evolution of human species and anthropogenic influences on different spheres are significant across the Holocene-Anthropocene boundary, the changes over longer time scales were controlled by the establishment of orographic barriers through the formation of major mountain chains, opening and closure of ocean basins, alteration of oceanic and atmospheric circulation patterns, different weathering trends, CO_2 degassing as well as sequestration of atmospheric CO_2, variations in volume of ice sheets and resultant sea-level variations, destabilization of gas hydrates, emplacement of large magmatic provinces, volcanic events, and multiple bolide impact events. The secular and anomalous physical and chemical changes have resulted in rapid and extensive evolution of mammals, diversity and geographic sprawl of rainforests, and coastal and offshore biota, including extensive reef systems. This chapter has exemplified all these natural dynamics (extended stratigraphic records, climatic and sea-level trends, biotic lineage and preservation, etc.) well preserved at various spatiotemporal scales in terms of chemostratigraphic markers. Though similar to other time slices, the establishment of synergy between various chemostratigraphic markers and enhancement of resolution during the Cenozoic are still tenuous. An entirely different problem is discrimination of natural and anthropogenic influences on markers, and it requires consideration and consensus.

13.1. INTRODUCTION

Sediments and sedimentary rocks provide useful records about past changes in the provenance, tectonic setting, depositional environmental condition, climate variation, and postdepositional process, and all these changes are deciphered from the mineralogical composition as well as chemical characterization of independent sedimentary units [Ramkumar, 1999]. It has been recognized that stratigraphy is the result of a system consisting of geological setting, climate, and biotic and abiotic processes of sediment production, deposition, and diagenesis. They produce distinct changes in mineralogical-geochemical compositions and provide useful signatures in order to distinguish the changes occurring at different temporal-spatial scales of interest [Ramkumar et al., 2006; Ramkumar, 2014]. Chemostratigraphy has already emerged as an important investigation tool for stratigraphic sequences, and it relies on the elemental and isotopic variations of stratigraphic records in space and time [Ramkumar, 2015b].

[1] Instituto de Geología, Universidad Nacional Autónoma de México, Ciudad de México, México

[2] Department of Geology, Periyar University, Salem, TN, India

[3] Department of Applied Geology, Curtin University, Sarawak, Malaysia

Chemostratigraphy Across Major Chronological Boundaries, Geophysical Monograph 240, First Edition.
Edited by Alcides N. Sial, Claudio Gaucher, Muthuvairavasamy Ramkumar, and Valderez Pinto Ferreira.

The temporal resolution of the sedimentary record is linked to the sediment transport and depositional processes ranging from few seconds (catastrophic events, mass wasting, avalanche, etc.), few hours (tidal cycle), or millions of years. The cycles can be of high frequency (diurnal, seasonal, annual, decadal, centennial, and millennial) or low frequency (orbital). Geochemical compositions are generally unique for stratigraphic changes that occurred at corresponding scales [*Ramkumar and Sathish,* 2007; *Ramkumar,* 2015a]. However, the stacking pattern of these cycles (cycle-in-cycle pattern), omission surfaces, stratigraphic gaps, and diagenetic overprinting can alter the pristine characteristics of the signatures significantly [*Ramkumar and Berner,* 2015]. An ability to characterize the depositional units in terms of unique geochemical signatures would help establish robust chemostratigraphy of the sedimentary record under study and correlate widely separated strata with measurable confidence levels [*Ramkumar et al.,* 2010, 2011]. It also helps to recognize completeness of the stratigraphic record [*Berggren et al.,* 1995; *Ramkumar et al.,* 2013, 2015] and is useful to distinguish lava flows and deposition of volcaniclastic sediments. Thus, chemostratigraphy is not restricted to sedimentary sequences alone, but found its usefulness in other lithologies as well.

The Cenozoic era, spanning the last 66 Ma [*Cohen et al.,* 2013], is one of the intensively studied time slices of Earth's history [*Ravizza and Zachos,* 2003; *Lyle et al.,* 2008 and references therein]. Given cognizance to the extensive studies available in the published literature, we attempt to review the available chronological and chemostratigraphic information in order to enlist the geochemical markers that documented geological, climatic, and paleobiotic events and perturbations, their duration and association, etc. These are then interpreted in terms of gaps in our existing understanding to point toward the future trends. Finally, these are discussed in terms of what has been done pertaining to the chronological boundaries of this era and the potential of chemostratigraphic markers in defining the events and boundaries [*Berggren et al.,* 1995; *Ramkumar,* 2015b].

13.2. CHRONOSTRATIGRAPHY OF THE CENOZOIC ERA

An abrupt boundary separates the Mesozoic era from the Cenozoic era virtually everywhere in global stratigraphic records [*Zachos and Arthur,* 1986; *Molina et al.,* 1998, 2006; *Keller et al.,* 2003, 2009, 2012; *Ravizza and Zachos,* 2003; *Ramkumar et al.,* 2005; *Lyle et al.,* 2008; *Meléndez and Molina,* 2008]. Thenceforth, our world assumed its modern shape over the last 66 Ma. According to the recent publication [*Cohen et al.,* 2013] of the International Commission on Stratigraphy (www.stratigraphy.org),

this time span is divided into three periods, namely, the Paleogene (66–23.03 Ma; comprising the Paleocene, the Eocene, and the Oligocene epochs), the Neogene (23.03–2.59 Ma; comprising the Miocene and the Pliocene epochs), and the Quaternary (last 2.59 Ma; comprising the Pleistocene and the Holocene epochs). The Quaternary was a period of environmental changes greater than any other time over the last 66 Ma [*Bradley,* 2015], and the stratigraphic records of latter parts of this epoch have been influenced by anthropogenic activities. As a result, there have been attempts to introduce a new epoch by subdividing the Holocene into the Holocene and the Anthropocene to mark the advent of anthropogenic impacts on stratigraphic record [*Crutzen,* 2002; *Andersson et al.,* 2005; *Walker et al.,* 2012; *Ruddiman,* 2013; *Monastersky,* 2015; *Waters et al.,* 2016]. Distinction of this new epoch by academic and research community through unique markers and establishment of standard global stratigraphic sections, however, awaits approval from the stratigraphy commission. Enhancement of the geologic time scale allows for better understanding of the Earth system; for example, it helps precise estimations and modeling of rates and feedbacks of climate change, evolution, and tectonic and geochemical processes [*Hinnov and Ogg,* 2007], which in turn provide for near-accurate understanding on deep past and better predictive capabilities on future. In this regard, though unsuccessful, the proposal of *Cramer et al.* [1999] in redefining the Paleocene-Eocene (P-E) boundary was based on the latest Paleocene thermal maximum (LPTM). This event lasted <20 ka and is characterized by negative carbon isotope excursion (CIE) [*Aubry,* 2000] of ~4.7‰ for terrestrial records and ~2.8‰ for marine records [*McInerney and Wing,* 2011]; 1–3‰ negative excursion of oxygen isotope and benthic foraminiferal extinction event in deep marine sedimentary records; lowest occurrences of short-lived and evolutionarily anomalous species of planktonic foraminifera such as *Acarinina africana, Acarinina sibaiyaensis,* and *Morozovella allisonensis;* and a mammalian turnover in terrestrial records. This event marks the Clarkforkian-Wasatchian boundary in the North America and the base of the Sparnacian "stage" in Europe [*Cramer et al.,* 1999]. Similarly, the advent of magnetic stratigraphy and $^{40}Ar/^{39}Ar$ dating of the sedimentary sections located in the North America has improved the resolution as well as our understanding of this era [*Prothero,* 2004]. Detailed stratigraphic studies have now become extremely sophisticated, especially the late Cenozoic [e.g., *Hodell et al.,* 2007], for which the marine stratigraphic record provides excellent undisturbed sections to which many separate techniques were applied (magnetostratigraphy [*Barendregt et al.,* 1996, 1998, 2010; *Barendregt and Duk-Rodkin,* 2004; *Gnibidenko,* 2007; *Dallanave et al.,* 2009, 2012; *Ogg,* 2012], carbon, oxygen, and strontium isotope stratigraphy [*Ravizza and Zachos*, 2003; *Becker et al.*, 2005; *Lyle et al.*, 2008; *Gradstein*, 2012; *Grossman*, 2012; *McArthur et al.*, 2012; *Saltzman and Thomas*, 2012; *Noorbergen et al.*, 2015; *Lacasse et al.*, 2017; *Yavuz et al.*, 2017], refined biostratigraphy [*Beiersdorf et al.*, 1995; *Fejfar et al.*, 1997; *Wade and Olsson*, 2009; *Wade et al.*, 2011; *Gradstein*, 2012; *Vilar et al.*, 2016; *Yavuz et al.*, 2017]), and correlations were carried out with the record of glacio-eustasy (facies and paleoecological changes in the sedimentary record) and the Milankovitch astronomical periodicities [*Miall*, 2010].

13.3. MAJOR GEOLOGICAL EVENTS AND SPATIOTEMPORAL SCALES

While efforts are taken to provide a comprehensive review on major geological events and their impacts on spatial and temporal scales, we are aware that it will not be prudent to present an inventory, given the multitudes of studies available in the literature and the length of the era. Hence, we focused on events that are globally recognizable through geochemical markers, which contributed to the definition of global boundaries within this period [*Cramer et al.*, 1999, 2009; *Kurtz et al.*, 2003; *Ravizza and Zachos*, 2003; *Lyle et al.*, 2008; *Griffith et al.*, 2011; *Hansen et al.*, 2013, 2017].

13.3.1. Tectonic Events

Plate tectonics changed shapes and positions of the ocean basins and continents and modified the oceanic circulation patterns and environmental parameters that influenced the atmospheric phenomena. For example, globe-circling current around the Antarctica developed in the Oligocene when tectonic movements opened critical seaways. Conversely, the Cenozoic closure of the seaways between the North and South America and between Europe and Africa blocked the globe-circling equatorial current that was existent during the late Jurassic and Cretaceous. Similarly, the Alpine orogeny exemplifies the dominant influence of tectonics over climate change caused through alteration of atmospheric circulation and precipitation patterns and therefore modified the topography and sediment influx into the oceans [*Barnes*, 1999]. These changes significantly produced the biotic and geochemical patterns of ensued sedimentary records, thus offering proxies to understand past tectonic, climatic, and biotic interactions and changes across chronostratigraphic boundaries and other time slices.

The Cenozoic era witnessed the northward directed, counterclockwise rotation of Africa, compression and fusing of many microplates of the Mediterranean, transformation of the Eurasian marine archipelago into continent, rise of major mountain chains including the Himalayas and the Alps, subduction of Africa beneath the European plate, closure of the Tethys Ocean, formation and evolution of the Indo-Pacific Ocean and the Mediterranean Sea, and closure of the Paratethys [*Rögl*, 1998; *Popov et al.*, 2004]. Paleogeographic changes also modified the areas subjected to weathering under differing climatic regimes and thus caused variations in cation fluxes through surface runoff. The present configuration of the continents was established over the Cenozoic, with India completing its drift to join the Asian landmass early in this period [*Ramkumar et al.*, 2017]. The development of major montane regions such as the Himalayan-Tibetan and Andean and associated regional uplift through tectonic processes might have enhanced the weathering [*Law et al.*, 2017] and resulted in the release of large volumes of calcium and magnesium into the oceans that later precipitated as marine carbonates [*Boucot and Gray*, 2001]. This enhanced carbonate precipitation during the Cenozoic in oceanic realms suggests the prevalent relationships between extreme climatic variations and mechanical-chemical weathering, tectonics and glaciation, and drawdown of atmospheric CO_2. *Boucot and Gray* [2001] opined that the relationship between the silicate weathering and the area of landmass in tropical latitudes has to be accounted in the global weathering and climatic models. As the influx of Ca and Mg from continental weathering and precipitation of carbonate minerals in the oceanic regimes are related to tectonic reorganization of plates, the large carbon deposits require both higher productivity and topographic and climatic situations conducive for carbon accumulation. The Cenozoic had well-known intervals of higher productivity and conducive climate. The carbon accumulation, however, did not occur in the absence of suitable topographic conditions. In such a case, it is necessary to document and link the periods of dynamic topographies and understand the factors that were responsible for this anomalous phenomenon. In the Eocene (~50 Ma), the Australian Plate began separating from the Antarctica, opening up an oceanic passage through which the Antarctic circumpolar current was established at ~30 Ma. Other changes in plate-ocean configurations were formation of isthmus that separated the Atlantic and Pacific oceans and a seaway that permitted tropical waters into the Arctic Ocean through the western Siberia [*Saltzman*, 2002]. These events of tectonic closings occurred between ~10 and 3.4 Ma.

Emplacement of large igneous provinces (LIPs) with opening of the North Atlantic and associated atmospheric and climatic changes mark the Cenozoic [*Molina*, 2015]. New intra-oceanic subduction zones were initiated over much of the western Pacific at ~52 Ma. They included the Aleutians and a broad stretch of subduction zones extending from Izu-Bonin-Mariana to Fiji and

Tonga-Kermadec. This resulted in the final shift from continental arcs to island arcs in the western Pacific and coincided with the termination of continental arc magmatism in southern Eurasia due to the collision between India and Eurasia. From this point on, the climate cooled considerably, culminating in development of the Antarctic ice sheets during ~35–40 Ma [*Lee et al.*, 2013].

13.3.2. Biotic Events

According to *Markov and Korotayev* [2007], overall shape of the biodiversity curve of the Earth's life history depends on differences in the mean rates of diversity growth in the Paleozoic (low), Mesozoic (moderate), and Cenozoic (high). In comparison, an increase during the Cenozoic was largely due to higher origination rate compared to the Mesozoic through different mechanisms of positive feedback between diversity and growth rate [*Markov and Korotayev*, 2007]. Despite this general contrast, the Cenozoic forms are more complex in terms of domination of detritus feeders and longer food chains. The eras of the geological time scale are characterized by distinct domination of certain fossil groups and separated by major extinction events that paved way for evolution and flourishing of newer fossil groups. For example, brachiopods, trilobites, and graptolites dominated the Paleozoic and the Mesozoic by ammonites, belemnites, marine reptiles, and dinosaurs. Similarly, mammals and molluskan groups such as the bivalves and gastropods dominated the Cenozoic [*Sessa et al.*, 2012]. Though the Cretaceous-Cenozoic global biodiversity trend was thought to be a time slice of major global diversity, it became questionable when new methodologies for tabulating, evaluating, and standardizing fossil collections showed that only a moderate rise occurred at all spatial scales. This observation has not thwarted [*Sessa et al.*, 2012] to opine that the ecological reorganization and innovation during the late Mesozoic through early Cenozoic was no less dramatic than the diversity drop at the Cretaceous-Paleogene (K-Pg) boundary. *Hallam and Wignall* [1999] opined that one of the most striking extinction events in the Cenozoic took place in the latest Paleocene, with extinction of 50% of deepwater benthic foraminifera, and this event has been widely recognized at bathyal-abyssal depths across the world. Based on the analysis of diversity trends of brachiopods, *Ruban* [2009] reported low turnover during the entire Phanerozoic, slightly stronger during the early Paleozoic, and close to zero during the Cenozoic, meaning stabilization of diversity structure of brachiopods. The P-E boundary also marks the commencement of mammal dispersion; evolution and diversity of larger mammals; rearrangement of geographic distribution of fauna and flora; expansion of terrestrial flora, especially rainforest

up to 50–60° toward poles; emplacement of large igneous-volcanic provinces with opening of the North Atlantic; and closure of the Tethys.

The ecosystems usually had positive feedback to the ancient climatic extremes. For example, the middle Mesozoic-Cenozoic radiation was generally associated with a greenhouse-dominated climatic regime. Warm climate facilitated the biodiversification and radiation by fostering greater population size and thus increased the extinction resistance [*Zhong-Qiang et al.*, 2014]. While comparing and enlisting the biotic events and extinctions that occurred during the Mesozoic and the Cenozoic, *Benton and Harper* [2009] stated that extinctions since the K-Pg event have been more modest in scope. The Eocene-Oligocene events at ~34 Ma were marked by extinctions among plankton and open-water bony fishes in the sea and by a major turnover among mammals in Europe and in the North America. These were followed by a dramatic extinction among mammals in the North America during the middle Oligocene and minor losses of plankton in the middle Miocene, but neither event was large. Planktonic extinctions occurred during the Pliocene due to the disappearances of bivalves and gastropods in tropical seas. As the great ice sheets withdrew from Europe and the North America, large mammals such as mammoths, mastodons, woolly rhinos, and giant ground sloths died at the end of the Pleistocene [*Benton and Harper*, 2009]. Both the climate and human-induced changes contributed to the extinction. However, the loss of these mammal species was negligible (<1%) and thus may not qualify to be considered as an extinction event.

Based on the dominance of photoautotrophic organisms in carbonate production and their abundance, *Schlager* [2005] stated that light had the most important control on skeletal carbonate precipitation during the Cenozoic. The role of phytoplankton and their effect on the marine ecosystem and global climate are reasonably understood throughout the Mesozoic and Cenozoic because dinoflagellates, diatoms, coccolithophorids, and phytoplankton groups that dominated during these eras are still present today. The worldwide rise of predatory gastropods during the early Cenozoic has been well documented [*Sessa et al.*, 2012]. Assemblages of the latest Cretaceous were dominated by suspension feeders, during the initial 0.5 Ma duration, and thenceforth the deposit-feeding mode became dominant. The early Paleocene settings were repopulated largely by species descended from the Cretaceous genera at time scales of 20–50 Ma. The number of dasycladalean algal species fluctuates within a wide range of 0–100, with a maximum occurring at the Paleocene, followed by the Permian and the early Cretaceous principally under the control of a combination of biological and environmental factors, in addition to the conditions of preservation [*Aguirre and*

Riding, 2005]. *Aguirre and Riding* [2005] also recorded rapid diversification during the Paleocene to the middle Eocene reaching a maximum of 39 genera, followed by a decline during the Eocene. From the Oligocene to the Pliocene, the decrease continued, reaching minima of 4–5 genera. The Cenozoic reef history was set in an icehouse world. It succeeded the warm, greenhouse interval of the Cretaceous and expanded the Cretaceous reef belts. Different tropical ecosystems emerged and were quick to regenerate following disturbances. Corals also formed associations in reefs with coralline algae and produced distinctive reef ecosystems on a global scale. Throughout much of the Cenozoic, sea grasses too have played a major role in trapping, stabilizing, and producing carbonate sediments. The calcareous algae genus *Halimeda* was rare in the early Cretaceous and became more common in the late Cretaceous. It did not become widely distributed until early in the Cenozoic. It emerged as a major sediment producer and reef builder only in the late Miocene-Pliocene interval [*Stanley et al.*, 2010]. Mangroves have similarly influenced sediment accumulation since the Miocene through the baffling action of their roots [*Reading*, 1996]. The rapid expansion of grasses into dryland and well-drained upland environments during the Cenozoic had a profound effect on fluvial styles, especially in ephemeral systems as it would baffle overland flows, and possibly enhanced rill development [*Strömberg*, 2011]. The origin and diversification of grasses in the Cenozoic might have supplemented the enhancement of weathering [*Boucot and Gray*, 2001].

A pulse of evolutionary turnover in terrestrial and marine biota happened during the East Antarctic ice sheet expansion, and it was associated with ~60 m sea-level drop as evidenced by changes in planktonic and benthic foraminiferal assemblages [*Tian et al.*, 2010]. Both surface and deep ocean circulation systems changed or reorganized with the southern component water production increasing during the middle Miocene climatic transition. A stepwise sea surface cooling of 6–7 °C and a deep ocean cooling of 2–3 °C were observed in high latitudes of the southwest Pacific during the middle Miocene climatic transition. This abrupt cooling event is marked by a global event of ~1‰ change in benthic foraminiferal $\delta^{18}O$ at ~13.8 Ma [*Tian et al.*, 2010]. The major climatic cooling of the past 30 Ma was associated with massive spread of grasslands on most continents and a matching rise in diversity of large grazing mammals, perhaps especially the ungulate mammals [*Benton*, 2013]. Based on the compilation of data on paleotemperature trends and diversity patterns of calcareous planktons for the past 230 Ma, *Prokoph et al.* [2004] stated that the Cenozoic shows a perfect coupling between ocean temperature and sea-level fall with that of diversity of calcareous nanoplankton. In addition to the influences of dramatic

climatic and environmental changes on speciation, diversification, dwindling, and population explosion events, the Cenozoic also depicted direct influence on the size of the organisms [*Ramkumar and Menier*, 2017]. According to *Schmidt et al.* [2004], the size of planktic foraminifera showed three major episodes of dwarfing during the last 70 Ma. *Benton and Harper* [2009] opined that the Cenozoic planktic foraminifera provide strong support for a stationary model of evolutionary change, with size changes being strongly correlated with extrinsic factors such as fluctuations in latitudinal and surface water temperature gradients [*Schmidt et al.*, 2004]. These authors documented sharp changes in sizes of low-latitude planktic taxa during 65–42 Ma (sharp decrease), 42–12 Ma (moderate decrease), and 12 Ma–present (gigantism). They also recorded gigantism during the intervals of global cooling (Eocene and Neogene) associated with latitudinal and temperature gradients, high diversity, and changes in size in association with periods of varying productivity (Paleocene and Oligocene). Similarly, the Cenozoic explosion of silicifying diatoms severely limited the abundance of silica for sponges and radiolarians [*Lazarus et al.*, 2009].

Prothero [2004] was of the opinion that the relationships between climate and organisms such as planktonic and benthic microfossils, benthic mollusks and echinoids, land plants, land snails, and reptiles and amphibians were well established. However, there are few discordant relationships such as the North American land mammal extinctions and major bolide impacts as well as major volcanic events during the Cenozoic. According to him, the absence of any relationship between land mammal extinction events and the bolide impacts (http://www.unb.ca/passc/ImpactDatabase) during this period is intriguing. Similarly, the eruptions of the North Atlantic Paleogene volcanic provinces at 61 and 56 Ma were documented in the middle and late Paleocene. They did not correspond to significant extinctions of either mammals or any other group of organisms. The Ethiopian and Yemeni traps, which were presumed to be responsible for the Eocene-Oligocene extinctions, were dated between 29.5 and 31 Ma, or the middle of the late Oligocene, when there were no extinctions of consequence in land mammals. Examination of four largest climatic events over the past 50 Ma (the 37 Ma cooling event at the end of the middle Eocene, the early Oligocene refrigeration at 33 Ma, the expansion of C4 grasslands at 7.5 Ma, and the glacial-interglacial cycles of the Pleistocene) found no association between the detailed records of these well-documented climatic changes and any significant extinctions [*Prothero*, 2004]. In addition, the mammalian faunas showed almost no change through all four of these intervals and showed much more faunal turnover at times when there was no evidence of climatic change.

Ravizza and Peucker-Ehrenbrink [2003] analyzed the osmium (Os) isotope trends of bulk sediments from the South Atlantic, Equatorial Pacific, and Italy to document a pronounced minimum in $^{187}Os/^{188}Os$ (0.22–0.27) in the late Eocene (34.5–34 Ma) and a subsequent rapid increase in $^{187}Os/^{188}Os$ (~32 Ma). They interpreted these variations as a result of ultramafic weathering event and an increased influx of extraterrestrial particles to the Earth. In addition to the Duchesnean-Chadronian (middle-late Eocene) and middle Orellan (early Oligocene) events, there were other important climatic events in the Cenozoic that might have caused faunal change in mammals. For example, the expansion of C4 grasslands at 7.5 Ma at almost all middle latitudes produced a dramatic isotopic signal in both tooth enamel and in soil carbonates of the late Miocene terrestrial records in the North America, Pakistan, South America, and East Africa.

All these parameters also influenced the marine habitat [*Ramkumar and Menier*, 2017]. Relatively few studies have tracked both the diversity and abundance structure of assemblages through the late Mesozoic to early Cenozoic at a finer stratigraphic scale or evaluated the influence of habitat on these trends [*Sessa et al.*, 2012]. Lower dasycladalean diversity was coincidental with higher sea level during 70–100 Ma, whereas the maxima accompanied falling sea level during the Selandian-Thanetian. Similar dwindling of genera was observed throughout the remainder of the Cenozoic [*Aguirre and Riding*, 2005].

This chapter on biotic evolution, extinction, diversity, and distribution during the Cenozoic unequivocally indicates:

1. Significant changes in the biosphere structure, distribution, and diversity

2. Continuation of the late Mesozoic trends for the extrinsic process-induced perturbations

3. Dynamic control over species-genera-family by the climatic-oceanographic parameters enforced by evolving continental-scale tectonic and landscape dynamics

4. Published/available studies restricted either to specific events, faunistic or floristic groups, or any specific location/chronological record

5. Necessity to understand biotic-climatic-oceano-graphic-tectonic evolution on a variety of scales

13.3.3. Climatic Events

The Phanerozoic paleotemperature history unravels three major glaciations during the Ordovician, Permo-Carboniferous, and late Cenozoic. Four significant warm events occurred during the late Cretaceous, Eocene, early Pliocene, and Holocene [*Saltzman*, 2002]. This observation suggests that the warmer events were more frequent during the Cenozoic. Studies over the last three decades

relate the change of skeletal carbonates from predominantly calcite to metastable carbonates from the Mesozoic to the Cenozoic with changes in oceanic and atmospheric chemistry, Mg/Ca ratio of seawater, dominant biotic composition, and vital effect of organisms [*Van de Poel and Schlager*, 1994; *Ravizza and Zachos*, 2003]. *Pomar and Hallock* [2008] reasoned that decline in atmospheric pCO_2 through the Cretaceous and scarce habitats for the neritic lime mud precipitation were some of the possible causes. A peak in oceanic Ca^{2+} concentrations at the same time promoted calcification through biotic control. The influences of temperature and Mg/Ca ratio in marine environments were also demonstrated by many studies [*Stanley and Hardie*, 1998, 1999; *Lear et al.*, 2000; *Ravizza and Zachos*, 2003]. Low-Mg calcite precipitation was promoted during the icehouse conditions, and precipitation of high-Mg calcite and metastable aragonite occurred during the increased greenhouse conditions. The algae *Halimeda* became a major sediment producer during the latter part of the Cenozoic, and it resulted from substantial rise of Mg/Ca ratio of the seawater [*Stanley et al.*, 2010]. Mg/Ca ratio of seawater also influenced the mineralogy, calcification, and primary production [*Stanley et al.*, 2010]. The oceans played a critical role in the chemical balance of the atmospheric system, particularly with respect to the atmospheric carbon dioxide levels. Because the oceans contain very large quantities of CO_2 in solution, even a small change in the oceanic CO_2 balance may have profound consequences for the radiation balance of the atmosphere and hence the climate [*Bradley*, 2015]. Across the K-Pg boundary, decrease in Ca^{2+} and pCO_2 and increase in global temperature gradients (i.e., high-latitude and deepwater cooling) enforced adoptative changes, extinctions, and turnovers in biotic realm. All these strategies efficiently linked photosynthesis and calcification and promoted successive changes of the dominant skeletal factory: larger benthic foraminifers (protist-protist symbiosis) during the Paleogene, red algae during the Miocene, and modern coral reefs (metazoan-protist symbiosis) since the late Miocene [*Pomar and Hallock*, 2008].

Stable isotopes extracted from foraminifer tests have provided valuable data on ancient sea temperatures through the Mesozoic and Cenozoic [*Van de Poel and Schlager*, 1994; *Morse et al.*, 1997]. In addition to the extreme global warmth during the Paleocene-Eocene Thermal Maximum (PETM) (~56 Ma), three more hyperthermal events occurred during the P-E [*Higgins and Schrag*, 2006]. Soon after the PETM, the climate remained unstable during the early-middle Eocene (including at the South Pole), and it was characterized by low latitudinal temperature gradient [*Zhong-Qiang et al.*, 2014]. The Climate Leaf Analysis Multivariate Program (CLAMP) experiment involving morphologic and ecologic analysis of the Antarctic region led to the proposition that the region experienced warm humid temperate climate with strong seasonality in temperature and precipitation [*Pomar and Hallock*, 2008]. This has also challenged the climatic models and improved our understanding on latitudinal heat transfer during increased greenhouse conditions [*Zhong-Qiang et al.*, 2014]. The stratigraphic records of the Cenozoic era preserve evidences for the Milankovitch cycles, in terms of sediment packages and also geochemical markers. *Boulila et al.* [2011] opined that high-resolution $\delta^{18}O$ records for the late Cenozoic icehouse (younger than 33.8 Ma) climatic variations mirror the precession, obliquity, and eccentricity cycles. The 41 ka cycle of obliquity forcing was the primary driver of glacial variability. The climatic, oceanographic, and environmental changes that occurred before, during, and after PETM were explained by *Molina* [2015] and *McInerney and Wing* [2011]. According to *Molina* [2015], the North Atlantic rifting generated volcanic activity and caused a global warming by injecting CO_2 into the atmosphere. This study also stated that the Indian rifting modified global marine currents from a thermohaline circulation system to a halothermal circulation by restricting the Tethys current and increased the seawater salinity. This increased the temperature in marine depths and destabilized methane hydrates. Climatic reversal-sea level change and followed large amount of methane gas and CO_2 released into the atmosphere, and strongly negative excursion of carbon isotopes ($\delta^{13}C$) represented the sedimentary records of the late Maastrichtian [*Ramkumar et al.*, 2005], and similar events can be surmised for PETM as well. Thus, methane and carbon dioxide might have produced an intense greenhouse effect and a sudden increase in temperature, giving rise to the PETM.

The warm Cretaceous climate continued into the early Paleogene with a distinct optimum near the P-E boundary (PETM) and the early Eocene. A gradual decrease in temperature during the late Eocene culminated in the formation of the first ice sheets in Antarctica around the Eocene-Oligocene boundary. The late Cenozoic global cooling was probably evidenced by three major expansions of the polar ice sheets, that is, the southern hemisphere ice sheet of Antarctica at ~34 Ma, the East Antarctic ice sheet expansion at ~13.8 Ma, and the northern hemisphere ice sheet at ~2.7 Ma. The onset of glaciation occurred first in Antarctica close to the Eocene-Oligocene boundary (~34 Ma). Major Arctic glaciation possibly initiated in the late Pliocene (~2.7 Ma). The middle Miocene East Antarctic ice sheet expansion marked the middle Miocene climatic transition during 14.2–13.8 Ma, and it was a major step in the Cenozoic climate evolution [*Tian et al.*, 2010]. A warming trend began during the late Oligocene and continued into the middle Miocene with a climax at the middle Miocene climatic optimum. Another similar trend was observed during the early Pliocene, and it reflected warming until 3.2 Ma.

The climatic extremes of the Cenozoic are relatively better studied as they are closer to the present-day analogs than those recorded in the deep past [*Zhong-Qiang et al.*, 2014]. Anthropogenic effects on the atmosphere, biosphere, and continental hydrology constitute an uncertain force comparable to the natural geologic forces that shaped the Earth's history. Our understanding of climatic variation during the most recent era of Earth's history is necessary to disentangle the underlying natural factors involved in climate change as well as to understand how environmental conditions may change in the future Anthropocene [*Bradley*, 2015].

13.3.4. Relative Sea-Level Trends

Relative changes in sea level altered the oceanographic circulation patterns, oxic-anoxic conditions, stability and movement of calcite compensation depths, dominant mineralogy and their precipitation patterns, and magnitude of global events such as Messinian salinity crisis [*Barnes*, 1999; *Bradley*, 2015; *Molina*, 2015]. New tectonic reconstructions of the seafloor allow quantification of the Cretaceous and the Cenozoic seafloor bathymetry changes on the eustatic sea level [*Conrad*, 2013]. Recently published sea-level curves for the late Mesozoic and the Paleogene (third- to fourth-order cycles) are primarily based on paleobathymetric reconstructions and backstripping data, as well as $\delta^{18}O$ records. Many of these can be correlated regionally and globally. It suggests that most of the observed sea-level fluctuations were indeed eustatic and not a result of local tectonics. Supplementary evidences for glacio-eustatic origin of these sea-level cycles were provided by the higher rates (~10–60 m/Ma) of sea-level change, and no other known mechanism could produce such rapid and continuous rise and fall in the sea level [*Conrad*, 2013].

Haq et al. [1988] reconstructed the relative sea-level cycles during the Cenozoic (Fig. 13.1). Major buildups of continental ice sheets occurred during the middle Cenozoic to the present, and durations of these sea-level fluctuations, however, are poorly constrained [*Reading*, 1996]. Global sea levels were lower in the Cenozoic compared to the expansive Middle and Late Cretaceous epicontinental seas, and it could have been driven by a combination of decreasing spreading rates and aging ocean basins [*Seton et al.*, 2009]. Reduction in sprawl of epicontinental seas over the Cenozoic perhaps was a result of cooling trend until the commencement of PETM [*Melott and Bambach*, 2011]. Stratigraphy, paleontology, outcrop geology, and seismic stratigraphy provide evidences that relatively rapid 1–10 Ma "third-order" cycles and even more rapid cycles of relative sea-level changes

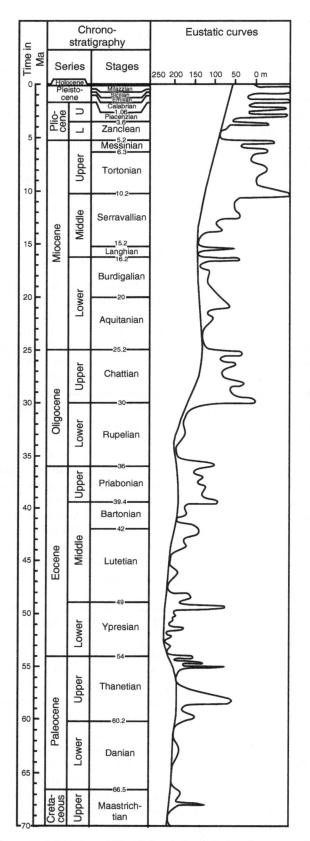

Figure 13.1 Relative sea-level fluctuations during the Cenozoic era. Modified after *Haq et al.* [1988].

occurred during the early Cenozoic. Glacial evidence came through ice-rafted debris in sediments from both southern and northern polar regions, and it dated back to 45 Ma and possibly to 58 Ma in Spitsbergen [*Dawber and Tripati*, 2011]. Conditions were warmer and the global sea level rose by more than 100 m during the PETM [*Segev et al.*, 2011]. Continental-scale glaciation was established in the Antarctica by 34 Ma, and ice sheets were established in northern high latitudes by 23 Ma. Several significant paleoceanographic events can be related to the global cooling [*Tian et al.*, 2010]. The global sea level also fluctuated during the middle Miocene. *Conrad* [2013] suggested that the sea-level variations of hundreds of meters cannot be explained by climatic effects alone as the Earth's present-day ice sheets contain the seawater equivalent to 64 m of sea level and an ice-free world should, however, exhibit an average of ~45 m eustatic sea level higher than present levels when isostatically compensated. Similarly, the seawater cooling by ~12 °C can lead to additional ~12 m sea-level drop, totaling ~57 m of sea-level fall. Variations in excess of ~57 m sea level during the late Cretaceous, and Cenozoic, must have resulted from geological changes in the "containing" volume of the ocean basins. According to *Conrad* [2013], the sea-level trends of the Quaternary also record a climate-induced glaciation history accompanied by spatial variations in sea level associated with the solid Earth's elastic and viscous responses to loading. The estimated sea-level drop associated with the East Antarctica ice sheet expansion (~60 m; *Tian et al.*, 2010) was large enough to facilitate growth of a part of ice sheet at this time. It is generally agreed that high-frequency fluctuations in the range of ~90–120, 60–80, and ~30–60 m during the Oligocene-Pliocene, Pliocene–early Pleistocene, and late Pleistocene icehouse periods were caused by freezing and thawing of polar ice sheets. Nevertheless, amplitudes of the Pliocene–Pleistocene sea-level fluctuations were unprecedented compared to those of the remaining part of the Cenozoic and Mesozoic eras [*Sømme et al.*, 2009]. The geographic extents of land-sea distribution were not proportional. For example, the relative proportions of land and sea changed little during the Quaternary on a global scale in spite of sea-level changes due to the growth and decay of continental ice sheets. The ocean area decreased by only 3% and land-surface area increased about 10% with the sea level 120–130 m below its present level.

The Cenozoic era witnessed relative sea-level variations, perhaps at an unprecedented magnitude and durations. However, this observation was possibly due to the presence and study of well-preserved stratigraphic records in the most recent era of Earth's history. Nevertheless, this does not preclude the uniqueness of dramatic changes that took place during this era and also the influences that it exerted on topographic evolution, source-sink interactions, and mineralogical-geochemical pathways and ultimately on the biosphere.

13.4. DISCUSSION

13.4.1. Chemostratigraphic Markers

According to the Cretaceous-Paleogene (K-Pg) Working Group, the boundary between the Mesozoic and Cenozoic eras is characterized by major planktonic foraminiferal extinction, first appearance datum of the Cenozoic species, and a major lithologic change characterized by the presence of a thin red layer at base of the boundary clay showing various geochemical anomalies [Canudo et al., 1991]. In addition, calcitic oceans of the Mesozoic were transformed into high-Mg and aragonite-dominated skeletal carbonate-rich seas as a function of glaciation, temperature, oceanic-atmospheric chemistry, and secular variation of seawater composition (i.e., Mg/Ca). Based on an overall increase in the rate of carbonate accumulation by severalfolds over the past 100 Ma and a shift in the locus of accumulation from continents to oceans, Schlager [2005] commented that the planktonic foraminifers and coccolithophorids progressively replaced shallow-water benthos in precipitating carbonate from the ocean. Dominant reef builders and sediment producers secreted/precipitated low-Mg calcite in calcite seas and aragonite or high-Mg calcite in aragonite seas throughout the Phanerozoic eon [Stanley et al., 2010]. Mg/Ca ratio of seawater exerted a strong influence over the carbonate biomineralization. Production and preservation trends of silica also changed significantly during the Cenozoic [Nilsen et al., 2003]. The sedimentation rates during the late Cenozoic were significantly higher than in the early Cenozoic as a consequence of the extensive glaciations [Schlager, 2005]. Coral reefs were scarce and larger benthic foraminifers were the dominant shallower-water carbonate producers during the early Cenozoic. The glacial facies were widespread across all the northern continents, on and around the Antarctica, and in the Andes. The ocean depths became anoxic or hypoxic, and the calcite compensation level rose by several hundred meters. These conditions promoted deposition of a thick layer of shale at the P-E boundary [Molina, 2015]. The locus of biogenic silica deposition shifted from the Atlantic Ocean to the North Pacific and the Antarctic oceans during the middle Miocene climatic transition. Large sedimentary deposits of organic carbon and phosphate occurred both in the marginal seas and in the open oceans, and they are termed as Monterey carbon excursion events [Tian et al., 2010]. These events were characterized by $\delta^{13}C$ maxima from ~17 to ~13.5 Ma and reflected episodic changes in organic carbon deposition relative to the carbonates.

The P-E boundary coincided with the PETM, and it showed negative excursion of $\delta^{13}C$ (Fig. 13.2b), a major crisis in smaller abyssal and bathyal foraminifera, and multiple environmental changes like decreased carbonate saturation, massive thermal enhancement, ocean acidification, lower oxygen levels, and globally reduced food supply and a massive carbon injection [McInerney and Wing, 2011; Molina, 2015]. Prominent negative CIEs recorded in marine (−3.5 to −4‰) and terrestrial (−5 to −6‰) carbonates were near synchronous with the PETM warming, and it indicated massive release of isotopically lighter carbon (4000–7000 GT; Dickens et al., 1995] into the ocean-atmosphere system [Jones et al., 2010]. The abrupt greenhouse gas-induced global warming caused a major disruption to the carbon cycle [Zhong-Qiang et al., 2014]. However, the structure and magnitudes of CIE during the PETM are not fully understood [Zhang et al., 2017]. The PETM was characterized by an increase of 5–6°C in temperature (Fig. 13.2a; Zachos et al., 2001, 2003), and it was associated with rise in sea surface temperature [Kennett and Stott, 1991], rise of about 2 km in the calcite compensation depth [Zachos et al., 2005], deep ocean acidification, widespread oceanic anoxia [Zhong-Qiang et al., 2014], massive release of methane hydrate from sediments on the continental slope [Dickens et al., 1995], turbiditic oxidation, global-scale conflagration of peatlands [Kurtz et al., 2003], and excessive accumulation of organic carbon. Another potential mechanism for the rapid release of large amounts of CO_2 was isolation of a large epicontinental seaway by tectonic uplift and associated volcanism or continental collision, followed by desiccation and bacterial respiration of the aerated organic matter [Higgins and Schrag, 2006]. The contact metamorphism associated with intrusion of LIPs into organic-rich sediments caused production of thermogenic CH_4 and CO_2 [Svensen et al., 2004] and accelerated continental weathering [Zhong-Qiang et al., 2014]. LIPs like the Andean and Deccan volcanic events were of brief duration (~2 Ma), and its frequency was higher during the K-Pg than the mid-Cenozoic. However, the frequencies were still far too low to sustain continuous and prolonged greenhouse conditions before, during, and after the PETM [Lee et al., 2013]. A dense vegetation cover is effective at protecting the bedrock and its overlying regolith from erosion. Even the steep mountain slopes can be effectively stabilized by plants [Nichols, 2009] thwarting erosion and sediment supply to streams that may in the long term affect the geochemical cycles. Hence, the intervals of extensive development of land plants and swamps during the Cenozoic need consideration in establishing geochemical trends [Nichols, 2009].

A long-term positive carbon excursion during the middle Miocene was coincident with warmer climate and higher sea level before the transition to the Neogene

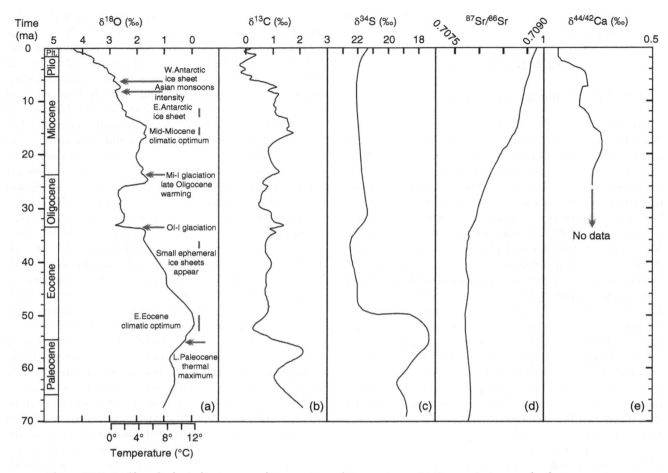

Figure 13.2 Profiles of selected isotopes and temperature of Cenozoic era. (a) Oxygen isotope and paleotemperature (After *Zachos et al.* [2001]). (b) Carbon isotope (After *Zachos et al.* [2001]). (c) Sulfur isotopic composition of Cenozoic seawater from marine barites (After *Paytan et al.* [2004]). (d) Strontium isotope (After *McArthur et al.* [2012]). (e) Ca isotopic composition of marine carbonates and phosphates (After *Farkas et al.* [2007]).

glacial-interglacial climatic mode and cryospheric circulation system [*Holbourn et al.,* 2004]. The benthic foraminiferal δ¹⁸O curves reflect extreme glacio-eustatic sea-level fluctuations, climate, and reorganization of oceanic circulation patterns during this interval (Fig. 13.2a). The mechanisms, timing, and duration of these events, however, are not yet understood completely. It was similar to another warm interval during the early Paleogene (65–41 Ma) that lacks understanding about carbon cycling, productivity, and export between systems [*Faul et al.,* 2003]. The general cooling trend of deep water and high-latitude seas and development of the Antarctic ice sheets during the Eocene-Oligocene transition (~34 Ma) were demarcated by proxies such as planktonic and benthic foraminiferal oxygen isotope compositions [*Nilsen et al.,* 2003]. Similarly, the δ¹³C of bulk calcium carbonate (CaCO₃) as high as 4‰ at around ~57 Ma with duration of 2–4 Ma represented the largest sustained isotopic excursion of the Cenozoic. The Paleocene δ¹³C maximum

was interpreted as an increase in absolute or relative organic carbon burial and export productivity.

Studies involving documentation of the changes in seawater ⁸⁷Sr/⁸⁶Sr provided information about roles of increasing physical and chemical weathering rates of crustal rocks, altering climatic patterns due to orographic effects, metamorphism and unroofing of relatively old crust, riverine influx into the ocean, silicate weathering and associated atmospheric drawdown of CO₂, and role and timing of uplift of the Himalayas and Tibetan Plateau regions during the Cenozoic [*Kashiwagi and Shikazono,* 2003; *Banner,* 2004; *Brand et al.,* 2004; *Pattanaik et al.,* 2013]. The strontium isotopes changed unidirectionally at a rate of 40×10^{-6} Ma⁻¹ in response to riverine input from newly uplifted mountain chains such as the Himalayas, Alps, Andes, and Cordillera (Fig. 13.2d; *Barnes,* 1999; *Ebneth et al.,* 2001; *McArthur et al.,* 2012). However, *Jacobson and Kaufman* [1999] opined that the Cenozoic trend is more influenced by continental glaciations than the heightened erosion enforced by orogeny. While discussing the secular trends of strontium isotopes, *Vollstaedt et al.* [2014] stated that the $^{87}Sr/^{86}Sr$ isotope ratio of seawater is not sensitive to temporal changes in output flux of marine strontium, primarily controlled by burial of calcium carbonate ($CaCO_3$) at the ocean floor. The stable isotopes of Sr are sensitive to the imbalances of oceanic Sr budget. Despite the widespread utility, there are discrepancies between modeled and observed seawater $^{87}Sr/^{86}Sr$ ratios [*Vollstaedt et al., 2014*]. These discrepancies were cumulative results of enhancement of weathering rates after deglaciations and tectonic uplifts and associated incongruent weathering of silicates, enhanced weathering of island arcs, release of Sr from riverine particulate matter, and uncertain estimates of the low-temperature alteration of the oceanic crust and groundwater discharge. Based on these observations, *Vollstaedt et al.* [2014] concluded that radiogenic Sr isotope systems provide information about Sr input fluxes into the ocean and do not quantify the output fluxes. Data of this Sr calibration curve were obtained from analysis of foraminiferal calcite collected from the DSDP/ODP sites [*McArthur et al., 2012*].

13.4.2. Cenozoic Era: A Continuum of the Mesozoic Trends

Changes in the global climate and continental collisions continued across the Cenozoic without much change from the late Mesozoic. The interval comprising the late Mesozoic to early Cenozoic was of significant biologic turnover and ecologic reorganization within marine assemblages, but the timing and causes of these changes remain poorly understood [*Sessa et al., 2012*]. Progressive continental reshuffling and associated changes in oceanic circulation played a large role in disrupting biotic distributions. Events such as breakup of the Gondwana also influenced the climate [*Stilwell, 2003*]. During the Paleocene, the southern Gondwana continued to split apart; the Indian plate was moving toward Asia and Africa was moving north toward Europe. The increase in total carbonate accumulation over the past 100 Ma was probably related to deposition of pelagic sediment on ocean crust and therefore more rapidly recycled than the carbonate rocks on the continents. *Mackenzie and Morse* [1992] supported this assertion and stated that the Cenozoic trend was essentially an extension of the late Cretaceous trend characterized by decline in cratonic carbonate deposition and its transportation to the deep sea. Replacement of Jurassic shallow-water calcareous organisms by planktonic shelled marine protists and a decline of sea level since the late Mesozoic decreasing the areas of warm carbonate-rich waters were the two reasons for the shift of principal loci of carbonate accumulation from continental to deep ocean [*Mackenzie and Morse, 1992*]. Continuation of the

Mesozoic biotic and therefore climatic and environmental trends could be gauged from a simple fact that the rainforests expanded and occupied widest geographic extent during the Cenozoic. However, it all started with diversification of angiosperms (that in turn marks the onset of profound changes in floras dominated by ferns, conifers, and cycads) in the early Cretaceous and angiosperm-dominated ecosystems in the Cenozoic era [*Heimhofer et al., 2005*]. Similarly, the PETM was exemplified by many biota. For example, the calcareous nanoplankton first appeared during the late Triassic, increased in abundance and diversity through the Jurassic and Cretaceous, and was severely affected by the K-Pg mass extinction. However, it subsequently radiated during the early Paleogene and remained a major component of the calcifying plankton throughout the Cenozoic. It is extremely abundant in surface waters of the modern oceans. Compilation of new and published data on paleotemperatures trends of the Cenozoic by *Cramer et al.* [2009] and *Friedrich et al.* [2012] suggested that the deep oceans started to cool, except for the two greenhouse intervals during the early Eocene (47–55 Ma) as well as the early to middle Miocene (15–20 Ma). Both of them had duration comparable with the middle Cretaceous thermal maximum.

Following the mass extinction across the K-Pg transition, another interval of exceptionally low coral diversity reduced carbonate deposition, and a virtual absence of tropical reefs occurred during the Paleocene [*Stanley, 2003*]. Oxygen isotope compositions of carbonate-rich deep-sea sediments provide a long-term perspective of deepwater temperature and global ice volume [*Bradley, 2015*]. In addition, the oxygen isotopic patterns also indicate a combination of factors including, but not limited to, long-term changes in continental position, ocean circulation, mountain-building episodes, atmospheric composition, and position of Earth's orbit relative to the Sun. High-resolution studies of carbon and oxygen isotopes documented extreme global warmth and an abrupt negative CIE during the PETM as well as three more hyperthermal events during the Paleocene and the Eocene [*Higgins and Schrag, 2006*]. Paleotemperature proxy data indicated a rapid surface ocean warming of ~5–9 °C within 1–10 ka during the PETM. *Scheibner and Speijer* [2008] stated that the PETM itself was part of the long-term warming trend that commenced at the early Paleogene. A shallower carbonate compensation depth allowed less deposition and preservation of carbonate in pelagic environments from the middle Cenozoic to the present. Pelagic carbonate deposition would be a consequence of higher inputs of CO_2 into the exogenic system rather than the driver of higher inputs into the exogenic system. In this regard, the early Paleogene cooling might have caused waning of continental arc magmatism of the late Cretaceous in the North America. A last spurt of CO_2 production possibly interrupted the cooling and gave rise to the early Eocene climatic optimum at 52 Ma. However, the overall trend in the Cenozoic appears to be that of cooling [*Lee et al.,* 2013]. Nevertheless, as presented in previous sections, consensus on the cooling trend yet eludes the researchers and warrants further investigation.

13.5. CONCLUSIONS

The Cenozoic era offers a unique opportunity to study climate variation, biodiversity, relative sea level, and topographic and geochemical evolution of the continental and oceanic systems in relatively higher resolution [*Ramkumar,* 2015b]. A review of available literature presents chemostratigraphic markers of tectonic, biotic, and climatic events along with relative sea-level trends and the associated forcings. The most recent time slice of Earth's history offers information to understand older deposits as well as the present and future trends. It can be summed up that the Cenozoic geodynamic changes in landscapes and environments were further amplified by drastic climate changes. The warm climate of the Cretaceous continued into the early Paleogene with a distinct optimum near the Paleocene-Eocene boundary and the early Eocene. A gradual decrease in temperature during the late Eocene culminated in formation of the first ice sheets in Antarctica around the Eocene-Oligocene boundary. A renewed warming trend that began during the late Oligocene continued into the middle Miocene with a climax at the middle Miocene climatic optimum. The turning point at around 14.2 Ma led to the onset of the middle Miocene climate transition indicated by the cooling of surface waters and the expansion of the East Antarctic ice sheet. A gentle warming during the early Pliocene at 3.2 Ma reflected the trend reversal. It is also revealed that except for the K-Pg perturbation, the biotic, environmental, climatic, and ecological trends of the Cenozoic were a continuation of the Mesozoic trends, which, in turn, exemplified the role of the Cenozoic stratigraphic records as a window to the pre-Cenozoic. Biodiversity of the Cenozoic biota was studied in terms of several subdivisions owing to the greater stratigraphic resolution: early and late Paleocene; early, middle, and late Eocene; early and late Oligocene; early, middle, and late Miocene; Pliocene; Pleistocene; and Holocene. Our review presents the geological, climatic, and paleobiotic events and the potential chemostratigraphic recognition according to these subdivisions.

ACKNOWLEDGMENTS

Authors thank the editors of this special volume on chemostratigraphy of major boundaries, especially Prof. A.N. Sial, for the invitation to submit this chapter about one of the interesting time slices of Earth's history. We thank the anonymous reviewers, Editor-in-Chief Prof. A.N. Sial, and handling Editor Prof. Valderez Ferreira for the constructive suggestions and comments, which helped us to see through the reader's perspective and make amends in the manuscript accordingly. This review would not have been possible without the predecessors who toiled hard to document various facets of the Cenozoic.

REFERENCES

Aguirre, J. and Riding, R.. 2005. Dasycladalean algal biodiversity compared with global variations in temperature and sea level over the past 350 Myr. Palaios *20*, 581–588.

Andersson, A.J., Mackenzie, F.T. and Lerman, A., 2005. Coastal ocean and carbonate systems in the high CO_2 world of the Anthropocene. American Journal of Science *305*, 875–918.

Aubry, M.P., 2000. Where should the Global Stratotype Section and Point (GSSP) for the Paleocene/Eocene boundary be located? Bulletin de la Société Géologique de France *171*, 461–76.

Banner, J.L., 2004. Radiogenic isotopes: systematics and applications to earth surface processes and chemical stratigraphy. Earth-Science Reviews *65*, 141–194.

Barendregt, R.W. and Duk-Rodkin, A., 2004. Chronology and extent of Late Cenozoic ice sheets in North America: a magnetostratigraphic assessment. In: Ehlers, J. and Gibbard, P.L. (Eds.), Quaternary Glaciations: Extent and Chronology. Part II: North America. Developments in Quaternary Science 2, series Editor Jim Rose, Elsevier, Amsterdam, pp. 1–7.

Barendregt, R.W., Enkin, R.J., Duk-Rodkin, A. and Baker, J., 1996. Paleomagnetic evidence for Late Cenozoic glaciations in the Mackenzie Mountains of the Northwest Territories, Canada. Canadian Journal of Earth Sciences *33*(6), 896–903.

Barendregt, RW., Vincent, J.S., Irving, E. and Baker, J., 1998. Magnetostratigraphy of Quaternary and Late Tertiary sediments on Banks Island, Canadian Arctic Archipelago. Canadian Journal of Earth Sciences *35*(2), 147–161.

Barendregt, R.W., Enkin, R.J., Duk-Rodkin, A. and Baker, J., 2010. Paleomagnetic evidence for multiple Late Cenozoic glaciations in the Tintina Trench, west-central Yukon, Canada. Canadian Journal of Earth Sciences *47*(7), 987–1002. doi:10.1139/E10-021.

Barnes, C.R., 1999. Paleoceanography and paleoclimatology: an Earth system perspective. Chemical Geology *161*, 17–35.

Becker, J., Lourens, L.J., Hilgen, F.J., Van der Laan, E., Kouwenhoven, T.J. and Reichart, G.J., 2005. Late Pliocene climate variability on Milankovitch to millennial time scales: a high-resolution study of MIS 100 from the Mediterranean. Palaeogeography, Palaeoclimatology, Palaeoecology *228*, 338–360. doi:10.1016/j.palaeo.2005.06.020.

Beiersdorf, H., Bickert, T., Cepek, P., Fenner, J., Petersen, N., Schijnfeld, J., Weiss, W. and Won, M.Z., 1995. High-resolution stratigraphy and the response of biota to Late Cenozoic environmental changes in the central equatorial Pacific Ocean (Manihiki Plateau). Marine Geology *125*, 29–59.

Benton, M.J., 2013. Origins of biodiversity. Palaeontology *56*, 1–7.

Benton, M.J. and Harper, D.A.T., 2009. Introduction to Paleobiology and the Fossil Record. Blackwell Publishing, Chicester, 592p.

Berggren, W.A., Kent, D.V., Swisher, C.C., III, and Aubry, M.P., 1995. A revised Cenozoic geochronology and chronostratigraphy. In: Berggren, W.A., Kent, D.V., Aubry, M.P., and Hardenbol, J. (Eds.), Geochronology, Time Scales and Global Stratigraphic Correlation: A Unified Temporal Framework for an Historical Geology. SEPM Special Publications, *54*, SEPM (Society for Sedimentary Geology), Tulsa, OK, pp. 129–212.

Boucot, A.J. and Gray, J., 2001. A critique of Phanerozoic climatic models involving changes in the CO content of the atmosphere. Earth-Science Reviews *56*, 1–159.

Boulila, S., Galbrun, B., Miller, K.G., Pekar, S.F., Browning, J.V., Laskar, J. and Wright, J.D., 2011. On the origin of Cenozoic and Mesozoic "third-order" eustatic sequences. Earth-Science Reviews *109*, 94–112.

Bradley, R.S., 2015. Paleoclimatology: Reconstructing Climates of the Quaternary, III Edition. Elsevier, Amsterdam. 675p.

Brand, U., Legrand-Blain, M. and Streel, M., 2004. Biochemostratigraphy of the Devonian-Carboniferous boundary global stratotype section and point, Griotte Formation, La Serre, Montagne Noire, France. Palaeogeography, Palaeoclimatology, Palaeoecology *205*, 337–357.

Canudo, J.I., Keller, G. and Molina, E., 1991. Cretaceous/tertiary boundary extinction pattern and faunal turnover at Agost and Caravaca, S.E. Spain. Marine Micropalaeontology *17*, 319–341.

Cohen, K.M., Finney, S.C., Gibbard, P.L. and Fan, J.-X., 2013. The ICS international chronostratigraphic chart. Episodes *36*, 199–204.

Conrad, C.P., 2013. The solid Earth's influence on sea level. Geological Society of America Bulletin *125*, 1027–1052.

Cramer, B.S., Aubry, M., Miller, K.G., Olsson, R.K., Wright, J.D. and Kent, D.V., 1999. An exceptional chronologic, isotopic, and clay mineralogic record of the latest Paleocene thermal maximum, Bass River, NJ, ODP 174AX. Bulletin of Geological Society of France *170*, 883–897.

Cramer, B.S., Toggweiler, J.R., Wright, J.D., Katz, M.E. and Miller, K.G., 2009. Ocean overturning since the Late Cretaceous: inferences from a new benthic foraminiferal isotope compilation. Paleoceanography *24*, PA4216. doi:10.1029/2008PA001683.

Crutzen, P.J., 2002. Geology of mankind. Nature *415*, 23.

Dallanave, E., Agnini, C., Muttoni, G. and Rio, D., 2009. Magneto-biostratigraphy of the Cicogna section (Italy): implications for the late Paleocene–early Eocene time scale. Earth and Planetary Science Letters *285*, 39–51. doi:10.1016/j.epsl.2009.05.033.

Dallanave, E., Agnini, C., Muttoni, G. and Rio, D., 2012. Paleocene magneto-biostratigraphy and climate-controlled rock magnetism from the Belluno Basin, Tethys Ocean, Italy. Palaeogeography, Palaeoclimatology, Palaeoecology *337–338*, 130–142.

Dawber, C. and Tripati, A., 2011. Constraints on glaciation in the middle Eocene (46–37 Ma) from Ocean drilling Program (ODP) Site 1209 in the tropical Pacific Ocean. Paleoceanography *26*, PA2208. doi:10.1029/2010PA002037.

Dickens, G.R., O'Neil, J.R., Rea, D.K. and Owen, R.M., 1995. Dissociation of oceanic methane hydrate as a cause of the carbon isotope excursion at the end of the Paleocene. Paleoceanography *10*, 965–971.

Ebneth, S., Shields, G., Veizer, J., Miller, J.F. and Shergold, J.H., 2001. High-resolution strontium isotope stratigraphy across the Cambrian-Ordovician transition. Geochimica et Cosmochimica Acta *65*, 2273–2292.

Farkas, J., Böhm, F., Wallmann, K., Blenkisop, J., Eisenhauer, A., van Geldern, R., Munnecke, A., Voigt, S. and Veizer J., 2007. Calcium isotope record of Phanerozoic oceans: implications for chemical evolution of seawater and its causative mechanisms. Geochemica et Cosmochimica Acta, *71*, 5117–5134.

Faul, K.L., Anderson, D. and Delaney, M.L., 2003. Cretaceous and early Paleogene nutrient and paleoproductivity Late records from Blake Nose, western North Atlantic Ocean. Paleoceanography *18*, 1042. doi:10.1029/2001PA000722.

Fejfar, O., Heinrich, W.-D., Pevzner, M.A. and Vangengeim, E.A., 1997. Late Cenozoic sequences of mammalian sites in Eurasia: an updated correlation. Palaeogeography, Palaeoclimatology, Palaeoecology *133*, 259–288.

Friedrich, O., Norris, R.D. and Erbacher, J., 2012. Evolution of middle to Late Cretaceous oceans: a 55 m.y. record of Earth's temperature and carbon cycle. Geology *40*, 107–110.

Gnibidenko, Z.N., 2007. Late Cenozoic paleomagnetism of West Siberian Plate. Russian Geology and Geophysics *48*, 337–348.

Gradstein, F.M., 2012. Biochronology. In: Gradstein, F., Ogg, J.G., Schmitz, M.D., and Ogg, G.M. (Eds.), The Geological Time Scale2012. Elsevier BV, Amsterdam, pp. 43–61. doi:10.1016/B978-0-444-59425-9.00003-2.

Griffith, E.M., Paytan, A., Eisenhauer, A., Bullen, T.D. and Thomas, E., 2011. Seawater calcium isotope ratios across the Eocene-Oligocene transition. Geology *39*(7), 683–686. doi:10.1130/G31872.

Grossman, E.L., 2012. Oxygen isotope stratigraphy. In: Gradstein, F.M., Ogg, J.G., Schmitz, M.D., and Ogg, G.M. (Eds.), The Geological Time Scale 2012. Elsevier BV, Amsterdam, pp. 181–206. doi:10.1016/B978-0-444-59425-9.00010-X.

Hallam, A. and Wignall, P.B., 1999. Mass extinctions and sea-level changes. Earth-Science Reviews *48*, 217–250.

Hansen, J., Sato, M., Russell, G. and Kharecha, P., 2013. Climate sensitivity, sea level and atmospheric carbon dioxide. Philosophical Transactions of the Royal Society A *371*, 20120294. doi:10.1098/rsta.2012.0294.

Hansen, J., Sato, M., Kharecha, P., von Schuckmann, K., Beerling, D.J., Cao, J., Marcott, S., Masson-Delmotte, V., Prather, M.J., Rohling, E.J., Shakun, J., Smith, P., Lacis, A., Russell, G. and Ruedy, R., 2017. Young people's burden: requirement of negative CO_2 emissions. Earth System Dynamics *8*, 577–616. doi:10.5194/esd-8-577-2017.

Haq, B.U., Hardenbol, J. and Vail, P.R., 1988. Mesozoic and Cenozoic chronostratigraphy and cycles of sea-level change. In: Wilgus, C.K., Hastings, B.S., Kendall, C.S., Posamentier, H.W., Ross, C.A., and Van Wagoner, J.C. (Eds.), Sea-Level Changes: An Integrated Approach. SEPM Special Publication *42*,

Society of Economic Paleontologists and Mineralogists, Tulsa, OK, pp. 71–108.

Heimhofer, U., Hochuli, P.A., Burla, S.S., Dinis, J.M.L. and Weissert, H., 2005. Timing of Early Cretaceous angiosperm diversification and possible links to major paleoenvironmental change. Geology 33, 141–144.

Higgins, J.A. and Schrag, D.P., 2006. Beyond methane: towards a theory for the Paleocene–Eocene thermal maximum. Earth and Planetary Science Letters 245, 523–537.

Hinnov, L.A. and Ogg, J.G., 2007. Cyclostratigraphy and the astronomical time scale. Stratigraphy 4, 239–251.

Hodell, D.A., Kamenov, G.D., Hathorne, E.C., Zachos, J.C., Rohl, U. and Westerhold T., 2007. Variations in the strontium isotope composition of seawater during the Paleocene and early Eocene from ODP Leg 208 (Walvis Ridge), Geochemistry Geophysics Geosystems 8, Q09001. doi: 10.1029/2007GC001607.

Holbourn, A., Kuhnt, W., Simo, J.A.T. and Li, Q., 2004. Middle Miocene isotope stratigraphy and paleoceanographic evolution of the northwest and southwest Australian margins(Wombat Plateau and Great Australian Bight). Palaeogeography, Palaeoclimatology, Palaeoecology 208, 1–22. doi:10.1016/j.palaeo.2004.02.003.

Jacobson, S.B. and Kaufman, A.J., 1999. The Sr, C and O isotopic evolution of Neoproterozoic seawater. Chemical Geology 161, 37–57.

Jones, D., Ridgwell, A., Lunt, D.J. and Maslin, M.A., 2010. A Palaeogene perspective on climate sensitivity and methane hydrate instability. Philosophical Transactions of the Royal Society 368, 2395–2415.

Kashiwagi, H. and Shikazono, N., 2003. Climate change during Cenozoic inferred from global carbon cycle model including igneous and hydrothermal activities. Palaeogeography, Palaeoclimatology, Palaeoecology 199, 167–185.

Keller, G., Stinnesbeck, W., Adatte, T. and Stüben, D., 2003. Multiple impacts across the Cretaceous-Tertiary boundary. Earth Science Review 62, 327–363.

Keller, G., Abramovich, S., Berner, Z. and Adatte, T., 2009. Biotic Effects of the Chicxulub Impact, K-T catastrophe and sea level change in Texas. Palaeogeography, Palaeoclimatology, Palaeoecology 271, 52–68.

Keller, G., Adatte, T., Bhowmick, P., Upadhyay, H., Dave, A., Reddy, A.N. and Jaiprakash, B.C., 2012. Nature and timing of extinctions in Cretaceous-Tertiary planktic foraminifera preserved in Deccan intertrappean sediments of the Krishna–Godavari Basin, India. Earth and Planetary Science Letters 341–344, 211–221.

Kennett, J.P. and Stott, L.D., 1991. Abrupt deep-sea warming, paleoceanographic changes and benthic extinctions at the end of the Paleocene. Nature 353, 225–228.

Kurtz, A.C., Kump, L.R., Arthur, M.A., Zachos, J.C. and Payton, A., 2003. Early Cenozoic decoupling of the global carbon and sulfur cycles. Paleoceanography 18(4), 1090. doi:10.1029/2003PA000908.

Lacasse, C.M., Santos, R.V., Dantas, E.L., Vigneron, Q., de Sousa, I.M.C., Harlamov, V., Lisniowski, M.A., Pessanha, I.B.M., Frazao, E.P. and Cavalcanti J.A.D., 2017. $^{87}Sr/^{86}Sr$ dating and preliminary interpretation of magnetic susceptibility logs of giant piston cores from the Rio Grande Rise in the South Atlantic. Journal of South American Earth Sciences 80, 244–254. doi:10.1016/j.jsames.2017.09.034.

Law, R.D., Thigpen, J.R., Merschat, A.J. and Stowell, H.H., 2017. Linkages and feedbacks in orogenic systems. GSA Memoir 213, 374 p. doi:10.1130/MEM213.

Lazarus, D.B., Kotrc, B., Wulf, G. and Schmidt, D.N., 2009. Radiolarians decreased silicification as an evolutionary response to reduced Cenozoic ocean silica availability. Proceedings of the National Academy of Sciences of the United States of America 106, 9333–9338.

Lear, C.H., Elderfield, H. and Wilson, P.A., 2000. Cenozoic deep sea temperatures and global ice volume from Mg/Ca in benthic foraminiferal calcite. Science 287, 269–272.

Lee, C.A., Shen, B., Slotnick, B.S., Liao, K., Dickens, G.R., Yokoyama, Y., Lenardic, A., Dasgupta, R., Jellinek, M., Lackey, J.S., Schneider, T. and Tice, M.M., 2013. Continental arc–island arc fluctuations, growth of crustal carbonates, and long-term climate change. Geosphere 9, 21–36.

Lyle, M., Barron, J., Bralower, T.J., Huber, M., Olivarez Lyle, A., Ravelo, A.C., Rea, D.K. and Wilson, P.A.. 2008. Pacific Ocean and Cenozoic evolution of climate. Review in Geophysics 46, RG2002. doi:10.1029/2005RG000190.

Mackenzie, F.T. and Morse, J.W., 1992. Sedimentary carbonates through Phanerozoic time. Geochimica et Cosmochimica Acta 56, 3281–3295.

Markov, A.V. and Korotayev, A.V., 2007. Phanerozoic marine biodiversity follows a hyperbolic trend. Palaeoworld 16, 311–318.

McArthur, J.M., Howarth, R.J. and Shields, G.A., 2012. Strontium isotope stratigraphy. In: Gradstein, F.M., Ogg, J.G., Schmitz, M., and Ogg, G. (Eds.), The Geological Time Scale 2012. Elsevier BV, Amsterdam, pp. 127–144. doi:10.1016/B978-0-444-59425-9.00007-X.

McInerney, F.A. and Wing, S.L., 2011. Thermal maximum: a perturbation of carbon cycle, climate, and biosphere with implications for the future. Annual Review of Earth and Planetary Sciences 39, 489–516.

Meléndez, A. and Molina, E., 2008. The Cretaceous-Tertiary (KT) boundary. In: García-Cortés, A., Villar, J.A., Suarez-Velgrade, J.P., and Salvador Gonzalez, C.I. (Eds.), Contextos Geológicos Españoles. Publicaciones del Instituto Geológico y Minero de España, Madrid, pp. 107–113.

Melott, A.L. and Bambach, R.K., 2011. A ubiquitous, 62-Myr periodic fluctuation superimposed on general trends in fossil biodiversity. II. Evolutionary dynamics associated with periodic fluctuation in marine diversity. Paleobiology 37, 383–408.

Miall, A.D., 2010. The Geology of Stratigraphic Sequences, II Edition. Springer-Verlag, Heidelberg, 522p.

Molina, E., 2015. Evidence and causes of the main extinction events in the Paleogene based on extinction and survival patterns of foraminifera. Earth-Science Reviews 140, 166–181.

Molina, E., Arenillas, I. and Arz J.A., 1998. Mass extinction in planktonic foraminifera at the Cretaceous/Tertiary boundary in subtropical to temperate latitudes. Bulletin de la Société Géologique de France 169, 351–372.

Molina, E., Alegret, L., Arenillas, I., Arz, J.A., Gallala, N., Hardenbol, J., von Salis, K., Steurbaut, E., Vandenberghe, N. and Zaghbibturki, D., 2006. The Global Boundary Stratotype

Section and Point for the base of the Danian Stage (Paleocene, Paleogene, "Tertiary", Cenozoic) at El Kef, Tunisia: original definition and revision. Episodes *29*(4), 263–273.

Monastersky, R., 2015. The human age. Nature *519*, 144–147.

Morse, J.W., Wang, Q. and Tsio, M.Y., 1997. Influences of temperature and Mg:Ca ratio on $CaCO_3$ precipitates from seawater. Geology *25*, 85–87.

Nichols, G., 2009. Sedimentology and Stratigraphy, II Edition. Wiley-Blackwell, Oxford, 419p.

Nilsen, E., Anderson, L. and Delaney, M., 2003. Paleo-productivity, nutrient burial, climate change and the carbon cycle in the western equatorial Atlantic across the Eocene/Oligocene boundary. Paleoceanography *18*(3), 1057. doi:10.1029/2002PA000804.

Noorbergen, L.J., Lourens, L.J., Munsterman, D.K. and Verreussel, R.M.C.H., 2015. Stable isotope stratigraphy of the early Quaternary of borehole Noordwijk, southern North Sea. Quaternary International *386*, 148–157. doi:10.1016/j.quaint.2015.02.045.

Ogg, J.G., 2012. Geomagnetic polarity time scale. In: Gradstein, F., Ogg, J.G., Schmitz, M.D., and Ogg, G.M. (Eds.), The Geological Time Scale. Elsevier BV, Amsterdam, pp. 85–113. doi:10.1016/B978-0-444-59425-9.00005-6.

Pattanaik, J.K., Balakrishnan, S., Bhutani, R. and Singh, P., 2013. Estimation of weathering rates and CO_2 drawdown based on solute load: significance of granulites and gneisses dominated weathering in the Kaveri River Basin, Southern India. Geochimica et Cosmochimica Acta *121*, 611–636.

Paytan, A., Kastner, M., Campbell, D. and Thiemens, M.H., 2004. Seawater sulfur isotope fluctuations in the Cretaceous. Science *304*, 1663–1665.

Pomar, L. and Hallock, P., 2008. Carbonate factories: a conundrum in sedimentary geology. Earth-Science Reviews *87*, 134–169.

Popov, S.V., Rögl, F., Rozanov, A.Y., Steininger, F.F., Shcherba, I. G. and Kováč, M., 2004. Lithological-Paleogeographic maps of Paratethys. 10 Maps Late Eocene to Pliocene. Courier Forschungsinstitut Senckenberg *250*, 1–46.

Prokoph, A., Rampino, M.R. and El Bilali, H., 2004. Periodic components in the diversity of calcareous plankton and geological events over the past 230 Myr. Palaeogeography, Palaeoclimatology, Palaeoecology *207*, 105–125.

Prothero, D.R., 2004. Did impacts, volcanic eruptions, or climate change affect mammalian evolution? Palaeogeography, Palaeoclimatology, Palaeoecology *214*, 283–294.

Ramkumar, M., 1999. Role of chemostratigraphic technique in reservoir characterisation and global stratigraphic correlation. Indian Journal of Geochemistry *14*, 33–45.

Ramkumar, M., 2014. Characterization of depositional units for stratigraphic correlation, petroleum exploration and reservoir characterization. In: Sinha, S. (Ed.), Advances in Petroleum Engineering. Studium Press L.L.C., Houston, pp. 1–13.

Ramkumar, M., 2015a. Discrimination of tectonic dynamism, quiescence, and third order relative sea level cycles of the Cauvery Basin, South India. Annales Géologiques de la Péninsule Balkanique *76*, 19–45.

Ramkumar, M., 2015b. Toward standardization of terminologies and recognition of chemostratigraphy as a formal stratigraphic method. In: Ramkumar, M. (Ed.), Chemostratigraphy: Concepts, Techniques and Applications. Elsevier, Amsterdam, pp. 1–21. doi:10.1016/B978-0-12-419968-2.00001-7.

Ramkumar, M. and Sathish, G., 2007. Integrated sequence and chemostratigraphic modelling: a sure-fire technique for stratigraphic correlation, petroleum exploration and reservoir characterization. In: Rajendran, S., Srinivasamoorty, K. and Aravindan, S. (Eds.), Mineral Exploration: Recent Strategies. New India Publishers, Delhi, pp. 21–40.

Ramkumar, M. and Berner, Z., 2015. Temporal trends of geochemistry, relative sea level and source area weathering in the Barremian-Danian strata of the Cauvery Basin, South India. In: Ramkumar, M. (Ed.), Chemostratigraphy: Concepts, Techniques and Applications. Elsevier, Amsterdam, pp. 273–308. doi:10.1016/B978-0-12-419968-2.00011-X.

Ramkumar, M. and Menier, D., 2017. Eustasy, High-Frequency Sea Level Cycles and Habitat Heterogeneity. Elsevier, Cambridge, MA, 91p.

Ramkumar, M., Harting, M. and Stüben, D., 2005. Barium anomaly preceding K/T boundary: plausible causes and implications on end Cretaceous events of K/T sections in Cauvery Basin (India), Israel, NE-Mexico and Guatemala. International Journal Earth Sciences *94*, 475–489.

Ramkumar, M., Stüben, D. and Berner, Z., 2006. Elemental interrelationships and depositional controls of Barremian-Danian strata of the Cauvery Basin, South India: implications on scales of chemostratigraphic modelling. Indian Journal of Geochemistry *21*, 341–367.

Ramkumar, M., Stüben, D. and Berner, Z., 2010. Hierarchical delineation and multivariate statistical discrimination of chemozones of the Cauvery Basin, South India: implications on Spatio-temporal scales of stratigraphic correlation. Petroleum Science *7*, 435–447.

Ramkumar, M., Stüben, D. and Berner, Z., 2011. Barremian-Danian chemostratigraphic sequences of the Cauvery Basin, South India: implications on scales of stratigraphic correlation. Gondwana Research *19*, 291–309.

Ramkumar, M., Alberti, M., Fürsich, F.T. and Pandey, D.K., 2013. Depositional and diagenetic environments of the Dhosa Oolite Member (Oxfordian), Kachchh Basin, India, based on Petrographic data: implications on the origin and occurrence of ooids and their correlation with global oolite peak. In: Ramkumar, M. (Ed.), On a Sustainable Future of the Earth's Natural Resources. Springer-Verlag, Heidelberg, pp.179–230.

Ramkumar, M., Alberti, M. and Fürsich, F.T., 2015. Chemostratigraphy of the Dhosa Oolite Member (Oxfordian), Kachchh Basin, western India: implications for completeness of the stratigraphic record and correlation with global oolite peak. In: Ramkumar, M. (Ed.), Chemostratigraphy: Concepts, Techniques and Applications. Elsevier, Amsterdam, pp. 309–340. doi:10.1016/B978-0-12-419968-2.00012-1.

Ramkumar, M., Menier, D., Manoj, M.J., Santosh, M. and Siddiqui, N.A., 2017. Early Cenozoic rapid flight enigma of the Indian subcontinent resolved: roles of topographic top loading and subcrustal erosion. Geoscience Frontiers *8*, 15–23.

Ravizza, G. and Peucker-Ehrenbrink, B., 2003. The marine $^{187}Os/^{188}Os$ record of the Eocene-Oligocene transition: the interplay of weathering and glaciations. Earth and Planetary Science Letters *210*, 151–165.

Ravizza, G.E. and Zachos, J.C., 2003. Records of Cenozoic ocean chemistry. In: Holland, H.D. and Turekian, K.K. (Eds.), Treatise on Geochemistry. Elsevier, Amsterdam, pp. 551–581. doi:10.1016/B0-08-043751-6/06121-1.

Reading, H.G., 1996. Sedimentary Environments: Processes, Facies, and Stratigraphy, III Edition, Blackwell Science, Cambridge, MA, 688p.

Rögl, F., 1998. Paleogeographic considerations for Mediterranean and Paratethys seaways (Oligocene to Miocene). Annalen Naturhistorisches Museum Wien 99A, 279–310.

Ruban, D.A., 2009. Phanerozoic changes in the high-rank suprageneric diversity structure of brachiopods: linear and non-linear effects. Palaeoworld 18, 263–277.

Ruddiman, W.F., 2013. The Anthropocene. Annual Reviews on Earth and Planetary Sciences 41, 45–68.

Saltzman, B., 2002. Dynamical Paleoclimatology: Generalized Theory of Global Climate Change. Academic Press, San Diego, 354p.

Saltzman, M.R. and Thomas, E., 2012. Carbon isotope stratigraphy. In: Gradstein, F., Ogg, J.G., Schmitz, M.D., and Ogg, G.M. (Eds.), The Geological Time Scale 2012. Elsevier BV, Amsterdam, pp. 207–232.

Scheibner, C. and Speijer, R.P., 2008. Late Paleocene–Early Eocene Tethyan carbonate platform evolution: a response to long- and short-term paleoclimatic change. Earth-Science Reviews 90, 71–102.

Schlager, W., 2005. Carbonate sedimentology and sequence stratigraphy. Concepts in Sedimentology and Paleontology 8, doi:10.2110/csp.05.08.

Schmidt, D.N., Thierstein, H.R. and Bollmann, J., 2004. The evolutionary history of size variation of planktic foraminiferal assemblages in the Cenozoic. doi:10.1594/PANGAEA. 694693,Supplement to: Schmidt, DN et al. (2004): the evolutionary history of size variation of planktic foraminiferal assemblages in the Cenozoic. Palaeogeography, Palaeoclimatology, Palaeoecology, 212, 159–180.

Segev, A., Schattner, U. and Lyakhovsky, V., 2011. Middle-Late Eocene structure of the southern Levant continental margin: tectonic motion versus global sea-level change. Tectonophysics 499, 165–177.

Sessa, J.A., Bralower, T.J., Patzkowsky, M.E., Handley, J.C. and Ivany, L.C., 2012. Environmental and biological controls on the diversity and ecology of Late Cretaceous through Early Paleogene marine ecosystems in the U.S. Gulf Coastal Plain. Paleobiology 38, 218–239.

Seton, M., Gaina, C., Muller, R.D. and Heine, C., 2009. Mid-Cretaceous seafloor spreading pulse: fact or fiction? Geology 37, 687–690.

Sømme, T., Helland-Hansen, H. and Granjeon, D., 2009. Impact of eustatic amplitude variations on shelf morphology, sediment dispersal, and sequence stratigraphic interpretation: Icehouse versus greenhouse systems. Geology 37, 587–590.

Stanley, G.D., Jr., 2003. The evolution of modern corals and their early history. Earth-Science Reviews 60, 195–225.

Stanley, S.M. and Hardie, L.A., 1998. Secular oscillations in the carbonate mineralogy of reef-building and sediment-producing organisms driven by tectonically forced shifts in seawater chemistry. Palaeogeography, Palaeoclimatology, Palaeoecology 144, 3–19.

Stanley, S.M. and Hardie, L.A., 1999. Hypercalcification: paleontology links plate tectonics and geochemistry to sedimentology. GSA Today 9, 2–7.

Stanley, S.M., Ries, J.B. and Hardie, L.A., 2010. Increased production of calcite and slower growth for the major sediment-producing algae Halimeda as the Mg/Ca ratio of sea water is lowered to a "calcite sea" level. Journal of Sedimentary Research 80, 6–16.

Stilwell, J.D., 2003. Patterns of biodiversity and faunal rebound following the K/T boundary extinction event in Austral Palaeocene molluscan faunas. Palaeogeography, Palaeoclimatology, Palaeoecology 195, 319–356.

Strömberg, C.A.E., 2011. Evolution of grasses and grassland ecosystems. Annual Review of Earth and Planetary Sciences 39, 517–544.

Svensen, H., Planke, S., Malthe-Sorenssen, A., Jamtyeit, B., Myklebust, R. Eidem, T.R. and Rey, S.S., 2004. Release of methane from a volcanic basin as a mechanism for initial Eocene global warming. Nature 429, 542–545.

Tian, J., Yang, M., Lyle, M.W., Wilkens, R. and Shackford, J.K., 2010. Obliquity and long eccentricity pacing of the Middle Miocene climate transition. Geochemistry, Geophysics, Geosystems 14, 1740–1755.

Van de Poel, H.M. and Schlager, W., 1994. Variations in Mesozoic–Cenozoic skeletal carbonate mineralogy. Geologie en Mijnbouw 73, 31–51.

Vilar I.C., Flynn, L. J. and Ostende, L.W.V.D.H., 2016. Windows into deep time: Cenozoic faunal change in long continental records of Eurasia. Comptes Rendus Palevol 15, 753–762.

Vollstaedt, H., Eisenhauer, A., Wallmann, K., Böhm, F., Fietzke, J., Liebetrau, V., Krabbenhöft, A., Frkas, J., Tomasŏvych, A., Raddatz, J. and Veizer, J., 2014. The Phanerozoic $\delta^{88/86}Sr$ record of seawater: new constraints on past changes in oceanic carbonate fluxes. Geochimica et Cosmochimica Acta 128, 249–265.

Wade, B.S. and Olsson, R.K., 2009. Investigation of pre-extinction dwarfing in Cenozoic planktonic foraminifera. Palaeogeography, Palaeoclimatology, Palaeoecology 284, 39–46.

Wade, B.S., Pearson, P.N., Berggren, W.A. and Pälike, H., 2011. Review and revision of Cenozoic tropical planktonic foraminiferal biostratigraphy and calibration to the geomagnetic polarity and astronomical time scale. Earth Science Reviews 104, 111–142. doi:10.1016/j.earscirev.2010.09.003.

Walker, M.J.C., Berkelhammer, M., Björck, S., Cwynar, L.C., Fisher, D.A., Long, A.J., Lowe, J.J., Newnham, R.M., Rasmussen S.O. and Weiss, H., 2012. Formal subdivision of the Holocene Series/Epoch: a Discussion Paper by a Working Group of INTIMATE (Integration of ice-core, marine and terrestrial records) and the Subcommission on Quaternary Stratigraphy (International Commission on Stratigraphy). Journal of Quaternary Science 27(7), 649–659. doi:10.1002/jqs.2565.

Waters, C. N., Zalasiewicz, J., Summerhayes, C., Barnosky, A.D., Poirier, C., Gałuszka, A., Cearreta, A., Edgeworth, M., Ellis, E.C., Ellis, M., Jeandel, C., Leinfelder, R., McNeill, J.R., Richter, D.B., Steffen, W., Syvitski, J., Vidas, D., Wagreich, M., Williams, M., Zhisheng, A., Grinevald, J., Odada, E., Oreskes, N. and Wolfe, A.P., 2016. The Anthropocene is functionally and stratigraphically distinct from the Holocene. Science *351*(6269), 137. doi:10.1126/science.aad2622.

Yavuz, N., Culha, G., Demirer, Ş.S., Utescher, T. and Aydın, A., 2017. Pollen, ostracod and stable isotope records of palaeoenvironment and climate: upper Miocene and Pliocene of the Çankırı Basin (Central Anatolia, Turkey). Palaeogeography, Palaeoclimatology, Palaeoecology *467*, 149–165. doi:10.1016/j.palaeo.2016.04.023.

Zachos, J.C. and Arthur, M.A., 1986. Paleoceanography of the Cretaceous/Tertiary boundary event: inferences from stable isotope and other data. Palaeoceanography *1*, 5–26.

Zachos, J., Pagani, M., Sloan, L., Thomas, E. and Billups, K., 2001. Trends, rhythms, and aberrations in global climate 65 Ma to present. Science *292*, 686–693

Zachos, J.C., Wara, M.W., Bohaty, S., Delaney, M.L., Petrizzo, M.R., Brill, A., Bralower, T.J. and Premoli-Silva, I., 2003. A transient rise in tropical sea surface temperature during the Paleocene–Eocene Thermal Maximum. Science *302*, 1551–1554.

Zachos, J.C., Rohl, U., Schellenberg, S.A., Hodell, D., Thomas, E., Sluijs, A., Kelly, C., McCarren, H., Kroon, D., Raffi, I., Lourens, L.J. and Nicolo, M., 2005. Rapid acidification of the ocean during the Paleocene–Eocene thermal maximum. Science *308*, 1611–1615.

Zhong-Qiang, C., Joachimski, J., Montañez, I. and Isbell, J., 2014. Deep time climatic and environmental extremes and ecosystem response: an introduction. Gondwana Research *25*, 1289–1293.

Zhang, Q., Wendler, I., Xu, X., Willems, H. and Ding, L., 2017. Structure and magnitude of the carbon isotope excursion during the Paleocene-Eocene thermal maximum. Gondwana Research *46*, 114–123.

INDEX

Chemostratigraphy Across Major Chronological Boundaries, Geophysical Monograph 240, First Edition.
Edited by Alcides N. Sial, Claudio Gaucher, Muthuvairavasamy Ramkumar, and Valderez Pinto Ferreira.
© 2019 the American Geophysical Union. Published 2019 by John Wiley & Sons, Inc.